S0-CRS-051

SELECTED PAPERS IN BIOCHEMISTRY

SELECTED PAPERS IN BIOCHEMISTRY

Volume 11

ENZYME MECHANISM

Edited by
Osamu Hayaishi
Kyoto University

UNIVERSITY PARK PRESS
Baltimore · London · Tokyo

UNIVERSITY PARK PRESS
Baltimore · London · Tokyo

Library of Congress Cataloging in Publication Data

Hayaishi, Osamu, 1920-   comp.
Enzyme mechanism.

(Selected papers in biochemistry, v. 11)
Bibliography : p.
1. Enzymes—Addresses, essays, lectures.
I. Title. [DNLM : 1. Enzymes—Metabolism—Collected works. QU 5 S464 v. 11 1973]
QP509. S45 vol. 11 [QP601] 574.1′92′08s [574.1′925′08]
ISBN 0-8391-0621-1   72-13088

UTP 3345-67517-5149
Printed in Japan

Originally published by
UNIVERSITY OF TOKYO PRESS

FOREWORD

This volume has been edited with the hope that it will serve as a useful text for students and research workers interested in the mechanisms underlying enzyme action. The study of enzyme mechanisms requires a multidisciplinary approach and hence this volume includes a broad survey of different approaches that deal with various aspects of enzyme functions, rather than detailed accounts of a specialized subject area.

It is a pleasure to acknowledge the interest and expert assistance of Dr. Yuzuru Ishimura during the preparation of this book.

Kyoto, May 1972

Osamu Hayaishi

BIBLIOGRAPHY

1. Dixon, M. and Webb, E. C., Enzymes, 2nd ed., Academic Press, Inc., New York, 1963.
2. Gutfreund, H., An Introduction to the Study of Enzymes, Blackwell Scientific Publications, Ltd., Oxford, 1965.
3. Bray, H. G. and White, K., Kinetics and Thermodynamics in Biochemistry, 2nd ed., Academic Press, Inc., New York, 1967.
4. Bruice, T. C. and Benkovic, S., Bioorganic Mechanisms, Vols. I, II, W. A. Benjamin, Inc., New York, 1966.
5. Goodwin, T. W., Harris, J. I. and Hartley, B. S., eds., Structure and Activity of Enzymes, Academic Press, Inc., New York, 1964.
6. Ingraham, L. L., Biochemical Mechanisms, John Wiley & Sons, Inc., New York, 1962.
7. Klotz, I. M., Energy Changes in Biochemical Reactions, Academic Press, Inc., New York, 1967.
8. Kosower, E. M., Molecular Biochemistry, McGraw-Hill Book Company, New York, 1962.
9. McElory, W. D. and Glass, B., eds., The Mechanism of Enzyme Action, The Johns Hopkins Press, Baltimore, 1954.

10. Waley, S. G., Mechanisms of Organic and Enzymic Reactions, Oxford University Press, London, 1962.
11. Webb, J. L., Enzyme and Metabolic Inhibitors, Vols. I ~ III, Academic Press, Inc., New York, 1963 ~ 1966.
12. Atkinson, D. E., Regulation of Enzyme Action, *Ann. Rev. Biochem.* **35**, 85, 1966.
13. Bender, M. L. and Breslow, R., "Mechanisms of Organic Reactions," *in* Comprehensive Biochemistry, M. Florkin and E. H. Stotz, eds., Vol. II, p. 1, Elsevier Publishing Company, Amsterdam, distributed by American Elsevier Publishing Company, New York, 1962.
14. Bender, M. L. and Kezdy, F. J., Mechanism of Action of Proteolytic Enzymes, *Ann. Rev. Biochem.* **34**, 49, 1965.
15. Chance, B., Enzyme-substrate Compounds, *Advances in Enzymol.* **12**, 153, 1951.
16. Eigen, M. and Hammes, G., Elementary Steps in Enzyme Reactions, *Advances in Enzymol.* **25**, 1, 1963.
17. Gutfreund, H. and Knowles, J. R., "The Foundations of Enzyme Action," *in* Essays in Biochemistry, P. N. Campbell and G. D. Greville, eds., Vol. 3, p. 25, Academic Press, Inc., New York, 1967.
18. Jencks, W. P., Mechanism of Enzyme Action, *Ann. Rev. Biochem.* **32**, 639, 1963.
19. Monod, J., Wyman, J. and Changeux, J. P., On the Nature of Allosteric Transitions: A Plausible Model, *J. Mol. Biol.* **12**, 88, 1965.
20. Phillips, D. C., The Three-dimensional Structure of an Enzyme Molecule, *Scientific American* **215** (Nov.), 78, 1966.
21. Rose, I. A., Mechanism of Enzyme Action, *Ann. Rev. Biochem.* **35**, 23, 1966.
22. Westheimer, F. H., Mechanisms Related to Enzyme Catalysis, *Advances in Enzymol.* **24**, 441, 1962.
23. Bernhard, S., The Structure and Function of Enzymes., W.A. Benjamin, Inc., New York, 1968.
24. Jencks, W. P., Catalysis in Chemistry and Enzymology, McGraw-Hill Series in Advanced Chemistry, McGraw-Hill, Inc., 1969.

まえがき

この巻は，酵素の反応機構に興味をもつ学生や研究者に有用なテキストを提供する意図をもって編集された．酵素反応機構の解明にはさまざまな技術を駆使した多方面からのアプローチが必要である．したがって，この巻では細分化した特定の分野についての詳細な説明を行なうことよりはむしろ，酵素のもつ機能の諸相に関する，種々の異なった方向からの研究を広く概観するという立場をとった．

なお，この巻を編集するにあたっての，石村巽博士のご関心と適切なご助力に対して感謝する．

京都にて
1972年5月

早石　修

CONTENTS

Volume
11
ENZYME MECHANISM

Reprinted from THE JOURNAL OF BIOLOGICAL CHEMISTRY, Vol. 151, 1943

THE KINETICS OF THE ENZYME-SUBSTRATE COMPOUND OF PEROXIDASE

BY BRITTON CHANCE

(From the Johnson Research Foundation, University of Pennsylvania, Philadelphia, and the Physiological Laboratory, University of Cambridge, Cambridge, England)

(Received for publication, May 26, 1943)

Studies on the over-all kinetics of enzyme action revealed in the majority of cases and over certain concentration ranges that the enzymatic activity was related linearly to the enzyme concentration and hyperbolically to the substrate concentration. On the basis of such evidence Michaelis and Menten (13) showed that such relationships were explained on the assumption that an intermediate compound of enzyme and substrate was formed: $E + S \rightarrow ES \rightarrow E + P$. As the rate of formation of such a compound was assumed to be quite rapid, the rate of breakdown was the rate-determining step. This theory was extended by Briggs and Haldane (2) who pointed out that the rate of formation of the intermediate compound could in certain cases be limited by the number of collisions of enzyme and substrate, and modified the Michaelis theory accordingly. The resulting theory has been extremely useful as a first approximation in the explanation of enzyme action and has given a basis for the comparison of different enzymes in terms of their affinity and activity.

The reaction velocity constants are, however, lumped into one term, the Michaelis constant, and are not separately determined. It is the purpose of this research to determine these constants separately, and to show whether the Michaelis theory is an adequate explanation of enzyme mechanism. Moreover, studies on the over-all enzyme activity do not permit a determination of whether the enzyme-substrate compound exists in fact and, if it exists, whether such a compound is responsible for the enzyme activity.

Several attempts have been made to identify enzyme-substrate compounds. Stern (16) made direct spectroscopic measurements of the compound of catalase and ethyl hydroperoxide and found that this compound was unstable and decomposed after several minutes in the presence of 1 M ethyl hydroperoxide. This was interpreted to indicate that the intermediate compound was responsible for the decomposition of all the ethyl hydroperoxide in this period. Although independent tests showed that ethyl hydroperoxide was decomposed by catalase, no data were given on the amount or rate of decomposition of ethyl hydroperoxide in the spectroscopic experiment (Green (8)).

Keilin and Mann (11) studied the compound of peroxidase and hydrogen peroxide by visual spectroscopy. Their observations include the fact that a spectroscopically defined compound of peroxidase and hydrogen peroxide is formed and that this compound rapidly decomposes in the presence of an oxygen acceptor. While these experiments indicate the existence of an unstable intermediate compound, no direct relation between this intermediate compound and the enzymatic activity is given. A conclusive proof of the Michaelis theory rests on such evidence.

This paper describes a detailed study of the compound of horseradish peroxidase and hydrogen peroxide, an enzyme-substrate compound. The enzyme activity in the presence of leucomalachite green, an acceptor, and hydrogen peroxide, a substrate, has been studied in the usual manner and the Michaelis constant determined. A new apparatus and a new method of studying the kinetics of rapid reactions have been developed and used to measure directly the reaction velocity constants which compose the Michaelis constant. These are the rates of formation and breakdown of the enzyme-substrate compound. The equilibrium of enzyme and substrate in the absence of an acceptor has also been studied. These new data have then been compared with the Michaelis constant which has been determined in the classical manner. A point by point comparison between experiment and theory has been made possible by solutions of the differential equations representing the Briggs and Haldane modifications of the Michaelis theory. In this way, the validity of the Michaelis theory has been clearly demonstrated, and the important relationship between the enzyme-substrate compound and its activity has been clearly shown. A preliminary report of this work was given earlier (Chance (4)).

Preparation and Standardization—The method of Elliott and Keilin (7) was used for the preparation of peroxidase. The first alcohol precipitate was usually discarded and in a particular case 1 gm. of enzyme, PZ[1] = 256, was obtained from 7 kilos of horseradish. The enzyme was kept in a volume of 75 cc. and was tested periodically for hematin iron and PZ. As neither the apparatus nor the information was available at the time, the peroxidase was not purified further in the manner recently indicated by Theorell (18).

A typical preparation contained 5×10^{-5} M hematin iron. The light absorption was measured at 640 and 400 mμ with a grating photoelectric spectrophotometer and it was found that $\epsilon_{640} = 12 \pm 2$ and $\epsilon_{410} = 125 \pm 12$

[1] PZ or purpurogallin number indicates peroxidase activity in terms of mg. of purpurogallin formed from pyrogallol in 5 minutes at 20° per mg. of dry weight of enzyme preparation. 12.5 mg. of H_2O_2 and 1.25 gm. of pyrogallol in 500 cc. of water are used.

(c = 1 mM, d = 1 cm.) at pH 6.2 in 0.01 M phosphate buffer on the basis of total hematin iron.[2] The extinction coefficients given do not represent those of a pure peroxidase.

Perhydrol, diluted to 1 M and kept at 0°, was tested periodically by permanganate titration. Further dilutions were freshly made up before each experiment.

A slightly oxidized saturated solution of leucomalachite green in 0.05 M acetic acid was standardized by oxidation in the presence of peroxidase and hydrogen peroxide. The light absorption at 610 mμ was measured and the concentration determined in terms of a standard solution of malachite green ($\epsilon_{614} \doteq 50$). The pH was maintained by 0.05 M acetate buffer at 4.1.

Method

This is set forth elsewhere (Chance (3, 5, 6)). The Hartridge-Roughton (10) flow method has been modified to give fluid economy and photoelectric resolution greatly exceeding the designs of Roughton and Millikan (15) and adequate for the direct measurement of the kinetics of the hematin compounds in a 1 mm. bore observation tube at concentrations of 1×10^{-6} mole of hematin Fe per liter. The apparatus is shown in Fig. 1, and details of the various parts may be obtained in the references above.

Controls—Detailed controls on the efficient mixing by this apparatus have been described in a previous paper (Chance (3)), indicating that the mixing was essentially complete in 2×10^{-4} second for the highest values of flow velocity. In these experiments the times were long compared to the minimum time range of the apparatus.

Controls on the linearity of the photoelectric system were carried out by plotting deflection of the recorder against concentration of the reactant and a linear relationship was obtained, as the light absorption was very small.

Under certain conditions, the production of malachite green may interfere with the measurement of the kinetics of the intermediate compound. The absorption of the dye is rather high at 420 mμ, as shown in Fig. 2, and would add to the absorption of the enzyme. A 4×10^{-6} M malachite green solution would cause a 3 per cent error in the measurement of 1×10^{-6} M hematin Fe peroxidase solution. This sets a limit to the amount of malachite green formed in the presence of a given amount of enzyme.

A compensation for the effect of malachite green absorption was effected by varying the relative amounts of light incident on the 370 and 430 mμ filter combinations so that the absorption of malachite green affected each photocell equally.

[2] $\epsilon \text{ (extinction coefficient)} = \dfrac{\log_{10} I_0/I}{d \text{ (cm.)} \times c \text{ (mM per liter)}}$.

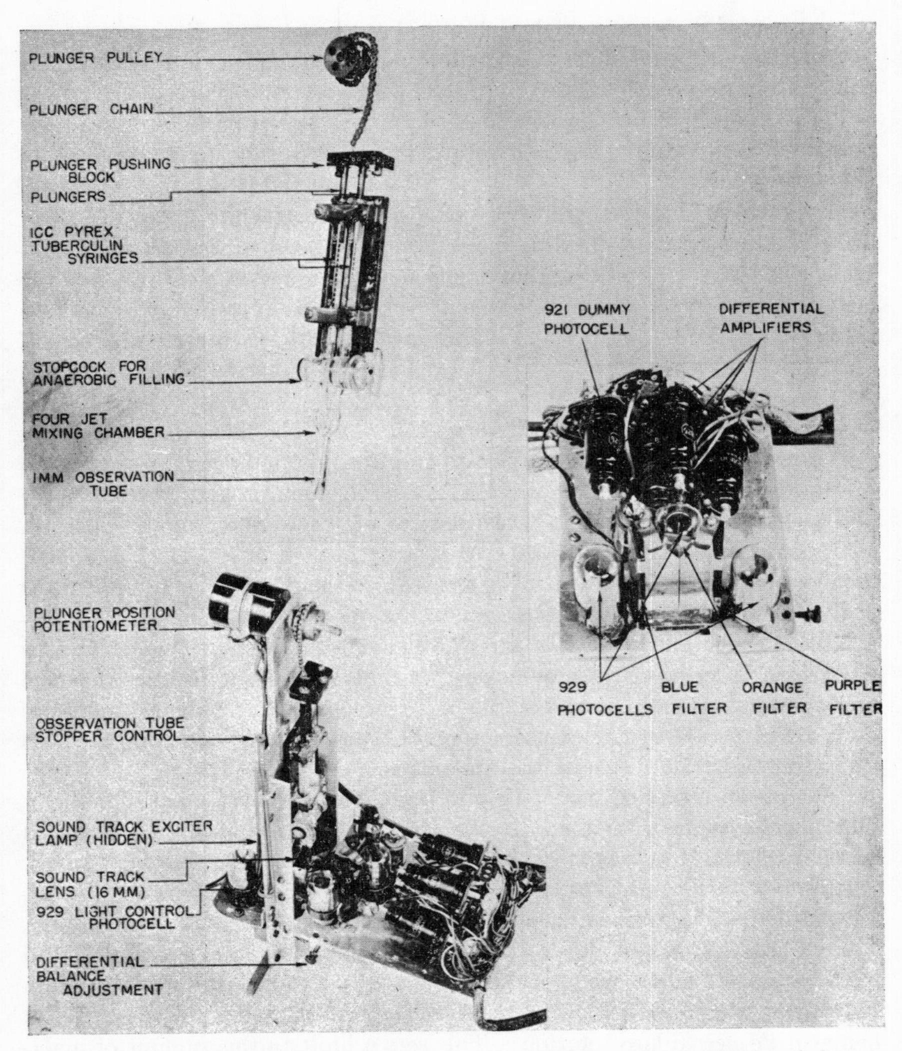

FIG. 1. Syringe unit, photocell unit, and assembled apparatus. Light and electrostatic shields are removed.

Procedure

In order to explain the experimental method more clearly the procedure used to obtain the data of Fig. 3 will be outlined. The enzyme solution was centrifuged before experiment in order to remove denatured protein and give a clear brown solution. Shortly before an experiment, the enzyme was diluted to 2×10^{-6} M hematin Fe. Hydrogen peroxide was diluted to 16×10^{-6} M just previous to an experiment. A saturated solution of leucomalachite green in 0.05 M acetic acid was diluted to 60×10^{-6} M in acetate buffer to make the final pH 4.0.

The syringes shown in Fig. 1 were thoroughly rinsed with cleaning solution and carefully flushed out with water in order that there might be no trace

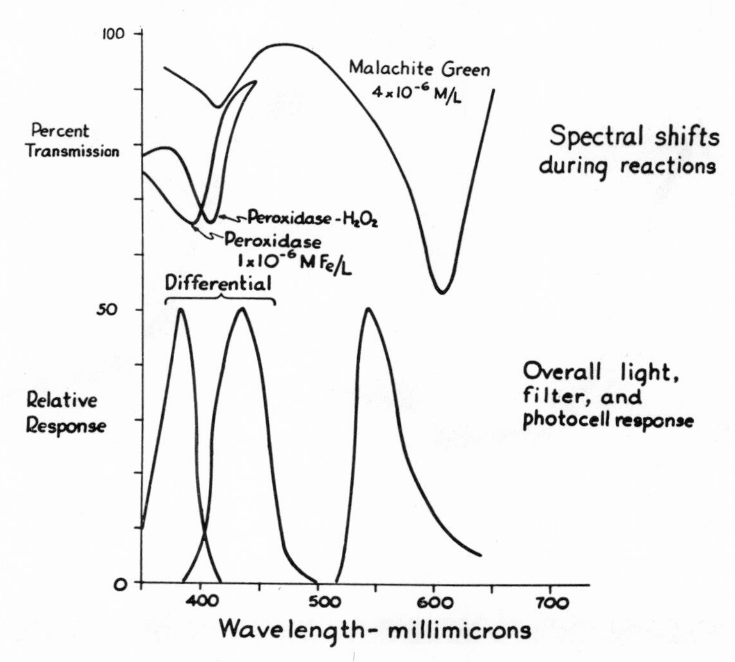

FIG. 2. The upper curves give the light transmission of enzyme, enzyme-substrate compound, and oxidized acceptor under the conditions of Fig. 3. The lower curves give the filter combinations used to measure the kinetics of the reactions. The trough depth was 16 times that of the 1 mm. observation tube of the rapid reaction apparatus. The spectral interval was approximately 8 mμ. The wave-length markers read 15 mμ low.

of the enzyme in the tube which was to be filled with substrate and acceptor. The right-hand syringe was then filled with a mixture of 8×10^{-6} M hydrogen peroxide and 30×10^{-6} M leucomalachite green in 0.05 M acetate buffer. These reactants were squirted into the top of the syringe while the outlet was held closed with a small rubber pad mounted on a lever shown in Fig. 1. The syringe plunger was then entered in the barrel and held in place at the top of the syringe by means of a plunger driving block. The left-hand syringe was flushed out with water and filled with 2×10^{-6} M enzyme solution while the outlet tube was again held closed by means of the stopper. The plunger for the left syringe was then entered and fitted into the driving block. Both plungers were carefully pushed a few mm. down their respective barrels to make sure that they were running smoothly and were accurately aligned. The zero point of the recording mirror oscillograph was checked and a trial run was made by sharply pushing the driving block approximately 1 cm. This caused the reactants to be mixed and to flow down the observation tube very rapidly and, at the end of the

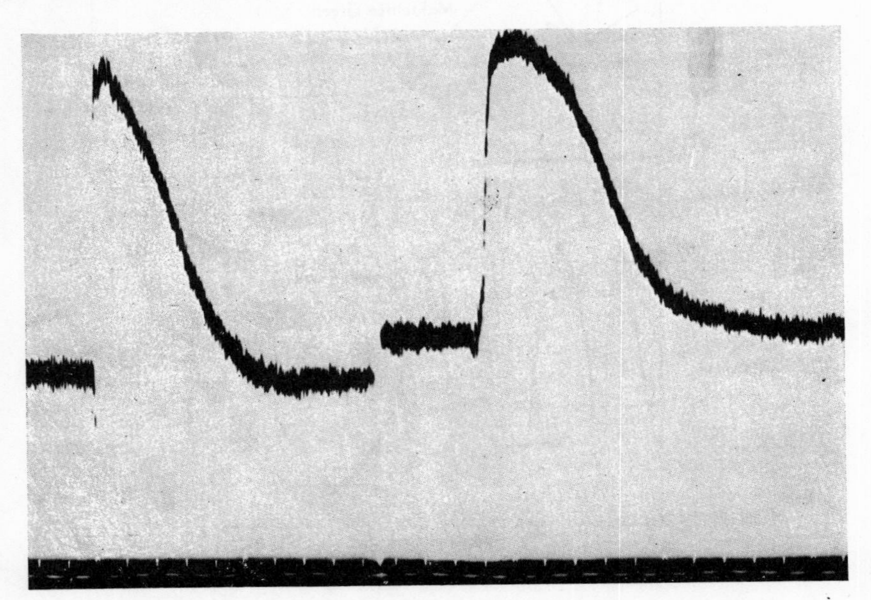

FIG. 3. Mirror oscillograph recording of the production of malachite green (left) and the corresponding kinetics of the enzyme-substrate compound (right). Time markers, 0.2 second. Peroxidase = 1×10^{-6} mole of hematin Fe per liter, H_2O_2 = 4×10^{-6} mole per liter, leucomalachite green = 15×10^{-6} mole per liter, pH = 4.0.

discharge, to stop before the photocell and light beam. The progress of the reaction that ensued in the portion of liquid stopped in the path of the light beam was measured directly by the photoelectric amplifiers. Either Amplifier 1 or 2 could be used, as shown in Chance (5). If the deflection was too large, the amplifier gain was readjusted so that the picture was approximately three-quarters of full linear scale. If it was then considered that the experiment was suitable for recording, the camera attached to the mirror oscillograph was set in operation, and the syringe plungers were given a second sharp push which caused the kinetic curves to repeat themselves. In this way the kinetics of the intermediate compound and the over-all reaction were recorded. This process was repeated until the syringes were completely discharged, and in most cases it was found that three to six curves could be obtained from one filling of the syringes.

A second experiment was carried out immediately to calibrate the maximum concentration of the enzyme-substrate compound. This was done in the same manner as the first experiment except that the leucomalachite green was omitted. Hence the substrate concentration would be sufficient to saturate the enzyme completely, as was indicated by independent experiment. This reaction was also recorded photographically. The deflection corresponded to 1×10^{-6} M hematin Fe enzyme-substrate compound and is marked on Fig. 10.

A third experiment was necessary to calibrate the amount of malachite green formed. Malachite green, formed by peroxidase action, was diluted to 4×10^{-6} mole per liter and used to calibrate the photoelectric amplifier of the system measuring the rapid reaction. The right-hand syringe was filled with the malachite green solution, and the left-hand syringe was filled with water. These two solutions were pushed down, not simultaneously, but alternately, so that the observation tube was filled first with malachite green and then water. The resulting deflection was recorded photographically and gave the deflection corresponding to 4×10^{-6} M malachite green. In this way, the amount of malachite green which had been formed in the experiment was accurately determined. This calibration point appears in Fig. 10. These calibrations were made so that it was unnecessary to rely upon any long time stability of the photoelectric amplifier or recording system.

Results

Equilibrium of Enzyme and Substrate

$$\text{Peroxidase} + H_2O_2 \underset{k_2}{\overset{k_1}{\rightleftharpoons}} \text{peroxidase}\cdot H_2O_2 \tag{1}$$

This reaction was studied by direct photoelectric measurements of the equilibrium concentration of enzyme-substrate compound as a function of substrate concentration. If hydrogen peroxide is mixed with peroxidase, the spectrum changes as in Fig. 2 and the compound denoted peroxidase-H_2O_2, Complex I (Keilin and Mann (11)), is formed, as the substrate is not in great excess.

In order to measure this equilibrium it is essential that k_3, the first order velocity constant for the enzymatic breakdown of the intermediate compound, be negligible compared to k_1, the second order constant for the combination of enzyme and substrate, and k_2, the first order constant for the reversible breakdown of the enzyme-substrate compound. As Keilin has pointed out, the small amount of acceptor present in the enzyme preparation may be oxidized by the addition of hydrogen peroxide and under these conditions the enzymatic breakdown of the enzyme-substrate compound is small. Under these conditions the intermediate compound appeared moderately stable at pH 6.2, although its concentration remained constant for only 5 to 10 seconds at pH 4.2. However, complete stability was not essential for measurements in the rapid reaction apparatus, and it was desired to carry out these reactions at the same pH as the other studies (4.0).

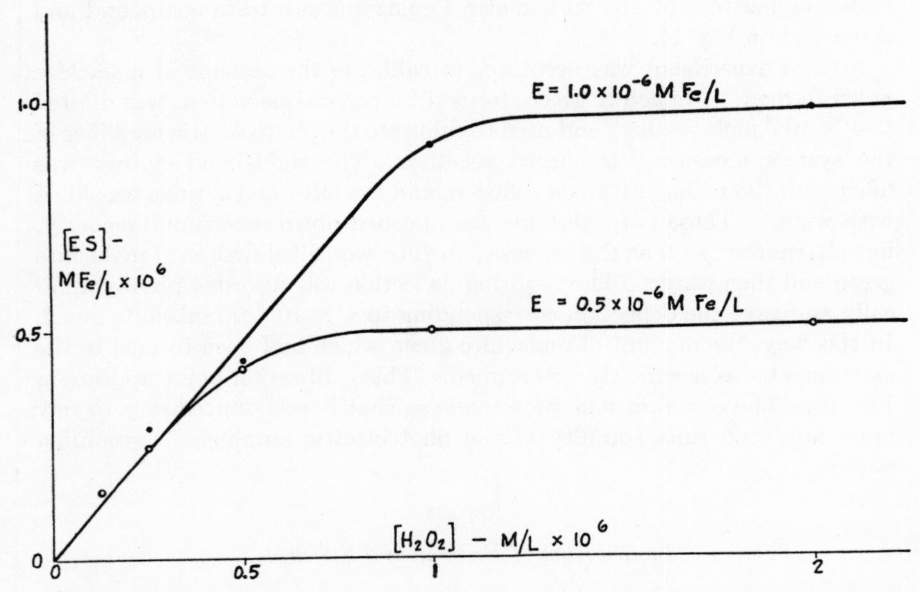

FIG. 4. Equilibrium of enzyme and substrate in absence of acceptor. Ordinate, intermediate compound as total hematin iron; abscissa, initial H_2O_2. pH = 4.2.

The experiments were carried out in this manner. The left-hand syringe was filled with varying concentrations of substrate, while the right-hand syringe was filled with a known concentration of enzyme. Both syringe plungers were then pushed downward in short, sharp pushes so that the observation tube was filled with mixed but unchanged enzyme and substrate, and, after the flow had stopped, the photoelectric system measured and recorded the rate of formation of the intermediate compound and the equilibrium concentration of enzyme-substrate compound. This experiment was repeated for different initial substrate concentrations, and the equilibrium value of the enzyme-substrate compound is plotted in Fig. 4 against initial substrate concentration. It is assumed that the maximum ordinate corresponds to complete conversion of enzyme into enzyme-substrate compound of concentration equal to the independently determined molar hematin iron.

The data of Fig. 4 indicate very small dissociation of the intermediate compound, and the equilibrium constant estimated from two points on Fig. 4 giving finite values is 2×10^{-8}. As the enzymatic breakdown of the enzyme-substrate compound was not zero, this figure should be regarded as a minimum value. Evidently the enzyme was nearly completely converted into its enzyme-substrate compound by an equimolal concentration of substrate. This indicates that all this hematin iron existed as compounds capable of reacting similarly with hydrogen peroxide, *i.e.* forming a spectroscopically defined intermediate compound.

Rate of Formation of Enzyme-Substrate Compound

$$\text{Peroxidase} + H_2O_2 \xrightarrow{k_1} \text{peroxidase} \cdot H_2O_2 \qquad (2)$$

The rate of this reaction has been determined in the manner described before; namely, the right-hand syringe is filled with a 2×10^{-6} M hydrogen peroxide solution, while the left-hand syringe is filled with a 2×10^{-6} M hematin iron enzyme solution. The syringe plungers are again pushed down rapidly, and the reaction was measured after the flow had stopped in the observation tube. The half time of this reaction was 0.1 second. The experiment was then repeated with substrate concentrations from 0.5 to 8×10^{-6} M. The half time and curve shapes of these data were measured, and it was found that a bimolecular equation approximately satisfied the variation of rate with substrate concentration. Higher substrate concentrations have not been used to a great extent, as there is some question whether or not a compound of different spectral absorption denoted peroxidase-H_2O_2, Complex II (Keilin and Mann (11)), might be formed. There is also slight evidence to lead one to believe that the reaction might not follow a bimolecular course at substrate concentrations greater than 10×10^{-6} mole per liter. Experiments in which concentrations of substrate lower than 0.5×10^{-6} mole per liter are employed involved larger experimental errors, owing to the small changes in light transmission.

The data fit a second order kinetic equation, as Fig. 5 shows. Over a range of enzyme concentrations from 1 to 2×10^{-6} mole of hematin Fe per liter and a range of substrate concentrations from 0.5 to 4×10^{-6} mole per liter the mean value of the second order velocity constant was 1.2×10^{7} liter mole^{-1} sec.$^{-1}$. The mean error is 0.4×10^{7}. The previous section gave the ratio of k_2 to k_1 as 2×10^{-8}, or larger; hence k_2 is 0.2 sec.$^{-1}$ or less.

It is now apparent that the enzyme and substrate unite with extreme rapidity to form a relatively tight complex, and it is interesting to note that the ratio of k_2/k_1 is considerably smaller than the Michaelis constant determined by measurement of the over-all enzyme action (5×10^{-6}, Mann (12)). k_3 is possibly far greater than k_2 in the case of peroxidase, and this will be shown to be true in the next section.

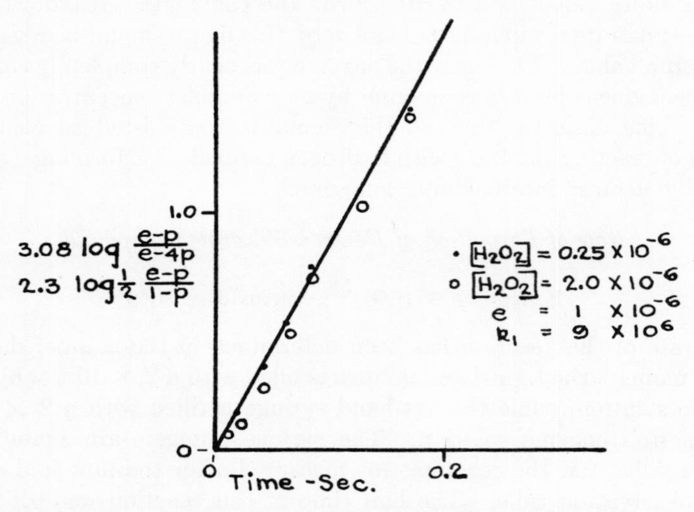

FIG. 5. Kinetics of formation of intermediate compound plotted for two values of substrate concentration according to the second order equation. $k_1 = 9 \times 10^6$ liter mole^{-1} sec.$^{-1}$, pH = 4.0.

Rate of Breakdown of Enzyme-Substrate Compound

$$A + \text{peroxidase}\cdot H_2O_2 \xrightarrow{k_3} \text{peroxidase} + H_2O + AO \qquad (3)$$

The decomposition of the intermediate compound in the presence of an oxygen acceptor is shown schematically by Equation 3. We will choose an oxygen acceptor in the presence of which peroxidase has a high activity. The oxidation products must not interfere with the measurement of the enzyme-substrate compound. This restriction eliminates acceptors like pyrogallol, hydroquinone, and guiacol, while leucomalachite green and ascorbic acid were found to be most satisfactory. In order to demonstrate the effect of such oxygen acceptors on the enzyme-substrate compound, the enzyme is mixed with substrate and acceptor, and the kinetics of the intermediate compound are observed. In Fig. 6 the concentration of the intermediate compound is recorded as a function of time for various concentrations of ascorbic acid. (In contrast to the results of Tauber (17) a polyphenol was not essential in this process.) The right-hand syringe is filled with a mixture containing 8×10^{-6} M H_2O_2, 0.05 M acetate buffer, pH 4.2, and varying concentrations of ascorbic acid. The left-hand syringe is filled with 2×10^{-6} M enzyme solution. The curves show that in the presence of 2.9×10^{-6} mole of ascorbic acid, the intermediate compound is stable for a long period of time. The stability of the compound is indicated, of course, by the length of time required for its concentration to fall to zero, for this is taken to mean that all the substrate has been con-

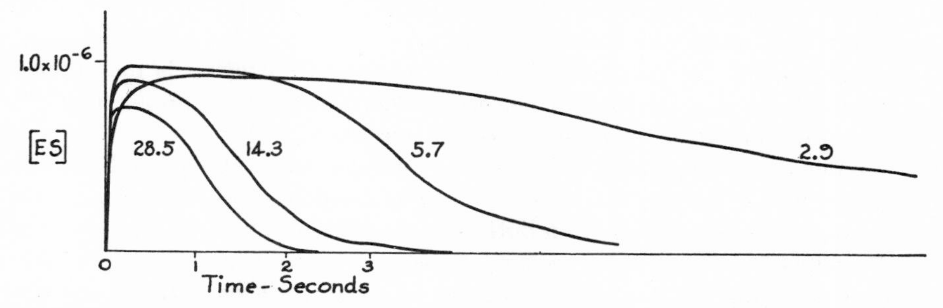

FIG. 6. The effect of an acceptor on the kinetics of the enzyme-substrate compound. $E = 1 \times 10^{-6}$ mole of hematin Fe per liter, $H_2O_2 = 4 \times 10^{-6}$ mole per liter, ascorbic acid as indicated in micromoles per liter, pH = 4.2.

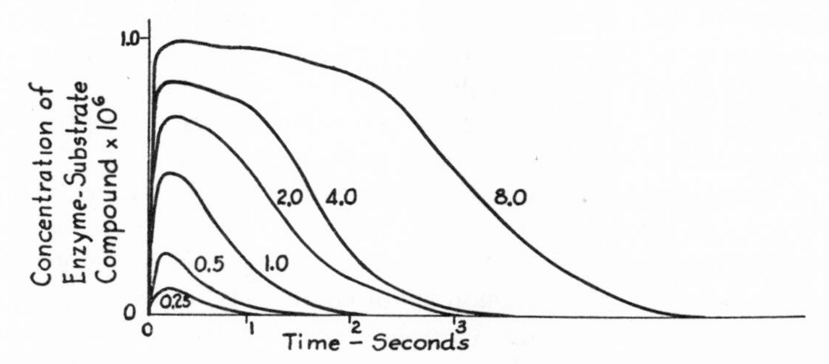

FIG. 7. The effect of substrate on the kinetics of the enzyme-substrate compound. $E = 1 \times 10^{-6}$ mole of hematin Fe per liter, ascorbic acid approximately 14×10^{-6} mole per liter, initial H_2O_2 as indicated in micromoles per liter, pH = 4.2.

sumed. The curves of Fig. 6 for higher concentrations of ascorbic acid clearly show a marked decrease in this interval. The curves also indicate a decrease in the maximum concentration of the enzyme-substrate compound, $p_{max.}$, with increasing ascorbic acid concentration. This decrease in $p_{max.}$ is due to the higher rate of breakdown of the intermediate compound. The low value of $p_{max.}$ in the 2.9×10^{-6} M ascorbic acid curve is believed due to experimental error.

The effect of the substrate concentration is shown in Fig. 7, when the acceptor concentration has been maintained in excess of the substrate concentration. The first interesting feature of this family of curves is the variation of height of the curves with substrate concentration, giving a method of directly studying enzyme-substrate affinity from measurements of the enzyme-substrate compound rather than from the over-all enzyme action. It is seen, for the particular value of ascorbic acid concentration, that the enzyme is one-half saturated by 1×10^{-6} M *initial* substrate concentration. It should also be noted that the area under each curve increases regularly with the initial substrate concentration. One would expect this, as k_3, the rate of breakdown of the enzyme-substrate compound, should be constant as the acceptor concentration is maintained constant and it is found that the area under the curve is proportional to the total amount of hydrogen peroxide consumed.

While k_3 can be determined from the kinetics shown above, we have yet to devise an experiment in which the rate of breakdown of the intermediate compound is determined from both enzyme-substrate kinetics and the rate of production of oxidized acceptor. This experiment is of great importance in determining the relation between the over-all reaction and the kinetics of the enzyme-substrate compound. The rate of disappearance of ascorbic acid could not be measured with this apparatus, as it was not adaptable for wave-lengths below 350 mμ. Leucomalachite green was used as an oxygen acceptor for the following reasons. (1) The mechanism of its oxidation appears simple compared to that of pyrogallol. (2) The absorption is quite strong and does not seriously interfere with the measurement of the enzyme absorption. (3) The linearity between enzyme concentration and rate of formation of malachite green is quite good.

One experimental difficulty in the use of leucomalachite green is a variation in the amount of the dye formed. Only when the leuco base is partially oxidized is the full amount realized and not even then at higher enzyme concentrations. This phenomenon is not completely understood.

On the right-hand side of Fig. 3 are shown the kinetics of the intermediate compound recorded by a photokymograph. The time is read from left to right with markers every 0.2 second. The break in the base-line corresponds to the moment when the syringe plungers were pushed downwards and, after 0.1 second, the flow stops and the reaction of enzyme, substrate, and acceptor proceeds. The formation of the intermediate compound occurs quite rapidly, as is indicated by the abrupt upward deflection of the tracing. Within 0.1 second the enzyme-substrate compound has reached its maximum concentration ($p_{max.}$), and it maintains a steady state for 0.2 second. After this time the substrate concentration has fallen to such a value that the rate of formation of the intermediate compound no longer balances its rate of breakdown. Hence its concentration decreases rapidly and in 1 second has fallen to zero, and the enzyme is all liberated.

The calibrations above indicated that $p_{max.} = 0.85 \times 10^{-6}$ mole of hematin Fe per liter in this experiment.

On the left side of Fig. 3 is shown the rate of production of malachite green by the enzyme system under identical conditions. Here again the break in the base-line indicates a push of the syringe plungers. However, the very rapid upward deflection in this case simply represents clearing out malachite green from the previous run. After 0.1 second the flow stops and the production of malachite green begins just as soon as the intermediate compound has formed. The reaction continues at nearly constant velocity as long as the concentration of the intermediate compound is constant. (The slight variation in slope is due to experimental error.) As this falls, so falls the rate of the over-all reaction, and both reach zero at approximately the same time. Calibrations given above indicated that 4×10^{-6} mole of malachite green was formed in this experiment.

This very simple experiment gives qualitative indication that the relationship between the kinetics of the enzyme-substrate compound and the over-all enzyme activity is that predicted by the Briggs and Haldane modifications of the Michaelis theory.

These experiments have been carried out for substrate concentrations ranging from 5×10^{-7} to 8×10^{-6} mole per liter. At the lower concentrations the error in recording was somewhat large, and at those higher than 6×10^{-6} mole per liter the transmission change due to the formation of the quantity of malachite green interfered with measurements of the enzyme kinetics (see "Controls" above). Enzyme concentrations ranged from 2.5 $\times 10^{-7}$ to 2×10^{-6} M hematin Fe. Lack of an adequate supply of enzyme limited the highest concentrations to 2×10^{-6} M hematin Fe.

Interpretation

Calculation of k_3—The "Appendix" gives methods for determining k_3 from the over-all reaction (Equations 9 and 12) and from the enzyme-substrate kinetics (Equations 11, 13, and 16).

The rate of the *over-all reaction* is 4.3×10^{-6} mole of malachite green per second and $p_{max.} = 0.85 \times 10^{-6}$ mole per liter. From Equations 9 and 12, $k_3 = 5.1$ sec.$^{-1}$.

From the *enzyme-substrate kinetics* there are available the following data for Equation 13.

$k_1 = 1 \times 10^7$ liter mole^{-1} sec.$^{-1}$; $x_0 = 4 \times 10^{-6}$ mole per liter.

$p_{max.} = 0.85 \times 10^{-6}$ mole per liter and $k_2 = 0.2$ sec.$^{-1}$.

$\int_0^t pdt$ is evaluated graphically at $t = 0.24$ second when $p = p_{max.}$ and found to be 0.17×10^{-6} mole second; hence $k_3 = 4.3$ sec.$^{-1}$ for $k_2 = 0$ and 4.2 sec.$^{-1}$ for $k_2 = 0.2$ sec.$^{-1}$.

$\int_0^t pdt$ also may be evaluated graphically at $t = \infty$ when $p = 0$ and $x = 0$. The integral is found to be 0.84×10^{-6} mole second and on substitution in Equation 11, $k_3 = 4.8$ sec.$^{-1}$.

According to Equation 16, the value of k_3 is given by $x_0/(p_{\text{max.}} \cdot t_{\frac{1}{2}})$. As $t_{\frac{1}{2}} = 0.9$ second, k_3 is calculated to be 5.2 sec.$^{-1}$.

The rate of breakdown of the enzyme-substrate compound in the presence of ascorbic acid is determined from the data of Fig. 6. Using convenient Equation 16, we find in Fig. 8 that the variation of k_3 with ascorbic acid is of such a nature that k_3 divided by the ascorbic acid concentration gives a constant indicative of a second order combination of acceptor and enzyme-substrate compound. The same relationship held for leucomalachite green, and the corresponding quotient is 3×10^5 liter mole^{-1} sec.$^{-1}$.

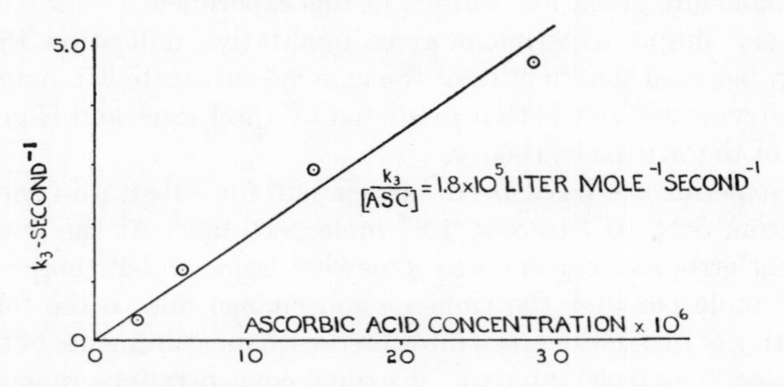

FIG. 8. Variation of k_3 with acceptor concentration. k_3 was obtained by Equation 16 from data of Fig. 6.

The constancy of k_3 for a given acceptor concentration is given in Fig. 9 for the data of Fig. 6 on the basis of Equation 16. The experimental check of the equation is satisfactory although the acceptor concentration was somewhat depleted in two reactions with higher substrate concentrations.

A particular curve for $x_0 = 1.0 \times 10^{-6}$ mole per liter has been examined and k_3 at 14×10^{-6} M ascorbic acid is found to be 2.2, 2.0, and 2.5 sec.$^{-1}$ from Equations 16, 11, and 13 respectively.

There is then substantial agreement between values of k_3 calculated from three different points of the enzyme-substrate kinetics corresponding to the times $p = p_{\text{max.}}$, $p = p_{\text{max.}}/2$, and $p = 0$ ($t = \infty$) and between values of k_3 determined from the over-all reaction.

Calculation of Michaelis Constant—There are three ways by which we can determine the Michaelis constant and thereby check the validity of the theory.

The first method is to calculate this constant from k_2, k_3, and k_1 which have all been experimentally determined. k_3, calculated solely *from the kinetics of the enzyme-substrate compound* above, is found to be 4.2 sec.$^{-1}$. k_1 is found to be 1×10^7 liter mole^{-1} sec.$^{-1}$ and k_2 a minimum value of 0.2 sec.$^{-1}$. The Michaelis constant is then calculated to be 0.44×10^{-6} from Equation 8.

This value may also be calculated according to Equation 8 from concentrations which obtain during the steady state. The saturation of the enzyme, $p_{max.}$, is known from the experiment and the corresponding value of x may be readily determined. It is important to note that the value of x is not the initial concentration of substrate as is usually

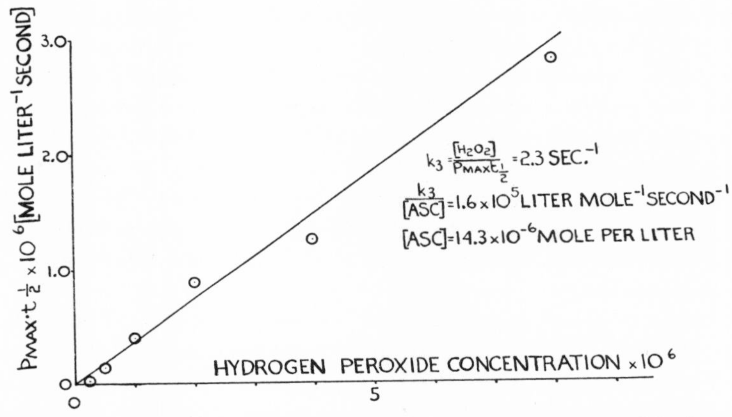

FIG. 9. Experimental test of Equation 16 indicating constancy of k_3 for varying substrate concentration and fixed acceptor concentration. From the data of Fig. 7

the case when this initial concentration is so large compared with the enzyme concentration that the amount of substrate combined with enzyme is relatively small. The value of x when p reaches $p_{max.}$ is calculated from the experimental data in three ways which follow: (a) x is readily calculated from Equation 11, as we have already determined the necessary quantities. $p_{max.}$ is equal to 0.85×10^{-6} mole per liter, $\int_0^{0.24} pdt$ is 0.17×10^{-6} mole second, and k_3 is 4.2 sec.$^{-1}$. x is 2.4×10^{-6} mole per liter and Km, calculated from Equation 8, is 0.43×10^{-6} mole per liter. This method is, of course, not independent of the calculation of k_3 shown previously; nevertheless, *all* the data used to determine Km in this manner are obtained from concentration measurements of the enzyme-substrate compound during the steady state. (b) The amount of substrate which has been consumed by the time p reaches its maximum value can be determined from Fig. 3 (right) by a second graphical method. The area under the whole curve for the kinetics of the enzyme-substrate compound represents the disappearance of 4×10^{-6} mole of substrate. In fact, this is true for $\int_0^\infty pdt = 0.84 \times 10^{-6}$ mole second and the mean value of $k_3 = 4.9$ sec.$^{-1}$, whence $k_3 \int_0^\infty pdt = 4.0 \times 10^{-6}$ mole, the initial substrate concentration. The area under the curve from time zero until p reaches its maximum value is representative of the amount of substrate which has been decomposed during that time and this is 0.9×10^{-6} mole of decomposed substrate. To this we must add the amount of substrate which is combined with the enzyme, $p_{max.}$. From this, x is readily calculated and the Michaelis constant is found to be 0.40×10^{-6}. This method is completely independent of a determination of k_3, as this quantity appears in both numerator and denominator. (c) If we assume that for each molecule of malachite green formed 1 molecule of substrate has been decomposed, we have directly the amount of substrate that disappeared enzymatically. At 0.24 second this is 0.9×10^{-6} mole. When $p_{max.}$ is added to this, the Michaelis constant is calculated to be 0.40×10^{-6}, which agrees very closely with the other values determined independently.

The classical determination of the Michaelis constant by Mann (12) gives 5×10^{-6} mole per liter at pH 4.0 and an acceptor concentration of 0.007 per cent. This constant varied linearly with acceptor concentration over this range. These data also indicated a linear relationship. Hence Mann's value of Km was reduced to our acceptor concentration by dividing by the concentration differential, 10. This gives 0.5×10^{-6}, which agrees fairly well with the above independently determined values in view of the widely different enzyme and substrate concentrations.

Correlation with Complete Solutions of Michaelis Theory—While previous data suggest the validity of the Michaelis theory, a much more convincing proof is furnished by the data on the superposition of the differential analyzer (see "Appendix") and direct experimental curves.

The solid curves in Fig. 10 show the kinetics of the enzyme-substrate compound (right) and the over-all reaction (left) for the following values of reaction velocity constants and concentrations: $e = 1 \times 10^{-6}$ mole per liter, $x_0 = 4 \times 10^{-6}$ mole per liter, $k_1 = 0.9 \times 10^7$ liter mole^{-1} sec.$^{-1}$, $k_2 = 0$ sec.$^{-1}$, $k_3 = 4.5$ sec.$^{-1}$. The experimental curves of Fig. 3 ($e = 1 \times 10^{-6}$ mole of hematin Fe per liter, $x_0 = 4 \times 10^{-6}$ mole per liter, leucomalachite green $= 15 \times 10^{-6}$ mole per liter, pH $= 4.0$) are plotted as circles to the proper scale in Fig. 10. The independently determined values of reaction velocity constants are $k_1 = 1.2 \times 10^7$ liter mole^{-1} sec.$^{-1}$, $k_2 \leqq 0.2$ sec.$^{-1}$, $k_3 = 4.9$ sec.$^{-1}$ (mean). Remarkably good agreement is obtained in view of the possible error in all experimental quantities required to determine the mathematical solution.

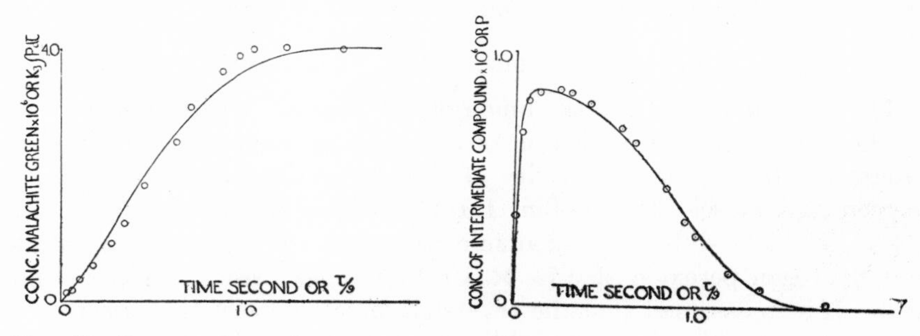

Fig. 10. A comparison of experimental enzyme-substrate and "over-all" kinetics (circles) with a mathematical solution of the Michaelis theory for experimentally determined reaction velocity constants and concentrations (solid lines).

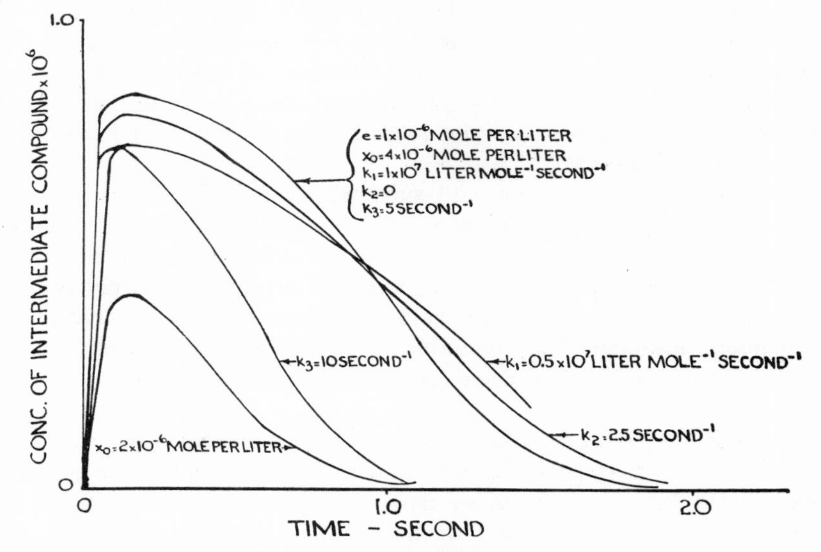

Fig. 11. Effect of variation of reaction velocity constants and concentrations on the shape of the mathematical solutions of the Michaelis theory.

The overshoot in the experimental points (Fig. 10, right) is possibly due to inadequate speed of response in the photocell amplifier. The scatter of points (Fig. 10, left) is thought to represent an instrumental rather than intrinsic irregularity.

Calculations show that the mathematical solutions are quite sensitive to changes in the experimental values. x_0 and k_3 cause large changes in $t_{\frac{1}{2}}$, while k_1 and k_2 affect $p_{max.}$ and the shape of the curve as shown in Fig. 11. The effect of enzyme and leucomalachite green concentration is not given by the mathematical solutions but would be large.

DISCUSSION

The extreme rapidity of the union of enzyme and substrate is indicated by the second order rate constant of 1×10^7 liter mole^{-1} sec.$^{-1}$. It is very interesting to note the similarity of this rate to the measured value for oxygen and muscle hemoglobin, 1.9×10^7 liter mole^{-1} sec.$^{-1}$ (Millikan (14)). Haldane's (9) calculated minimum rate for the union of catalase and hydrogen peroxide of 0.76×10^7 liter mole^{-1} sec.$^{-1}$ is quite similar also. If it is assumed that the reversible breakdown of the enzyme-substrate compound is also slow in the case of catalase and hydrogen peroxide, the similarity between the catalase and peroxidase values is more striking. Such concordance in the rates of union of small molecules and proteins would lead one to question whether or not these reaction velocities were limited by the number of collisions. While experiments made at 0° indicate but little change in the rate of formation of the enzyme-substrate compound, these results are preliminary and should not be used to substantiate the conclusion that the joining of enzyme and substrate is a collision-limited process.

The equilibrium of enzyme and substrate was directly studied as the irreversible breakdown of the intermediate compound (k_3) was quite small in the absence of acceptor. The equilibrium constant was found to have a minimum value of 2×10^{-8}. This indicates an extremely tight complex of enzyme and substrate, and this value is of the same order as that for CO hemoglobin, although the individual rates differ considerably. Cytochrome *c* peroxidase-hydrogen peroxide evidently dissociates more readily (1).

The studies on the enzymatic function of the enzyme-substrate compound were also carried out with ascorbic acid and leucomalachite green as acceptor, the latter over a rather narrow range, as the absorption of dye interfered with the measurement of the enzyme-substrate compound. The Michaelis theory has been checked by various determinations of the Michaelis constant. The first method is from *kinetic* data; namely, the rate of formation of the enzyme-substrate compound, the rate of reversible breakdown, and the rate of its irreversible breakdown into free enzyme and altered substrate. The Michaelis constant was determined from the sum of the last two divided by the first. This was also determined from *concentration measurements* at the steady state, when the concentration of the intermediate compound passed through its maximum value. From this maximum value and the corresponding substrate concentration we can again directly calculate the Michaelis constant. The third method is the classical one wherein the rate of the over-all reaction is measured, and the concentration of substrate giving one-half maximal activity is determined. To these three methods a fourth one has been added to take advantage of the fact that the data are complete kinetic curves of the enzyme-substrate compound, and therefore, with complete solutions of the Michaelis equation we may compare, point by point, theory and experiment.

In all cases in the range of experimental concentrations the kinetics of the intermediate compound were related to the kinetics of the over-all reaction in a manner explained by the Michaelis theory, substantiating the conclusion that the mechanism of a second order combination of enzyme and substrate followed by a first order decomposition is essentially correct for peroxidase action at this particular acceptor concentration.

The rate of breakdown of the intermediate compound of peroxidase and hydrogen peroxide is very small compared to that of catalase $k_3 = 3 \times 10^5$ sec.$^{-1}$. The value for catalase assumes that the Michaelis theory holds and that a chain mechanism is not responsible for the enzyme action. The peroxidase kinetics indicate that a chain mechanism plays no prominent part, if any, as the induction period in the production of dye is no longer than is required by the formation of the enzyme-substrate compound and there is also no further production of dye after the enzyme-substrate compound has disappeared. It is possible that the difference between these two enzymes lies mainly in the slower breakdown of the peroxidase intermediate compound.

It is of considerable interest to know whether there is a bimolecular combination of the enzyme-substrate compound and the acceptor. No spectroscopic evidence of such compound formation from 360 to 600 mμ was found. However, kinetic evidence for such a combination is given by the variation of the enzyme activity with acceptor concentration. The rate of production of malachite green and the effect of ascorbic acid on the kinetics of the enzyme-substrate compound strongly suggest a bimolecular combination with acceptor in accordance with Mann (12).

The mechanism by which the acceptor is oxidized is still obscure. As this may take place through single electron changes involving the formation of a free radical of the triphenylmethyl type in the case of malachite green, studies were made[3] to find spectroscopic evidence for such intermediates. While no data were obtained in the visible spectrum, the question is still open.

[3] Dr. Fred Karush collaborated in this study.

SUMMARY

Under the narrow range of experimental conditions, and at a temperature of approximately 25°, the following data were obtained.

1. The equilibrium constant of peroxidase and hydrogen peroxide has a minimum value of 2×10^{-8}.

2. The velocity constant for the formation of peroxidase-H_2O_2 Complex I is 1.2×10^7 liter mole^{-1} sec.$^{-1}$, $\pm 0.4 \times 10^7$.

3. The velocity constant for the reversible breakdown of peroxidase-H_2O_2 Complex I is a negligible factor in the enzyme-substrate kinetics and is calculated to be less than 0.2 sec.$^{-1}$.

4. The velocity constant, k_3, for the enzymatic breakdown of peroxidase-H_2O_2 Complex I varies from nearly zero to higher than 5 sec.$^{-1}$, depending upon the acceptor and its concentration. The quotient of k_3 and the leucomalachite green concentration is 3.0×10^5 liter mole^{-1} sec.$^{-1}$. For ascorbic acid this has a value of 1.8×10^5 liter mole^{-1} sec.$^{-1}$.

5. For a particular acceptor concentration, k_3 is determined solely from the enzyme-substrate kinetics and is found to be 4.2 sec.$^{-1}$.

6. For the same conditions, k_3 is determined from a simple relationship derived from mathematical solutions of the Michaelis theory and is found to be 5.2 sec.$^{-1}$.

7. For the same conditions, k_3 is determined from the over-all enzyme action and is found to be 5.1 sec.$^{-1}$.

8. The Michaelis constant determined from kinetic data alone is found to be 0.44×10^{-6}.

9. The Michaelis constant determined from steady state measurements is found to be 0.41×10^{-6}.

10. The Michaelis constant determined from measurement of the overall enzyme reaction is found to be 0.50×10^{-6}.

11. The kinetics of the enzyme-substrate compound closely agree with mathematical solutions of an extension of the Michaelis theory obtained for experimental values of concentrations and reaction velocity constants.

12. The adequacy of the criteria by which experiment and theory were correlated has been examined critically and the mathematical solutions have been found to be sensitive to variations in the experimental conditions.

13. The critical features of the enzyme-substrate kinetics are $p_{max.}$ and curve shape, rather than $t_{\frac{1}{2}}$. $t_{\frac{1}{2}}$ serves as a simple measure of dx/dt.

14. A second order combination of enzyme and substrate to form the enzyme-substrate compound, followed by a first order breakdown of the compound, describes the activity of peroxidase for a particular acceptor concentration.

15. The kinetic data indicate a bimolecular combination of acceptor and enzyme-substrate compound.

It is a very great pleasure to acknowledge the aid of Dr. F. J. W. Roughton, Dr. F. A. Cajori, Dr. G. A. Millikan, and Dr. J. G. Brainerd, and the keen interest of Dr. D. W. Bronk in this research. The aid of the American Philosophical Society is gratefully acknowledged. It is also a source of regret that the problem could not be concluded where it was initiated.

Appendix

Extension of Michaelis Theory

These reactions are represented by Briggs and Haldane as the bimolecular combination of the enzyme, E, and substrate, S, to form an intermediate compound, ES, followed by a monomolecular decomposition into free enzyme and activated or altered substrate, Q, representative of the products of the "over-all" enzyme action.

$$E + S \underset{k_2}{\overset{k_1}{\rightleftharpoons}} ES \xrightarrow{k_3} E + Q \tag{4}$$

If e is the total molar enzyme concentration, x the molar substrate concentration, p the molar concentration of ES, k_1 the second order rate constant, and k_2 and k_3 the first order rate constants, then

$$\frac{dp}{dt} = k_1 x(e - p) - (k_2 + k_3)p \tag{5}$$

$$\frac{dx}{dt} = -k_1 x(e - p) + k_2 p \tag{6}$$

These two equations represent the rate of formation of the intermediate compound and the rate of disappearance of the substrate.

The solution of these equations has been already obtained by Briggs and Haldane for the special conditions of the steady state, when

$$p = p_{max.}, \qquad \frac{dp}{dt} = 0, \qquad \text{and} \qquad \frac{p_{max.}}{e} = \frac{x}{x - Km} \tag{7}$$

where

$$Km = \frac{k_2 + k_3}{k_1} = x\,\frac{(e - p_{max.})}{p_{max.}} \tag{8}$$

A further solution valid during the steady state is obtained by adding Equations 5 and 6,

$$\frac{dx}{dt} = -k_3 p_{max.} \tag{9}$$

where dx/dt is the rate of disappearance of substrate. This equation is useful for determining k_3.

In addition to these solutions for the steady state the general solution of these differential equations can be indicated thus:

$$\frac{d(p + x)}{dt} = -k_3 p \tag{10}$$

and is obtained by adding Equations 5 and 6. As $p = 0$ and $x = x_0$ when $t = 0$, we have, on integrating,

$$x = x_0 - p - k_3 \int_0^t pdt \tag{11}$$

When $dp/dt = 0$,

$$\frac{dx}{dt} = -k_3 \frac{d\int_0^t pdt}{dt} \tag{12}$$

The right-hand member may represent the rate of appearance of oxidized substrate such as malachite green. We may then use Equation 9 to calculate k_3.

Also when $dp/dt = 0$, $p = p_{\text{max.}}$ and at this time Equation 7 is valid. On solving Equation 7 for x and substituting in Equation 12, we have, after simplification,

$$k_3 = \frac{k_1(x_0 - p_{\text{max.}})(e - p_{\text{max.}})}{p_{\text{max.}} + k_1(e - p_{\text{max.}})\int_0^t pdt} - \frac{k_2\, p_{\text{max.}}}{p_{\text{max.}} + k_1(e - p_{\text{max.}})\int_0^t pdt} \tag{13}$$

This equation is useful to calculate k_3 when the curve of p against t is known as in the case of Fig. 3.

As a check on the mathematics, let us substitute in Equation 13 the condition that $p_{\text{max.}}$ is nearly equal to e, t is small, and x is nearly equal to x_0. It will be seen that Equation 7 is obtained as would be expected.

Complete Solutions for Michaelis Theory[4]—Under the experimental conditions it was found that the steady state existed only for a fraction of a second. In order to determine whether the transient portions of the curves satisfied the Michaelis theory, solutions of differential Equations 5 and 6 were required. For satisfactory solutions from the differential analyzer, the following substitutions were necessary. Let

$$K_2 = 10^6 \frac{k_2}{k_1}, \qquad X = 10^6 x, \qquad \tau = k_1 et$$

$$K_3 = 10^6 \frac{k_3}{k_1}, \qquad P = \frac{p}{e}, \qquad \Theta = k_1 t 10^{-6}$$

[4] These solutions were obtained with the aid of Dr. J. G. Brainerd, Moore School, University of Pennsylvania.

$$\therefore \frac{dP}{d\Theta} = (1 - P)X - (K_2 + K_3)P \tag{14}$$

$$\frac{dX}{d\tau} = -(1 - P)X + K_2 P \tag{15}$$

When e is equal to 1×10^{-6} mole per liter, the experimental value, then τ is equal to Θ and a series of solutions of these equations may be obtained for this particular value of the enzyme concentration. The solutions have been carried out in this manner, and it should be noted that they are valid only for this enzyme concentration.

The mathematical solutions are given in Fig. 12, and solutions have been obtained for $K_2 = 0$; $K_3 = 0$ to 2; and $X_0 = 1$ to 8. Solutions for $K_2 = 0$ to 8 and $K_3 = 0$ to 8 were also made.

The upper portions of Fig. 12 show the disappearance of substrate, X, and production of "over-all" products, Q. The substrate concentration starts at its initial value, $X = X_0$, and falls to zero. The "over-all" production, represented by $K_3 \int_0^\tau P d\tau$, begins at zero and continues until the substrate is exhausted.

The lower left portion of Fig. 12 gives the ordinary solutions of the bimolecular reaction of enzyme and substrate, when K_2 and K_3 are zero, for four values of the initial substrate concentration, X_0. 1 unit of the ordinate corresponds to $p = e$; *i.e.*, complete conversion of the enzyme into the intermediate compound. The abscissae, plotted in units of τ, are converted into time units by the appropriate values of k_1. Directly above this is the disappearance of substrate due to the bimolecular reaction.

In the remaining lower portions of Fig. 12, K_2, the velocity constant of the reversible reaction is zero and K_3 has a finite value. The concentration of the intermediate compound increases, passes through as a maximum, and then falls to zero. The maximum concentration and area under the curve increase with the substrate concentration. Directly above are the corresponding curves for the over-all reaction. The initial rush in the kinetics of the disappearance of substrate and the induction period in the formation of "over-all" products are significant features.

The abscissae in all cases are represented in units of τ and therefore a wide range of values of k_1 can be used. The range of values of K_2, K_3, and X_0 is that corresponding to the values for which solutions have been obtained. However, to determine k_2 or k_3, K_2 or K_3 is multiplied by k_1.

For larger values of k_1 and X_0, and smaller values of K_3, dp/dt will become quite small for a considerable time and solutions corresponding to $dp/dt = 0$ can easily be obtained, as shown by Briggs and Haldane (2).

The families of curves obtained from the differential analyzer are obviously applicable to any reversible bimolecular combination and a consecutive monomolecular breakdown of the intermediate compound. It should be noted that these mathematical solutions do not completely describe peroxidase action, as they do not include the acceptor process. Hence they are valid only for fixed acceptor concentrations giving constant k_3.

A useful feature of these mathematical solutions is that they reveal arbitrary relationships between members of the families of curves. As an example it is found from measurements of the mathematical solutions

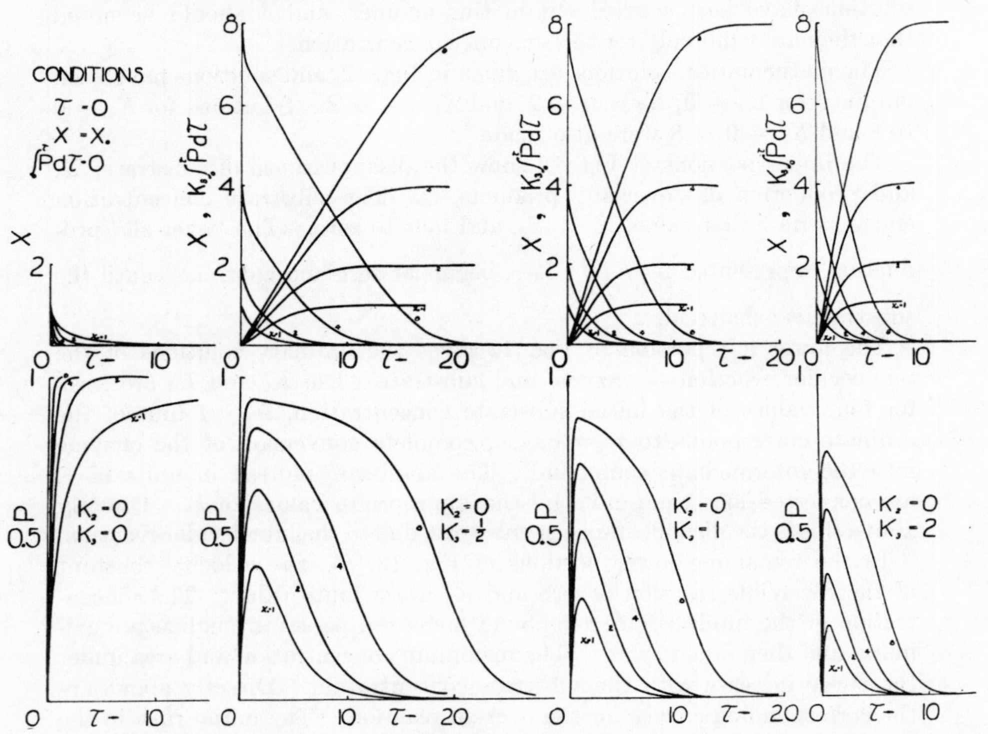

FIG. 12. Differential analyzer solutions of the Michaelis theory for the over-all reaction (upper group) and the kinetics of the enzyme-substrate compound (lower group).

for the kinetics of the enzyme-substrate compound that k_3 can be determined directly by this formula if the acceptor concentration is constant.

$$k_3 = \frac{x_0}{p_{\max.}\, t_{\frac{1}{2}}} \tag{16}$$

$p_{\max.}$ is the maximum value of p. $t_{\frac{1}{2}}$ is the time required for p to fall from $p_{\max.}$ to $p_{\max.}/2$. This relationship has been tested for experimental curves and found very satisfactory, as shown in Fig. 9.

If Equation 16 is combined with Equation 9, we find

$$t_{\frac{1}{2}} = -\frac{x_0}{dx/dt} \tag{17}$$

This indicates that there are a variety of curves of the same half width that would satisfy our experimental data. The shape and $p_{max.}$ would be quite different.

BIBLIOGRAPHY

1. Abrams, R., Altschul, A. M., and Hogness, T. R., *J. Biol. Chem.*, **142,** 303 (1942).
2. Briggs, G. E., and Haldane, J. B. S., *Biochem. J.*, **19,** 338 (1925).
3. Chance, B., *J. Franklin Inst.*, **229,** 455 (1940).
4. Chance, B., *Science*, **92,** 455 (1940).
5. Chance, B., *Rev. Scient. Instruments*, **13,** 158 (1942).
6. Chance, B., Harvey, E. N., Johnson, F., and Millikan, G. A., *J. Cell. and Comp. Physiol.*, **15,** 195 (1940).
7. Elliott, K. A. C., and Keilin, D., *Proc. Roy. Soc. London, Series B*, **114,** 210 (1934).
8. Green, D. E., Mechanisms of biological oxidations, Cambridge (1940).
9. Haldane, J. B. S., *Proc. Roy. Soc. London, Series B*, **108,** 559 (1931).
10. Hartridge, H., and Roughton, F. J. W., *Proc. Roy. Soc. London, Series B*, **104,** 376 (1923).
11. Keilin, D., and Mann, T., *Proc. Roy. Soc. London, Series B*, **122,** 119 (1937).
12. Mann, P. J. G., *Biochem. J.*, **24,** 918 (1931).
13. Michaelis, L., and Menten, M. L., *Biochem. Z.*, **49,** 333 (1913).
14. Millikan, G. A., *Proc. Roy. Soc. London, Series B*, **120,** 366 (1936).
15. Roughton, F. J. W., and Millikan, G. A., *Proc. Roy. Soc. London, Series A*, **155,** 258, 269, 277 (1936).
16. Stern, K. G., *Enzymologia*, **4,** 145 (1937).
17. Tauber, H., *Enzymologia*, **1,** 209 (1936).
18. Theorell, H., *Ark. Kemi, Mineral. o. Geol.*, **14 B,** 1 (1940).

Reprinted from ACTA CHEMICA SCANDINAVICA, Vol. 5, 1951

Studies on Liver Alcohol Dehydrogenase

II. The Kinetics of the Compound of Horse Liver Alcohol Dehydrogenase and Reduced Diphosphopyridine Nucleotide

HUGO THEORELL and BRITTON CHANCE

Biochemical Department, Medical Nobel Institute, Stockholm, Sweden, and Johnson Research Foundation, University of Pennsylvania, Philadelphia, U. S. A.

The elucidation of the mechanism of enzyme action by means of direct measurements of enzyme-substrate compounds has previously been limited to the peroxide compounds of the hemoproteins [1]. In catalases and peroxidases, the addition of substrate causes changes of absorption spectra that are readily measurable with rapid and sensitive spectrophotometric methods [2]. Theorell and Bonnichsen [3] have recently found a small shift in the spectrum of reduced diphosphopyridine nucleotide (DPNH) upon the addition of liver alcohol dehydrogenase (ADH) **. This paper describes kinetics and equilibrium studies of the enzyme-substrate compound, ADH-DPNH, in the reduction of aldehyde (Ald) to alcohol (Alc). The reactions of this enzyme-substrate complex are outlined by the following equations,

$$\text{ADH} + \text{DPNH} \underset{k_2}{\overset{k_1}{\rightleftharpoons}} \text{ADH-DPNH} \tag{1}$$

$$\text{H}^+ + \text{ADH-DPNH} + \text{Ald} \underset{k_3}{\overset{k_4}{\rightleftharpoons}} \text{ADH-DPN}^+ + \text{Alc} \tag{2}$$

$$\text{ADH-DPN}^+ \underset{k_5}{\overset{k_3}{\rightleftharpoons}} \text{ADH} + \text{DPN}^+ \tag{3}$$

* This work was supported in part by the Office of Naval Research and by the National Institutes of Health, United States Public Health Service, and by a travelling grant from the *Statens medicinska forskningsråd* for one of us (H. T.).

** The symbol ADH is here taken to represent the portion of the alcohol dehydrogenase molecule that binds one molecule of DPNH.

and the assumptions and simplifications represented by these equations are discussed below. The values of the equilibrium constant for Eqn. 1, and for the values of the reaction velocity constants, k_1 and k_4, are determined experimentally and are compared with values computed from studies of the overall activity of this enzyme.

PREPARATIONS

ADH and DPNH were prepared as described by Bonnichsen[4, 5] and had purities of 70—100 % and 69 % respectively. The actual concentration of ADH was computed according to part I.

The concentration of acetaldehyde was tested enzymatically with ADH and excess DPNH at pH 7.0. The concentration of formaldehyde was tested by the method of MacFadyen [6].

METHODS

The spectroscopic data of Theorell and Bonnichsen show that the largest changes of molecular extinction coefficient ($\Delta\varepsilon$) of the DPNH spectrum caused by the addition of ADH occur at 310 and 350 mμ. The actual values of $\Delta\varepsilon$ are small, 1.5 and 2.4 $cm^{-1} \times mM^{-1}$ at 310 and 350 mμ respectively. There is an isosbestic point between the DPNH and ADH-DPNH spectra (after subtraction of the ADH absorption) at 328 mμ. But there is no single wavelength where the formation of ADH-DPNH may be recorded without recording changes in [DPNH] at the same time.

In equilibrium studies of the ADH-DPNH complex (part I) it is possible to correct for the DPNH or ADH absorption, but it is neither practical nor accurate to do this in kinetic studies.

Thus the problem of measuring the reaction kinetics of the ADH-DPNH complex is a much more formidable one that that of measuring the hydrogen peroxide compounds of catalases and peroxidases. In the latter case the substrate has negligible absorption in the region of the enzyme and thus the measurements of the enzyme kinetics are not interfered with. In addition the molecular extinction coefficients of catalases and peroxidases change by about 50 $cm^{-1} \times mM^{-1}$ at 405 mμ on combination with peroxide. In the case of the ADH · DPNH complex, the change of extinction coefficient is much less than that of the hemoproteins, namely 2.4 $cm^{-1} \times mM^{-1}$ at 350 mμ and the DPNH absorption is relatively large at this wavelength ($\varepsilon_{350} = 5.7\ cm^{-1} \times mM^{-1}$). It has been necessary to develop a spectrophotometric method that will respond only to the formation and disappearance of the ADH-DPNH complex and reject the changes of light absorption caused by DPNH alone.

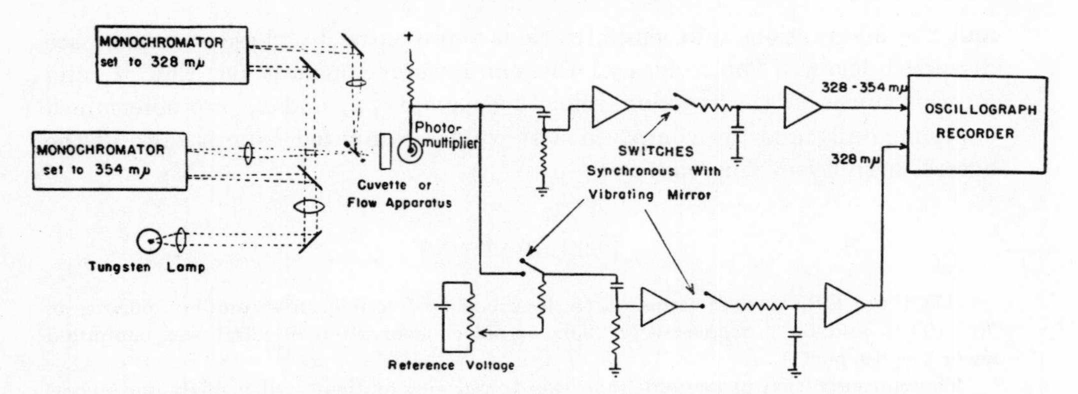

Fig. 1. A block diagram of the double beam system for measuring the difference of light absorption at 328 and 354 mμ as well as the light absorption at 328 mμ only. All parts of the optical system, except the lamp bulb, are quartz. The mirror and the switches vibrate at 60 cps. The symbol —▷— represents an amplifier. (MD-13.)

A consideration of the spectrum of DPNH shows that the extinction coefficients of DPNH are very nearly equal at the wavelengths 328 and 354 mμ. Thus a spectrophotometric method that measures only the difference of light absorption at these two wavelengths effectively rejects changes in concentration of DPNH. On the other hand the difference of extinction coefficients of the ADH-DPNH complex at these two wavelengths is measurable (about 1.4 $cm^{-1} \times mM^{-1}$ — see Table 1 of part I).

By alternately flickering light of a wavelength of 328 and 354 mμ through the ADH + DPNH solution and thence to a photocell, an alternating current wave is obtained whose amplitude represents only the difference of light absorption caused by the ADH-DPNH complex. And by more complex electronic circuits [2], it is possible to record simultaneously the change of light absorption at 328 mμ which is a measure of the [DPNH] since the ADH-bound and the free DPNH spectra have an isosbestic point at this wavelength. Fig. 2 illustrates the performance of this circuit. The lower trace indicates the deflection recorded by the circuit operating at 328 mμ when buffer alone and buffer plus 8.4 μM DPNH are alternately placed in the light path. The upper trace shows that the change of [DPNH] causes a change of only 1 % in the circuit measuring the difference of light absorption at 328 and 354 mμ. The wavelength readings for optimum rejection vary about ± 0.5 mμ. Since the wavelengths must be set accurately in order to achieve good rejection of DPNH, an experimental test similar to that of Fig. 2 is made before each experiment and the longer wavelength is adjusted to give optimum rejection (see Fig. 3). Thereby thermal drift of the monochromators is eliminated.

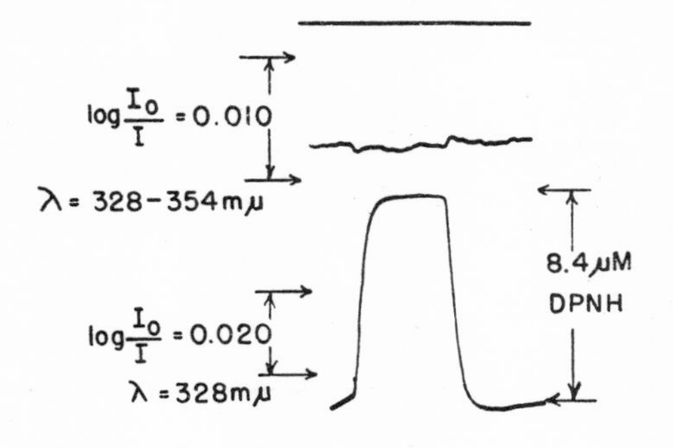

Fig. 2. The measurement of 8.4 μM DPNH by the optical density change at 328 mμ (lower trace) and the nearly complete rejection of this optical density change by a differential measurement between 328 and 354 mμ (upper trace) (Expt. 7b).

Under actual experimental conditions the ratio [DPNH] : [ADH] rarely exceeds 2 and the error in the measurement of [ADH · DPNH] caused by [DPNH] is about 2 %.

The sensitivities used in Fig. 2 are representative of those necessary for the measurement of the equilibrium and kinetics of the reactions of ADH and DPNH. The random fluctuations of the upper trace correspond to an optical density increment ($\log I_0/I$) of 5×10^{-4}. In view of the small errors caused by random fluctuations and by the change of [DPNH], it is possible to use concentrations of ADH and DPNH as low as a few micromolar.

A typical record of the titration of dilute ADH with DPNH is given in Fig. 3. In this case an open 3 ml cuvette is used and 0.01 ml additions of 0.63 *mM* DPNH are made with a stirring rod. The first deflection of the traces only represents the stirring of the solution and it is seen that no net change of optical density results. Each of the successive deflections (except the last

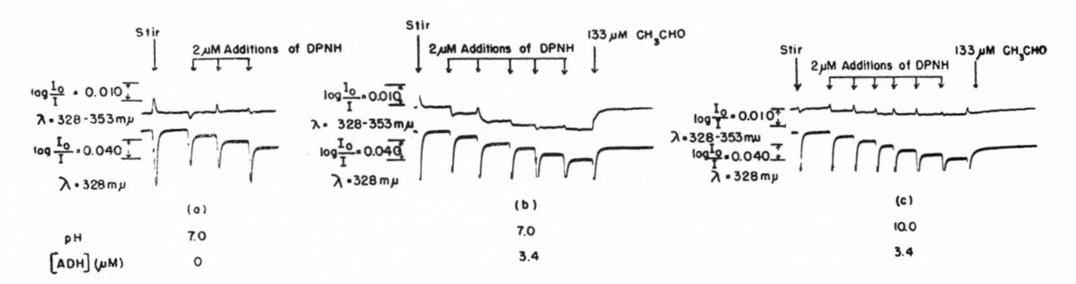

Fig. 3. Illustrating the titration of ADH with DPNH. The difference between the absorption at 328 and 354 mμ is recorded on the upper trace and the absorption at 328 mμ is recorded on the lower trace. Record a is a control experiment by which the rejection of DPNH by the circuit operating at 328—354 mμ is tested with [ADH] = 0. A downward deflection represents an increase in light absorption at 328 mμ and a decrease of absorption at 328—354 mμ. 3.4 μM ADH, pH = 7.0, 0.01 M PO_4''', pH = 10, 0.01 M glycine-NaOH buffers. (Expt. 5c, 33c—35.)

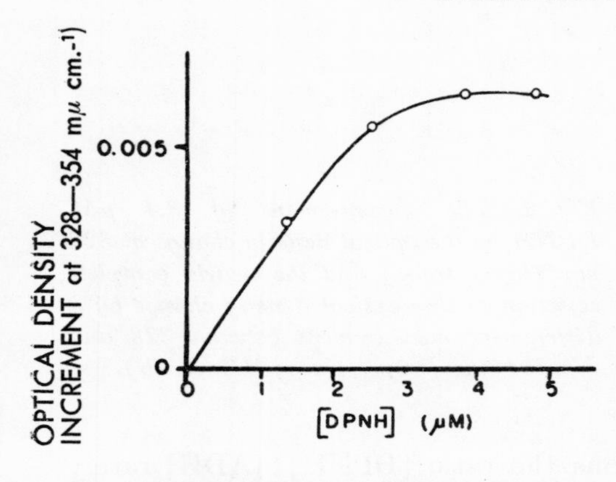

Fig. 4. The results of the titration of very dilute ADH with DPNH at pH = 7.0. The performance of the apparatus used in this experiment is given by Fig. 2. 1.17 μM ADH, pH = 7.0, 0.01 M PO_4''' buffer. (Expt. 6f.)

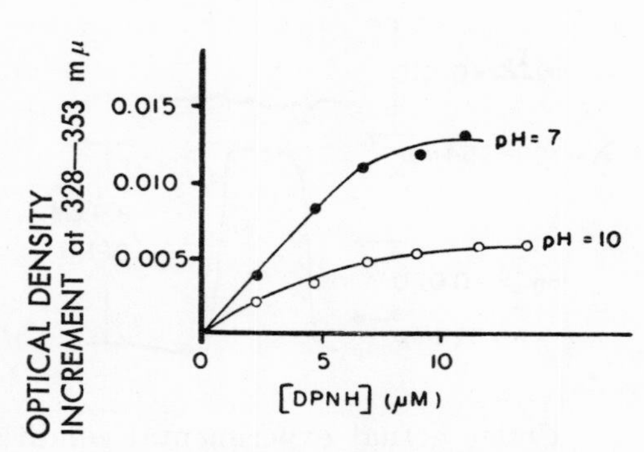

Fig. 5. The results of a titration of ADH with DPNH at pH = 7 and 10 plotted from the experimental data of Fig. 3. (Expt. 5c, 33c—35.)

one) correspond to the addition of 2 *μM* DPNH (0.01 ml of 0.6 *mM*) and it is seen that the deflections increase regularly at 328 mμ. In fact the exact amount of DPNH added is determined from this trace. In record (a) no ADH is present and the addition of DPNH causes no effect upon the circuit recording at 328—353 mμ. In record (b) 3.4 *μM* is present and differences between the absorption at 328 and 353 mμ are now recorded. On the first few additions of DPNH, a relatively large decrease of optical density is measured, but further additions cause relatively smaller increases. On the last addition, acetaldehyde (133 *μM*) is added which reacts with the ADH-DPNH complex and liberates ADH, causing the trace at 328—353 mμ to return to the original base-line. These records are discussed in detail below.

It was found that the combination of ADH with DPNH is too fast to be measured by ordinary mixing methods and a special flow apparatus having a rectangular observation tube with an optical path of 1 cm was used. Control experiments on the rate of formation of the peroxidase hydrogen peroxide complex show that the mixing is adequate for the measurement of reactions whose half-times are greater than 0.01 sec [2]. By using very dilute ADH and DPNH the half-times of the reactions actually measured were of the order of 0.1—0.05 sec for which this type of apparatus performs satisfactorily.

In summary, it may be said that the measurement of the kinetics and equilibrium of ADH and DPNH has been made possible by new spectrophotometric and rapid reaction techniques that have been used up to the limit of their sensitivity and time resolution. The accuracy of the results is therefore not as great as might be desired but is surely adequate for a preliminary exploration of the mechanism of the reaction.

THE EQUILIBRIUM OF DPNH AND ADH

The titrations reported earlier in part I were necessarily carried out at such high concentrations of ADH and DPNH that no measurable dissociation of DPNH occurred. Those experiments were, however, excellently suited to a determination of the stoichiometry of the ADH-DPNH reaction. With the more sensitive techniques described above it is possible to reduce the [ADH] by a factor of 10 to 100 and Fig. 4 shows that a measurable dissociation constant can be obtained when 1.17 μM ADH (see footnote p. 1127) is titrated with DPNH. The average of the two values of dissociation constant computed from Fig. 4 is about 10^{-7} *M* (based on the fact that ADH binds 2 DPNH at pH = 7.0 (part I). The affinity of ADH for DPNH is indeed very high.

The effect of pH upon the optical density change caused by the formation of the ADH-DPNH complex is shown by the original data of Fig. 3 which are plotted in Fig. 5. Since somewhat larger [ADH] must be used to give adequte deflections at pH = 10, the titration at pH = 7.0 gives a scarcely measurable dissociation constant. But the value of the dissociation constant at pH = 10 can be calculated to be roughly 3×10^{-6} *M* (based on the fact that ADH binds 1 DPNH at pH = 10 (part I). Thus there is an increased dissociation of DPNH from ADH in alkaline solutions.

THE RELATIONSHIP BETWEEN THE ADH-DPNH COMPLEX AND THE REDUCTION OF ALDEHYDE

The experiment in Fig. 3 at pH = 7 shows that the ADH-DPNH complex rapidly disappears upon the addition of acetaldehyde. In fact, the reaction with acetaldehyde is so rapid that it has been difficult to increase the acetaldehyde concentration much above a stoichiometric equivalent of DPNH without causing the life-time of the ADH-DPNH complex to be too short for accurate measurement. Fortunately, it has beens found that formaldehyde reacts with the ADH-DPNH complex much more slowly than acetaldehyde and therefore the most satisfactory kinetic experiments have been carried out on the reduction of formaldehyde by the system ADH + DPNH.

A typical example of the formation and disappearance of the ADH-DPNH complex in the presence of excess formaldehyde is given in Fig. 6. As usual, the disappearance of DPNH is measured simultaneously at 328 mμ. These

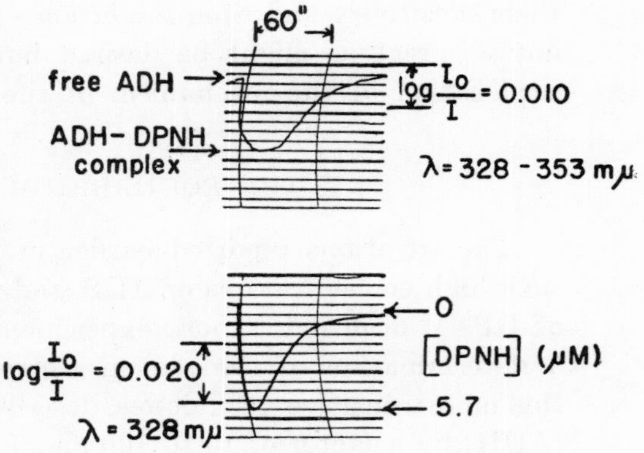

Fig. 6. The kinetics of formation and disappearance of the ADH · DPNH complex measured at 328—354 mμ (upper trace) and the simultaneous disappearance of DPNH measured at 328 mμ (lower trace). 1 cm optical path flow apparatus. 1.17 μM ADH, 8.8 μM DPNH (initial), 66 μM HCHO pH = 7.0, 0.01 M PO$_4'''$. (Expt. 7c—5.)

experiments were carried out in the flow apparatus by mixing ADH with DPNH plus HCHO. Since the experiment shown in Fig. 6 is the second of a duplicate set, the record begins with "free ADH" (+DPN) remaining from the previous test. On initiating the flow of reactants, the fairly rapid formation of the ADH-DPNH complex starts immediately as is indicated by the fall of the 328—354 mμ trace. The complex exists in a steady state for some seconds, and then decomposes into "free ADH". At 328 mμ, the initiation of the flow replaces the spent DPNH solution and causes the abrupt downward deflection (increase in optical density). The chemical reaction causing the disappearance of DPNH then proceeds and the linear upward sweep of the trace is sustained until the [DPNH] falls to a very low value.

One of the obvious criteria of an enzyme-substrate complex that follows the theory of Michaelis and Menten is that the half-time for the cycle of the enzyme-substrate complex should be twice the half-time for the overall reaction[7]. The values of the half times are 51 and 25 sec. for the cycle and the overall reaction respectively and, according to this criterion, verify the role of the ADH-DPNH complex to be that indicated by Eqns. 1 and 2.

The velocity constant for the reaction of ADH-DPNH with formaldehyde (k_4) is computed directly from experiments similar to those of Fig. 6 by measurement of the kinetics of the ADH-DPNH complex. The value of k_4 is computed according to the following equation (4)

$$k_4 = \frac{[\mathrm{DPNH}]_0}{[\mathrm{ADH} \cdot \mathrm{DPNH}]_m \; t_{\frac{1}{2}\,\mathrm{off}} \; [\mathrm{HCHO}]_0} \tag{4}$$

$[\mathrm{DPNH}]_0$ is the initial [DPNH] (M)
$[\mathrm{HCHO}]_0$ is the initial [HCHO] (M)

Table 1. A summary of values for the velocity of the reaction of formaldehyde and acetaldehyde with the ADH-DPNH complex. 1.17 μM ADH, pH = 7.0, 0.01 M PO_4''' 27° C, (Expt. 7c)
(note: 1.17 μM ADH gives 2.34 μM complex.)

$[DPNH]_0$ μM	8.8	3.7	7.4	7.4	$[DPNH]_0$ μM	7.5	15
$[HCHO]_0$ μM	66	66	132	330	$[CH_3CHO]_0$ μM	33	66
$[ADH\text{-}DPNH]_m$ μM	2.3	2.0	2.3	2.0	$[ADH\text{-}DPNH]_m$ μM	0.56	0.89
$t\frac{1}{2}$ off (sec.)	52	22	19	8.5	$t\frac{1}{2}$ off (sec.)	2	3
$k_4 \times 10^{-3}$ ($M^{-1} \times sec.^{-1}$)	1.1	1.3	1.3	1.3	$k_4 \times 10^{-5}$ ($M^{-1} \times sec.^{-1}$)	2	0.9

$[ADH\text{-}DPNH]_m$ is the maximum concentration of the complex formed in the particular experiment (M).* $t\frac{1}{2}$ off is the time interval between formation and half-disappearance of the ADH-DPNH complex (sec).

The value of k_4 for several experimental conditions is given in Table 1. The average value for formaldehyde is 1.3×10^3 $M^{-1} \times sec^{-1}$ at pH = 7.0.

Considerable difficulty was experienced in obtaining satisfactory "cycles" of the ADH-DPNH complex in the presence of a reasonable excess of acetaldehyde over DPNH for under those conditions the complex concentration ($[ADH\text{-}DPNH]_m$) is too small for accurate measurement. The preliminary value of k_4 for acetaldehyde based on the data of Table 1 is 10^5 $M^{-1} \times sec^{-1}$.

THE VELOCITY CONSTANT FOR THE FORMATION OF THE ADH-DPNH COMPLEX

The rapid formation of the ADH-DPNH complex is verified by records such as Fig. 6 which show that ADH and DPNH even in extremely dilute solution combine in much less than 1 sec. The flow method has therefore been used and a typical record is shown in Fig. 7. Formaldehyde is present so that the baseline before starting the flow corresponds to the optical density of free ADH. The flow is then started momentarily and the formation of the complex is observed. The flow is then restarted and is held near its maximum velocity for over one second in order to allow the photoelectric circuits to respond

* The concentration of ADH used in this computation is twice the actual molarity of ADH because two molecules of DPNH are bound to one molecule of ADH at pH = 7.0[3].

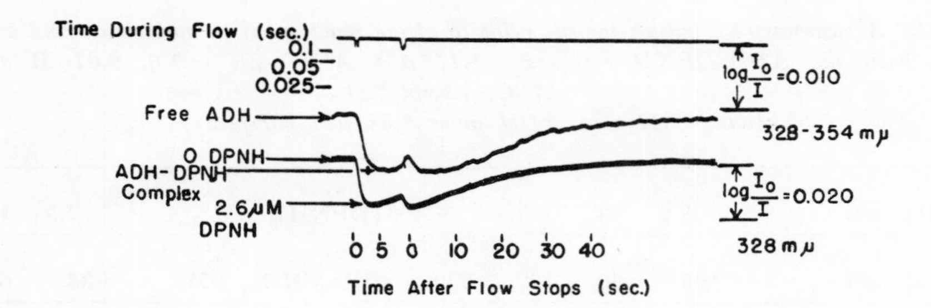

Fig. 7. A measurement of the speed of combination of ADH and DPNH. 1 cm path flow apparatus. 1.17 μM ADH, 3.7 μM DPNH, 66 μM HCHO, pH = 7.0, 0.01 M PO_4'''. *(Expt. 7c—10.)*

fully to the change of optical density. The trace clearly indicates that the maximum amount of complex is not reached during the flow. As soon as the flow stops, the maximum amount of complex is formed.

The trace at 328 mμ serves as a good control, in this case the second initiation of the flow should only restore the [DPNH] to its full value — and so it does.

The velocity constant for the formation of the ADH-DPNH complex (k_1) may be computed according to the second order equation on the assumption that the two DPNH molecules are bound independently and that the effective molarity of ADH is twice its actual concentration

$$k_1 = \frac{2.3}{t\,([\mathrm{DPNH}]_0 - [\mathrm{ADH}]_0)} \log \frac{[\mathrm{ADH}]_0\,([\mathrm{DPNH}]_0 - [\mathrm{ADH\text{-}DPNH}]_t)}{[\mathrm{DPNH}]_0\,([\mathrm{ADH}]_0 - [\mathrm{ADH\text{-}DPNH}]_t)} \tag{5}$$

The subscript 0 denotes initial concentrations and t denotes the concentration at time t.

A summary of the values of k_1 obtained with several values of $[\mathrm{DPNH}]_0$ is given in Table 2. The average of all values is $k_1 = 4 \times 10^6\ \mathrm{M}^{-1}, \times \mathrm{sec}^{-1}$, although the values obtained at higher [DPNH] are considered less accurate.

DISCUSSION

These experiments show that the combination of DPNH with liver alcohol dehydrogenase is a rapid reaction and that the DPNH is tightly bound to the protein. An estimate of the value of the velocity constant for the dissociation of DPNH molecules from the protein is given by the product of the dissociation constant and the velocity constant for the combination of DPNH and ADH

Table 2. A summary of data on the velocity constant for the combination of ADH and DPNH. 1.17 μM ADH, pH = 7.0, 0.01 M PO_4''', 27° C, (Expt. 7c) (note: 1.17 μM ADH gives 2.34 μM complex).

$[DPNH]_0$ (μM)	4.3	3.7	7.4	7.4
$[ADH\text{-}DPNH]_m$ (μM)	1.6	1.9	1.7	2.0
t (sec.)	0.067	0.12	0.12	0.095
$k_1 \times 10^{-6}$ ($M^{-1} \times sec.^{-1}$)	5.2	5.4	1.6	3.2

($4 \times 10^6 \times 10^{-7} = 0.4$ sec.$^{-1}$). Thus the half-time for the dissociation reaction is about 1.7 sec. $(\frac{0.693}{0.4})$.

The complex of ADH and DPNH also reacts fairly rapidly with aldehydes, the velocity constants are 1.3×10^3 and 10^5 $M^{-1} \times sec^{-1}$ for formaldehyde and acetaldehyde respectively.

The results of these kinetic tests may be summarized according to the following equations for the case of acetaldehyde:

$$\text{ADH} + \text{DPNH} \underset{k_2 = 0.4 \text{ sec}^{-1}}{\overset{k_1 = 4 \times 10^6 \text{M}^{-1} \times \text{sec}^{-1}}{\rightleftharpoons}} \text{ADH-DPNH} \tag{6}$$

$$\text{ADH} \cdot \text{DPNH} + \text{Ald} + \text{H}^+ \overset{k_4 \approx 10^5 \text{ M}^{-1} \times \text{sec}^{-1}}{\rightleftharpoons} \text{ADH-DPN}^+ + \text{Alc} \tag{7}$$

The velocity constant k_4 represents the value for the reaction of ADH-DPNH and aldehyde at pH = 7.0.

At pH = 7, the reaction of alcohol and ADH-DPN is very slow and is neglected.

In the presence of various small formaldehyde concentrations, the values of k_4 are reasonably constant and it is probable that the dissociation of DPN from the complex is not a rate limiting step with dilute formaldehyde. In this case the reaction mechanism is identical to that previously studied for catalases and peroxidases:

$$\text{E} + \text{S} \rightleftharpoons \text{ES} \tag{8}$$
$$\text{ES} + \text{A} \rightleftharpoons \text{E} + \text{P} \tag{9}$$

and the methods of computing k_4 derived previously are valid [8].

When more concentrated aldehyde solutions are used an accumulation of the ADH-DPN complex is to be expected. Studies of the overall reaction are in accord with this explanation because a definite maximum in the aldehyde-activity relationship is obtained [3]. Although such a maximum might be attributed to an ADH-aldehyde complex, the accumulation of the ADH-DPN complex would provide an equally satisfactory explanation. The next step in the reaction is then assumed to be

$$\text{ADH-DPN}^{+} \underset{k_5}{\overset{k_3}{\rightleftharpoons}} \text{ADH} + \text{DPN}^{+} \qquad (10)$$

In order to determine whether the mechanism outlined by the Eqns. 6, 7 and 10 is compatible with the observed data on the overall activity of ADH, it is necessary to determine the relationships between the reaction velocity constants as defined by these equations and the values of activity and the "Michaelis constants" for DPNH and acetaldehyde. Eqns. 6, 7, and 10 are comverted into the usual form for computations of enzyme-substrate kinetics:

$$\underset{e - p_1 - p_2}{\text{E}} + \underset{x_1}{\text{S}_{\text{I}}} \underset{k_2}{\overset{k_1}{\rightleftharpoons}} \underset{p_1}{\text{ES}_{\text{I}}} \qquad (11)$$

$$\underset{p_1}{\text{ES}_{\text{I}}} + \underset{a_1}{\text{A}} \overset{k_4}{\rightleftharpoons} \underset{p_2}{\text{ES}_{\text{II}}} + \underset{a_2}{\text{AH}} \qquad (12)$$

$$\underset{p_2}{\text{ES}_{\text{II}}} \underset{k_5}{\overset{k_3}{\rightleftharpoons}} \underset{e - p_1 - p_2}{\text{E}} + \underset{x_2}{\text{S}_{\text{II}}} \qquad (13)$$

In this form the first two steps of the reaction mechanism are seen to be a form of the simple mechanism shown to apply, for example, to catalase, alkylhydrogen peroxides, and alcohol [9]. But in this case the enzyme is not released until ES_{II} is dissociated. As usual, the concentrations of the various reactants at any time are written underneath the appropriate symbols. Since only a steady state analysis is required, only the differential equations for p_1, p_2, and a are required.

$$\frac{dp_1}{dt} = k_1 \ x_1 \ (e - p_1 - p_2) - (k_2 + k_4 a_1) \ p_1 \tag{14}$$

$$\frac{dp_2}{dt} = k_4 a_1 p_1 - k_3 p_2 + k_5 x_2 \ (e - p_1 - p_2) \tag{15}$$

$$\frac{da_1}{dt} = k_4 a_1 p_1 \tag{16}$$

In the steady-state both $\frac{dp_1}{dt}$ and $\frac{dp_2}{dt}$ are negligible and p_1 and p_2 can be computed.

$$p_1 = \frac{k_1 \ x_1 \ (e - p_2)}{k_1 x_1 + k_2 + k_4 a_1} = \frac{e - p_2}{1 + \frac{K_{m_1}}{x_1}} \tag{17}$$

$$K_{m1} = \frac{k_2 + k_4 a_1}{k_1} \tag{18)*}$$

$$p_2 = \frac{k_4 \ a_1 p_1 + k_5 \ x_2 \ (e - p_1)}{k_3 + k_5 x_2} = \frac{e - p_1 \left(1 - \frac{k_4 a_1}{k_5 x_2}\right)}{1 + \frac{K_{m_2}}{x_2}} \tag{19}$$

$$K_{m_2} = \frac{k_3}{k_5} \tag{20}$$

K_{m_2} is actually the dissociation constant of the ADH-DPN complex and is about 2×10^{-5} M. See below p. 1141.

$$\frac{p_1}{e} = \frac{1}{\left(1 + \frac{x_2}{K_{m_2}}\right) \left(1 + \frac{K_{m_1}}{x_1}\right) - \frac{x_2}{K_{m_2}} + \frac{k_4 a_1}{k_3}} \tag{21}$$

The overall activity is usually measured by the rate of disappearance of DPNH, $-\frac{da_1}{dt}$, and the turnover number is $\frac{1}{e} \times \frac{da_1}{dt}$

From Eqn. 16,

* K_{m_1} is usually regarded as a Michaelis constant.

$$\frac{1}{e} \times \frac{da_1}{dt} = k_4 a_1 \times \frac{p_1}{e} \qquad (22)$$

and on substituting for $\frac{p_1}{e}$ its value given in Eqn. 21,

$$\frac{1}{e}\frac{da_1}{dt} = \frac{1}{k_4 a_1 \left[\left(1 + \frac{x_2}{K_{m_2}}\right)\left(1 + \frac{K_{m_1}}{x_1}\right)\right] - \frac{x_2}{K_{m_2}} + \frac{1}{k_3}} \qquad (23)$$

Under the conditions of these experiments a very small [DPN] is formed since the initial [DPNH] is very low. And in studies of the overall activity, the initial rate is measured (part I) and again very little [DPN] forms. Thus the amount of DPN under both these conditions is probably negligible compared to K_{m_2} ($\sim 1 \times 10^{-5}$ M). Eqn. 23 is therefore simplified as follows

$$\frac{1}{e}\frac{da_1}{dt} = \frac{1}{\frac{1}{k_1 x_1} + \frac{1}{k_4 a_1}\left(1 + \frac{k_2}{k_1 x_1}\right) + \frac{1}{k_3}} \qquad (24)$$

The values of [DPNH] giving maximal turnover number then depends upon the [aldehyde]. For large [aldehyde],

$$\frac{1}{e}\frac{da_1}{dt} = \frac{1}{\frac{1}{k_1 x_1} + \frac{1}{k_3}} \qquad (25)$$

and the maximum turnover number for large [DPNH] is

$$\frac{1}{e}\frac{da_1}{dt} = k_3 \qquad (26$$

The [DPNH] giving half maximal activity in the presence of excess aldehyde is

$$(x_1)_{\frac{1}{2}} = \frac{k_3}{k_1} \qquad (27)$$

Thus at *high* [aldehyde], the "Michaelis constant"for DPNH depends upon the rate of dissociation of DPN from ADH, and not at all upon the rate of dissociation of DPNH from ADH.

For *low* [aldehyde] $(k_3 \gg k_4 a)$

$$\frac{1}{e}\frac{da_1}{dt} = \frac{1}{\dfrac{1}{k_1 x_1} + \dfrac{1}{k_4 a_1}\left(1 + \dfrac{k_2}{k_1 x_1}\right)} \tag{28}$$

and the maximum turnover number is

$$\frac{1}{e}\frac{da_1}{dt} = k_4 a_1 \tag{29}$$

The [DPNH] giving half maximal activity in the presence of low [aldehyde] is

$$(x_1)_{\frac{1}{2}} = \frac{k_2 + k_4 a_1}{k_1} = K_{m_1} \tag{30}$$

Thus only at very low [aldehyde] will the true dissociation constant for the ADH-DPNH complex be measured (when $k_2 \gg k_4 a_1$).

At large [DPNH], the effect of aldehyde upon the activity will be as follows:

$$\frac{1}{e}\frac{da_1}{dt} = \frac{1}{\dfrac{1}{k_4 a_1} + \dfrac{1}{k_3}} \tag{31}$$

For large [aldehyde] and [DPNH], the maximum turnover number is k_3 as already given by Eqn. 26. The [aldehyde] giving half maximal activity is

$$(a_1)_{\frac{1}{2}} = \frac{k_3}{k_4} \tag{32}$$

These formulas and the relevant assumptions are summarized in Table 3.

The values $(a_1)_{\frac{1}{2}}$ and $(x_1)_{\frac{1}{2}}$ corresponding to the conditions of Eqns. 27 and 32 as well as the value of k_3, have already been published by Theorell and Bonnichsen, see Table 7 part I.

Since no overall data are available on the reactions of ADH, DPNH, and formaldehyde, some new data are plotted in Fig. 8 and are summarized

Table 3. Summary of equations for calculation of reaction velocity constants (x_2, $a_2 \approx 0$).

Equation number	26	27	29	30	31	32
x_1	∞	$\frac{k_3}{k_1}$	∞	$\frac{k_2 + k_4 a_1}{k_1}$	∞	∞
a_1	∞	∞	small	small	variable	$\frac{k_3}{k_4}$
$\frac{1}{e}\frac{da_1}{dt}$	k_3	$\frac{k_3}{2}$	$k_4 a_1$	$\frac{k_4 a_1}{2}$	$\frac{1}{\frac{1}{k_4 a_1} + \frac{1}{k_3}}$	$\frac{k_3}{2}$

in Table 4. The values of the reaction velocity constants k_1 and k_4 are computed from the overall data on acetaldehyde and formaldehyde and are compared with the values obtained by direct measurements of the reaction kinetics of ADH-DPNH in Table 5.

The agreement of the overall and direct data is especially close in the tests in which dilute formaldehyde was used. And this corresponds most closely to the conditions used in the direct studies of the ADH-DPNH complex. In view of the fact that there is over 100 fold difference in the [ADH] for the two studies, the agreement is considered to be satisfactory.

If the mechanism described here applies to the oxidation of alcohol by ADH and DPN, the formulae derived for the computation of the reaction velocity constants can be used. Although no direct kinetic measurements have been made on the ADH-DPN complex, the value of k'_3 for the alcohol-DPN system (the velocity constants for this system are designated by primes) should be the same as the value of k_2 for the ADH-DPNH. The value of k'_3 computed from the data of Theorell and Bonnichsen at pH = 6.8 is 1.1 sec^{-1} and is in fair agreement with our calculated value of k_2 (0.4 sec^{-1}).

The value of K_m of Eqn. 20 ($K_m = \frac{k'_3}{k'_5}$) has already been determined by Theorell and Bonnichsen (part I) to be about 200 times greater than the dissociation constant of the ADH-DPNH complex at pH = 7.0, thus $\sim 2 \times 10^{-5}$. Since k_3 is already known (see Table 7, part I, k'_3 = 39—45 sec^{-1} at pH = 7.0), k'_5, the velocity constant for the combination of ADH and DPN, is computed to be 2×10^{6} M^{-1} $\times$ sec^{-1}. This value is in remarkably good agreement with the value of k_1, the velocity constant for the combination of ADH and

11

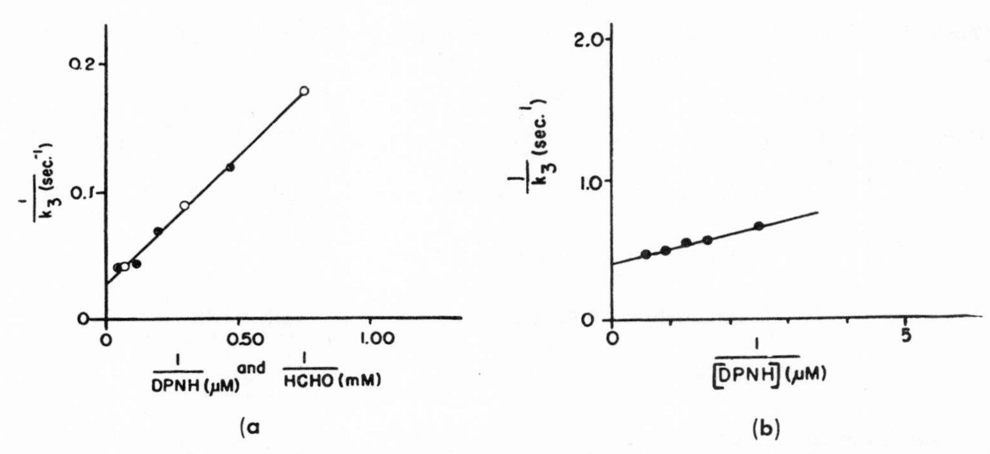

Fig. 8. The effect of [DPNH] and [HCHO] upon the activity of ADH. In Fig. 8 a the solid circles represent the values obtained with varying [DPNH] for [HCHO] = 10 mM. The open circles represent the values obtained with varying [HCHO] and [DPNH] = 8.6 μM. [ADH] = 3 × 10^{-9} M. In Fig. 8b, the effect of varying [DPNH] for 0.67 mM HCHO is shown. [ADH] = 15 × 10^{-9} M. pH = 7.22, 0.01 M PO_4''' for all experiments. (Expts. 9e, 9g.)

Table 4. Summary of overall data on the reaction of ADH, DPNH, and HCHO (pH = 7.22, 26°).

Substance	K_m (M)	k_3 (sec.$^{-1}$)	k_4a_1 (sec.$^{-1}$)	[HCHO] (mM)	[DPNH] (μM)
Formaldehyde	8 × 10^{-3}	19	—	—	8.6
DPNH	8 × 10^{-6}	19	—	10	—
DPNH	2.6 × 10^{-7}	—	1.25	0.67	—

DPNH, in view of the uncertainties of the various quantitites. The combination of ADH with DPN or DPNH apparently occurs at about the same speed, but their dissociation velocities differ considerably, DPNH being bound much more tightly. And on this basis the different activities of ADH towards alcohol and aldehyde are readily explained.

The close agreement of the reaction velocity constants for the combination of ADH with DPN and DPNH is reasonable in view of the similarity of the latter two molecules and in view of the probability that they combine at the

Table 5. Velocity constants for the reactions of ADH computed on the basis of the mechanism of equations 6, 7, and 10.

	Substrate used	k_1 ($M^{-1} \times sec.^{-1}$)	k_4 ($M^{-1} \times sec.^{-1}$)	k_2 ($sec.^{-1}$)
Computed from overall data	Acetaldehyde	1.6×10^6	2.2×10^5	—
	Formaldehyde (10 μM)	2.4×10^6	2.4×10^3	—
	Formaldehyde (0.67 μM)	4.8×10^6	1.9×10^3	—
	Alcohol	—	—	1.1
Measured directly from the kinetics of ADH-DPNH	Formaldehyde	4×10^6	—	0.4
	Formaldehyde	—	1.3×10^3	—
	Acetaldehyde	—	10^5	—

same position on the protein. The latter supposition is supported by preliminary experiments on the competition between DPN and DPNH for ADH.

There are many aspects of these reactions that require further study but these preliminary results encourage us to believe that the detailed analysis of the mechanism of action of catalases and peroxidase based on direct studies of enzyme-substrate compounds may be applied to the reactions of many enzyme systems.

SUMMARY *

1. A rapid spectrophotometric method for measuring the formation and disappearance of the compound of ADH and DPNH without appreciable interference from the absorption of DPNH has been developed.

2. A titration of very dilute ADH (1.17 μM) with DPNH gives a dissociation constant for the ADH-DPNH complex of 10^{-7} M at pH = 7.0.

3. The velocity constant for the formation of the ADH-DPNH complex is $4 \times 10^6 M^{-1} \times sec^{-1}$ at pH = 7.0.

4. The velocity constant for the dissociation of DPNH from ADH is computed to be 0.4 sec^{-1}.

5. The velocity constant for the reaction of the ADH-DPNH complex with formaldehyde is $1.3 \times 10^3 M^{-1} \times sec^{-1}$ at pH = 7.0. The values of k_4

* The experiments were carried out at 26° C.

are constant over a reasonable range of experimental conditions. A preliminary value for the velocity constant for acetaldehyde is 10^5 $M^{-1} \times sec^{-1}$ at pH = 7.0.

6. The ADH-DPNH complex in the presence of dilute formaldehyde fulfils the requirements for a Michaelis intermediate.

7. On the basis of the following mechanism for the action of ADH,

$$ADH + DPNH \underset{k_2}{\overset{k_1}{\rightleftharpoons}} ADH \cdot DPNH$$

$$H^+ + ADH \cdot DPNH + Aldehyde \overset{k_4}{\rightleftharpoons} ADH\text{-}DPN^+ + alcohol$$

$$ADH\text{-}DPN^+ \underset{k_5}{\overset{k_3}{\rightleftharpoons}} ADH + DPN^+,$$

the velocity constants k_1, k_4, and k_2 have been computed from data on the overall activity of the enzyme in very dilute solutions and values of 3×10^6 $M^{-1} \times sec^{-1}$, 2.2×10^5 $M^{-1} \times sec^{-1}$ (for acetaldehyde), 1.3×10^3 $M^{-1} \times sec^{-1}$ (for formaldehyde) and 1.1 sec^{-1} respectively are obtained and agree reasonably well with the values obtained by direct measurements of the kinetics of the ADH · DPNH complex. This agreement has been obtained without assuming the formation of compounds of ADH with aldehyde or alcohol.

8. The velocity constant for the combination of ADH and DPN is calculated to be about 2×10^{-6} M^{-1} sec^{-1} at pH = 7.0 and it is concluded that DPN and DPNH are bound by ADH at about the same speed and on the same place.

REFERENCES

1. Chance, B. In *Advances in Enzymology* **12** (1951) 153.
2. Chance, B. *Rev. of Sci. Inst.* **22**, (1951) 619—638.
3. Theorell, H. *8e Conseil de Chimie de l'Institut International de Solvay*, Bruxelles 1950, p. 395; Theorell and Bonnichsen, Part I. *Acta Chem. Scand.* **5** (1951) 1105.
4. Bonnichsen, R. K. *Acta Chem. Scand.* **4** (1950) 714.
5. Bonnichsen, R. K. *Acta Chem. Scand.* **4** (1950) 715.
6. MacFadyen, D. A. *J. Biol. Chem.* **17** (1945) 107.
7. Chance, B. In *Modern trends in physiology and biochemistry* (in press) (Interscience, N. Y.)
8. Chance, B. *J. Biol. Chem.* **151** (1943) 553.
9. Chance, B. *J. Biol. Chem.* **179** (1949) 1341.

Received April 28, 1951.

Reprinted from THE JOURNAL OF BIOLOGICAL CHEMISTRY, Vol. 245, 1970

The Oxygenated Form of L-Tryptophan 2,3-Dioxygenase as Reaction Intermediate*

(Received for publication, September 26, 1969)

YUZURU ISHIMURA, MITSUHIRO NOZAKI, AND OSAMU HAYAISHI

From the Department of Medical Chemistry, Kyoto University Faculty of Medicine, Kyoto, Japan

TAKAO NAKAMURA,‡ MAMORU TAMURA, AND ISAO YAMAZAKI

From the Biophysics Division, Research Institute of Applied Electricity, Hokkaido University, Sapporo, Japan

SUMMARY

In order to clarify the reaction mechanism of L-tryptophan 2,3-dioxygenase (L-tryptophan: oxygen oxidoreductase, EC 1.13.1.12), a hemoprotein, spectral and kinetic studies were carried out with highly purified enzyme preparations from *Pseudomonas fluorescens* (ATCC 11299).

A new spectrally distinct species of the enzyme heme (λ_{max}; 418, 545, and 580 mμ) was observed during the steady state of the catalytic reaction. The formation of the new spectral species was absolutely dependent on the simultaneous presence of both oxygen and L-tryptophan with the ferrous enzyme. In the absence of L-tryptophan, the ferrous heme in the enzyme was oxidized to a ferric state by molecular oxygen. Available evidence indicated that the observed spectrum was due to a ternary complex of oxygen, the enzyme, and L-tryptophan and represented the oxygenated form.

By means of rapid reaction spectrophotometry, the oxygenated form was shown to be an obligatory intermediate of the reaction. The rate of the over-all reaction, as judged by the accumulation of L-formylkynurenine, was always proportional to the amount of the oxygenated form present during the entire course of the reaction. The rate constant for the decomposition of the oxygenated form was estimated to be 19 sec^{-1}, which agreed well with the turnover number of the enzyme (18 sec^{-1}). The rate constant for the binding of oxygen with the ferrous enzyme in the presence of L-tryptophan and that for the reverse reaction were also determined to be 5×10^6 M^{-1} sec^{-1} and 230 sec^{-1}, respectively. Thus, the binding of oxygen with the ferrous enzyme in the presence of L-tryptophan is a reversible process, as are those with hemoglobin and myoglobin.

Some properties of the *Pseudomonas* enzyme are also described.

* This investigation was supported in part by Public Health Service Research Grants CA-04222, from the National Cancer Institute, and AM-10333, from the National Institute of Arthritis and Metabolic Diseases, and by grants from the Squibb Institute for Medical Research and the Scientific Research Fund of the Ministry of Education of Japan. The data were taken from a dissertation submitted by Yuzuru Ishimura in November 1968 to the Graduate School of Kyoto University in partial fulfillment of the requirements for the degree of Doctor of Medical Science.

‡ Present address, Department of Biology, Faculty of Science, Osaka University, Toyonaka, Osaka, Japan.

L-Tryptophan 2,3-dioxygenase, commonly known as tryptophan pyrrolase or tryptophan oxygenase,[1] is an enzyme which catalyzes the oxidative ring cleavage of L-tryptophan to L-formylkynurenine (1). The enzyme has been isolated both from animal liver and *Pseudomonas* (2, 3).

Since the enzyme activity was first observed in rat liver extracts by Kotake and Masayama in 1936 (4), the catalytic reaction mechanism as well as its regulatory control mechanism has been the subject of extensive investigations. With the use of the heavy isotope of oxygen ($^{18}O_2$), atoms of oxygen incorporated into L-formylkynurenine were found to be derived from molecular oxygen and not from water (5). Tanaka and Knox (6), using partially purified preparations both from *Pseudomonas* and from rat liver, identified this enzyme as an iron porphyrin protein. Thus, the enzyme was shown to be a unique dioxygenase containing heme as prosthetic group. Later, Feigelson and Greengard (7, 8), using liver enzyme, showed that the apoenzyme could be obtained in an inactive form and that the catalytic activity was restored upon addition of exogenous hematin. Although it has been suggested that the heme iron in the enzyme might be the oxygen-binding site (9), no direct evidence has so far been available either to prove or disprove it. Thus, there has been a considerable discussion in the literature as to the reaction mechanism of the catalytic process, especially the role of heme iron and its valence state during the catalysis (10–14).

The present paper describes the detection and characterization of a new spectral species of L-tryptophan 2,3-dioxygenase which was observed during the steady state of the reaction. Spectral and kinetic analyses of the new species revealed that it is the oxygenated state of heme in the enzyme, and an obligatory intermediate of the reaction. The role of substrate, L-tryptophan, in the oxygenated form is also discussed, together with some other properties of the enzyme. Preliminary accounts of a portion of this work have appeared (15–17).

EXPERIMENTAL PROCEDURE

Materials and Methods

Reagents—L-Tryptophan was purchased from Yoneyama Chemical Industries, Ltd. Tryptophan analogues, including

[1] L-Tryptophan 2,3-dioxygenase is preferred because the term L-tryptophan oxygenase may be confused with L-tryptophan 5-monooxygenase and L-tryptophan oxygenase (decarboxylating).

2,3-dihydro-DL-tryptophan, 4-methyl-, and 5-methyl-DL-tryptophan, were generous gifts of Dr. B. Witkop, Laboratory of Chemistry, National Institute of Arthritis and Metabolic Diseases, National Institutes of Health. α-Methyl-DL-tryptophan from Sigma was kindly donated by Dr. J. B. Wittenberg of the Albert Einstein College of Medicine. Purified preparations of protohematin IX were kindly provided by Dr. P. Feigelson of Columbia University. 5-Hydroxy-L-tryptophan and 5-hydroxy-D-tryptophan were products of Calbiochem. Streptomycin sulfate was obtained from Meiji-Seika Company, Ltd., Tokyo. Crystalline catalase was a product of Sigma. Horseradish peroxidase was prepared according to the method of Paul (18). All other chemicals were of analytical grade.

Instruments—Usual spectrophotometric measurements were carried out either with a Cary model 15 recording spectrophotometer or with a Hitachi model EPS-2 recording spectrophotometer. For rapid scan spectrophotometry, a Hitachi rapid scan spectrophotometer,[2] model RSP-2 equipped with a Hitachi memoriscope V-104 and an automatic scanner, was used. The wave length scale of the spectrophotometer was calibrated with a Didymium glass filter. The equipment used for the stopped flow experiments (20) was a modified Hitachi spectrophotometer model 139 combined with a flow apparatus, based on the design by Chance (21). Details of the equipment have been described previously (22). Oxygen consumption was measured polarographically with a rotating platinum electrode (23).

Determinations—Protein was measured either by the biuret method (24) with crude enzyme preparations or by the method of Warburg and Christian (25). The hematin content of the enzyme preparation was determined as the pyridine-ferrohemochrome according to the method of Keilin and Hartree (26).

Enzyme Assay—The standard assay system is a modification of that originally described by Tanaka and Knox (6) and contained 25 μmoles of L-tryptophan, 10 μmoles of L-ascorbate, 250 μmoles of potassium phosphate buffer (pH 7.0), and the enzyme in a final volume of 2.5 ml. The assays were carried out at 24° under normal atmospheric conditions in a cuvette with 1-cm light path. The reaction was initiated by the addition of enzyme and the increase in optical density at 321 mμ due to the formation of formylkynurenine (ϵ = 3750) was followed continuously with a recording spectrophotometer. After a short lag period (approximately 1 min), the reaction proceeded linearly with time until optical density reached 0.6, and the rate was proportional to the amount of enzyme up to 0.1 unit. When the reduced form of enzyme was used for the assay instead of the oxidized form, ascorbate was omitted from the system. The enzyme activity was also determined by the measurement of oxygen consumption polarographically with the same reaction mixture. One unit of enzyme activity is defined as that amount which catalyzes the conversion of 1 μmole of L-tryptophan to L-formylkynurenine per min under the standard assay conditions. Specific activity is expressed as units per mg of protein.

Preparations of Reduced Form of Enzyme and Anaerobic Conditions—The heme in tryptophan 2,3-dioxygenase is autoxidizable and hence the purified enzyme was usually obtained as the ferric form (native). The ferric enzyme was reduced by either one of the following procedures: (*a*) quantitative additions of sodium dithionite under anaerobic conditions or (*b*) illumination of light both in the presence and absence of L-tryptophan under anaerobic conditions (14, 27, 28). Anaerobic conditions were obtained either by gassing the reaction mixture with purified nitrogen for 15 min or by the use of Thunberg type cuvettes.

[2] This was originally constructed by Baba and Shindo (19) of Research Institute of Applied Electricity, Hokkaido University, Sapporo, Japan, and has been developed by Hitachi, Ltd., Japan.

Preparation of Enzyme

Growth of Bacteria—*Pseudomonas fluorescens* (ATCC 11299) was used as the source of enzyme. The organism was subcultured on an agar slant and then grown in a liquid medium containing 0.2% L-tryptophan, 0.13% yeast extract, 0.3% NH_4Cl, 0.15% K_2HPO_4, 0.05% KH_2PO_4, and 0.02% $MgSO_4 \cdot 7H_2O$. Inoculations were carried out by transferring the cells from a slant to 5-liter Erlenmeyer flasks containing 1.5 liters of the sterilized liquid medium. Cells were grown at 23° for 20 to 24 hours with vigorous mechanical shaking and were harvested with a Sharples centrifuge, shortly before they reached maximal growth. The yield of wet packed cells was about 2.5 g per liter of medium. The packed cells could be stored for several weeks with little loss of activity at −15°.

Crude Extracts—All subsequent operations were carried out at about 5°, centrifugations were carried out at 10,000 × *g* for 20 min, and all of the phosphate buffers contained L-tryptophan in a final concentration of 1 mM to stabilize the enzyme, unless otherwise noted.

Wet packed cells, 200 g, were ground mechanically with 400 g of aluminum oxide (Wako W-800) in a chilled porcelain mortar for about 10 min, until the mixture became viscous. Two liters of 0.01 M potassium phosphate buffer, pH 7.0 were added to the mixture and the resulting slurry was centrifuged at 13,000 × *g* for 30 min to remove cell debris and alumina.

Streptomycin Treatment—To the viscid crude extracts (1,800 ml) were added 100 ml of a 10% streptomycin sulfate solution with mechanical shaking. After further stirring for 15 min, the massive precipitate was removed by centrifugation at 13,000 × *g* for 20 min.

Ammonium Sulfate Fractionation—To the streptomycin supernatant solution (1850 ml) were added 450 g of ammonium sulfate with stirring. After 30 min, the resulting precipitate was collected by centrifugation and dissolved in 0.05 M phosphate buffer, pH 7.0, in a final volume of 200 ml.

Heat Treatment under Anaerobic Conditions—To the ammonium sulfate fraction, 20 ml of 0.04 M L-tryptophan were added, and the mixture was kept at room temperature for 10 min under a nitrogen stream. The dissolved oxygen in the medium was almost completely eliminated by this process. Then the solution was heated to 53–55° for 5 min in a water bath under nitrogen. After cooling to 5° in an ice bath, denatured protein was removed by centrifugation. To the supernatant solution (195 ml), 28 g of ammonium sulfate were added with stirring, and, after 30 min, the resulting precipitate was collected by centrifugation. The red precipitate was dissolved in a minimum volume of 0.05 M phosphate buffer (approximately 10 ml), pH 7.0. Ammonium sulfate was removed with a Sephadex G-25 column (4 × 15 cm) which had been equilibrated with the same buffer. All of the colored eluate was collected until the presence of ammonium sulfate was detected.

DEAE-cellulose Chromatography—The eluate from a Sephadex G-25 column was diluted five times with distilled water and then placed on a DEAE-cellulose column (4.5 × 15 cm), which had previously been equilibrated with 0.01 M potassium phosphate buffer, pH 7.0. The column was washed successively with each 200 ml of 0.01 M and 0.05 M phosphate buffer, pH 7.0. The enzyme was then eluted with 200 ml of 0.15 M phosphate buffer, concentrated by the addition of ammonium sulfate (40% saturation), and dissolved in a minimum volume of 0.05 M phosphate buffer, pH 7.0.

Sephadex G-200 Column Chromatography—The concentrated solution (2 to 3 ml) was passed through a Sephadex G-200 column (2.5 × 50 cm) equilibrated with 0.05 M phosphate buffer, pH 6.5, free of L-tryptophan. Usually, two protein fractions were obtained as judged by absorbance at 280 mμ; one appeared at the void volume of the column and the other was eluted a little later. The middle three-fifths of the total volume of the second peak was collected for use.

The purification procedure resulted in an over-all purification of about 180-fold with 5% in yield. The enzyme thus obtained had a specific activity of 5.3 units per mg of protein (13). More recently, Poillon *et al.* (29) described a method of purification which yielded homogeneous enzyme preparations. However, our preparations after DEAE-cellulose chromatography were essentially free of other hemoproteins as well as other chromophores such as flavoprotein and were therefore used for spectral and kinetic studies. Catalase and peroxidase activities were found to be negligible as judged by the spectrophotometric method of Chance (30) and the pyrogallol test (31), respectively. Further attempts to purify the enzyme were usually accompanied by a decrease in activity per hematin molecule. The enzyme can be stored for several days without an appreciable loss of activity if it is kept at 0° under anaerobic conditions in the presence of 1 mM tryptophan.

RESULTS

Properties of Enzyme

Absorption Spectra—The absorption spectra of both the oxidized (native) and the dithionite-reduced forms of the purified tryptophan 2,3-dioxygenase in the presence and absence of L-tryptophan are shown in Fig. 1. These represent typical absorption spectra of a high spin protohemoprotein having absorption maxima at 404.5, 500, and 635 mμ in the ferric state and at 432, 553, and 588 mμ in the ferrous state, respectively. The highest ratio of the absorbance at 404.5 mμ to that at 280 mμ so far obtained was 1.5 in the ferric state of enzyme. Almost no change of the spectra was observed between pH 6.0 and 9.0. The ferric enzyme was instantaneously reduced by the addition of either sodium dithionite or borohydride both in the presence and absence of L-tryptophan. The reduced heme was autoxidizable in the absence of L-tryptophan. The spectral changes in the Soret region caused by substrate were first reported by Maeno and Feigelson (14) and were confirmed by us (15). As shown in the *inset* of the figure, addition of L-tryptophan to either form of enzyme under anaerobic conditions caused slight but significant changes of their spectra not only in the Soret region but also in the entire visible region. These results suggest that L-tryptophan combines with the enzyme irrespective of the valence state of its heme in the absence of oxygen. No such spectral changes were observed either when other aromatic amino acids such as tyrosine and histidine were added to the enzyme solution or when L-tryptophan was added to other high spin protohemoproteins such as catalase and peroxidase under similar conditions.

In Fig. 2 is shown the absorption spectrum of the pyridine ferrohemochrome of the enzyme as well as that of the authentic hematin. Coincidence of the absorption maxima and troughs indicates that the iron porphyrin in the enzyme is protohematin IX, as reported previously (13). Extinction coefficients of the Soret absorption of the enzyme in the absence of L-tryptophan were estimated to be 175 × 10³ and 145 × 10³ per mole of heme in the ferric and ferrous state of enzyme, respectively.

Substrate Specificity and Effects of pH—In good agreement with the results of previous investigators (6), the enzyme was absolutely specific for L-tryptophan. None of the following tryptophan analogues was active as substrate under the standard

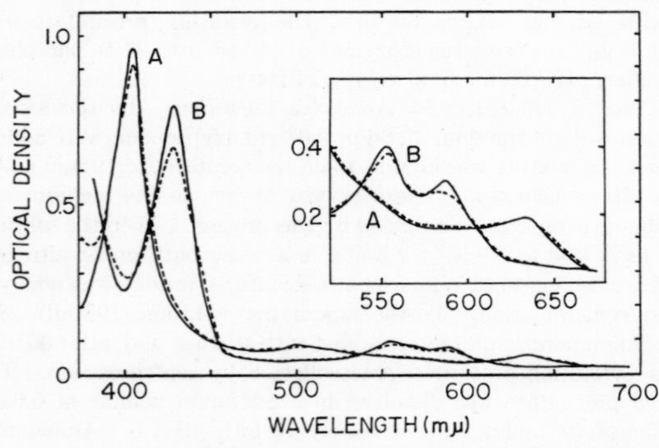

FIG. 1. Absorption spectra of L-tryptophan 2,3-dioxygenase. *A*, the ferric form; *B*, the ferrous form, obtained by the addition of sodium dithionite (5 μmoles) to *A*. The *solid* and *dotted lines* represent the spectra before and after the addition of L-tryptophan (approximately 2 mg), respectively. All of the solution contained 3.9 mg of L-tryptophan 2,3-dioxygenase (specific activity, 1.5), 5.4 mμmoles in terms of heme content, and 50 μmoles of potassium phosphate buffer, pH 6.5, in a final volume of 1.0 ml at 24°. The *inset* represents the spectra taken with four times concentrated enzyme solutions under identical conditions.

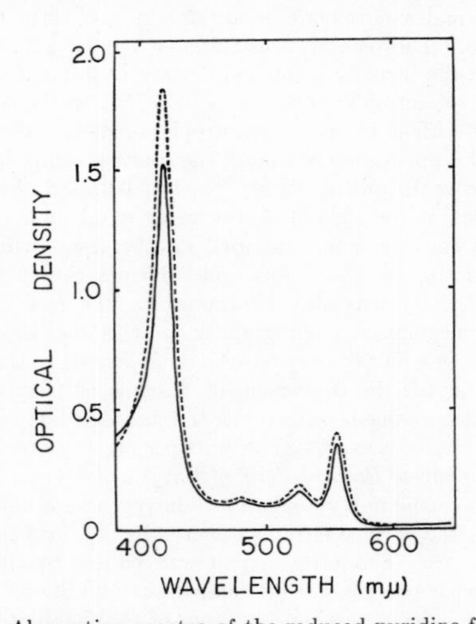

FIG. 2. Absorption spectra of the reduced pyridine-ferrohemochrome of L-tryptophan 2,3-dioxygenase (——, 7.5 mg of enzyme with specific activity 3.6) in a final volume of 2.5 ml and of authentic hematin (- - -, 12 μM).

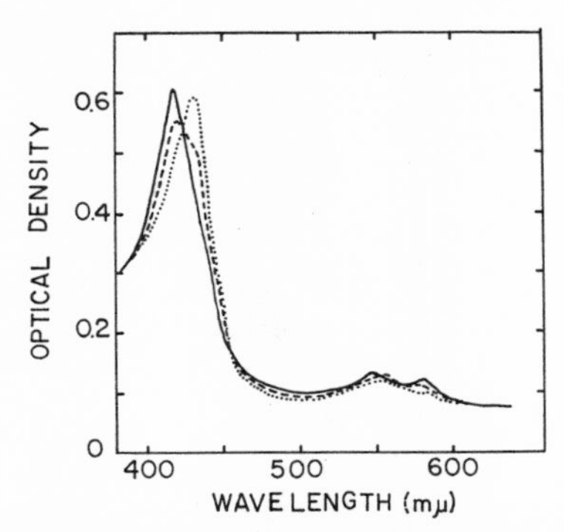

FIG. 3. Spectrum of L-tryptophan 2,3-dioxygenase obtained at 5° by continuous bubbling of oxygen. The reaction mixture contained 22.7 mμmoles of the reduced form of L-tryptophan 2,3-dioxygenase (specific activity, 3.7), 500 μmoles of potassium phosphate buffer (pH 7.0), and 50 μmoles of L-tryptophan in a final volume of 5.0 ml. The reaction was initiated by introducing gaseous oxygen into a cuvette by bubbling (35). ——, recorded during the continuous bubbling of oxygen; – – –, recorded when the supply of oxygen was stopped; · · · ·, ferrous enzyme before and after the reaction.

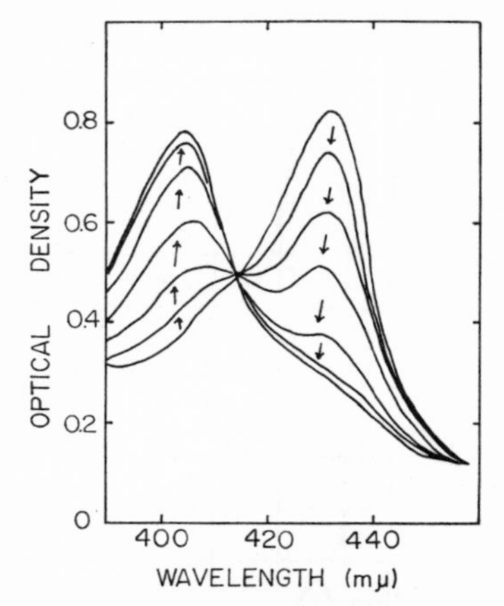

FIG. 4. Spectral changes during the autoxidation of ferrous L-tryptophan 2,3-dioxygenase in 5% oxygen. The reaction mixture contained 17.1 mμmoles of ferric L-tryptophan 2,3-dioxygenase (specific activity, 2.3) and 300 μmoles of potassium phosphate buffer, pH 7.0, in a final volume of 3.0 ml under anaerobic conditions. The ferric enzyme was reduced by the addition of 30 mμmoles of sodium dithionite under anaerobic conditions. Excess dithionite was removed by bubbling commercial nitrogen which contained a trace amount of oxygen. The reaction was then initiated by the addition of 0.15 ml of 0.1 M potassium phosphate buffer, pH 7.0, which had previously been equilibrated with pure oxygen. The spectra were recorded at 1-min intervals.

assay conditions: 2,3-dihydro-DL-tryptophan, 4-methyl-D,L-tryptophan, 5-methyl-D,L-tryptophan, L-tryptophan amide, L- and D-5-hydroxytryptophan, and α-methyl-DL-tryptophan. However, some of these analogues were found to act as inhibitors at higher concentrations. Among those, 5-hydroxy-L-tryptophan was the most potent inhibitor, while 5-hydroxy-D-tryptophan was not. On the other hand, α-methyl-DL-tryptophan activated the enzyme at low L-tryptophan concentrations, as has been described already (32, 33). Details of these studies will be described elsewhere.

The pH optimum for enzyme activity was 7.0 to 7.3. The enzyme was most stable at pH 6.5 in 0.05 M potassium phosphate buffer. The isoelectric point of the enzyme protein was determined to be pH 5.2 by the isoelectric focusing method of Vesterberg and Svenson (34).

K_m Values for Oxygen and L-Tryptophan and Turnover Number of Enzyme—Reciprocal plots of the over-all reaction rate against oxygen concentrations at fixed tryptophan concentrations were linear within analytical limits and the K_m value for oxygen with a saturated concentration of L-tryptophan (5 mM) was found to be 60 μM. In contrast, plots of reciprocal velocity against reciprocal tryptophan concentrations were nonlinear at all levels of oxygen and the apparent K_m value for L-tryptophan determined in a concentration range between 50 μM and 10 mM under an air atmosphere was 240 μM. Mutual effects of oxygen and L-tryptophan concentrations were observed on each K_m value and the apparent dissociation constants for each substrate were increased by decreasing the concentration of the other substrate. These results are in agreement with those reported previously (6, 33).

The maximal turnover number of the enzyme per hematin molecule was estimated to be 1100 min^{-1} under the standard assay conditions.

Absorption Spectra of Enzyme during Reaction

Detection of Oxygenated Form of Enzyme at 5°—We have previously reported that, when the valence state of heme in the enzyme was examined during the steady state of the reaction, a new spectrum appeared in the Soret region, which was indicative of neither ferric heme nor ferrous heme, nor a mixture of these, suggesting the possible presence of a transient intermediate (27, 28). This finding was further confirmed and extended in the following experiments. Incubations were carried out at 5° to slow down the reaction rate in a wide cuvette with 1-cm light path into which a continuous supply of gas can be added through bubbling without interference with optical measurements (35). When oxygen was introduced to a solution containing both the ferrous enzyme and tryptophan, a new absorption spectrum appeared immediately, having its absorption maxima at 418,[3] 545, and 580 mμ (Fig. 3, *solid line*). These peak positions are very close to those of known oxygenated heme proteins such as oxyhemoglobin, oxymyoglobin, and Compound III of horseradish peroxidase (15). When the supply of oxygen was stopped, the spectrum reverted to that of the original ferrous enzyme, with an isosbestic point at 422 mμ (Fig. 3). When oxygen was reintroduced, the spectrum again changed to that of the new species. This process could be repeated until all of the tryptophan in the system was consumed. With the ferric enzyme under the same conditions, no such spectrum was observed. In the absence of L-tryptophan, the ferrous enzyme was converted very slowly to a ferric form, as shown in Fig. 4. This slow oxidation of the enzyme took several minutes for completion and no species of heme other than the ferric and ferrous state could be detected during the process. Thus, the formation of the new species is dependent on the presence of

[3] The position of the γ band was reported in the preliminary communication (15) to be 415 mμ. This could have been due to the contamination of a small amount of the ferric form.

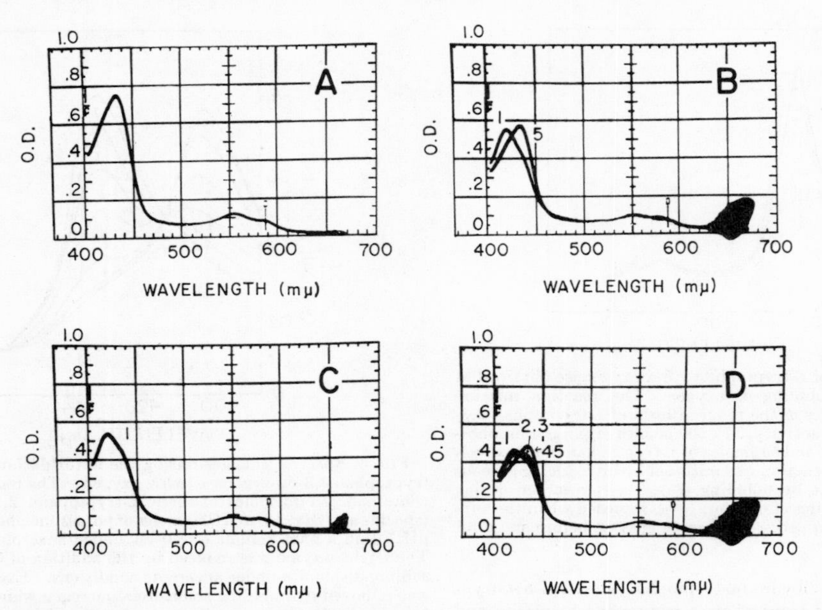

FIG. 5. Spectra of the oxygenated form of L-tryptophan 2,3-dioxygenase obtained by the use of rapid scan spectrophotometry at 24°. The incubation mixture contained 14.5 mμmoles of reduced L-tryptophan 2,3-dioxygenase (specific activity, 3.7), 500 μmoles of potassium phosphate buffer (pH 7.0), and 30 μmoles of L-tryptophan in a final volume of 2.4 ml under anaerobic conditions. The reaction was initiated by the injection of 0.5 ml of 0.1 M potassium phosphate buffer, pH 7.0, which had previously been equilibrated with pure oxygen. *A*, before addition of oxygen; *B*, at 1 and 5 sec after the injection of oxygen; *C*, at only 1 sec after the injection of oxygen obtained by a parallel experiment under conditions identical with *B*; *D*, at 1, 2.3, and 45 sec after the additional injection of 0.5 ml of the oxygen-equilibrated buffer to *B*.

L-tryptophan. These results, together with the kinetic evidence described below, indicate that the observed spectrum is due to a ternary complex of oxygen, the enzyme, and tryptophan, and represents the oxygenated form.

Rapid Scan Spectrophotometry of Oxygenated Form—A spectral species similar to that described above was also shown to exist under the standard assay conditions (20% oxygen at 24°) by the following experiments with the use of a Hitachi rapid scan spectrophotometer. This instrument permits full spectral scans of the entire visible region three times per sec with a scanning time of 0.15 sec. In Fig. 5 are shown results of these experiments in which the reaction was initiated by the addition of oxygen to a solution containing ferrous enzyme and tryptophan. It can be seen that the main portion of enzyme had been converted to the new species after 1 sec but it reverted to the original ferrous enzyme after 5 sec. A lowering of the peak height in Fig. 5*B* was due to a dilution caused by the addition of the buffer solution containing oxygen. Fig. 5*C* shows a record only at 1 sec after the initiation of the reaction. In addition to the changes in the Soret region, the formation of two new peaks at 545 and 580 mμ is evident. These processes could be observed repeatedly by the addition of oxygen, as shown in Fig. 5*D*.

Stopped Flow Experiments

Spectrum of Tryptophan 2,3-Dioxygenase in Steady State of Reaction—A typical record of a stopped flow experiment is shown in Fig. 6, in which a solution containing both the ferrous enzyme and L-tryptophan was allowed to react with molecular

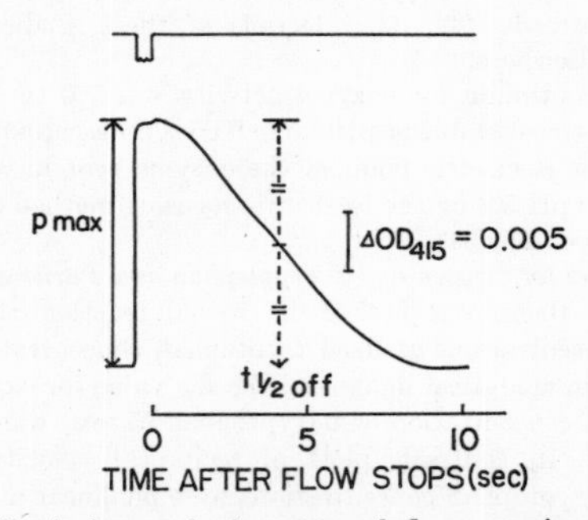

FIG. 6. Typical record of a stopped flow experiment at 25°. Concentrations of ferrous L-tryptophan 2,3-dioxygenase (specific activity, 3.7), L-tryptophan, oxygen, and potassium phosphate buffer, pH 7.0, were 1.1 μM, 5 mM, 60 μM, and 0.1 M, respectively. The reaction was started by mixing the buffer solution containing oxygen with the mixture of both enzyme and L-tryptophan. Changes in absorbance were followed at 415 mμ. The time during the flow was calculated to be 6 msec from the flow velocity trace (*upper curve*) (21).

oxygen and changes in absorbance were followed at 415 mμ. An almost instantaneous increase in optical density was observed followed by its gradual disappearance. The maximum change in absorbance as represented by the peak height is denoted by p_{max} and the time for changes in optical density to fall from p_{max} to $p_{max/2}$ is designated as $t_{\frac{1}{2}}$ off according to the Chance's

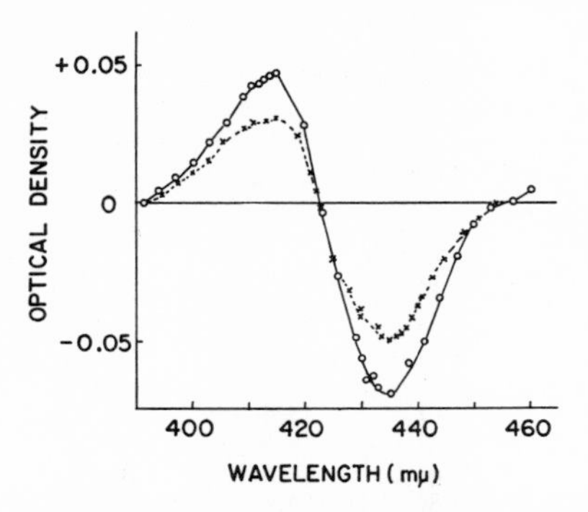

FIG. 7. Difference spectra of the steady state and ferrous enzyme in the Soret region. Concentrations of ferrous L-tryptophan 2,3-dioxygenase, L-tryptophan, and potassium phosphate buffer, pH 7.0, were 1.4 μM, 5 mM, and 0.1 M, respectively. The initial concentrations of oxygen were 600 μM (O——O) and 60 μM (X- - -X), respectively.

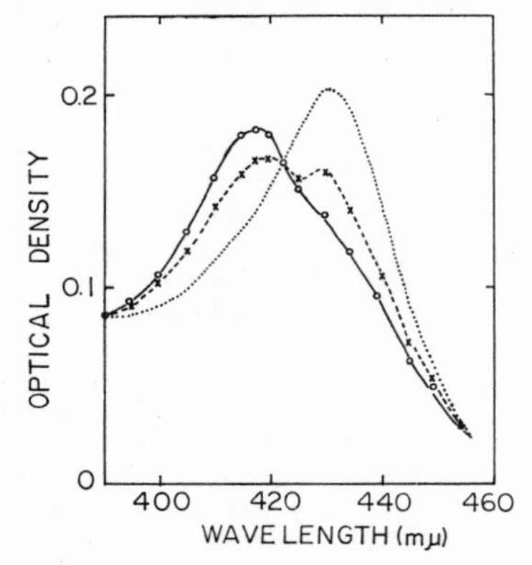

FIG. 8. Absolute spectra of L-tryptophan 2,3-dioxygenase during the steady state of the reaction reconstructed from the data in Fig. 7. ····, the ferrous form of enzyme; O——O and X- - -X, the same as in Fig. 7.

notation (36). By following the reaction at different wave lengths in the entire Soret region, it was found that the values for $t_{\frac{1}{2}}$ off were constant and independent of the wave length at which the measurement was made. This indicates that the changes in optical densities are attributable to the appearance and disappearance of only one spectral species. When the values for p_{max} were plotted against the wave length, a difference spectrum was obtained which corresponded to the spectrum of the steady state minus the ferrous state of the enzyme. In Fig. 7 are shown these difference spectra which were obtained with 600 and 60 μM of initial oxygen concentrations, respectively. It can be seen that the same spectral species was obtained with different concentrations of oxygen. The absolute spectra of the enzyme which were reconstructed from curves in Fig. 7 and the spectrum of ferrous enzyme are shown in Fig. 8. The reconstructed spectra correspond to either the oxygenated form of the enzyme or its mixture with the ferrous enzyme. Thus, the changes in optical density observed in the stopped flow experiments are due to the formation and decomposition of the oxygenated form of tryptophan 2,3-dioxygenase.

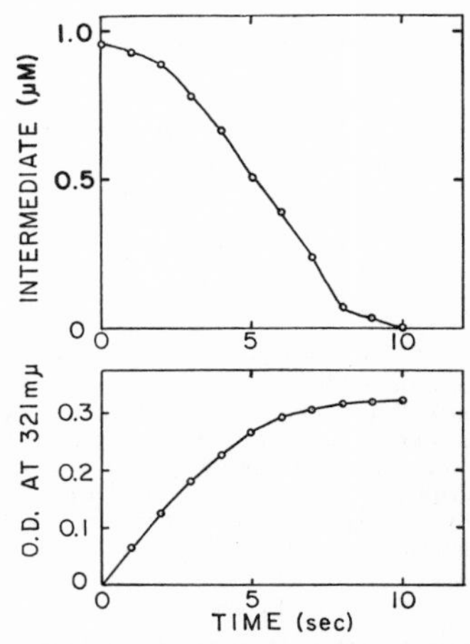

FIG. 9. Time course of the accumulation of the reaction product and the amount of the oxygenated enzyme. Concentrations of ferrous L-tryptophan 2,3-dioxygenase (specific activity, 3.7), L-tryptophan, and potassium phosphate buffer, pH 7.0, were 1.4 μM, 5 mM, and 0.1 M, respectively. Initial concentration of oxygen was 120 μM. Accumulation of the reaction product, L-formylkynurenine (*lower figure*), and the amount of the oxygenated form (*upper figure*) were recorded in parallel experiments as described in the text under identical experimental conditions.

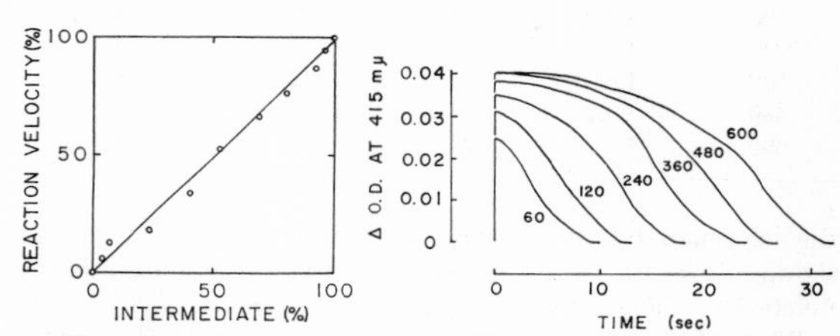

FIG. 10 (*left*). Linear relationship between the rate of over-all reaction and the amount of the oxygenated form replotted from the data in Fig. 9.

FIG. 11 (*right*). Effects of oxygen concentrations on the kinetics of the oxygenated form of L-tryptophan 2,3-dioxygenase. Concentrations of ferrous L-tryptophan 2,3-dioxygenase (specific activity, 3.7), L-tryptophan, and potassium phosphate buffer, pH 7.0, were 1.4 μM, 5 mM, and 0.1 M, respectively. *Numerals* in the figure represent the initial micromolar concentrations of oxygen.

Relationship between Rate of Over-all Reaction and Amounts of Oxygenated Form of Enzyme—Since 415 mμ is the most prominent peak of the difference spectra, all of the following stopped flow experiments were carried out at this wave length unless otherwise noted. In Fig. 9 are shown the relative amounts of the oxygenated form during the entire course of the reaction (*upper curve*) as well as the accumulation of the reaction product, L-formylkynurenine, as determined by the increase in optical density at 321 mμ (*lower curve*). There were recorded in parallel experiments under identical conditions in which the initial concentration of oxygen was 120 μM. When the rates of over-all reaction at different time intervals calculated from the tangents of the *lower curve* were plotted against the relative amounts of

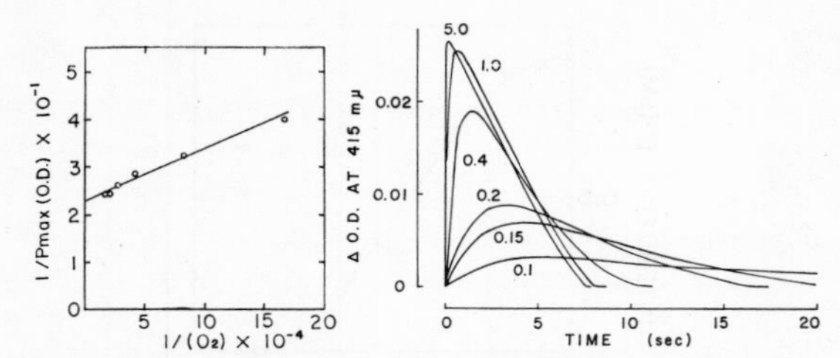

FIG. 12 (*left*). A reciprocal plot of p_{max} values against oxygen concentrations obtained from the data in Fig. 11.

FIG. 13 (*right*). Effects of L-tryptophan concentrations on the kinetics of the oxygenated form. Concentrations of ferrous L-tryptophan 2,3-dioxygenase (specific activity, 3.7), oxygen, and potassium phosphate buffer, pH 7.0, were 1.4 μM, 60 μM, and 0.1 M, respectively. *Numerals* in the figure represent the initial micromolar concentrations of L-tryptophan.

TABLE I

Rate constants for decomposition of oxygenated form of L-tryptophan 2,3-dioxygenase (K_d) obtained at different initial concentrations of oxygen

For experimental details, see the legend for Fig. 11.

Oxygen concentration	p_{max}	$t_{1/2}$ off	$p_{max} \cdot t_{1/2}$ off	K_d
μM	μM	sec	μM·sec	sec⁻¹
60	0.80	3.9	3.12	19.2
120	1.00	6.15	6.15	19.5
240	1.13	11.0	12.4	19.4
360	1.23	15.0	18.4	19.6
480	1.32	18.6	24.6	19.5
600	1.32	23.6	31.2	19.2

the oxygenated form, a linear relationship was obtained, as shown in Fig. 10. Thus, the rate of over-all reaction is a linear function of the amount of oxygenated enzyme.

Effect of Oxygen Concentrations on Kinetics of Oxygenated Form—While the above experiments were performed at a fixed concentration of oxygen, the experiments shown in Fig. 11 were carried out at different initial concentrations of oxygen in the presence of 5 mM L-tryptophan. It should be noted that the p_{max} increased as oxygen concentration increased and finally reached a saturated level. A reciprocal plot of p_{max} values against oxygen concentrations gave a straight line, as shown in Fig. 12, from which an apparent affinity of oxygen for the enzyme as well as p_{max} value at infinite concentration of oxygen was obtained. The former was estimated to be 50 μM, in good agreement with that obtained from the over-all reaction with the same concentration of L-tryptophan (see above). The latter was 0.31 optical density with the enzyme concentration of 10 μM in terms of heme. Assuming that the ferrous enzyme was totally converted into the oxygenated form at infinite oxygen concentrations, the molar extinction coefficient of the oxygenated heme in the enzyme was determined to be 133×10^3 at 418 mμ. The value is approximately the same as that of oxyhemoglobin and oxymyoglobin (37, 38).

It was noted that the area under each curve increased regularly with the initial oxygen concentration. When the area calculated by multiplying p_{max} by $t_{\frac{1}{2}}$ off was plotted against oxygen concentration, a linear relationship was again obtained. From these

TABLE II

Effect of L-tryptophan concentrations on K_d

For experimental details, see the legend for Fig. 13. K_d is the rate constant for the decomposition of the oxygenated form of L-tryptophan 2,3-dioxygenase.

L-Tryptophan concentration	p_{max}	$t_{1/2}$ off	$p_{max} \cdot t_{1/2}$ off	K_d
mM	μM	sec	μM·sce	sec⁻¹
0.1	0.13	20	2.6	23
0.15	0.23	11	2.5	24
0.20	0.29	8.5	2.46	24.4
0.40	0.61	4.5	2.75	21.8
1.0	0.82	4.0	3.28	18.3
2.0	0.83	3.8	3.15	19.0
5.0	0.86	3.6	3.09	19.4

data, the rate constants for the decomposition of the oxygenated form (product formation) calculated by Chance's equation are shown in Table I. The values (19.2 − 19.6 sec⁻¹) agree quite well with the turnover number of the enzyme determined by the conventional method (18 sec⁻¹) and the rate constant for the decomposition obtained from the data in Fig. 9 by dividing the rate of formylkynurenine formation by the amount of oxygenated form (18 sec⁻¹).

Effect of Tryptophan Concentrations—The next experiments were carried out in order to learn the effect of tryptophan concentration on the formation and decomposition of the oxygenated form. The results of typical experiments in which an anaerobic solution of ferrous enzyme was mixed with solutions containing varying amounts of L-tryptophan and 60 μM of oxygen are shown in Fig. 13.

The ferrous enzyme solution contained 50 μM L-tryptophan to eliminate a trace amount of dissolved oxygen and to prevent the oxidation of enzyme heme. When the solutions in both syringes of the flow apparatus were allowed to react, the tryptophan was diluted to below 25 μM, which was approximately one-tenth of its K_m value under these conditions and was therefore disregarded for the calculations. It is seen from the figure that the rate of formation of the oxygenated form as well as p_{max} decreased as L-tryptophan concentration decreased.[4] When values for p_{max} were plotted against L-tryptophan concentrations, a slightly sigmoidal curve was obtained, in agreement with the results of over-all reaction kinetics (32, 33). On the other hand, the area under each curve was found to be roughly constant, indicating that the rate constant for the decomposition of the oxygenated form was independent of L-tryptophan concentration within this range. The results were summarized in Table II.

Determinations of Rate Constant for Formation of Oxygenated Form and Reverse Rate Constant—These values were calculated from the measurement of the initial slope of the changes in optical density at 415 mμ when the ferrous enzyme was mixed with oxygen in the presence of L-tryptophan. As expected from the data in Fig. 11, the rate of formation was so rapid that it was not possible for us to use our ordinary flow apparatus (minimum dead time is about 4 msec). Through the kind collaboration of

[4] The time required for reaching p_{max} in Fig. 13 is probably due to the conformational change of the enzyme induced by L-tryptophan since the formation of the oxygenated form was accelerated by the preliminary incubation of enzyme with high concentrations of L-tryptophan.

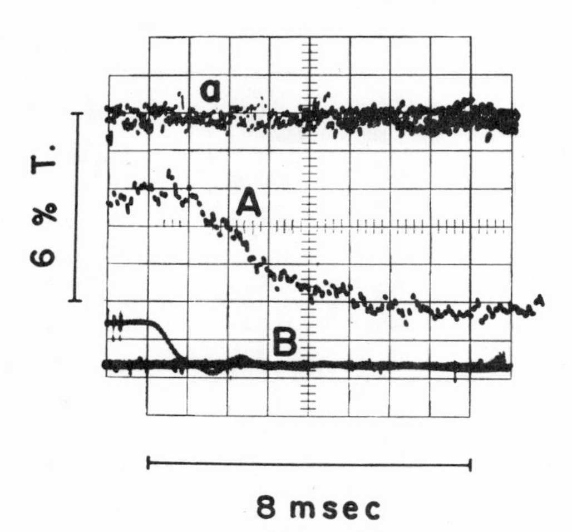

FIG. 14. A typical oscillograph trace of the changes in optical transmission at 415 mμ due to the formation of the oxygenated form. Concentrations of ferrous L-tryptophan 2,3-dioxygenase (specific activity, 2.3), oxygen, L-tryptophan, and potassium phosphate buffer, pH 7.0, were 6 μM, 120 μM, 5 mM, and 0.1 M, respectively. *A*, changes in optical transmission at 415 mμ due to the formation of the oxygenated form; *a*, level of the ferrous enzyme at 415 mμ, recorded after the oxygen was completely used up. *B*, flow velocity trace (21 ml per sec; the dead space of the flow cell was 15 μl); *ordinate*, for optical transmission, 1.2% per division; *abscissa*, for scanning time, 1 msec per division. The light path was 2.0 mm in length. The reaction was initiated by mixing oxygen with enzyme solution containing L-tryptophan.

Dr. K. Hiromi of Osaka Prefectural University, it was possible to observe reaction rates from 0.6 msec after mixing. A typical trace of a stopped flow experiment with Dr. Hiromi's instrument, in which final concentrations of oxygen and L-tryptophan were 120 μM and 5 mM, respectively, and that of the enzyme as heme was 6 μM, is shown in Fig. 14. Experiments were also carried out at several different oxygen concentrations (30 to 240 μM) and the rate of reaction was directly proportional to the oxygen concentration, indicating that the reaction was first order with respect to oxygen. The average rate constant calculated from all of our experiments is 5.0×10^6 M^{-1} sec^{-1}. Based on this value as well as other kinetic constants described earlier, the reverse rate constant is determined to be 231 sec^{-1}.

DISCUSSION

There has been considerable discussion in the literature as to the reaction mechanism of L-tryptophan 2,3-dioxygenase. Tanaka and Knox (6) and Tokuyama and Knox (11) presented evidence that the enzyme with its heme in the ferric state was catalytically inactive but could be reduced to an active ferrous form by the addition of either ascorbate or hydrogen peroxide in the presence of L-tryptophan. In their hypothetical reaction mechanism, Tanaka and Knox assumed that oxygen binds first with ferrous heme and then reacts with L-tryptophan, giving L-formylkynurenine as product. On the other hand, Feigelson, Ishimura, and Hayaishi (39, 40) and subsequently Feigelson and Maeno (12) and Maeno and Feigelson (14) reported that the ferric as well as the ferrous enzyme was catalytically active. A reaction mechanism was proposed by these authors in which the heme iron in the enzyme oscillates in charge, being sequentially reduced by L-tryptophan and reoxidized by oxygen.

Our interest in the general mechanism of dioxygenase reaction led us to undertake a study of these discrepancies. As cited above (39, 40), we once reported in collaboration with Feigelson that both ferric and ferrous enzyme could be regarded as the active form of L-tryptophan 2,3-dioxygenase. Later, however, it was noticed that the enzyme preparations used at that time contained some endogenous reductants which could reduce the ferric enzyme to active ferrous state in the presence of L-tryptophan (13). With the use of more purified preparations, we further demonstrated that, when the ferric enzyme was gradually reduced by illumination under anaerobic conditions, the process resulted in a proportional increase in the activity (27, 28, 41). Likewise, a gradual oxidation of the ferrous enzyme by air resulted in a parallel decrease in the activity. These results, together with the fact that the ferrous enzyme was fully active in the absence of the activator whereas the ferric form was almost inactive (41), indicated that the enzyme must be in the ferrous state to initiate the reaction. Thus, the original view of Tanaka and Knox (6) was revitalized. However, we also found that a new spectrum of the enzyme, which was due to neither the ferric nor the ferrous state of the heme, appeared in the Soret region during the steady state of the reaction (27, 28). The new spectral species was observed only in the presence of both L-tryptophan and oxygen, and therefore was interpreted to be due to a ternary complex of oxygen, L-tryptophan, and the enzyme. These findings were now confirmed and extended in this paper.

The spectrum of the new spectral species over the entire visible region was obtained and found to be almost identical with those of known oxygenated hemoproteins (15). These spectral characteristics together with the kinetic evidence that the binding of oxygen to the ferrous heme is a reversible process strongly support our view that the new spectral species is due to an "oxygenated" form of heme in L-tryptophan 2,3-dioxygenase. These interpretations are in accord with the previous observation of Maeno and Feigelson (9) that the photochemical action spectrum, with a maximum at 421 mμ, is identical with the Soret absorption spectrum of the tryptophan-carboxyferroprotoporphyrin-enzyme complex.

The oxygenated form of enzyme was observed whenever the reaction proceeded. It was detected under various experimental conditions with several different techniques, *i.e.* at different temperatures, varying concentrations of oxygen, L-tryptophan, and the enzyme by an ordinary spectrophotometric method, rapid scan spectrophotometry, and the stopped flow method. Furthermore, we were able to demonstrate that the rate constant for the decomposition of the oxygenated form was in good agreement with the maximal turnover number of the enzyme. Thus, the oxygenated form was proved to be an obligatory intermediate of the reaction.

Recently, Feigelson and Maeno (12) and Maeno and Feigelson (14) proposed a reaction mechanism which postulates the participation of ferric enzyme in the catalytic process. Their postulate is based mainly on the observation that, when the reaction was initiated with a ferric enzyme in the presence of ascorbate, no increase in optical densities at 432 mμ due to the formation of ferrous enzyme was observed during the steady state of catalysis. However, this does not necessarily mean that the enzyme is in the ferric state during the process. The present data as well as those reported previously (27, 28) clearly indicate that the enzyme is mainly in an oxygenated state and, therefore, almost no free ferrous enzyme can be detected during the steady state of the reaction.

It is to be noted that the oxygenation spectrum of enzyme heme was observed only in the presence of L-tryptophan. On the other hand, additions of L-tryptophan either to the ferrous or the ferric enzyme in the absence of oxygen caused modifications in their spectra over the entire visible region. Furthermore, the rate of decomposition of the oxygenated form was independent of L-tryptophan concentrations. Therefore, it seems reasonable to assume that first L-tryptophan and then oxygen binds to the ferrous enzyme to form a tryptophan-enzyme-oxygen complex, the oxygenated intermediate. If the observed intermediate does not contain the L-tryptophan that reacts with oxygen, the rate constant for its degradation should be affected by the L-tryptophan concentrations, but the data in Table II show that this is not the case.

The only enzymatically active species of oxygenated heme so far known is the Compound III of peroxidase, in which case divalent iron in the heme is oxygenated in the absence of its organic substrate (42). In contrast, in the case of L-tryptophan 2,3-dioxygenase, the organic substrate is necessary in order to form the oxygenated intermediate in a detectable quantity. In the absence of L-tryptophan, the ferrous form of enzyme is oxidized to the ferric form very slowly. Presumably tryptophan, by binding with the enzyme, evokes a conformational change in the enzyme protein and an increase in reactivity of the heme moiety towards oxygen. This interpretation is in accord with the findings that the dissociation constants of heme-binding substances such as cyanide and carbon monoxide with the enzyme were greatly altered by L-tryptophan (9, 15, 41, 43, 44)

The following reaction mechanism is therefore compatible with the above data. Tryptophan first combines with the ferrous enzyme and activates the heme in the enzyme. The activated enzyme then reacts with oxygen to form an intermediary ternary complex. Both substrates, tryptophan and oxygen, are activated in the complex and interact, yielding formylkynurenine as the product.

The oxygenated iron compound reported here requires much more detailed characterization. Nevertheless, our findings are quite consistent with the view that dioxygenase reactions in general involve a ternary complex of enzyme, oxygen, and organic substrate (16, 17, 27, 45). The role of tryptophan(s) in the oxygenation of the enzyme heme is especially noteworthy and is currently under investigation.

Acknowledgments—The authors are grateful to Dr. K. Hiromi for his kind collaboration in the determination of the rate constant for the formation of the oxygenated form. We are also indebted to Dr. C. A. Tyson for his aid in the preparation of this paper, to Mr. M. Fujita for his assistance in the use of the rapid scan spectrophotometer, and to Mrs. T. Kizu for her excellent technical assistance.

REFERENCES

1. KNOX, W. E., AND MEHLER, A. H., *J. Biol. Chem.*, **187**, 419 (1950).
2. KNOX, W. E., in S. P. COLOWICK AND N. O. KAPLAN (Editors), *Methods in enzymology, Vol. II*, Academic Press, New York, 1955, p. 242.
3. HAYAISHI, O., AND STANIER, R. Y., *J. Bacteriol.*, **62**, 691 (1951).
4. KOTAKE, Y., AND MASAYAMA, T., *Z. Physiol. Chem.*, **243**, 237 (1936).
5. HAYAISHI, O., ROTHBERG, S., MEHLER, A. H., AND SAITO, Y., *J. Biol. Chem.*, **229**, 889 (1957).
6. TANAKA, T., AND KNOX, W. E., *J. Biol. Chem.*, **234**, 1162 (1959).
7. FEIGELSON, P., AND GREENGARD, O., *Biochim. Biophys. Acta*, **50**, 200 (1961).
8. GREENGARD, O., AND FEIGELSON, P., *J. Biol. Chem.*, **237**, 1903 (1962).
9. MAENO, H., AND FEIGELSON, P., *J. Biol. Chem.*, **243**, 301 (1968).
10. KNOX, W. E., AND TOKUYAMA, K., in T. E. KING, H. S. MASON, AND M. MORRISON (Editors), *Oxidases and related redox systems, Vol. 1*, John Wiley and Sons, Inc., New York, 1965, p. 514.
11. TOKUYAMA, K., AND KNOX, W. E., *Biochim. Biophys. Acta*, **81**, 201 (1964).
12. FEIGELSON, P., AND MAENO, H., in K. BLOCH AND O. HAYAISHI (Editors), *Biological and chemical aspects of oxygenases*, Maruzen Company, Ltd., Tokyo, 1966, p. 411.
13. ISHIMURA, Y., OKAZAKI, T., NAKAZAWA, T., ONO, K., NOZAKI, M., AND HAYAISHI, O., in K. BLOCH AND O. HAYAISHI (Editors), *Biological and chemical aspects of oxygenases*, Maruzen Company, Ltd., Tokyo, 1966, p. 416.
14. MAENO, H., AND FEIGELSON, P., *J. Biol. Chem.*, **242**, 596 (1967).
15. ISHIMURA, Y., NOZAKI, M., HAYAISHI, O., TAMURA, M., AND YAMAZAKI, I., *J. Biol. Chem.*, **242**, 2574 (1967).
16. HAYAISHI, O., ISHIMURA, Y., NAKAZAWA, T., AND NOZAKI, M., in B. HESS AND H. STAUDINGER (Editors), *Biochemie des Sauerstoffs, 19. Colloquium der Gesellschaft für Biologische Chemie, 24–27 April 1968 in Mosbach, Baden*, Springer-Verlag, New York, 1968, p. 196.
17. HAYAISHI, O., *Ann. N. Y. Acad. Sci.*, **158**, 318 (1969).
18. PAUL, K. G., *Acta Chem. Scand.*, **12**, 1312 (1958).
19. BABA, H., AND SHINDO, Y., *Jap. Anal.*, **16**, 653 (1967).
20. CHANCE, B., in A. WEISSBERGER (Editor), *Techniques of organic chemistry, Vol. 8*, Part II, Interscience Publishers, New York, 1963, p. 1314.
21. CHANCE, B., AND LEGALLAIS, V., *Rev. Sci. Instrum.*, **22**, 627 (1951).
22. NAKAMURA, T., *Seikagaku*, **39**, 855 (1967).
23. HAGIHARA, B., *Biochim. Biophys. Acta*, **46**, 134 (1961).
24. LAYNE, E., in S. P. COLOWICK AND N. O. KAPLAN (Editors), *Methods in enzymology, Vol. III*, Academic Press, New York, 1957, p. 450.
25. WARBURG, O., AND CHRISTIAN, W., *Biochem. Z.*, **298**, 150 (1938).
26. KEILIN, D., AND HARTREE, E. F., *Biochem. J.*, **49**, 88 (1951).
27. HAYAISHI, O., *Proceedings of the Plenary Sessions of the Sixth International Congress of Biochemistry, New York, 1964, IUB Vol. 33*, p. 31.
28. ISHIMURA, Y., AND HAYAISHI, O., *Seikagaku*, **36**, 502 (1964).
29. POILLON, W. N., MAENO, H., KOIKE, K., AND FEIGELSON, P., *J. Biol. Chem.*, **244**, 3447 (1969).
30. CHANCE, B., *Methods Biochem. Anal.* **1**, 408 (1954).
31. CHANCE, B., AND MAEHLY, A. C., in S. P. COLOWICK AND N. O. KAPLAN (Editors), *Methods in enzymology, Vol. II*, Academic Press, New York, 1955, p. 773.
32. SCHIMKE, R. T., SWEENEY, E. W., AND BERLIN, C. M., *J. Biol. Chem.*, **240**, 4609 (1965).
33. FEIGELSON, P., AND MAENO, H., *Biochem. Biophys. Res. Commun.*, **28**, 289 (1967).
34. VESTERBERG, O., AND SVENSSON, H., *Acta Chem. Scand.*, **20**, 820 (1966).
35. YAMAZAKI, I., AND YOKOTA, K.-N., *Biochim. Biophys. Acta*, **132**, 310 (1967).
36. CHANCE, B., *Arch. Biochem. Biophys.*, **71**, 130 (1957).
37. SIDWELL, A. E., JR., MUNCH, R. H., BARRON, E. S. G., AND HOGNESS, T. R., *J. Biol. Chem.*, **123**, 335 (1938).
38. YAMAZAKI, I., YOKOTA, K., AND SHIKAMA, K., *J. Biol. Chem.*, **239**, 4151 (1964).
39. FEIGELSON, P., ISHIMURA, Y., AND HAYAISHI, O., *Biochem. Biophys. Res. Commun.*, **14**, 96 (1964).

40. Feigelson, P., Ishimura, Y., and Hayaishi, O., *Biochim. Biophys. Acta*, **96**, 283 (1965).
41. Ishimura, Y., Nozaki, M., Hayaishi, O., Tamura, M., and Yamazaki, I., *Advan. Chem. Ser.*, **77**, 235 (1968).
42. Yokota, K., and Yamazaki, I., *Biochem. Biophys. Res. Commun.*, **18**, 48 (1965).
43. Ishimura, Y., Nozaki, M., Hayaishi, O., Tamura, M., and Yamazaki, I., in K. Okunuki, M. D. Kamen, and I. Sezuku (Editors), *Structure and function of cytochromes*, University of Tokyo Press, Tokyo, 1968, p. 188.
44. Maeno, H., and Feigelson, P., *Biochemistry*, **7**, 968 (1968).
45. Hayaishi, O., and Nozaki, M., *Science*, **164**, 389 (1969).

Reprinted from THE JOURNAL OF THE AMERICAN CHEMICAL SOCIETY, Vol. 84, 1962

[CONTRIBUTION FROM THE DEPARTMENT OF CHEMISTRY AND RESEARCH LABORATORY OF ELECTRONICS, MASSACHUSETTS INSTITUTE OF TECHNOLOGY, CAMBRIDGE, MASSACHUSETTS]

A Kinetic Study of Glutamic–Aspartic Transaminase[1]

BY GORDON G. HAMMES AND PAOLO FASELLA

RECEIVED JULY 18, 1962

Kinetic studies of glutamic–aspartic transaminase have been made at high enzyme concentrations ($>10^{-5}$ M) using the temperature jump method. The observed relaxation times ranged from less than 50 μsec. to about 20 msec. The minimal mechanism consistent with spectral and kinetic data is

$$E_L + As = X_1 = X_2 = E_M + Oa \quad (1) \qquad E_M + Kg = Y_2 = Y_1 = E_L + Gm \quad (2)$$

Here E_L is the aldehydic form of the enzyme, E_M is the aminic form, Gm, Kg, Oa, and As are the substrates glutamate, ketoglutarate, oxalacetate, and aspartate, respectively, and the X's and Y's are enzyme–substrate intermediates. Individual rate constants or lower bounds thereof were obtained for all of the steps; in addition approximate spectra were found for the intermediates which are consistent with their being Schiff bases. The binding constants for Schiff base formation are quite high—10^3 to 10^5 M^{-1} depending on the specific substrate. The measured bimolecular rate constants are all about 10^7 M^{-1} sec.$^{-1}$, and the rate constants for the interconversion of intermediates are between 10 and 10^2 sec.$^{-1}$. These results are also in accord with stopped-flow and equilibrium dialysis experiments.

Introduction

The mechanism of enzymatic transamination has been the subject of many investigations,[2] but very little direct information has been obtained concerning the number and nature of the enzyme-substrate intermediates involved. This is due primarily to the fact that most kinetic studies have been carried out at very low enzyme concentrations (as is true with most enzyme systems) in order to make the time scale of the reaction experimentally accessible. Of course, at low enzyme concentrations direct observation of the intermediates is virtually impossible. On the other hand, spectral observations at high enzyme concentrations cannot be interpreted reliably because of the spectral overlap of the intermediates and native enzyme. The recent development of the temperature jump technique[3–5] for studying fast reactions in solution offers the attractive possibility of carrying out kinetic studies at high concentrations of enzyme, thus permitting direct observation of the intermediates. The principle of the temperature jump method is to perturb an equilibrium reaction mixture by rapidly raising the temperature and then to observe the rate with which the system attains its new equilibrium state. This rate is characterized by a spectrum of relaxation times which are functions of the rate constants and equilibrium concentrations.[6] Determination of the relaxation spectrum, therefore, permits the evaluation of kinetic parameters. The analysis is simplified by the fact that all rate processes near equilibrium are first order. The presence of a coenzyme in most transaminases endows this type of enzyme with spectral properties which are very convenient for following the reaction progress. Glutamic–aspartic transaminase was selected for study since it can now be obtained pure in relatively large quantities.[7,8]

Purified glutamic–aspartic transaminase has been shown to react with its substrates according to the general scheme[9,10]

$$E_L + As \rightleftharpoons X_1 \rightleftharpoons \ldots X_n \rightleftharpoons E_M + Oa \quad (1)$$

$$E_M + Kg \rightleftharpoons Y_n \ldots \rightleftharpoons Y_1 \rightleftharpoons E_L + Gm \quad (2)$$

Here E_L is the aldehydic form of the enzyme, E_M is the aminic form, X_n and Y_n designate enzyme-substrate intermediates, As is aspartate, Oa is oxalacetate, Gm is glutamate, and Kg is ketoglutarate. The number and nature of the intermediates X_n and Y_n, however, could not be determined. Steady state kinetic studies are in agreement with the above mechanism,[11,12] but the resultant Michaelis constants and maximum velocities are in general complicated functions of the rate constants and thus permit no conclusions to be drawn concerning the intermediate steps. Attempts to study the reaction of the aldehydic form of the enzyme with glutamate and aspartate by use of the stopped flow method[13] showed that the half time of the reaction (as measured by a decrease in absorbance at 360 mμ) is about 5 msec., which is very close to the resolution time of this method. Furthermore, ignorance of the number and absorption spectra of the various enzyme–substrate complexes makes the interpretation of such experiments in terms of a detailed mechanism extremely difficult.

In the work presented here, the kinetics of each half of the reaction was studied separately by the temperature jump method. Three relaxation times were observed for each of the two systems ranging from less than 50 μsec. to about 20 msec. The concentration dependence of the relaxation times indicates that each half of the reaction proceeds through the formation of at least two intermediates (X_1, X_2, and Y_1, Y_2). Furthermore ten of the twelve rate constants and all of the individual equilibrium constants could be calculated. Coupling these results with equilibrium measurements of the

(1) Supported in part by U. S. Army Signal Corps, Air Force, Office of Scientific Research, and Office of Naval Research and in part by the National Institutes of Health.

(2) (a) E. E. Snell, *Vitamins and Hormones*, **16**, 77 (1958); (b) A. E. Braunstein, "The Enzymes," Vol. II, Academic Press, New York, N. Y., 1960, p. 113.

(3) G. Czerlinski and M. Eigen, *Z. Elektrochem.*, **63**, 652 (1959).

(4) H. Diebler, Dissertation, Universität Göttingen, 1960.

(5) L. de Maeyer, *Z. Elektrochem.*, **64**, 65 (1960).

(6) M. Eigen, *ibid.*, **64**, 115 (1960).

(7) H. Lis, *Biochim. et Biophys. Acta*, **28**, 191 (1958).

(8) W. T. Jenkins, D. A. Yphantis and I. W. Sizer, *J. Biol. Chem.*, **234**, 51 (1959).

(9) W. T. Jenkins and I. W. Sizer, *ibid.*, **235**, 620 (1960).

(10) H. Lis, P. Fasella, C. Turano and P. Vecchini, *Biochim. et Biophys. Acta*, **45**, 529 (1960).

(11) C. Turano, P. Fasella and A. Giartosio, *ibid.*, **58**, 255 (1962).

(12) S. F. Velick and J. Varra, *J. Biol. Chem.*, **237**, 2109 (1962)

(13) H. Gutfreund, *Nature*, **192**, 820 (1961).

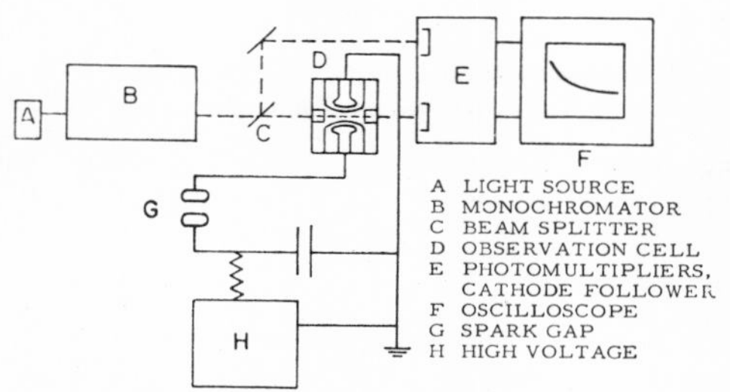

Fig. 1.—Block diagram of the temperature jump apparatus.

absorption spectra of various reaction mixtures permitted some spectral characteristics of the enzyme–substrate complexes to be evaluated. These results allow tentative conclusions to be made concerning the detailed mechanism of the transamination. In addition, stopped flow and equilibrium dialysis experiments are in accord with these conclusions.

Experimental

The temperature jump apparatus used was a modification of that described by Czerlinski and Eigen.[3] A schematic sketch of the system is shown in Fig. 1. In actual operation, the 0.1 μfd. condenser (Plastic Capacitors, Inc.) is charged to 30,000 volt, and then discharged through a suitably adjusted air gap. The observation cell, constructed from Plexiglas, is connected in parallel with the condenser *via* two brass platinized electrodes, and the discharge heats the volume of solution ($\sim$1.2 cc.) between the two electrodes. The temperature jump is about 8°. The entire discharge circuit is constructed concentrically and lined with mu metal to eliminate the effect of the magnetic field on the photomultipliers; the observation cell is surrounded by a thermostatted jacket. The relaxation time (reciprocal first order rate constant) of the heating period is $RC/2$[3] where R is the resistance of the cell and C is 0.1 μfd. For the apparatus being described here, $RC/2$ is about 5 μsec. so that chemical events occurring in times longer than 5–10 μsec. can be observed. Monochromatic light is obtained from a 100-watt tungsten lamp (Osram, West Germany) coupled with a Bausch and Lomb monochromator. The light is split into two beams by a half surfaced mirror; one beam passes through the cell to a photomultiplier (RCA 1P28), the other through air to a second photomultiplier. The path length through the solution is 1 cm. The intensities of the parallel beams of light are adjusted through use of diaphragms so that initially the photomultiplier outputs are equal; deviations from this null point produced by the temperature jump are observed on a Tektronix 545A oscilloscope equipped with a differential preamplifier. The oscilloscope is triggered with the voltage induced in a small coil near the spark gap by the discharge. Absorption changes as small as 0.1% can be measured in this manner. The relaxation effects were photographed with a Polaroid Land Camera.

The purified enzyme was prepared as previously described[7] and the amino and ketoacids were obtained from Calibiochem. All other chemicals were standard analytical grade reagents.

The general procedure was to mix the pyridoxal (aldehydic) form of the enzyme with varying proportions of glutamate or aspartate. All solutions were 0.16 M in phosphate buffer and were adjusted to pH 8.0 ± 0.1. Since the temperature jump cell requires about 25 ml. of solution, a substantial amount of enzyme is needed for each experiment. The equilibrium mixtures containing the various forms of the enzymes and substrates were thermostatted at 17° in the temperature jump cell and temperature jumps of 8° were applied to the solution as previously described. About four minutes were allowed to elapse between pulses in order to reestablish thermal equilibrium in the system. Each relaxation effect was photographed several times and the relaxation times were evaluated from the slopes of plots of the logarithm of the amplitude *versus* time. An actual oscilloscope trace of a system with two observable relaxation is shown in Fig. 2.

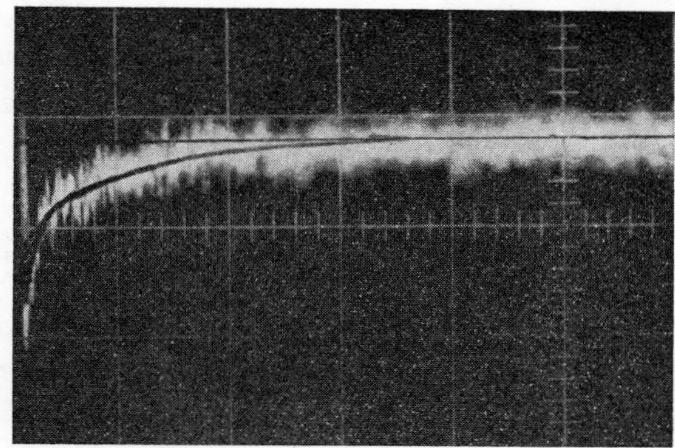

Fig. 2.—Relaxation effect in a transaminase system beginning with glutamate and pyridoxal enzyme: $(S_0) = 2 \times 10^{-4}$ M and $(E_0) = 5 \times 10^{-5}$ M, pH 7.9, λ = 430 mμ. The abscissa scale is 10 msec. per major division and the vertical scale is in arbitrary units of absorbency. The relaxation effect corresponds to a decrease in absorbance with time.

The wave length of the monochromatic light used was 430 mμ. This wave length was found to give optimal effects over the spectral range of 350 to 510 mμ. There are several possible reasons for the fact that the optimal effect is observed at this wave length. The sensitivity of the apparatus is not good at lower wave lengths (<400 mμ) because of the sharp decrease in lamp intensity so that the experiments could not be carried out at 360 mμ, the absorption maximum of E_L, or at 330 mμ, the absorption maximum of E_M. One possibility is that the spectral tail of the 360 peak is being observed; however, it is more probable that the presence of very small amounts of either the protonated form of the E_L enzyme, (HE_L), or its complex with the keto acid (both have absorption maxima at 430 mμ) is serving as an indicator. In keeping with this hypothesis the reaction of HE_L and keto acid to form a complex and the protonation of E_L have been shown to occur faster (under similar conditions) than any of the observed relaxation times through independent experiments.[14] The absorption spectrum and catalytic activity of the enzyme were determined with a Beckman DU Spectrophotometer before and after temperature jump experiments to check for possible enzyme denaturation. Approximately twenty high voltage pulses could be applied without any sign of denaturation.

Stopped flow experiments were carried out in an apparatus constructed from Plexiglas and glass syringes. A four jet mixer was used and the path length of the observation chamber was 1 cm. The detection system for measurement of absorbency changes in the sample was similar to that employed in the temperature jump. The resolution time of the apparatus was 5–10 msec. and the sensitivity was such that 1% changes in absorption could be detected. A detailed description of the apparatus is available elsewhere.[15] Four reactions were studied at a wave length of 390 mμ: E_L + Gm, E_L + As, E_M + Oa, and E_M + Kg. In all experiments, either E_L or E_M was present at a concentration of 2×10^{-5} M, while the substrate concentrations were made as large as possible—the limit being set by the resolution time of the apparatus.

The equilibrium dialysis experiments were carried out with 1.3-cm. Visking tubing which was first soaked 3 min. in 5×10^{-3} N NaOH and washed with distilled water according to the method of Scatchard and Pigliacampi.[16] Approximately 2 ml. of 0.16 M phosphate buffer, pH 8, was placed in the sack and dialyzed against 5 ml. of enzyme and substrate solution in the same buffer. Four initial combinations of enzyme and substrate were investigated. E_L + Gm, E_L + As, E_M + Kg, and E_M + Oa. Blank runs of the enzyme and substrate alone also were made. In order to prevent enzyme denaturation, the experiments had to be carried out at 4° and the dialysis time had to be minimized.

(14) G. G. Hammes and P. Fasella, to be published.

(15) G. Becker, B.S. Thesis, M.I.T., 1962.

(16) G. Scatchard and J. Pigliacampi, *J. Am. Chem. Soc.*, **84**, 127 (1962).

TABLE I

REACTION BETWEEN TRANSAMINASE AND ASPARTATE–OXALACETATE[a]

Concentrations, M						Relaxation times, msec. I		II		III	
S_0	E_0	E_L	X_1	X_2	$E_M(=Oa)$	Exp.	Calcd.	Exp.	Calcd.	Exp.	Calcd.
1.90×10^{-4}	1.90×10^{-5}	6.80×10^{-6}	2.10×10^{-6}	6.50×10^{-6}	3.60×10^{-6}	<0.100	...	..	..	..	..
1.60×10^{-4}	2.60×10^{-5}	1.06×10^{-5}	2.85×10^{-6}	8.80×10^{-6}	4.20×10^{-6}	...	...	..	..	1.3	1.39
8.00×10^{-5}	2.65×10^{-5}	1.55×10^{-5}	1.86×10^{-6}	5.74×10^{-6}	3.40×10^{-6}	...	...	..	..	1.8	1.62
3.73×10^{-4}	3.68×10^{-5}	8.80×10^{-6}	5.50×10^{-6}	1.67×10^{-5}	5.80×10^{-6}	...	...	18	18.7	..	..
7.18×10^{-4}	3.40×10^{-5}	4.65×10^{-6}	5.70×10^{-6}	1.77×10^{-5}	5.95×10^{-6}	...	...	15	15.8	1.07	1.02

[a] pH 8.0, 25°.

TABLE II

REACTION BETWEEN TRANSAMINASE AND GLUTAMATE–KETOGLUTARATE[a]

Concentrations, M						Relaxation times, msec. I		II		III	
S_0	E_0	E_L	X_1	X_2	$E_M(=Kg)$	Exp.	Calcd.	Exp.	Calcd.	Exp.	Calcd.
2.60×10^{-5}	2.60×10^{-5}	1.43×10^{-5}	2.52×10^{-6}	5.04×10^{-6}	4.10×10^{-6}	0.300	0.270	..	..	4.1	3.96
7.80×10^{-5}	2.60×10^{-5}	6.90×10^{-6}	4.53×10^{-6}	9.07×10^{-6}	5.50×10^{-6}	...	...	..	..	3.1	3.22
2.00×10^{-4}	5.30×10^{-5}	6.70×10^{-6}	1.24×10^{-5}	2.48×10^{-5}	9.10×10^{-5}	0.110	0.122	16	14.5	2.1	2.19
4.00×10^{-4}	5.30×10^{-5}	3.30×10^{-6}	1.34×10^{-5}	2.68×10^{-5}	9.50×10^{-6}	...	...	13	13.3	1.5	2.10

[a] pH 8.0, 25°.

The minimum time for equilibration without denaturation was found to be about 6 hours for experiments beginning with the E_L enzyme and 3 hours for those beginning with the E_M enzyme. The solutions were slowly but continuously stirred during the dialysis. At the end of the dialysis the contents of the dialysis bag were analyzed quantitatively for keto acids with 2,4-dinitrophenylhydrazine[17] and for amino acids with ninhydrin according to Moore and Stein.[18]

Results

For each of the reactions (1 and 2), three relaxation times were observed. No relaxation effects were found without substrate being present. In the case of the glutamate–enzyme system, all three relaxation times could be measured, while in the case of the aspartate–enzyme system only two of the relaxation times could be measured, although an upper bound could be obtained for the third. The results of these experiments are reported in Tables I and II together with the total enzyme (E_0) and substrate (S_0) concentrations employed. The relaxation times are averages obtained from several photographs and probably are only precise to ± 15% due to the small absolute magnitude of the effects. In reaction mixtures containing lower total concentrations of enzyme and substrate, the relaxation effects become too small for quantitative evaluation. At higher substrate concentrations, substrate inhibition occurred. This latter phenomenon will be discussed more fully a little later.

Each of the four reactions studied in the stopped flow requires some special comment. For E_L + Gm, the concentration of Gm was much greater than that of E_L in an attempt to make the reaction pseudo first order; unfortunately the observable change in absorption was quite small at times greater than 5–10 msec. However, an approximate pseudo first order rate constant was determined by plotting the logarithm of the signal amplitude *versus* time. In the case of E_L + As, no reliable first order constant could be obtained since the reaction became too fast when the concentration of As was high enough to cause a measurable absorption change. A lower bound of this rate constant can be calculated by making the conservative estimate that the reciprocal rate constant is less than 10 msec. if the reaction is too fast to measure. The reaction of E_M + Kg was found to be initially second order when E_M = Kg = 2×10^{-5} M; at higher concentrations the reaction was too fast to measure. For E_M + Oa, a lower bound for the second order rate constant could be obtained by assuming $t_{1/2} < 10$ msec. when the reaction was too fast to study with E_M = Oa = 2×10^{-5} M. The results obtained are given in Table III.

Not too much significance can be attached to the stopped flow measurements since the rates involved are really too fast for this method. The measured rate constants are precise only to ± 40%, while the lower bounds could easily be a factor of two larger because of the conservative estimate of the apparatus resolution time.

TABLE III

STOPPED FLOW RESULTS

Concentration, M		k, M^{-1} sec.$^{-1}$	
Enzyme	Substrate	Stopped flow	Temp. jump
$E_L = 2 \times 10^{-5}$	Gm $= 2 \times 10^{-4}$	1.2×10^5	7.2×10^5
$E_L = 2 \times 10^{-5}$	As $= 3 \times 10^{-3}$	$>4.5 \times 10^4$	1.5×10^5
$E_M = 2 \times 10^{-5}$	Kg $= 2 \times 10^{-5}$	4.5×10^6	9.0×10^6
$E_M = 2 \times 10^{-5}$	Oa $= 2 \times 10^{-5}$	$> 5 \times 10^6$	1.3×10^7

The equilibrium dialysis experiments were intended to yield some information on the magnitude of the enzyme–substrate binding constants. The results of four experiments are given in Table IV. The concentrations of amino and keto acids found are often close to the limit of the analytical methods employed so that experimental errors of ± 5–10% are expected.

Discussion

Since no relaxation effects were observed in the absence of substrates, it seems quite certain that the time constants reported are characteristic of the enzymatic reaction. Because three relaxation times are observed, a minimal mechanism requires three steps in each half reaction; the simplest possibility is

$$E_L + As \underset{k_{-1}}{\overset{k_1}{\rightleftharpoons}} X_1 \underset{k_{-2}}{\overset{k_2}{\rightleftharpoons}} X_2 \underset{k_{-3}}{\overset{k_3}{\rightleftharpoons}} E_M + Oa \quad (3)$$

(17) H. J. Koepsell and E. S. Sharp, *Arch. Biochem. Biophys.*, **38**, 443 (1952).

(18) S. Moore and W. H. Stein, *J. Biol. Chem.*, **211**, 909 (1954).

TABLE IV

EQUILIBRIUM DIALYSIS[a]

Initial enzyme	Initial substrate	Final unbound ketoacid Exp.	Final unbound ketoacid Calcd.	Final unbound amino acid Exp.	Final unbound amino acid Calcd.
$E_L = 2.37 \times 10^{-5}$	$Gm = 3.04 \times 10^{-5}$	$<2 \times 10^{-6}$	3.58×10^{-6}	1.61×10^{-5}	2.10×10^{-5}
$E_L = 2.37 \times 10^{-5}$	$As = 3.00 \times 10^{-5}$	$<2 \times 10^{-6}$	2.00×10^{-6}	2.50×10^{-5}	2.50×10^{-5}
$E_M = 2.44 \times 10^{-5}$	$Kg = 2.88 \times 10^{-5}$	1.4×10^{-5}	1.23×10^{-5}	1.00×10^{-5}	1.16×10^{-5}
$E_M = 3.25 \times 10^{-5}$	$Oa = 3.57 \times 10^{-5}$	1.60×10^{-5}	1.43×10^{-5}	2.00×10^{-5}	1.86×10^{-5}

[a] pH 8, 4°, all concentrations in M.

$$E_M + Kg \underset{k_3'}{\overset{k'_{-3}}{\rightleftarrows}} Y_2 \underset{k_2'}{\overset{k'_{-2}}{\rightleftarrows}} Y_1 \underset{k_1'}{\overset{k'_{-1}}{\rightleftarrows}} E_M + Gm \quad (4)$$

One of the relaxation times (τ_1) is much shorter than the other two so that the restoration of equilibrium for the step it characterizes can be assumed to be essentially independent of the other steps. Since the relaxation time of an uncoupled unimolecular reaction (e.g., $A \rightleftarrows B$) should be independent of concentration, τ_1 must be associated with one of the bimolecular reactions in the mechanism. Furthermore since glutamate and aspartate are present in concentrations much higher than ketoglutarate and oxalacetate, the most logical assumption is that τ_1 and τ_1' respectively characterize the steps

$$As + E_L \underset{k_{-1}}{\overset{k_1'}{\rightleftarrows}} X_1 \quad (5)$$

$$Gm + E_L \underset{k'_{-1}}{\overset{k_1'}{\rightleftarrows}} Y_1 \quad (6)$$

In this case the relaxation times can be shown readily to be[6]

$$1/\tau_1 = k_1(\overline{As} + \overline{E}_L) + k_{-1} \quad (7)$$

$$1/\tau_1' = k_1'(\overline{Gm} + \overline{E}_L) + k'_{-1} \quad (8)$$

where the bars designate equilibrium concentrations. If step 5 is assumed to be equilibrated much faster than the other steps (this is actually a direct consequence of the fact that $\tau_1 << \tau_2$ and τ_3)

$$-\delta X_1 = \frac{\delta X_2 + \delta E_M}{1 + [K_1(\overline{E}_L + \overline{As})]^{-1}}$$

Mechanism 3 can then be characterized by rate laws (9) and (10) in the neighborhood of equilibrium[6,19,20]

$$-d\delta E_M/dt = k_{-3}(\overline{E}_M + \overline{Oa})\delta E_M - k_3\delta X_2 = a_{11}\delta E_M + a_{12}\delta X_2 \quad (9)$$

$$-d\delta X_2/dt = \left[\frac{k_2}{1 + [K_1(\overline{E}_L + \overline{As})]^{-1}} - k_{-3}(\overline{E}_M + \overline{Oa})\right]\delta E_M + \left(k_{-2} + \frac{k_2}{1 + [K_1(\overline{E}_L + \overline{As})]^{-1}} + k_3\right)\delta X_2 = a_{21}\delta E_M + a_{22}\delta X_2 \quad (10)$$

The coefficients a_{11}, a_{12}, a_{21} and a_{22} are defined by the above equations, δ represents the difference between the actual and equilibrium concentration of the species under consideration, and $K_1 = k_1/k_{-1}$. The relaxation times can now be calculated (see ref. 6, 19 and 20 for details) by solving for the two roots of the determinant

$$\begin{vmatrix} a_{11} - 1/\tau & a_{12} \\ a_{21} & a_{22} - 1/\tau \end{vmatrix} = 0$$

The results are

$$\frac{1}{\tau_{2,3}} = \frac{(a_{11} + a_{22}) \pm \sqrt{(a_{11} + a_{22})^2 - 4(a_{11}a_{22} - a_{12}a_{21})}}{2} \quad (11)$$

where the positive square root corresponds to τ_3 and the negative square root to τ_2. In the limit where the equilibrium involving the bimolecular reaction is adjusted much more quickly than the other step, the relaxation times are

$$\frac{1}{\tau_3} = k_3 + k_{-3}(\overline{Oa} + \overline{E}_M) \quad (12)$$

$$\frac{1}{\tau_2} = \frac{k_2}{1 + [K_1(\overline{E}_L + \overline{As})]^{-1}} + \frac{k_{-2}}{1 + [K_3(\overline{E}_M + \overline{Oa})]^{-1}} \quad (13)$$

where $K_3 = k_{-3}/k_3$. In the present case equations 12 and 13 were found to be sufficient within 10 or 15% and were used as a first approximation. All of the final calculations were carried out using equation 11. Equations identical to 9–13 can be obtained for mechanism 4 except the equilibrium and rate constants are primed and the equilibrium concentrations of As and Oa are replaced by Gm and Kg, respectively.

The relaxation times contain 12 unknown rate constants (actually 10 rate constants plus one equilibrium constant, since τ_1 could not be measured due to its shortness and the small absolute effect). To determine these eleven constants are 13 measured relaxation times plus the mass conservation of substrate and enzyme for each experiment (16 equations). In addition the equilibrium constant for the over-all transmination is known to be about $^1/_7$.[21] An exact solution of such a system of equations is virtually hopeless; however a method of successive approximations was used, and it quickly became apparent that the order of magnitude of the various rate constants was quite rigidly fixed by the large number of equations which had to be satisfied.

A further refinement of the calculations was made by making use of the equilibrium spectra of the reaction mixtures. Since the substrate absorption is negligible above 300 mμ, the total absorbancy, a, of a solution containing all of the reactants in mechanism 3 in the spectral region 300–600 mμ is

$$a = \epsilon_L(E_L) + \epsilon_M(E_M) + \epsilon_{X_1}(X_1) + \epsilon_{X_2}(X_2)$$

Here the epsilons are the extinction coefficients of the various species and a 1 cm. path length has been assumed. The analogous relationship for a reac-

(19) G. G. Hammes and R. A. Alberty, *J. Am. Chem. Soc.*, **82**, 1564 (1960).

(20) R. A. Alberty and G. G. Hammes, *Z. Elektrochem.*, **64**, 124 (1960).

(21) B. E. C. Banks, K. C. Oldham, E. H. Thain and C. A. Vernon, *Nature*, **183**, 1084 (1959).

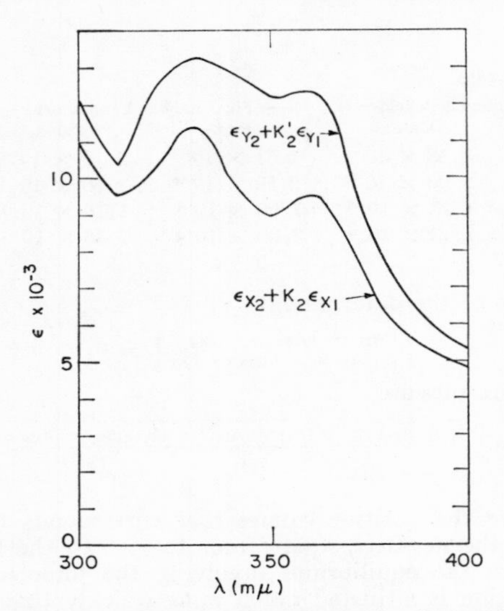

Fig. 3.—Spectra of transamination reaction intermediates. See text for explanation of symbols.

tion mixture involving the mechanism in equation 4 is obvious. Since ϵ_L and ϵ_M can be measured independently $\epsilon_{X_1}(X_1) + \epsilon_{X_2}(X_2)$ can be determined for each reaction mixture. Furthermore this quantity divided by X_2 should be a constant at any given wave length equal to $\epsilon_{X_2} + K_2\epsilon_{X_1}$ where $K_2 = k_{-2}/k_2$. Thus the spectra provide another check on the consistency of the data and in addition give some information about the spectral properties of the intermediates.

The best fit of all the data is obtained by using the parameters

$k_1/k_{-1} = 1.8 \times 10^3\ M^{-1}$	$k_1' = 3.3 \times 10^7\ M^{-1}$ sec.$^{-1}$
	$k'_{-1} = 2.8 \times 10^3$ sec.$^{-1}$
$k_2 = 80$ sec.$^{-1}$	$k_2' = 61$ sec.$^{-1}$
$k_{-2} = 26$ sec.$^{-1}$	$k'_{-2} = 30$ sec.$^{-1}$
$k_3 = 1.4 \times 10^2$ sec.$^{-1}$	$k_3' = 70$ sec.$^{-1}$
$k_{-3} = 7 \times 10^7\ M^{-1}$ sec.$^{-1}$	$k'_{-3} = 2.1 \times 10^7\ M^{-1}$ sec.$^{-1}$

Although the measurement of the fastest relaxation time was too uncertain for calculation of k_1 and k_{-1}, these lower bounds could be obtained

$$k_1 > 10^7\ M^{-1}\ \text{sec.}^{-1}$$

$$k_{-1} > 5 \times 10^3\ \text{sec.}^{-1}$$

Using the above constants, all of the relaxation times (with one exception) can be calculated to $\pm 10\%$, which is well within the experimental error. The one exception is at the highest concentration used in the glutamate–enzyme system where the difference between the calculated and observed values is about 30%. A summary of the observed and calculated relaxation times is given in Tables I and II. Also included are the calculated concentrations of the various enzyme species present in each reaction mixture. The equilibrium constants of the various individual steps, the half reactions and the over-all reaction are easily calculable from the rate constants. Using the above rate constants, values of $\epsilon_{X_2} + K_2\epsilon_{X_1}$ and $\epsilon_{Y_2} + K'_2\epsilon_{Y_1}$ can be calculated to better than $\pm 15\%$ in the concentration range under consideration; these parameters are shown as a function of wave length in Fig. 3. At higher concentrations of substrate than those reported both the spectra and relaxation times deviate from their predicted values. This probably is due to several different effects. At high concentrations of glutamate ($>5 \times 10^{-4}$) and aspartate ($>1 \times 10^{-3}$ M), combination of these two substrates with the aminic form of the enzyme probably occurs. Assuming a value of this binding constant of $10^3\ M^{-1}$ for glutamate and E_M, and $10^2\ M^{-1}$ for aspartate and E_M, all of the equilibrium spectral observations can be accounted for well within experimental error. An alternative possibility which is consistent with the experimental results is that the ketoglutarate formed begins to react with E_L at higher concentrations, thus lowering the effective enzyme concentration. Such an interaction is known to occur at lower pH's.[8] Whether or not the corresponding temperature jump data also are explained so simply is difficult to say since calculations cannot be made without knowing all of the rate constants involved. However, certainly the combined effects of coupling and superimposing relaxation times could account easily for the excessive shortening of the relaxation time observed at high concentrations. Finally a possibly explanation of the data at higher concentrations is that other intermediates become important and the simple mechanism proposed is not sufficient. A word of caution should be inserted here: although the rate constants given probably are precise to about $\pm$ 25% within the framework of the mechanism proposed, a mechanism involving more intermediates could certainly also be made consistent with the data. However, in such a complex case the assignment of the relaxation times to any of the various mechanistic steps is impossible. As is true with any kinetic investigation, only a minimal mechanism consistent with all known facts should be given. In the present situation, a very large number of kinetic and spectral observations are correlated within experimental error over the accessible concentration range.

Although the rather crude results of the stopped flow experiments cannot be used to postulate a mechanism, at least it can be seen if the data is consistent with the mechanism proposed on the basis of the temperature jump experiments. At the concentration of substrates employed the rate of equilibration of the enzyme–substrate reactions is fast compared to the rate of interconversion of intermediates; therefore the measured second order rate constants for the over-all half reaction are equal to an equilibrium constant (for enzyme–substrate formation) times a first order rate constant (for interconversion of intermediates) in terms of the proposed mechanism. The values of these over-all rate constants calculated from temperature jump data are presented in Table III together with the stopped flow results. In the cases where only lower bounds are obtained from the stopped flow data, the results from the two methods are consistent. Otherwise the rate constants obtained from the stopped flow are of the same order of magnitude, but slightly less than expected. Actually this is not too surprising because this simple analysis assumed all of the reactions to be irreversible, while in point of fact this is a very poor approximation. An exact analysis of the full rate equations is not profitable, but qualitatively the effect of introducing the back reaction would be to make the apparent rate constant smaller, in agreement with the observed results. However, it should be emphasized that the only conclusion that can be stated with certitude is that the stopped flow results do not contradict the analysis of the temperature jump data.

Unfortunately the equilibrium dialysis experiments are also not conclusive. The agreement between the measured values of the final substrate concentrations and those predicted on the basis of the temperature jump experiments are given in Table IV; the agreement is certainly as good as one can expect considering the experimental errors and the fact that the two experiments were necessarily done at different temperatures. The desirability of carrying out the equilibrium dialyses at concentrations similar to those employed in the temperature jump experiments had two important consequences: (1) the concentration of substrates was such that often detection was barely within the limits of the analytical methods employed and (2) the per cent of substrate bound to the enzyme often was small. Thus in the cases where asparate and oxalacetate were the initial substrates, the results really provided a good check of the over-all equilibrium of that particular half reaction, but did not provide a critical test of the proposed enzyme-substrate binding constants. For the other half reaction, on the other hand, a considerable fraction of the enzyme is present as intermediates according to the calculations. Again it can only be said that the two experiments are consistent within experimental error.

In principle, the results of these experiments should be able to predict the results of steady state kinetic studies. However, in the present case a complicating factor exists, namely, a monomer–dimer equilibrium is thought to exist for this enzyme.[22] In fact essentially all of the steady state experiments have been carried out with monomer enzyme, while in these experiments the enzyme is probably all dimer. Therefore it is not too surprising that the calculated Michaelis constants and maximum velocities differ considerably from those actually measured.[11,12] In fact, the values of both the maximum velocities and Michaelis constants are much greater for the monomer enzyme. This means the stabilities of the enzyme–substrate complexes are much greater with the dimer enzyme, but the interconversion of intermediates is slower. This trend is consistent with preliminary steady state experiments with the dimer enzyme[22] although a quantitative comparison is not yet possible.

The spectra in Fig. 3 both show two peaks, one at 330 mμ and another of lower intensity at 360 mμ. Since the difference in height between the two peaks is larger in the enzyme-4 carbon substrate system (aspartate–oxalacetate) where the equilibrium is more in favor of the second enzyme–substrate intermediate, X_2, the intermediates, X_2 and Y_2 probably can be identified with the peaks at 330 while the first enzyme-substrate intermediates (X_1 and Y_1) probably absorb around 360. Absorption peaks around 360–370 mμ are characteristic of Schiff bases between pyridoxal phosphate and amino acids,[23] while absorption peaks around 310–330 mμ are found for pyridoxine derivatives with no conjugated double bond involving the carbon atom attached to position 4 of the ring. Thus the enzyme reaction in terms of the known parts of the active site can be written schematically for each half reaction as shown in Fig. 4.

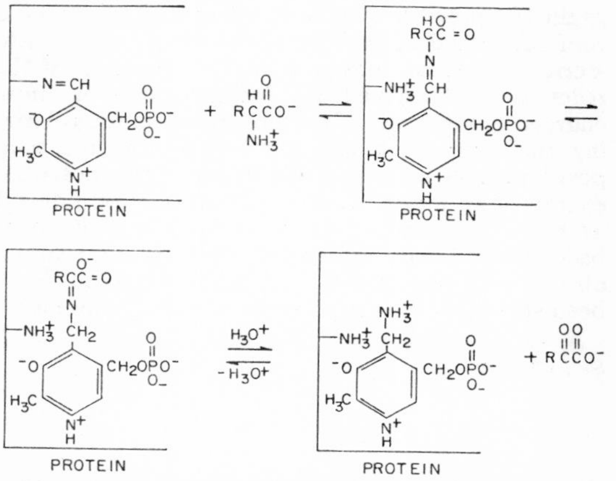

Fig. 4.—A schematic mechanism for enzymatic transamination. For aspartate–oxalacetate R is CH_2COO^-, while for glutamate–ketoglutarate it is $CH_2CH_2COO^-$.

Such a mechanism is consistent with that proposed for model reactions[2]; however to our knowledge this is the first direct kinetic evidence of such intermediates in the enzymatic transamination. As in the case of model reactions,[2] the rate limiting step at normal substrate concentrations is the conversion of one Schiff base into the other.

As usual the role of the protein moiety in the over-all mechanism is not clear. All of the bimolecular rate constants are of the order of 10^7 M^{-1} sec.$^{-1}$, while a typical value for model reactions (pyridoxal plus valine) is about 1 M^{-1} sec.$^{-1}$.[24] Therefore the enzyme catalyzes Schiff base formation by a factor of 10^6. This probably is due to two effects: binding of the substrate in a critical configuration and concerted local acid–base catalysis. The value of 10^7 M^{-1} sec.$^{-1}$ for the rate constant of enzyme–substrate formation is about 100 times smaller than that predicted for a diffusion controlled reaction.[25,26] Thus at least two more intermediates, the initial enzyme–substrate complexes, should be present in each half reaction. An interesting possibility is that a conformational change of the enzyme after the initial enzyme–substrate interaction is the rate controlling step in the Schiff base formation. The transformation of one Schiff base into another has a first-order rate constant about 10^8 times larger than typical model reactions.[27] Here

(22) P. Fasella, C. Turano, A. Giartosio, Comm. to the First International Congress of Biophysics, Stockholm, 1961.

(23) Y. J. Matsuo, *J. Am. Chem. Soc.*, **79**, 2011 (1957).

(24) G. G. Hammes and P. Fasella, unpublished results.

(25) R. A. Alberty and G. G. Hammes, *J. Phys. Chem.*, **62**, 154 (1958).

(26) G. G. Hammes and R. A. Alberty, *ibid.*, **63**, 274 (1959).

again the possible rate controlling step might be a configurational change bringing the substrate into a critical position for reaction. For example, if the reaction is catalyzed by bringing one or more charged groups into a favorable position for polarizing the chemical bonds involved in the reaction, positioning would be critical since the potential energy of an ion-induced dipole interaction falls off as r^{-4}. Of course, if the interconversion of Schiff bases is acid–base catalyzed, conformation would also be very important. Although metal ions have been shown to be very effective in catalyzing model reactions,[2] no significant amounts of metal ions have been found in purified glutamic–aspartic transaminase.[28] At the present time we are trying to measure protein conformational changes during the course of the reaction in order to ascertain whether or not the protein molecule is dynamically involved in the catalytic process.

The authors are indebted to Dr. M. Eigen and Dr. L. de Maeyer for helpful advice concerning construction of the temperature jump and to Mr. Gerald Becker for carrying out the stopped flow experiments.

(27) B. E. C. Banks, A. A. Diamantis and C. A. Vernon, *J. Chem. Soc.*, 4235 (1961).

(28) P. Fasella, G. G. Hammes and B. Vallee, *Biochim. et Biophys. Acta*, **65**, 142 (1962).

Reprinted from BIOCHEMISTRY, Vol. 8, 1969

The Significance of Intermediary Plateau Regions in Enzyme Saturation Curves*

John Teipel and D. E. Koshland, Jr.

ABSTRACT: An investigation has been made of the catalytic and binding parameters required to generate kinetic saturation curves possessing an intermediary plateau region. General rate equations were derived for the case in which the rate of equilibration between substrate and enzyme is rapid relative to the rate of catalysis. Analysis of the second derivative of these expressions revealed that kinetic or binding curves with pronounced intermediary plateau regions will only be produced when (1) the enzyme or enzymes present possess a total of *more than two* substrate binding sites and (2) the relative magnitude of the intrinsic catalytic or binding constants of these sites first decreases, then increases as the enzyme is saturated.

Multisite enzymes will not yield these undulating kinetic curves when their intrinsic catalytic, or binding constants progressively increase, progressively decrease or remain constant with saturation. Applying these and other diagnostic criteria, certain enzyme models could be unequivocably excluded as possible explanations for the complex kinetic curves experimentally observed.

Analysis of the shape of substrate saturation curves has often provided valuable information concerning the structure–function relationships of enzymes. Examination of the hyperbolic function observed in plots of initial velocity *vs.* substrate concentration, for example, led to the concept of an enzyme–substrate complex. (Michaelis and Menten, 1913). Later, the observation of sigmoidal-shaped binding curves suggested the presence of cooperative effects between binding sites (Bohr, 1903). Attempts to interpret the latter phenomenon in terms of intraprotein interactions provided several useful models (Monod *et al.*, 1965; Koshland *et al.*, 1966) relating subunit structure to enzyme function. One of these models (Koshland *et al.*, 1966) led to the prediction of negative cooperativity, a new mode of subunit interaction which has since been verified experimentally (Conway and Koshland, 1968).

Recent investigations of several enzymes, including phosphoenolpyruvate carboxylase (Corwin and Fanning, 1968), B ADP-glucose pyrophosphorylase (Gentner and Preiss, 1968), and cytidine triphosphate synthetase (Levitzki and Koshland, 1969) from *Escherichia coli*, glutamate dehydrogenase (LeJohn and Jackson, 1968) from Blastocladiella and glyceraldehyde 3-phosphate dehydrogenase (Gelb and Nordin, 1969) from insects, have revealed a new and anomalous type of kinetic behavior (see Figure 1). Plots of initial velocity *vs.* substrate concentration for these enzymes are hyperbolic at low substrate concentrations and sigmoidal at higher substrate levels. This transition from hyperbolic to sigmoidal behavior yields curves which contain a pronounced intermediary plateau region or "bump." Thus, unlike the purely hyperbolic curve, which possesses no points of inflection, or the purely sigmoidal curve, which possesses one point of inflection, the curves described above possess two points of inflection.

The observation of these bumpy curves by several independent investigators with a variety of enzymes suggests that this phenomenon might be widespread and indicative of some common structural properties. A study was therefore initiated to investigate the catalytic and binding parameters required to generate saturation curves with two points of inflection. The analysis which follows provides some general diagnostic aids by which the shapes of bumpy saturation curves might be related to structural features. These aids include a determination of the minimum number of binding sites and the nature of the cooperative interactions between sites which are consistent with this type of kinetic behavior.

Theory

One Ligand. The binding of a single substrate to a multisite enzyme may be expressed by the following equilibria

$$\begin{aligned} &E + S \rightleftharpoons ES & K_1 &= (ES)/(E)(S) \\ &ES + S \rightleftharpoons ES_2 & K_2 &= (ES_2)/(ES)(S) \\ &ES_{i-1} + S \rightleftharpoons ES_i & K_i &= (ES_i)/(ES_{i-1})(S) \\ &ES_{n-1} + S \rightleftharpoons ES_n & K_n &= (ES_n)/(ES_{n-1})(S) \end{aligned} \qquad (1)$$

where (S) and (E) refer to the concentration of free substrate and free enzyme, respectively, and n is the total number of substrate binding sites per molecule. In describing the binding of substrate to a multisite enzyme it will be valuable here to replace the association constants $K_1, K_2 \ldots K_n$, which include a statistical factor dependent upon the total number of binding sites, with the intrinsic binding constants $K_1', K_2' \ldots K_n'$, which express directly the inherent affinity of the individual binding sites for substrate. The relationship between the two sets of constants is given in eq 2.

* From the Department of Biochemistry, University of California, Berkeley, California 94720. *Received July 3, 1969.* The authors would like to acknowledge the support of the U. S. Public Health Service (Grant AM09765) and the American Cancer Society (local and national).

$$K_i = \frac{(n + 1 - i)}{i} K_i' \qquad (2)$$

The initial velocity, v_0, of a reaction catalyzed by an enzyme with *one* binding site is given by $v_0 = k(\text{ES})$, where k is a first-order rate constant. In the case of a multisite enzyme, v_0 maybe expressed as the sum of the individual turnover rates for each ES complex as shown in eq 3, where the values of k_i represent the average catalytic rate constants per site for a particular complex. If it is assumed that the rate of

$$v_0 = \sum_1^n i k_i(\text{ES}_i) \qquad (3)$$

equilibration between substrate and enzyme is rapid relative to the rate of catalysis, then (ES_i) in eq 3 is related to substrate concentration by the equilibrium expressions in eq 1. From eq 1, 2, and 3 one can thus readily derive a rate expression (eq 4), similar in form to the classical Adair equation (Adair, 1925) which describes initial velocity, v_0, as a function of substrate concentration(s). In eq 4 E_t is the total enzyme

$$v_0 = \frac{E_t \sum_1^n i k_i \psi_i (\text{S})^i}{1 + \sum_1^n \psi_i (\text{S})^i} \qquad (4)$$

concentration and ψ_i represents a product of the intrinsic binding constants.

$$\psi_i = \prod_1^i \left(\frac{n + 1 - j}{j}\right) K_j' \qquad (5)$$

For an enzyme possessing four binding sites eq 4, for example, would yield

$$v_0 = \frac{E_t[4k_1 K_1'(\text{S}) + 12k_2 K_1'K_2'(\text{S})^2 + 12k_3 K_1'K_2'K_3'(\text{S})^3 + 4k_4 K_1'K_2'K_3'K_4'(\text{S})^4]}{1 + 4K_1'(\text{S}) + 6K_1'K_2'(\text{S})^2 + 4K_1'K_2'K_3'(\text{S})^3 + K_1'K_2'K_3'K_4'(\text{S})^4} \qquad (6)$$

It is apparent that when the intrinsic catalytic constants are all equal, *i.e.*, $k_1 = k_2 = k_i = k_n$, then $nk(E_t) = V_{\max}$ and eq 4 reduces to the binding equation originally derived by Adair (1925), eq 7, where N_s is equal to the average number of

$$n\frac{v_0}{V_{\max}} = N_s = \frac{\sum_1^n i \psi_i (\text{S})^i}{1 + \sum_1^n \psi_i (\text{S})^i} \qquad (7)$$

moles of ligand bound per mole of enzyme.

More Than One Ligand. For the reaction of an enzyme with more than one substrate, or with reversible inhibitors or activators, a more generalized rate equation (eq 8) may be derived by the same treatment as described above. In this

$$v_0 = \frac{\sum_1^n \theta_i (\text{S})^i}{1 + \sum_1^n \phi_i (\text{S})^i} \qquad (8)$$

case θ and ϕ are not as simply related to the intrinsic catalytic and binding parameters as are the constants appearing in eq 4 and 7. θ and ϕ are complex constants which will generally include terms describing the interactions of both substrate and effectors with the enzyme. If, for example, an enzyme, possessing one binding site for substrate and one binding site for the ligand, X, reacts according to the following mechanisms

$$\begin{array}{ccccc} \text{E} & \underset{}{\overset{K_1(\text{S})}{\rightleftharpoons}} & \text{ES} & \xrightarrow{k_1(\text{slow})} & \text{product} \\ K_2(\text{X}) \downarrow\uparrow & & \uparrow\downarrow K_3(X) & & \\ \text{EX} & \overset{K_4(\text{S})}{\rightleftharpoons} & \text{ESX} & \xrightarrow{k_2(\text{slow})} & \text{product} \end{array} \qquad (9)$$

the rate equation will be given by eq 10

$$v_0 = \frac{E_t[k_1 K_1 + k_2 K_1 K_3(\text{X})](\text{S})}{1 + K_2(\text{X}) + [K_1 + K_1 K_3(\text{X})](\text{S})} \qquad (10)$$

If this rate equation is rewritten in the generalized form of eq 8, then θ and ϕ will have the values

$$\theta = \frac{E_t[k_1 K_1 + k_2 K_1 K_3(\text{X})]}{1 + K_2(\text{X})} \qquad (11)$$

$$\phi = \frac{K_1 + K_1 K_3(\text{X})}{1 + K_2(\text{X})} \qquad (12)$$

The important similarities to note between eq 4 and 8 are that (1) each is in the form of the classical Adair equation and (2) the degree of the equation, *i.e.*, the highest exponential power to which (S) is raised, is determined by the total number of substrate binding sites. It should also be noted that eq 8 may describe the kinetic behavior of a mixture of enzymes, where n, in this case, is equal to the total number of substrate binding sites contributed by all the species present.

As noted above, rate eq 4 and 8 were derived by assuming that the rates of equilibrium between substrate and enzyme were rapid relative to the rates of catalysis. If, however, the rate of the binding of substrate to enzyme is of the same order of magnitude as catalysis, then the concentrations of the ES complexes in these equations must be related to substrate concentration through a combination of kinetic constants rather than equilibrium constants. Under these conditions the rate expressions derived need not assume the form of a generalized Adair equation nor does the degree of the equation necessarily have to reflect the total number of substrate binding sites. Sanwal and Cook (1966) and Sweeney and Fisher (1968) have shown for bisubstrate reactions that when the relative rates of binding and catalysis lie within a certain range, the degree of the rate equation, with respect to one of the substrates, often exceeds that substrate's number of binding sites. The results of their analyses also pertain to the reaction of a single substrate plus effector with an enzyme possessing one substrate binding site and to the reaction of two or more molecules of the same substrate with an enzyme possessing multiple substrate binding sites. Therefore, if the rates of binding and catalyses are comparable, the kinetic

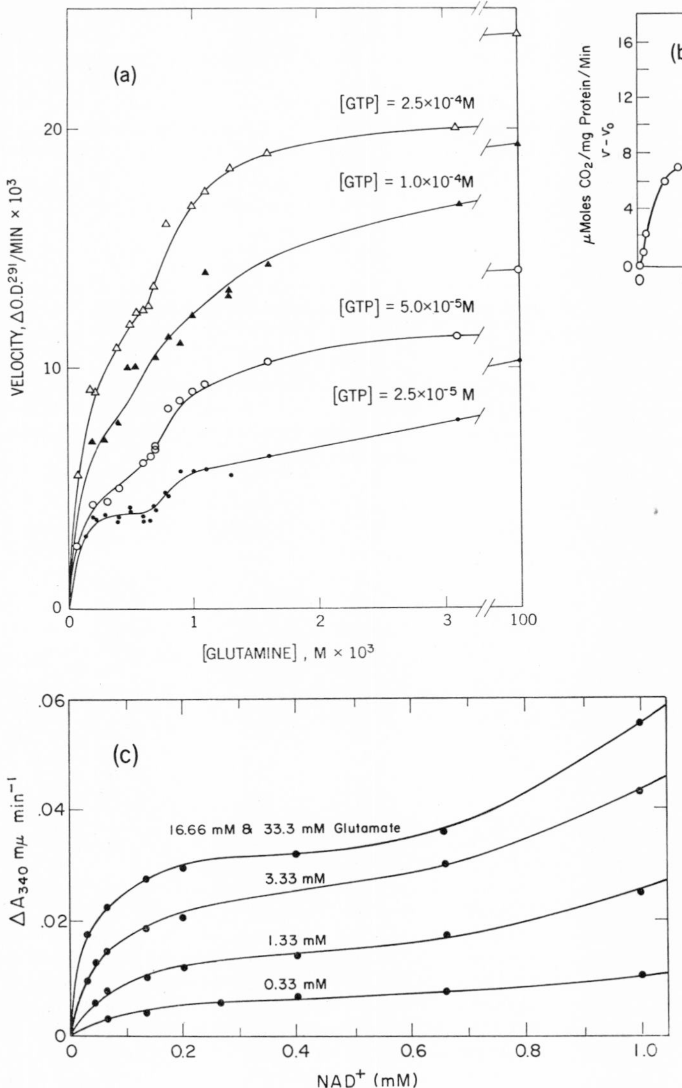

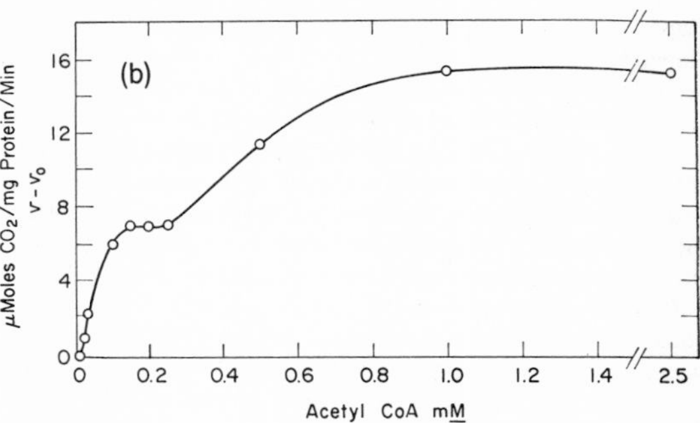

FIGURE 1: Experimental examples of saturation curves with intermediate plateau regions. (a) CTP synthetase: velocity of formation of cytidine triphosphate as a function of glutamine concentration at the concentrations of GTP, an allosteric effector, indicated. Data taken from Levitzki and Koshland (1969). (b) Phosphoenolpyruvate carboxylase: velocity of formation of oxalacetate as a function of acetyl-CoA concentration, an allosteric effector at a phosphoenolpyruvate concentration of 1×10^{-2} M. The ordinate represents the difference in initial velocity in the presence, v, and absence, v_0, of acetyl-CoA. Data taken from Corwin and Fanning (1968). (c) Glutamate dehydrogenase: velocity of the oxidative deamination of glutamate as a function of NAD^+, at the concentrations of glutamate indicated. Data taken from LeJohn and Jackson (1968).

behavior of an enzyme with a small number of substrate binding sites may resemble that expected of an enzyme possessing a larger number of such sites. The rate equations for this situation however will not be developed here since the formulation of these expressions depends upon a detailed knowledge of the catalytic mechanism of the enzyme.

Results

Inflection Point Analysis. In Figure 1 are shown initial velocity *vs.* substrate concentration plots, taken from the literature, for the enzymes cytidine triphosphate synthetase (Levitzki and Koshland, 1969), phosphoenolpyruvate carboxylase (Corwin and Fanning, 1968), and glutamate dehydrogenase (LeJohn and Jackson, 1968). The kinetic behavior of these enzymes is representative of a class of proteins whose saturation curves display an intermediary plateau region. Although the absolute shapes of these curves differ, all have one feature in common. Each curve is characterized by a decreasing slope at low substrate levels, followed by an increasing slope at intermediate substrate levels, and finally a decreasing slope at high substrate levels. The changing slope of these curves generates two points of inflection, one located at the point of transition from decreasing to increasing slope and the other at the point of transition from increasing to decreasing slope.

The second derivative, with respect to (S) of the rate equations for these curves, will yield the rate of the change in slope as a function of substrate concentration. Since the slopes of the experimental curves in Figure 1 are decreasing at both very low and very high substrate concentrations, the second derivative of these functions must be negative as $S \rightarrow 0$ and $S \rightarrow \infty$. (Although the slopes of the kinetic curves for glutamate dehydrogenase appear to be increasing at the highest substrate concentration shown in Figure 1C, kinetic measurements, reported in the same communication, show that these slopes decrease at higher substrate levels.)

In addition, the second derivative of these curves must equal zero at the two values of S which give inflection points, because for any continuous function $y = f(x)$, the values of x which give inflection points must be among the roots of $f''(x) = 0$. The type of function thus generated by the second derivative of the curves in Figure 1 is shown in a general form in Figure 2.

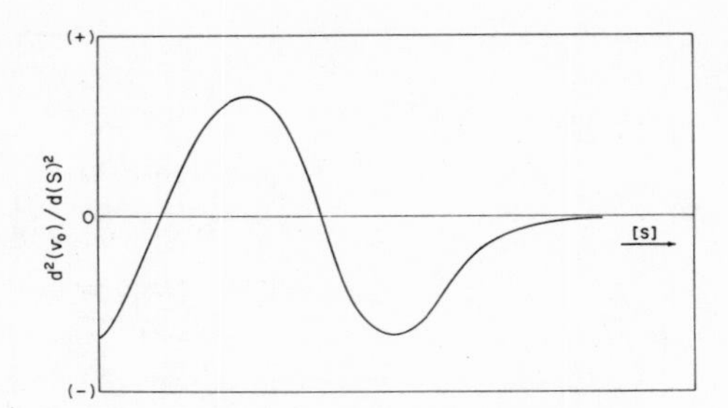

FIGURE 2: General form of the second derivative of a saturation curve with an intermediate plateau whose slope decreases at both low substrate levels and high substrate levels.

From the above analysis it follows that a saturation curve possessing an intermediary plateau or two points of inflection will be generated by eq 4 or 8 only if the second derivative of these expressions yields a function of the type seen in Figure 2. The second derivative of the general rate equations for a two-site and a four-site model were therefore examined.

Two-Site Model. For an enzyme or enzymes possessing a total of two substrate binding sites, eq 8 reduces to

$$v_0 = \frac{\theta_1(S) + \theta_2(S)^2}{1 + \phi_1(S) + \phi_2(S)^2} \quad (13)$$

The second derivative, with respect to S of eq 13 is given by

$$\frac{d^2v_0}{d(S)^2} = \frac{-2[\phi_2(\theta_2\phi_1 - \theta_1\phi_2)(S)^3 + 3\theta_2\phi_2(S)^2 + 3\theta_1\phi_2(S) + \theta_1\phi_1 - \theta_2]}{[1 + \phi_1(S) + \phi_2(S)^2]^3} \quad (14)$$

As discussed above, a curve of the type seen in Figure 2 will be generated by eq 14 only if (a) the value of the equation is less than zero as $S \rightarrow 0$ and $S \rightarrow \infty$, and (b) the value is equal to zero at two different values of S. Inspection of eq 14 reveals that the first criteria is satisfied when $\theta_1\phi_1 > \theta_2$ and $\theta_2\phi_1 > \theta_1\phi_2$. When condition (a) is met however, it is apparent that all of the coefficients of (S) are negative. Consequently the value of the function is less than zero at all (S) and condition (b) is not satisfied. Thus, regardless of the values selected for θ_1, θ_2, ϕ_1, and ϕ_2, eq 8 is incapable of generating a curve containing two points of inflection. Furthermore, since eq 8 is the most general rate expression for a two-site system, it may be concluded that neither a single enzyme with two substrate binding sites nor a mixture of two enzymes, each with one substrate binding site, in the presence or absence of effectors, will yield curves possessing an intermediary plateau region, provided that the rate of binding is rapid relative to the rate of catalysis.

Four-Site Model. For an enzyme possessing four substrate binding sites, rate eq 4 and 8 yield expressions containing eight independent parameters. The second derivatives of these expressions were too complex to be analyzed in the simple manner discussed above for the two-site model.

A second derivative analysis was therefore made of eq 7, which is both a binding expression and, when catalytic activity is directly proportional to saturation, a rate equation. The second derivative of eq 7 with respect to S for a four-site model is still a complicated expression and, since for the discussion to follow only the first and last few terms in the numerator of the equation are relevant, it is presented below in eq 15. The missing terms consist of the series $a_8(S)^8 + \ldots +$

$$\frac{d^2N_s}{d(S)^2} = \frac{-8K_1[K_1'^2K_2'^3K_3'^3K_4'^2(S)^9 + \ldots + 9K_2'(2K_1' - K_3')(S) + 4K_1' - 3K_2']}{[1 + 4K_1'(S) + 6K_1'K_2'(S)^2 + 4K_1'K_2'K_3'(S)^3 + K_1'K_2'K_3'K_4'(S)^4]^3} \quad (15)$$

$a_i(S)^i + \ldots + a_2(S)^2$, where the coefficients a_i consist of combinations of the equilibrium constants K_1', K_2', K_3', and K_4'.

The behavior of eq 15 was examined for cases where (1) the affinity for substrate progressively increased with saturation, *i.e.*, $K_4' > K_3' > K_2' > K_1'$ and (2) the affinity for substrate progressively decreased with saturation, *i.e.*, $K_1' > K_2' > K_3' > K_4'$. For the case where the affinity for substrate remains constant, *i.e.*, $K_1' = K_2' = K_3' = K_4'$, eq 7 will generate a Michaelis–Menten curve which is, of course, hyperbolic and without points of inflection.

In the first case, where affinity for substrate progressively increased, a function of the type presented in Figure 2 may be generated by eq 15. This conclusion is reached by the following argument. Inspection of eq 15 reveals that at very low (S) the term $4K_1' - 3K_2'$ will predominate. At slightly higher (S) the term $9K_2'(2K_1' - K_3')(S)$ will predominate. At very high (S) the term $K_1'^2K_2'^3K_3'^3K_4'^2(S)^9$ will predominate. The value of the function can be negative at very low (S) when $^4/_3K_1' > K_2'$, positive at slightly higher (S) when $K_3' > 2K_2'$, and again negative at very high (S) with any values of K_1', K_2', K_3', and K_4'. Values of intrinsic binding constants may be selected such that both $K_4' > K_3' > K_2' > K_1'$ and the above conditions for two inflection points are satisfied.

Although, as shown above, a saturation curve possessing two points of inflection may, in theory, be generated when $K_4' > K_3' > K_2' > K_1'$ substitution into eq 7 of appropriate values of the intrinsic binding constants did not yield curves with an intermediary plateau regions which were visibly detectable. This result is not contradictory to the conclusion reached above, but rather indicates that the inflection point test for the detection of intermediary plateau regions is a very sensitive criterion and thus represents a necessary but not sufficient condition for demonstrating the existence of pronounced "bumps" such as seen in Figure 1.

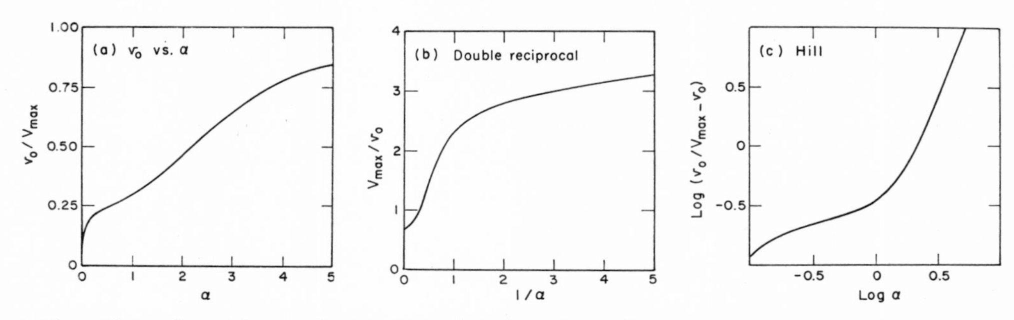

FIGURE 3: Three different plots of the same data for a curve displaying an intermediate plateau in a v_0 *vs.* (S) plot. (a) A v *vs.* (S) plot. (b) A Lineweaver–Burk plot. (c) A Hill plot. Data generated by eq 7 when $K_1' = 100$, $K_2' = 1$, $K_3' = 10$, $K_4' = 100$, and $n = 4$.

In the second case, where the affinity for substrate decreases with saturation, a situation might exist where $K_1' \gg K_2' \gg K_3' \gg K_4'$. Under these conditions the progressive binding of substrate would result in the nearly complete saturation of one binding site before an appreciable portion of another site with a lower affinity was occupied. One might intuitively reason in this case that intermediary plateau regions, associated with the saturation of each site, would be observed in a plot of N_s *vs.* (S). However, when $K_1' > K_2' > K_3' > K_4'$ all the coefficients of (S) in eq 15, including those indicated by the dots, are positive. Because the value of eq 15 is thus always less than zero, the curves generated by eq 7 in this case possess no points of inflection and are hyperbolic in shape. This result, which pertains to plots of N_s *vs.* (S) should not be confused with plots of N_s *vs.* log (S) (Koshland *et al.*, 1966). In the latter case plateau regions are produced when substrate affinity progressively decreases.

By a treatment similar to that outlined above, it may also be demonstrated that if the association constants are all equal, *i.e.*, $K_1' = K_2' = K_3' = K_4'$, saturation curves possessing an intermediary plateau region will *not* be generated when the magnitudes of the catalytic rate constants progressively increase, progressively decrease or remain constant with saturation.

Generation of Bumpy Curves. Curves displaying pronounced plateau regions or bumps could be generated by eq 4 provided that the binding or kinetic constants chosen were such that $K_i > K_{i+1} < K_{i+2}$ or $k_i > k_{i+1} < k_{i+2}$. Under these conditions the rate equation for an enzyme possessing as little as three substrate binding sites may yield curves with intermediary plateau regions. In Figure 3 are shown a v_0 *vs.* (S) plot, together with corresponding double reciprocal and Hill plots, for an enzyme possessing four binding sites where $K_1' = 1 \times 10^2$, $K_2' = 1 \times 10^0$, $K_3' = 1 \times 10^1$, and $K_4' = 1 \times 10^2$, and $k_1 = k_2 = k_3 = k_4$. The substrate concentration range for the graphs in Figures 3, 4, 5, and 6 have been normalized such that for a given set of association constants

$$\alpha = \left(\prod_1^n K_i\right)^{\frac{1}{n}} (\mathrm{S}) \qquad (16)$$

It may be seen that in each of the three classical methods for

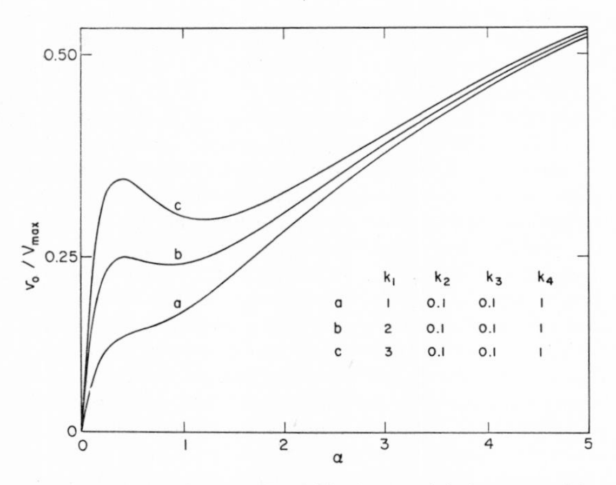

FIGURE 4: Plots of v_0 *vs.* (S) which show undulations caused by varying catalytic constants. Data generated by eq 4 when $n = 4$, $K_1' = K_2' = K_3' = K_4'$, and the values of the intrinsic catalytic constants, k, are as described in the figure.

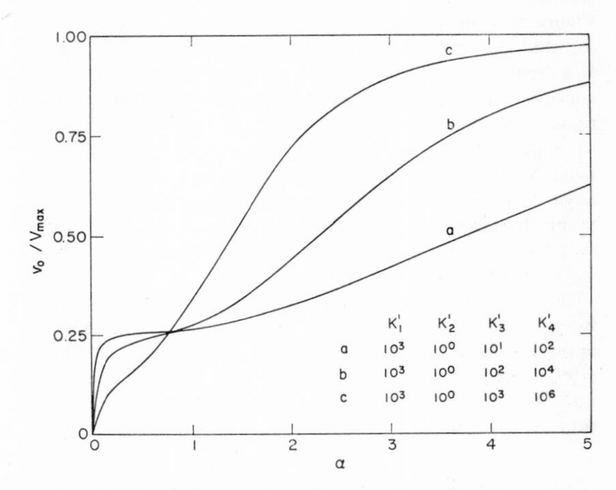

FIGURE 5: Effect on intermediate plateau regions of varying intrinsic binding constants. Data generated by eq 7 when $n = 4$, and the values of the intrinsic binding constants, K', are as described in the figure.

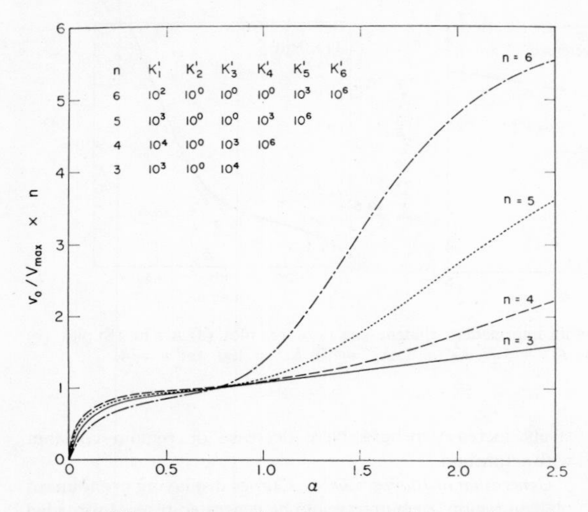

FIGURE 6: Effect on intermediate plateau regions on varying n, the number of binding sites. Data generated by eq 7 when the number of binding sites, n, and the values of the intrinsic binding constants, K', are as described in the figure.

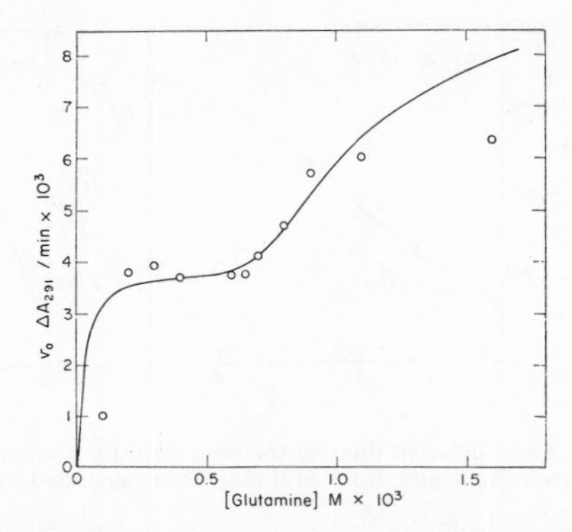

FIGURE 7: Comparison of theory and experiment for CTP synthetase. Experimental points represent rate of formation of cytidine triphosphate as a function of glutamine in the presence of 2.5×10^{-5} M GTP compared with the best theoretical curve based on eq 8 and consistent with the limitations imposed by eq 16 of Weber and Anderson (1965).

representing kinetic data the characteristic plateau region or bump is expressed. In the case of the Hill plot, however, the plateau region is not necessarily indicative of only bumpy saturation curves since quite similar curves may be generated by enzymes displaying a progressively decreasing affinity for substrate. Curves possessing intermediary plateau regions may also be generated by varying the catalytic constants k_1, k_2, k_3, and k_4 and setting $K_1' = K_2' = K_3' = K_4'$ as shown by the curves in Figure 4.

The sharpness of the plateau region, that is the rate of increase of the slope from the plateau to sigmoidal region of the curve, is dependent upon both the choice of kinetic and binding constants and the degree of the rate equation. In Figure 5 is shown the effect of increasing the magnitude of K_3' and K_4' relative to K_1' and K_2' for a four-site model. It is apparent that the slope of the sigmoidal region of the curve is increased only at the expense of reducing the plateau region. The effect of increasing the value of n, the total number of binding sites, on the shape of the saturation curve is dramatized in Figure 6. For each value of n the association constants have been selected so as to maximize the sharpness of the bump. It is evident from the shapes of the curves in Figure 6 that the rate of change in slope from plateau to sigmoidal region increases with increasing n.

The maximum rate of change in slope of a curve generated by eq 7 for a particular value of n may be determined by applying an expression derived by Weber and Anderson (1965). This expression, which sets limits on the maximum rate of change in slope as a function of the degree of saturation, is presented as

$$\frac{-n-Z}{n-N_s} < \frac{dZ}{dN_s} < \frac{n-Z}{n-N_s}\left(\frac{n}{Z} + \frac{n}{N_s} - 2\right) \quad (17)$$

Z represents the slope of the curve in a Hill plot of the binding data. It may be seen that both the upper and lower limits of eq 17 increase with increasing n.

The limitations imposed by eq 17 were applied to one of the kinetic curves of cytidine triphosphate synthetase (Figure 1a) in order to determine whether this data could be explained by eq 7. The value of n was set equal to four, since the enzyme has been shown to be composed of four subunits (Long *et al.*, 1969). The experimental points and allowable theoretical curve are presented in Figure 7. The theoretical curve was generated by conforming as closely as possible to the experimental data yet maintaining the change of slope limitations defined by eq 17. It may be seen that while the theoretical curve approximates fairly well the plateau and early sigmoidal region of the data, the maximum rate of decrease of the curve at higher substrate levels is not consistent with the experimental points. One may, therefore, conclude that the kinetic behavior of cytidine triphosphate synthetase under certain conditions cannot be represented by eq 7 when $n = 4$ regardless of the values selected for the intrinsic binding constants K_1', K_2', K_3', and K_4'.

The above observation at first suggested that differences in the magnitude of the intrinsic catalytic constants, k_1, k_2, k_3, and k_4 might play the prominent role in determining the kinetic behavior of the enzyme. An attempt to fit the curve for cytidine triphosphate synthetase (and the kinetic curve for phosphoenolpyruvate carboxylase, Figure 1b) using eq 4, $n = 4$, in which values for the intrinsic catalytic constants were now also incorporated, was, however, largely unsuccessful. The large number of variable parameters encountered for $n > 4$ prohibited an analysis of more complex expressions. The inability to fit those saturation curves possessing the more pronounced plateau regions with eq 4, which was derived by assuming the rate of equilibration between ligand and enzyme was rapid relative to the rate of catalysis, implies that, for these enzymes, either the rate of ligand binding or the rate of conformational changes within the enzyme are comparable with or slower than the catalytic process. The rate equations for the latter situation, which would undoubtedly contain

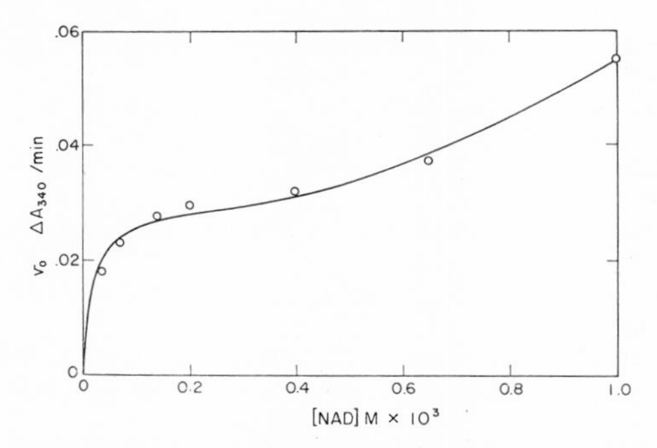

FIGURE 8: Comparison of theory and experiment for glutamate dehydrogenase. Experimental points measured for the rate of deamination of glutamate by glutamate dehydrogenase as a function of NAD^+ in the presence of 33.3 mM glutamate compared with the theoretical curve generated by eq 4, when $n = 4$, $K_1' = 100$, $K_2' = 1$, $K_3' = 10$, $K_4' = 100$ and $k_1 = k_2 = k_3 = k_4 = 2.4 \times 10^{-2}\ \Delta A_{340}\ \text{min}^{-1}\ E_t^{-1}$.

powers of (S) greater than four, could account for the rapidly changing slopes of these kinetic curves.

A fairly good fit of the experimental data was obtained in the case of glutamate dehydrogenase (Figure 1c) with eq 7, $n = 4$. A comparison between the theoretical curve and the experimental points, measured at one concentration of glutamate, is shown in Figure 8. This result suggests that, for glutamate dehydrogenase, an initial decreasing affinity for ligand followed by an increasing affinity for ligand may account for the intermediary plateau regions experimentally observed.

Discussion

An analysis has been presented of saturation curves which possess an intermediary plateau region. This analysis has consisted of an examination of the types of functions generated by general rate equations for multisite enzymes. These equations were derived by assuming that the rate of equilibration between substrate and enzyme was rapid relative to the rate of catalysis. This assumption, in the absence of detailed information on the catalytic mechanisms of the enzymes involved, appeared to be a reasonable starting proposition since (1) the rate of binding of substrate to enzyme has, in fact, been shown to be rapid relative to the rate of catalysis for many enzymes and (2) the rate equations derived for this case are quite general and thus capable of describing a wide range of kinetic behavior.

By differentiating the general rate equations, it was found that certain enzyme models could not account for the bumpy kinetic curves experimentally observed. Firstly, it was shown that enzymes displaying saturation curves with an intermediary plateau region, must possess *more than two* substrate binding sites. Interestingly, among the enzymes reported here, whose gross physical properties are known (glutamate dehydrogenase, cytidine triphosphate synthetase, and glyceraldehyde 3-phosphate dehydrogenase), all are composed of four or more subunits and are thus likely to possess at least this number of binding sites. Secondly, it was demonstrated that saturation curves with intermediary plateau regions will *not* be produced by an enzyme possessing more than two binding sites, when the magnitude of the catalytic or binding constants of these sites progressively increases, progressively decreases, or remains constant with saturation. Bumpy curves of the type experimentally observed could, however, be generated by multisite enzymes when the relative magnitude of the catalytic or binding constants first decreased, then increased as the enzyme was saturated.

If the intermediary plateau region of these curves results from an initial decreasing affinity for ligand followed by an increasing affinity for ligand, then protein models may be postulated which can account for this anomolous behavior. An increasing affinity for ligand is usually attributed to positively cooperative interactions of the allosteric type. A decreasing affinity for ligand, however, may result from one of two causes. For a single protein species the effect may arise from ligand-induced conformational changes. These forces, which may be termed negatively cooperative interactions, have been shown to be operative in glyceraldehyde 3-phosphate from rabbit muscle (Conway and Koshland, 1968). Alternatively, a decreasing affinity for ligand may result from a mixture of two or more proteins, or subunits (as in isozymes), each with a different intrinsic binding constant. It should be noted that from an analysis of binding curves alone, the two alternative explanations are indistinguishable. Determination of protein homogeneity, however, could resolve this ambiguity.

In cases in which there is first a decrease in affinity for ligand followed by an increase in affinity for ligand, it is apparent that this behavior may also be explained in one of two ways. Firstly, if only one protein is responsible for the observed phenomena, the curve may result from ligand induced cooperative interactions, in which negatively cooperative interactions predominate at low substrate concentrations and positively cooperative interactions predominant at higher substrate levels. Secondly, the apparently fluctuating affinity for ligand may result from a mixture of two proteins, one displaying Michaelis–Menten or negatively cooperative behavior and the other positively cooperative behavior. In the latter case it is necessary that the magnitude of intrinsic binding constant(s) of the Michaelis–Menten or negatively cooperative enzyme be greater than that of the first intrinsic binding constant of the positively cooperative enzyme.

Finally it is noted that the rate of change in slope in some cases, *i.e.*, cytidine triphosphate synthetase and phosphoenalpyruvate carboxylase, cannot be explained by even the most general rate equation based upon a rapid rate of equilibration between substrate and enzyme relative to the rate of catalysis. In such cases the data provide presumptive evidence that rate-determining conformational changes or rate-determining binding steps are involved in the over-all enzymic process. In addition, it should be noted that saturation curves with intermediary plateau regions may also be generated under certain conditions for single-site enzymes when there is interaction between the substrate and a modifier (as exists, for example, between ATP and magnesium ion) (London and Steck, 1969; Atkinson, private communication).

In summary, the mathematical analysis presented above reveals that this new phenomena involving intermediary plateau regions can only be accommodated by certain models of enzyme action. The existence of such plateaus restricts the potential models and thus can be used as a diagnostic test which provides information concerning structural and kinetic characteristics of the protein under study. Furthermore the analysis suggests that curves with even more complex relationships, *i.e.*, a pronounced maximum in the v_0 *vs.* (S) plots at intermediate ligand levels (Figure 4), are possible. These and other types of bumpy curves would most likely have been dismissed in the past due to the large number of closely spaced points required to verify such curves and due to the tendency to average out such bumps assuming they reflect experimental error. It will be of interest to see if further examples of these unusual saturation curves are discovered in the future.

References

Adair, G. S. (1925), *J. Biol. Chem. 63*, 529.

Bohr, C. (1903), *Z. Physiol. 17*, 682.

Conway, A., and Koshland, D. E., Jr. (1968), *Biochemistry 7*, 4011.

Corwin, L. M., and Fanning, G. K. (1968), *J. Biol. Chem. 243*, 3517.

Gelb, W., and Nordin, J. H. (1969), *Fed. Proc. 28*, 853.

Gentner, N., and Preiss, J. (1968), *J. Biol. Chem. 243*, 5882.

Koshland, D. E., Jr., Nemethy, G., and Filmer, D. (1966), *Biochemistry 5*, 365.

LeJohn, H. B., and Jackson, A. (1968), *J. Biol. Chem. 243*, 3447.

Levitzki, A., and Koshland, D. E., Jr. (1969), *Proc. Natl. Acad. Sci. U. S.*

London, W. P., and Steck, T. L. (1965), *Biochemistry 8*, 1767.

Long, C. W., Levitzki, A., Houston, L. L., and Koshland, D. E., Jr. (1969), *Fed. Proc. 28*, 342.

Michaelis, L., and Menten, M. (1913), *Biochem. Z. 49*, 333.

Monod, J., Wyman, J., and Changeux, J.-P. (1965), *J. Mol. Biol. 12*, 88 (1965).

Sanwal, B. D., and Cook, R. A. (1966), *Biochemistry 5*, 886.

Sweeny, J. R., and Fisher, J. R. (1968), *Biochemistry 7*, 561.

Weber, G., and Anderson, S. R. (1965), *Biochemistry 4*, 1942.

Reprinted from European Journal of Biochemistry, Vol. 7, 1969

Comments on the Kinetics and Mechanism of Yeast Hexokinase Action

Is the Binding Sequence of Substrates to the Enzyme Ordered or Random?

H. J. Fromm

Department of Biochemistry and Biophysics, Iowa State University, Ames

(Received August 6, 1968)

The mechanism of action of yeast hexokinase was reinvestigated in light of a recent report in which it was suggested that substrates add to the enzyme in an ordered manner with the binding of glucose required to precede the binding of ATP. Initial velocity experiments were undertaken with AMP which is a competitive inhibitor for ATP, and with the product inhibitor glucose-6-phosphate. The results of these studies serve to exclude the ordered mechanism for yeast hexokinase from consideration. A discussion is presented in an attempt to show that yeast hexokinase may react in a random fashion with its substrates before forming a ternary complex of enzyme-ATP-glucose.

In 1968 Noat, Ricard, Borel, and Got made initial rate studies with yeast hexokinase and concluded that the enzyme reaction mechanism was ordered-sequential with glucose the obligatory initial substrate (according to the nomenclature of Cleland [1,2]). These results are similar to those published by Hammes and Kochavi [3,4]; however, they are in fundamental disagreement with the findings of Fromm and his coworkers [5—7]. The latter group have suggested that substrates add randomly to hexokinase to form one or more enzyme-substrate ternary complexes which then break down randomly to form products. Furthermore in this mechanism, all steps equilibrate rapidly relative to the interconversion of the ternary complexes [8]. Evidence for the rapid-equilibrium random mechanism is provided not only from kinetic studies [5—7], but from investigations involving quenching of hexokinase fluorescence by both glucose and ATP [6] and from experiments on the ATPase activity of highly purified hexokinase [9,10]. It has been demonstrated that hexokinase which had been recrystallized many times exhibits ATPase activity and that this activity can be blocked by competitive inhibitors of glucose such as *N*-acetylglucosamine. It is important to note that ATP can, indeed, add to hexokinase in the absence of glucose in these experiments.

Noat *et al.* [2] list a number of arguments to support their view that the mechanism for yeast hexokinase is ordered [2]. These are:

a) The Michaelis constant for ATP changes with different sugars.

b) Alternative sugar substrate studies support the ordered mechanism and are at variance with random substrate addition to the enzyme.

c) The product inhibition studies of Fromm [5] involving ADP are incorrect because of ionic strength effects. Presumably the same can be said of this author's experiments with the inhibitor adenine [5].

d) Glucose-6-phosphate, a product, is a competitive inhibitor of glucose; and thus the sugar substrates react with free enzyme.

e) The isotope exchange studies of Fromm *et al.* [7] do not permit one to exclude ordered mechanisms because of the experimental protocol.

f) Cleland, in reviewing the status of the hexokinase problem [11], has suggested that the mechanism of yeast hexokinase is not of the rapid equilibrium random type.

This writer hopes to show, not only from the experiments presented in this report, but also from studies of Noat *et al.* [2] and others, that the mechanism for yeast hexokinase is clearly not ordered.

EXPERIMENTAL PROCEDURE

Materials

D-Glucose was obtained from Pfanstiehl Laboratories, Inc. Yeast hexokinase, glucose-6-phosphate dehydrogenase, pyruvate kinase, and rabbit muscle lactate dehydrogenase were products of C. G. Boehringer & Soehne (Mannheim) and obtained from Calbiochem. ATP, AMP, NADP, NADH, phosphoenolpyruvate, and glucose-6-phosphate were purchased from Sigma Chemical Co. All other reagents were of the highest grade available commercially.

Enzymes. Hexokinase or ATP : D-hexose-6-phosphotransferase (EC 2.7.1.1); pyruvate kinase or ATP : pyruvate phosphotransferase (EC 2.7.1.40); glucose-6-phosphate dehydrogenase or D-glucose-6-phosphate : NADP oxidoreductase (EC 1.1.1.49); lactate dehydrogenase or L-lactate : NAD oxidoreductase (EC 1.1.1.27); ATPase or ATP phosphohydrolase (EC 3.6.1.4).

Ion-low water, obtained by passing distilled water through a Rohm and Haas Amberlite MB-3 resin, was used to prepare all reagents. Yeast hexokinase, specific activity 206, was diluted with ice-cold 0.1 M Tris-Cl buffer, pH 7.6 containing 1 mg bovine serum albumin per ml immediately prior to each kinetic experiment. The diluted enzyme preparation showed no evidence of inactivation in the course of any complete kinetic study.

Methods

ATP was assayed enzymatically with hexokinase and glucose-6-phosphate dehydrogenase and excess glucose and NADP. A similar assay was employed to determine glucose concentration; however, in this case limiting glucose was used. AMP and glucose-6-phosphate were determined spectrophotometrically, and enzymatically (with excess NADP and glucose-6-phosphate dehydrogenase), respectively.

Initial velocity measurements were made in a Cary recording spectrophotometer, Model 15 (0—0.1 slide wire) in cells of 1 cm light path. The reation mixture samples were incubated at 29°, and the temperature of the spectrophotometer cell housing was maintained at 29° by circulating water from a temperature-controlled bath through thermospacers. Reactions were initiated with 0.1 μg enzyme, which had been maintained at 3°, and initial rates were determined with a coupled assay system containing glucose-6-phosphate dehydrogenase and NADP as described elsewhere [12] in those experiments involving AMP inhibition. In the kinetic experiments in which glucose-6-phosphate was present as a product inhibitor, velocities were recorded continuously using pyruvate kinase and lactate dehydrogenase as suggested by Wellner *et al.* [13] and as modified by Copley and Fromm [14].

Velocity in all kinetic experiments to be reported here is expressed as molarity of product formed per minute. Reactions were carried out in 2.5 ml assay mixtures containing 0.3 M Tris-Cl buffer, pH 7.7.

The concentration of Mg^{2+} (as $MgSO_4$) was adjusted in each reaction mixture so that the concentration of free Mg^{2+} was essentially 1 mM. Calculations used to determine the proper Mg^{2+}:ATP ratio were made using a value of 20,000 for the stability constant of $MgATP^{2-}$ [15]. When AMP was used as an inhibitor, additional Mg^{2+} was added assuming a stability constant of 100 for MgAMP [16] so as to maintain the free Mg^{2+} concentration at 1 mM. A similar adjustment was made for glucose-6-phosphate.

RESULTS

The fundamental question which this report addresses is whether the mechanism of yeast hexokinase involves ordered or random substrate addition to the enzyme. There seemed to be little point in attempting to repeat the studies of Noat *et al.* [2] completely, as most of their experiments are already in print elsewhere, but with different results [5,6]. Two experiments were carried out in an attempt to make a choice between the two possible mechanisms outlined above. One involves the use of AMP as a competitive inhibitor of ATP and the other involves the use of glucose-6-phosphate as a product inhibitor.

We suggested in 1962 [5] that competitive inhibitors of substrates could be used to make a choice between enzyme mechanisms involving ordered and random pathways. Indeed, this kinetic approach has been suggested for two [17] and three [18] substrate systems and used, we had hoped successfully, in the former case with yeast hexokinase [6] and also muscle lactate dehydrogenase [19]. The idea involved when using competitive inhibitors is that for a random pathway a competitive inhibitor for either substrate will produce mixed or non-competitive inhibition for the other substrate. Thus mannose, a competitive inhibitor for glucose was found to be a non-competitive inhibitor for ATP [5]. Similarly, for a random mechanism, a competitive inhibitor for ATP should show mixed or non-competitive inhibition with respect to glucose. Because both adenine and adenosine acted in this latter manner, it was concluded that the yeast hexokinase mechanism involved random substrate addition to the enzyme [6].

The equations for an ordered mechanism in which competitive inhibitors of substrates are used are available elsewhere [6,19]. In this case a competitive inhibitor for the substrate which adds to the enzyme first will cause mixed or non-competitive inhibition relative to the second substrate. On the other hand, a competitive inhibitor for the second substrate will produce uncompetitive inhibition relative to the initially adding substrate.

In Fig. 1 and 2 are shown data in which AMP was used to make a choice between the ordered and random mechanisms. In the proposal of Noat *et al.* AMP, a competitive inhibitor for ATP, should cause uncompetitive inhibition relative to glucose. It is quite clear that these findings are not consistent with an ordered mechanism in which glucose adds to the enzyme first followed by ATP. The data do, however, agree with an ordered mechanism in which ATP adds to hexokinase before glucose becomes bound to the enzyme as well as the random case.

In Scheme 1 is presented the pathway which we believe best represents the mechanism of action of yeast hexokinase.

$$\begin{array}{llll} E + A = EA, & K_1 & ECD = ED + C, & K_5 \\ E + B = EB, & K_2 & ECD = EC + D, & K_6 \\ EA + B = EAB, & K_3 & ED = E + D, & K_7 \\ EB + A = EAB, & K_4 & EC = E + C, & K_8 \end{array}$$

$$EAB \underset{k_2}{\overset{k_1}{\rightleftharpoons}} ECD$$

Scheme 1

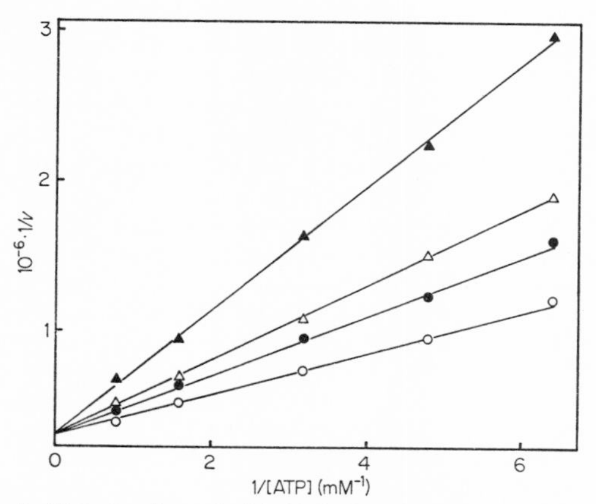

Fig. 1. *Plot of reciprocal of initial reaction velocity (v) versus reciprocal of the molar concentration of ATP in the absence and presence of AMP.* The concentrations of AMP are: (○), none; (●), 3.3 mM; (△), 4.9 mM; (▲), 9.9 mM. v was determined as a function of ATP concentration which was varied in the range from 0.1 mM to 0.8 mM. The glucose concentration was held constant at 0.25 mM. The reaction mixture samples contained 0.3 M Tris-Cl buffer, pH 7.7, 60 μM NADP and 0.2 units of glucose-6-phosphate dehydrogenase. Other details can be found under Experimental Procedure

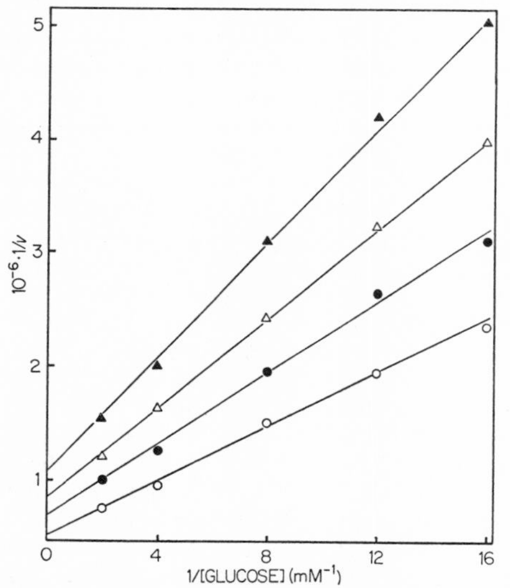

Fig. 2. *Plot of reciprocal of initial reaction velocity (v) versus reciprocal of the molar concentration of glucose in the absence and presence of AMP.* The concentrations of AMP are: (○), none; (●), 3.3 mM; (△), 6.6 mM; (▲) 9.9 mM. v was determined as a function of glucose concentration which was varied in the range from 62.5 μM to 500 μM. The ATP concentration was held constant at 0.16 mM. Other details are presented in the legend to Fig. 1

In Scheme 1 A, B, C, D, and E represent ATP, glucose, ADP, glucose-6-phosphate, and enzyme, respectively. It is assumed that all steps are very rapid relative to the interconversion of the ternary complexes.

For a competitive inhibitor of ATP, such as AMP (I), the following reactions might be expected to occur:

$$\mathrm{E} + \mathrm{I} = \mathrm{EI}, K_i; \ \mathrm{EB} + \mathrm{I} = \mathrm{EBI}, K_{ii}; \ \mathrm{EI} + \mathrm{B} = \mathrm{EBI}, K_{iii}.$$

The rate equation for this case is:

$$v = \frac{V_{\max}}{1 + \frac{K_4}{[\mathrm{A}]}\left(1 + \frac{[\mathrm{I}]}{K_{ii}}\right) + \frac{K_3}{[\mathrm{B}]} + \frac{K_1 K_3}{[\mathrm{A}][\mathrm{B}]}\left(1 + \frac{[\mathrm{I}]}{K_i}\right)} \quad (1)$$

where v and $V_{\max}$ are initial velocity and maximal velocity, respectively. The value calculated for $K_i = 5.7$ mM and the value determined for $K_{ii} = 7.5$ mM from the data of Fig. 1 and 2.

In 1962 we reported that mannose-6-phoshate was a mixed inhibitor of both glucose and ATP with yeast hexokinase [5]. We used this alternative product because our assay system utilized glucose-6-phosphate and a good continuous means was not available at that time to measure ADP production. These studies not only supported the suggested random mechanism but also indicated that mannose-6-phosphate binds primarily at the ATP site on the enzyme and only very weakly at the glucose site. Since that time other workers have also found that binding of glucose-6-phosphate is essentially at the ATP site, and data for brain, muscle and liver hexokinase indicate the product is a competitive inhibitor of ATP and a weak mixed inhibitor for glucose [20]. On the other hand, Parry and Walker [21] using liver glucokinase have shown that glucose-6-phosphate is a competitive inhibitor of glucose while Wettermark *et al.* [22] have demonstrated that the sugar phosphate is a competitive inhibitor of ATP with yeast hexokinase. Noat *et al.* [2] have reported that mannose-6-phosphate is a competitive inhibitor of mannose and fructose-6-phosphate is a competitive inhibitor of fructose in the yeast system.

It was first shown by Alberty [23] that in the case of an ordered system, when one studies initial rates in the presence of a reaction product, the product of the obligatory initial substrate will act as a competitive inhibitor for the substrate. This will be true even if abortive ternary complexes are formed [24] as both the substrate and product compete for the free enzyme. Thus in the studies of Noat and his coworkers [2] the sugar phosphate reaction product must act as a competitive inhibitor of the sugar if the mechanism for hexokinase is really ordered.

Fig. 3 and 4 depict data from experiments carried out with glucose-6-phosphate as product inhibitor. If these data, which indicate that glucose-6-phosphate is a mixed inhibitor of both ATP and glucose are correct, then the ordered mechanism proposed by

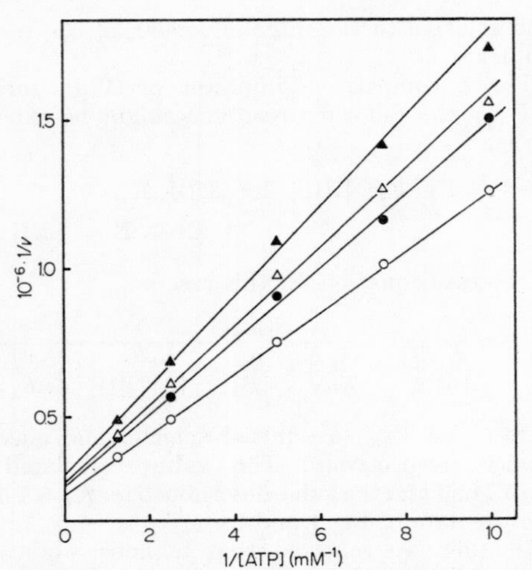

Fig. 3. *Plot of the reciprocal of initial reaction velocity (v) versus reciprocal of the molar concentration of ATP in the absence and presence of glucose-6-phosphate.* The concentrations of glucose-6-phosphate are: (○), none; (●), 10 mM; (△), 15 mM; (▲), 20 mM. v was determined as a function of ATP concentration which was varied in the range from 0.1 mM to 0.8 mM. The glucose concentration was held constant at 0.25 mM. The reaction mixture samples which also contained 0.3 M Tris-Cl buffer, pH 7.7, dialyzed lactate dehydrogenase (33 units), dialyzed pyruvate kinase (20 units), 0.5 mM phosphoenolpyruvate, 50 μM NADH, and 10 mM KCl were preincubated 10 min before adding hexokinase so as to utilize the small amount of ADP contained in the ATP. Other details can be found under Experimental Procedure

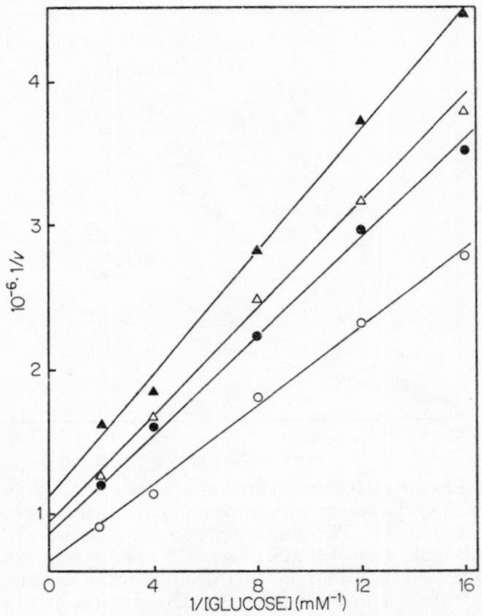

Fig. 4. *Plot of reciprocal of initial reaction velocity (v) versus reciprocal of the molar concentration of glucose in the absence and presence of glucose-6-phosphate.* The concentrations of glucose-6-phosphate are: (○), none; (●), 10 mM; (△), 15 mM; (▲), 20 mM. v was determined as a function of glucose concentration which was varied in the range from 62.5 μM to 500 μM. The ATP concentration was held constant at 0.147 mM. Other details are presented in the legend to Fig. 3

Noat *et al.* [2] is not valid. Furthermore, while these workers report that the sugar phosphate is a competitive inhibitor of the free sugar the results of Fig. 3 and 4 indicate that binding of the product at the sugar site on hexokinase is relatively weak compared to interaction at the ATP site.

If one assumes a random mechanism for the yeast enzyme as depicted in Scheme I, then in the presence of glucose-6-phosphate (D), the following reactions might be expected to occur:

$$\mathrm{E} + \mathrm{D} = \mathrm{ED},\ K_7;\ \mathrm{ED} + \mathrm{B} = \mathrm{EDB},\ K_i;\ \mathrm{ED} + \mathrm{A} = \mathrm{EDA},\ K_{ii}$$

and the resulting rate equation is:

$$v = \frac{V_{\max}}{1 + \frac{K_4}{[\mathrm{A}]}\left(1 + \frac{K_1 K_3 [\mathrm{D}]}{K_4 K_7 K_i}\right) + \frac{K_3}{[\mathrm{B}]}\left(1 + \frac{K_1 [\mathrm{D}]}{K_7 K_{ii}}\right) + \frac{K_1 K_3}{[\mathrm{A}][\mathrm{B}]}\left(1 + \frac{[\mathrm{D}]}{K_7}\right)} \qquad (2)$$

It is quite clear that this rate equation is at variance with the results of Noat *et al.* [2], however, it is consistent with the findings of Fig. 3 and 4, the observations of Wettermark *et al.* [22] for yeast hexokinase, and the data of Parry and Walker [21] for glucokinase which is also believed to involve random substrate binding [25].

Wettermark *et al.* [22] have shown with the enzyme from yeast that glucose-6-phosphate is a competitive inhibitor of ATP when the concentration of the non-varied substrate glucose is 220 mM. This level of glucose is approximately 1,000 times its K_m and thus the $K_3/[\mathrm{B}]$ term in Equation (2) should become insignificant. This could be expected to result in the type of effect reported by the Swedish workers. The results of Parry and Walker [21, 25] can be rationalized in a similar manner. These investigators held ATP at 5 mM or at 10 times the K_m for ATP. This high level of ATP could reasonably be expected to minimize the effect of the $K_4/[\mathrm{A}]$ term of Equation (2) which could in turn appear to indicate the glucose-6-phosphate is a competitive inhibitor for glucose.

At any rate it is clear that the independent studies of Wettermark *et al.* [22] as well as those presented in this report are at variance with an ordered mechanism for yeast hexokinase in which glucose adds to the enzyme before ATP.

DISCUSSION

In order to place the divergent views on the mechanism of action of yeast hexokinase into proper perspective, it might be useful to consider a few comments contained in a review article by Cleland [11]: "Hammes and Kochavi [3] showed that the initial velocity pattern at pH 8 was intersecting and the mechanism thus sequential; and that free ATP was an inhibitor apparently competitive *versus* MgATP. The authors considered that ATP combined only with the E-glucose complex and not with free enzyme, but it is doubtful that the precision of their data, which were obtained by following H^+ production in a pH-stat, would allow such a distinction to be made."

This author believes that the results of Noat *et al.* [2] must be evaluated in light of these comments, as these investigators, like Hammes and Kochavi [3], used a pH-stat in their experiments. Their results are qualitatively very similar to those of Hammes and Kochavi.

The initial rate experiments of Fromm and his coworkers [5—7] involving the forward yeast hexokinase reaction were carried out using a coupled assay with glucose-6-phosphate dehydrogenase. The validity and superiority of this procedure relative to other methods of assaying for hexokinase activity are documented in an article by Bennett *et al.* [26]. A cursory review of the hexokinase literature since 1962 reveals that the coupled assay method is used essentially to the exclusion of all other procedures for the measurement of initial velocity. It is important to note that in the studies of Fromm and coworkers [5—7] the amount of substrate utilized was always so small that initial reaction velocities could be determined with great precision.

Noat and Ricard [27] have presented some initial velocity data which are useful in the following discussion. They show that with one-eleventh the amount of enzyme used for their kinetic experiments [2] at 0.5 mM glucose and ATP initial velocity is maintained at most for 5 min and represents approximately 6 % of substrate utilization. At 5 mM glucose and 2 mM ATP initial velocity conditions are maintained for 10 min and represents about 5 % substrate utilization. It is in light of these statements that this writer wishes to comment on the results of Noat *et al.* [2].

In the introduction to this paper six points are listed which are the basis for the conclusions alluded to by Noat and his coworkers [2] regarding the sequence of substrate binding to yeast hexokinase. The following is a consideration of these points in light of the present and previous [5—7] investigations:

a) For a random mechanism the dissociation constant for ATP should remain the same regardless of the sugar substrate. The results of Noat *et al.* [2] with fructose and mannose (Fig. 4A and B) indicate that the dissociation constants are different with these two sugars thus weakening the argument for the random mechanism. We reported that the K_m for ATP was 0.16, 0.20, and 0.22 mM with glucose, fructose, and mannose, respectively [5,6]. The K_m values for the sugars as determined from initial rate experiments are 84 μM (glucose), 628 μM (fructose), and 109 μM (mannose) [5,6]. Actually, these data were presented as evidence in support of the random mechanism [6].

Rough calculation of the findings of Noat *et al.* [2] gives a K_m for ATP of 0.37 mM with fructose (Fig. 4A) and 0.07 mM with mannose (Fig. 4B). It is also possible to calculate the K_m for ATP with glucose to be about 0.4 mM (Fig. 1B). It is interesting to note that even though the K_m values for mannose and glucose are known to be very similar, both are very different from that reported for fructose [28]. According to Noat *et al.* [2] the K_m values for ATP with fructose and glucose are very similar; however, that for mannose is quite different from the K_m values for ATP determined with fructose and glucose.

One can raise questions concerning the validity of the initial velocity data in the studies of Noat *et al.* [2] by citing the results of Fig. 4B. At a mannose concentration of 0.5 mM the concentration of $MgATP^{2-}$ is 15.6 μM and the velocity 0.4 μM/sec. It can also be shown from Fig. 4B that the velocity is 0.3 μM/sec. where the mannose concentration is 25 μM. From the report of Noat and Ricard [27] cited above, initial velocity conditions should be maintained until only about 10 % of the substrate has been utilized [1]. Thus in the experiments where $MgATP^{2-}$ is 15.6 μM and mannose is 25 μM, initial rate conditions are maintained for only 4 sec and 8.3 sec, respectively. Even in the concentration range of 100 μM $MgATP^{2-}$ where initial velocity could persist at most until 10 μM $MgATP^{2-}$ had been utilized, the velocities reported are in the range of 0.4 μM/sec—1.0 μM/sec. It is difficult indeed to see how one can make a serious value judgement on initial velocities and thus K_m values for ATP with various sugar substrates from data of this type.

b) It had been suggested that a choice of mechanism for two substrate enzyme systems could be made by using alternative substrates [17]. Using mixtures

[1] In this discussion it is assumed that initial velocity conditions prevail until 10 % of the substrate is utilized in the hexokinase reaction. Actually, Noat and Ricard [27] present data which indicate that the initial velocity phase of the reaction persists until 5—6 % of the substrate has been consumed (Fig. 1A). Their data demonstrate that the linear portion of the velocity progress curve may last until 15 % of the substrate has been utilized in the presence of product at low concentrations of substrates (Fig. 1B). It is noteworthy that these experiments which were carried out with 0.09 nM hexokinase give linear velocities for 5—10 min. In the kinetic experiments of Noat *et al.* [2] the initial velocity phase of the hexokinase reaction would be expected to be much shorter as 1 nM or 11-times this amount of enzyme was used.

of glucose and fructose we concluded that the mechanism of yeast hexokinase is random [6]. The data presented in Fig. 3A of reference [6] indicates that in the presence of glucose and fructose, double reciprocal plots of 1/velocity *versus* 1/[ATP] are linear, where the ATP is varied from 0.1—1.0 mM. On the other hand Noat *et al.* [2] show that from 16.7—200 μM $MgATP^{2-}$ these plots are non-linear (Fig. 4C). They correctly point out that such data serve to exclude the random mechanism, however, we believe that these results suffer from the same limitations cited above. A fairly good case could be made for drawing a straight line through the point at the highest concentration of $MgATP^{2-}$ and the point at 50 μM $MgATP^{2-}$ in the non-linear curve in Fig. 4C. The only points which are not on such a straight line would be those at 25 μM and 16.7 μM $MgATP^{2-}$. The velocity at the former concentration of ATP is 0.25 μM/sec and at the latter 0.167 μM/sec. Considering that initial velocity conditions should prevail for only 10 sec at most in these experiments, and considering that if these points in Fig. 4C were lowered by only 10% they would fall on the straight line described above, it seems rather unreasonable to exclude the random mechanism because of the reasons suggested by Noat *et al.* [2].

c) We reported that ADP is a competitive inhibitor for ATP in the yeast hexokinase system [5]; however, Noat *et al.* [2] have suggested that this is an artifact caused by variations in ionic strength when Mg^{2+} and ADP are added in the product inhibition experiments. If ADP is indeed a competitive inhibitor for the nucleotide substrate, an ordered mechanism in which ATP is the second substrate to add to the enzyme may be ruled out [23,24].

Noat and his coworkers [2] show that although ADP is actually a competitive inhibitor of ATP at an ionic strength of 0.05—0.06, the inhibition becomes mixed as the ionic strength is increased to 0.3 (Fig. 7). The rationale for this phenomenon according to these workers is: "It is easy to calculate from his data [5] that, in any given experiment, the ionic strength of the reaction mixture is very small and not constant. But for low ionic strengths, the stability constant of complexes $MgATP^{2-}$ and $MgADP^{-}$ decreases considerably with increase in I (Fig. 6). It follows that when the concentration of $MgATP^{2-}$ varies, the concentration of $MgADP^{-}$ (Fig. 6) also necessarily varies. This provokes a systematic deformation of the Lineweaver-Burk plots."

A number of points can be made in answering this criticism by Noat *et al.* [2]. Firstly, one could present a good case for drawing the lines in Fig. 7B so that they converge at a single point; *i. e.*, for the reasons stated above the initial velocity data for these experiments may not be correct. In fact a small upward displacement of the point representing 2 mM ADP — 0.1 mM ATP could achieve this end.

Secondly, Fig. 6, which shows that the stability constants for $MgATP^{2-}$ and $MgADP^{-}$ decrease with ionic strength also indicate that the most desirable ionic strength to carry out kinetic experiments with enzymes such as hexokinase is in the ionic strength range of 0—0.15. At an ionic strength 0.10, as shown in Fig. 6 of Noat *et al.* [2], the stability constants for $MgATP^{2-}$ and $MgADP^{1-}$ are about 100,000 and 6,500 M^{-1}. These constants increase with decreasing ionic strength and if one works in the range 0.08 to 0.13 most of the nucleotides will be chelated to Mg^{2+}.

To illustrate this point note that in Fig. 7A of Noat *et al.* [2] $[ATP^{4-}]$ = 1.25 mM; $[Mg^{2+}]$ for the ATP = 3.12 mM; $[ADP^{3-}]$maximum = 2 mM; $[Mg^{2+}]$ for the ADP = 2 mM; and tetramethyl ammonium chloride concentration = 0.05 M. If all these species were totally ionic, *i. e.*, no $MgATP^{2-}$ *etc.* was formed, the maximum ionic strength one would calculate is 0.10. In the absence of ADP^{3-} and Mg^{2+} which were added in a ratio of 1, this value would fall to 0.08. It is quite clear from Fig. 6 that at an ionic strength of 0.10, 100% of the ATP would exist as $MgATP^{2-}$ while 92% of the ADP would be in the form Mg $MgADP^{1-}$. Presumably, an even higher percentage of the nucleotides would be bound at the lower ionic strengths.

We feel that the variation in the results between Fig. 7A and B is not due to ionic strength effects, but rather suggests the true precision of the kinetic data.

Thirdly, the kinetic data with adenine would not cause any alterations in ionic strength as it has no formal charge at pH 7.6.

Fourthly, if ionic strength did indeed cause a dissociation of $MgATP^{2-}$ complex the data would not be expected to follow Michaelis-Menten kinetics. This would occur as the concentrations for ATP plotted on the abscissa would be incorrect.

Fifthly, although Cohn did report that glucose was required for the binding of $MnADP^{1-}$ to yeast hexokinase from nuclear magnetic resonance experiments [29], she has not been able demonstrate this effect with more recent enzyme preparations [29a].

Sixthly, in the kinetic experiments presented in this report 0.3 M Tris-Cl buffer was used. The ordered mechanism appears to be excluded even at high ionic strength.

d) The data presented in this report for inhibition by glucose-6-phosphate merely serve to confirm earlier studies made with mannose-6-phosphate. These findings are at variance with the results, and the conclusions alluded to from the results, of Noat and coworkers [2].

e) and f) According to Noat and his colleagues [2] the isotope exchange experiments at equilibrium carried out by Fromm *et al.* [7] do not permit a choice to be made between ordered and random mechanisms. The statement of Noat *et al.* may be found in [2] (pp. 68—70) and is, "the study of the rate of exchange at equilibrium between ATP and

ADP and between glucose and glucose-6-phosphate, according to Boyer's method [30], had led Fromm [7] to conclude the existence of a rapid equilibrium random mechanism. As shown by Cleland [11], the results of Fromm [7] are not in agreement with such a mechanism. Moreover these results were obtained at pH 6.5, a condition very different from that in which we worked. Further the analysis of the equations presented by Boyer [30], using values of velocity constants that we worked out, shows that the ADP and ATP concentrations used by Fromm [7] are insufficient to allow the observation of a decrease in the exchange rate. As with the kinetic studies [5,7,31], the application [7] of Boyer's method [30] to the study of hexokinase has not, therefore, lent support for a random mechanism."

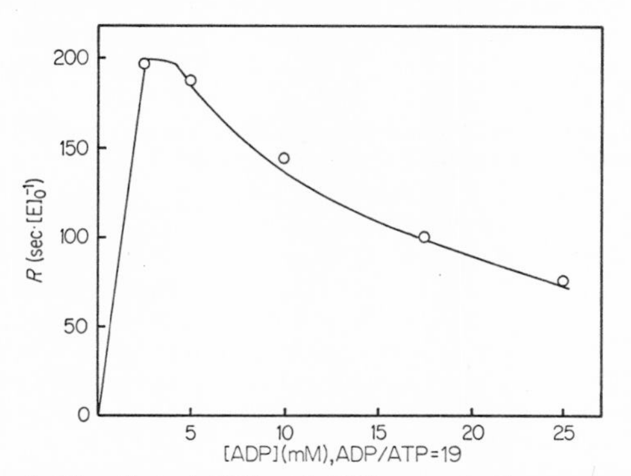

Fig. 5. *The effect of ATP and ADP concentrations on equilibrium reaction rates (R) catalyzed by the enzyme yeast hexokinase.* The points on the graph were calculated using Equation (3) for an ordered mechanism, the rate constants provided by Noat *et al.* [2], and the data of Fromm *et al.* [7] (Fig. 1). The concentration of glucose required for the solution of Equation (3) is 2.5 mM [7]. [ADP]/[ATP] = 19. Glucose-6-phosphate is 38.5 mM

It should be pointed out that the isotope exchange studies at equilibrium were carried out by Silverstein [32] and later incorporated into the paper by Fromm *et al.* [7]. In this writer's opinion there is little validity in the statement by Noat *et al.* [2] that the ADP and ATP concentrations were insufficient to allow the observation of a decrease in the glucose—glucose-6-phosphate exchange rate. In Fig. 5 is presented a theoretical curve using the data of [7] (Fig. 1) for glucose, ADP, and ATP, with Equation (5) of Alberty *et al.* [33] which may be written in the following form:

$$R = \frac{k_1k_2k_3k_4\,[\text{Glc}]\,[\text{ATP}]\,[\text{E}]_0}{\left[1 + \frac{k_1}{k_{-1}}[\text{Glc}] + \frac{k_1k_2\,[\text{Glc}]\,[\text{ATP}]}{k_{-1}k_{-2}} + \frac{k_{-4}\,[\text{G-6-P}]}{k_4}\right]} \times \frac{1}{[k_{-1}k_{-2}k_4 + k_{-1}k_3k_4 + k_2k_3k_4[\text{ATP}] + k_{-1}k_{-2}k_{-3}[\text{ADP}]]} \quad (3)$$

where R is the isotope exchange rate at equilibrium between glucose and glucose-6-phosphate.

The values for the rate constants were taken from the paper by Noat *et al.* [2] (Table 8).

Equation (3) is the equation for an ordered pathway with a single ternary complex in which glucose is the first substrate to add to the enzyme and glucose-6-phosphate the last product to leave. It is clear from Fig. 5 that there is a depression of the glucose—glucose-6-phosphate exchange, in contradiction to the statement of Noat *et al.* [2]. The experimental points for this study describe a rectangular hyperbola (Fig. 1 of [7]) and thus rule out the ordered mechanism proposed by Noat and coworkers [2]. The type of isotope exchange result to be expected for an ordered pathway is presented elsewhere [34].

It might be well to again allude to a portion of the review article by Cleland relative to yeast hexokinase [11]. He says that although isotope exchange studies at equilibrium indicate that the mechanism is not truly rapid equilibrium random it is undoubtedly random by the criteria of the initial velocity experiments. Furthermore, random mechanism in most cases will appear to be of the rapid equilibrium type when subjected to theoretical analysis. Finally, it should be pointed out that a similar statement regarding the mechanism of yeast hexokinase was made previously by Fromm *et al.* [7].

The results and comments contained in the present study suggest that substrates ATP and glucose add randomly to the enzyme yeast hexokinase. We feel that the method for measuring initial velocity employed by Noat *et al.* [2] which is from 10—100 times less sensitive than the glucose-6-phosphate dehydrogenase procedure which we have employed [5—7] is the basis for the divergent kinetic results reported from these laboratories. Although it is not possible to rule a mechanism in from initial rate studies it is, however, possible to exclude mechanisms. For this reason we feel that the proposed ordered mechanism of Noat *et al.* [2] is not valid for yeast hexokinase.

Journal Paper J-6040 and Project No. 1666 of the Iowa Agricultural and Home Economics Experiment Station, Ames, Iowa. This research was supported in part by Research Grant AM-11041 from the National Institutes of Health, U.S. Public Health Service.

REFERENCES

1. Cleland, W. W., *Biochim. Biophys. Acta*, 67 (1963) 104.
2. Noat, G., Ricard, J., Borel, M., and Got, C., *European J. Biochem.* 5 (1968) 55.
3. Hammes, G. G., and Kochavi, D., *J. Am. Chem. Soc.* 84 (1962) 2069.
4. Hammes, G. G., and Kochavi, D., *J. Am. Chem. Soc.* 84 (1962) 2073.
5. Fromm, H. J., and Zewe, V., *J. Biol. Chem.* 237 (1962) 3027.
6. Zewe, V., Fromm, H. J., and Fabiano, R., *J. Biol. Chem.* 239 (1964) 1625.

7. Fromm, H. J., Silverstein, E., and Boyer, P. D., *J. Biol. Chem.* 239 (1964) 3645.
8. Alberty, R. A., *J. Am. Chem. Soc.* 75 (1953) 1928.
9. Kaji, A., and Colowick, S. P., *J. Biol. Chem.* 240 (1965) 4454.
10. De La Fuente, G., and Sols, A., *Biochem. J.* 89 (1963) 36p.
11. Cleland, W. W., *Ann. Rev. Biochem.* 36 (1967) 77.
12. Fromm, H. J., and Zewe, V., *J. Biol. Chem.* 237 (1962) 1661.
13. Wellner, V. P., Zoukis, M., and Meister, A., *Biochemistry,* 5 (1966) 3509.
14. Copley, M., and Fromm, H. J., *Biochemistry,* 6 (1967) 3503.
15. O'Sullivan, W. J., and Perrin, D. D., *Biochemistry,* 3 (1964) 18.
16. Bock, R. M., *Enzymes,* 2 (1960) 3.
17. Fromm, H. J., *Biochim. Biophys. Acta,* 81 (1964) 413.
18. Fromm, H. J., *Biochim. Biophys. Acta,* 139 (1967) 221.
19. Zewe, V., and Fromm, H. J., *Biochemistry,* 4 (1965) 782.
20. Grossbard, L., and Schimke, R. T., *J. Biol. Chem.* 241 (1966) 3546.
21. Parry, M. J., and Walker, D. G., *Biochem. J.* 99 (1966) 266.
22. Wettermark, K. G., Borglund, E., and Brolin, S. E., *Anal. Biochem.* 22 (1968) 211.
23. Alberty, R. A., *J. Am. Chem. Soc.* 80 (1958) 1777.
24. Fromm, H. J., and Nelson, D. R., *J. Biol. Chem.* 237 (1962) 215.
25. Parry, M. J., and Walker, D. G., *Biochem. J.* 105 (1967) 473.
26. Bennett, E. L., Drori, J. B., Krech, D., Rosenzweig, M. R., and Abraham, S., *J. Biol. Chem.* 237 (1962) 1758.
27. Noat, G., and Ricard, J., *European J. Biochem.* 5 (1968) 71.
28. Crane, R. K., *Enzymes,* 6 (1962) 58.
29. Cohn, M., *Biochemistry,* 2 (1963) 623.
29a. Cohn, M., personal communication.
30. Boyer, P. D., *Arch. Biochem. Biophys.* 82 (1959) 387.
31. Robbins, E. A., and Boyer, P. D., *J. Biol. Chem.* 224 (1957) 121.
32. Silverstein, E., Ph. D. thesis, University of Minnesota, 1963.
33. Alberty, R. A., Bloomfield, V., Peller, L., and King, E. L., *J. Am. Chem. Soc.* 84 (1962) 4381.
34. Silverstein, E., and Boyer, P. D., *J. Biol. Chem.* 239 (1964) 3901.

H. J. Fromm
Department of Biochemistry and Biophysics
Iowa State University, Ames, Iowa 50010, U.S.A.

Reprinted from THE JOURNAL OF THE AMERICAN CHEMICAL SOCIETY, Vol. 85, 1963

[CONTRIBUTION FROM THE DEPARTMENT OF BIOLOGICAL SCIENCES, PURDUE UNIVERSITY, LAFAYETTE, IND., AND THE BIOLOGY DEPARTMENT, BROOKHAVEN NATIONAL LABORATORY, UPTON, N. Y.]

An All-or-None Assay for Assessing the Role of Amino Acid Residues in Enzyme Action—Application to Phosphoglucomutase[1,2]

BY W. J. RAY, JR., AND D. E. KOSHLAND, JR.

RECEIVED DECEMBER 12, 1962

An "all-or-none" assay, which distinguishes fully active and partially active enzymes from inert enzymes in a mixture of all of these, is described. Unlike the usual "efficiency" assay, the "all-or-none" assay gives equal weight to active and partially active enzymes. The difference in the two assays is exploited in interpreting the effect on phosphoglucomutase of the methylene blue-catalyzed photoöxidation reaction, and provides evidence for a formation of partially active enzyme intermediates, probably *via* oxidation of a single "surface" histidine residue. A correlation of the activity loss as determined by the "all-or-none" assay with previous data on amino acid modification during photoöxidation indicates that production of an inert enzyme having an efficiency less than $^1/_{200}$ of the native enzyme parallels the modification of a single surface methionine residue. The general applicability of such "all-or-none" assays to enzymes which form a measurable intermediate of reasonable stability is discussed.

Introduction

Methods utilizing chemical modification as a means of identifying amino acid residues involved in enzyme action have many pitfalls, not the least of which arises from the possibility of producing a partially active enzyme by virtue of the modification reaction under study. For example, if the observed activity of an enzyme has been reduced 10-fold by treatment with a given reagent, the remaining activity might be the result of (a) a small amount (10%) of fully active enzyme in the presence of inert enzyme (90%), (b) the reduced efficiency of a modified enzyme, 10% as active as the native enzyme and (c) complex mixtures of active, partially active and inert enzyme. To aid in the detection and analysis of such situations, an "all-or-none" assay has been devised which, in conjunction with the conventional "efficiency" assay, may also be used to determine whether modification of a given residue produces partially active or inert enzyme.

The conventional "efficiency" assay for assessing enzymic activity measures the amount of substrate converted, or product produced per unit time, under standard conditions. It therefore gives a weighted average of the catalytic efficiencies of the various enzymic species present in the assay. For example, an enzyme preparation which, by virtue of chemical modification, contained fully active enzyme (15%),

(1) A preliminary account of some of this work has been published.[2]

(2) W. J. Ray, Jr., J. J. Ruscica and D. E. Koshland, Jr., *J. Am. Chem. Soc.*, **82**, 4739 (1960).

modified enzyme with 10% of its native activity (30%) and inert enzyme (55%), would by such an "efficiency" assay be 18% as active as an equivalent amount of native enzyme.

On the other hand, an "all-or-none" assay, as herein defined, is designed to measure the *number* of enzymic sites possessing a pre-determined minimum efficiency rather than the *rate* of product formation. This is accomplished by quantitating some property, which is an essential feature of the over-all assay reaction, after a sufficient time interval has elapsed to ensure that the time dependence of this property has essentially become zero, even with respect to enzymic species that are able to function only at greatly reduced efficiency. For example, all the phosphoglucomutase in a complex mixture of active, partially active and inactive enzyme which retained its ability to transfer its phosphate to a substrate could be measured in such an assay by (a) using phosphoglucomutase previously labeled with P^{32}-phosphate, (b) treating the enzyme mixture to be assayed with a large excess of substrate (allowing sufficient time for all the enzyme capable of phosphate transfer to react), (c) determining the amount of phosphate remaining attached to the enzyme. Only inert enzyme would fail to lose its phosphate label in such an assay, and in the case cited above where 18% of the initial activity would be observed by the conventional "efficiency" assay, the "all-or-none" assay would indicate that 45% of the enzyme retained at least a small fraction of its original catalytic activity.

Thus, in such cases, the difference in enzyme activity remaining, as determined by these two assays, depends on the difference in the quantity measured. By exploiting this difference in conjunction with the measured modification rates of the various amino acid residues, the relative importance to the enzyme of individual amino acid residues may be assessed. In the present paper the application of this approach to the chemical modification produced by methylene blue-sensitized photoöxidation of phosphoglucomutase will be described.

Experimental

Reagents and Procedures.—The apparatus and general procedure for the photoöxidation reaction have been described previously.[3] Phosphoglucomutase was prepared from rabbit muscle by a modification of the method of Najjar.[4] The efficiency assay, measuring the rate of conversion[5] of G-1-P to G-6-P during a 10-minute time interval, utilizes conditions similar to those of Najjar[4] and has been described in a previous publication.[3]

P^{32}-Labeled glucose phosphates were prepared by adding trace amounts of sucrose phosphorylase and phosphoglucomutase to a solution, 10^{-3} M in sucrose and 10^{-4} M in $KH_2P^{32}O_4$, buffered at pH 6.0 with 0.1 M histidine. When the P_i had decreased to about 5% of its initial value, the reaction was terminated by passing the mixture through a small Dowex 50W-8% column in the acid form. The eluate, which contained G-1-P^{32} and G-6-P^{32} in an equilibrium ratio, was neutralized and subsequently used for labeling the enzyme.

P^{32}-Labeled phosphoglucomutase was prepared by treating a mixture of labeled glucose phosphates (above) with a 10-fold excess of enzyme at 0° for a few minutes, at pH's varying from 6.0 to 7.5. No added cofactors were required. Residual P^{32}-labeled substrate was removed either by chromatography,[3] or by a batch-wise procedure in which the enzyme in 10^{-3} M phosphate buffer, pH 6.0, was absorbed on a 2 × 2 cm. carboxymethylcellulose column, previously equilibrated against the same buffer. After washing the column with the dilute buffer until the radioactivity in the effluent had reached an insignificant value, the enzyme was eluted with a small volume of 0.05 M phosphate buffer, pH 7. Up to 90% of the substrate radioactivity could thus be incorporated into the enzyme.

All-or-None Assays.—All-or-none assays were carried out in three ways: (I) by measuring loss of P^{32}-phosphate from P^{32}-labeled enzyme on treatment with G-1-P, (II) by measuring loss of P^{32}-phosphate from P^{32}-labeled enzyme in treatment with G-6-P, (III) by measuring labeling of the enzyme on treating unlabeled enzyme with a mixture of G-1-P^{32} and G-6-P^{32}.

Assays I and II, which involved treatment of photoöxidized samples of P^{32}-labeled enzyme with 0.01 M G-1-P or G-6-P, were carried out at 30° for 5 minutes in 0.07 M potassium hydrogen phthalate buffer, pH 5.5, with an enzyme concentration of 0.1 mg./ml. After addition of 1 mg./ml. of carrier protein (bovine serum albumin) and trichloroacetic acid to a final concentration of 5%, the tubes were allowed to stand for 2 hours in ice. Aliquots of the supernatant liquid were then assayed, both for total P^{32} content and also for acid-labile phosphate, using the isobutyl alcohol extraction procedure for P_i. Other aliquots were subjected to paper electrophoresis (pyridine–acetic acid–water, 150:6:1350) for 2 hours at 900 v. As a control, phosphoglucomutase was used which had been photoöxidized until no detectable enzyme activity remained by the efficiency assay (40 minutes). When such material was used in all-or-none assays I and II, about 2% of the total radioactivity appeared in the trichloroacetic supernatant. This quantity was subtracted from the measured value of each assay in these series as a "constant" which was attributed to traces of P^{32}-glucose phosphates in the labeled enzyme preparation and to incomplete precipitation of the protein under the conditions employed.

Assay III, in which unlabeled enzyme, previously subjected to photoöxidation, was treated with radioactive substrate, was carried out in a solution 1.2×10^{-3} M in $MgSO_4$, 1.0×10^{-3} M in P^{32}-labeled glucose phosphates (an equilibrium mixture of G-1-P^{32} and G-6-P^{32}), 0.05 M in histidine, and 0.01 M in trishydroxymethylamine at pH 7.4, with sufficient enzyme to incorporate *ca.* 15% of the total radioactivity into the protein. (The amount of enzyme, *i.e.*, the dilution of photoöxidized enzyme required for this degree of incorporation, was determined by trial and error in pilot runs.) In each case, serum albumin was added to the assay mixture in sufficient amounts to maintain a protein concentration of 0.1 mg./ml. After 25 minutes at 30°, additional serum albumin and trichloroacetic acid were added as above and the mixture allowed to stand for 2 hours in ice. After centrifugation and washing (5% trichloroacetic acid) the protein was dissolved in 0.1 N NaOH and reprecipitated with trichloroacetic acid. The second precipitate was subsequently dissolved in 6 N ammonium hydroxide and plated for counting. As a control, phosphoglucomutase was used which had been photoöxidized 40 minutes. When such material was used in this assay, only about 0.3% of the total radioactivity was carried down by the protein. The quantity was subtracted from each assay tube as a constant which was attributed to non-specific adsorption by precipitated protein.

Results and Discussion

Comparison of Efficiency and All-or-None Assays.—The effect of dye-sensitized photoöxidation on phosphoglucomutase, as determined by the efficiency assay and all-or-none assay I, is shown in Fig. 1. At all points, the quantity registered, relative to the same quantity at $t = 0$, is substantially greater by the all-or-none assay than by the efficiency assay. This difference results from the fact that these assays were designed to take advantage of different aspects of phosphoglucomutase action[6] (eq. 1 and 2) and therefore measure different quantities.

$$\text{G-1-P} + \text{EP} \rightleftharpoons \text{G-1,6-P}_2 + \text{E} \quad (1)$$

$$\text{G-1,6-P}_2 + \text{E} \rightleftharpoons \text{G-6-P} + \text{EP} \quad (2)$$

The efficiency assay measures the rate of conversion of G-1-P to G-6-P in the presence of catalytic amounts of enzyme. A decrease in the efficiency of the enzyme will therefore be reflected in a decreased amount of G-1-P converted in the limited assay interval. On the other hand, all-or-none assay I (and also II, *vide infra*) measures the amount of enzyme which retains its capacity to participate in a phosphate transfer reaction with substrate even when the efficiency of this reaction is low. The labeled enzyme in the latter case is measured directly, rather than indirectly (*i.e.*, by assessing its catalytic capabilities), and its reaction

(3) W. J. Ray, Jr., and D. E. Koshland, Jr., *J. Biol. Chem.*, **237**, 2493 (1962).

(4) V. A. Najjar, in "Methods in Enzymology," ed. by S. P. Colowick and N. O. Kaplan, Vol. 1, Academic Press, Inc., New York, N. Y., 1955, p. 294.

(5) Abbreviations: G-1-P, glucose-1-phosphate; G-6-P, glucose-6-phosphate; G-1,6-P_2, glucose-1,6-diphosphate; P_i, inorganic phosphate; P_{10}, labile phosphate, *i.e.*, phosphate produced on 10-min. hydrolysis in 1 N acid at 100°; EP, phospho-form of phosphoglucomutase; E, dephospho enzyme.

(6) V. A. Najjar and M. E. Pullman, *Science*, **119**, 631 (1954).

efficiency enters only in limiting cases. Thus if native enzyme were able to transfer completely its phosphate to G-1-P in one second, an assay time of 17 minutes (*i.e.*, 1000 seconds) would be sufficient for a modified enzyme operating at 1/1000th the efficiency of native enzyme to undergo the same reaction. By labeling the phospho-enzyme with P^{32}-phosphate the extent of phosphate transfer can be measured quantitatively, and by using a 17-minute assay interval, an enzyme whose efficiency was lowered by a factor of 1000 would thus register in the all-or-none assay to the same degree as native enzyme. Only inert enzyme would retain its radioactive label and fail to register.

The term "inert" is used here to refer to enzyme which does not lose its P^{32}-label under the designated conditions of the various all-or-none assays. Enzyme with zero activity would of course be inert, but enzyme with a very low efficiency might also fail to register in the assay. In the above example, enzyme with an activity reduced by 10^{-6}-fold relative to the native enzyme would thus fail to undergo phosphate transfer in this assay. Hence inert as used here means not detectable under the assay conditions. If these assay conditions are correctly designed the detectable activity limits can be extremely low, but inert or "zero activity" in an enzyme system must always tacitly assume the phrase, "within experimental error."

Since the measured turnover number (G-1-P to G-6-P conversion) under the conditions of all-or-none assay I was 100 min.$^{-1}$ per 77,000 grams with native enzyme and since a positive effect in this assay might result from only a part of one turnover, the conditions utilized for this assay would allow at least 90% of a sample of native enzyme to undergo at least one phosphate transfer in 1/200th of the total time. Thus, a partially active enzyme with a 200-fold reduction in efficiency (*i.e.*, a turnover number of 0.5 min.$^{-1}$) would have a 90% probability of participating in at least one phosphate transfer and therefore registering in the assay. Enzyme which is inactive by this assay therefore must retain less than 0.5% of its original catalytic efficiency and may indeed be less active by several orders of magnitude.

Evidence for a Partially Active Species.—That the quantity measured at the various reaction times, relative to the same quantity at $t = 0$, was invariably greater by all-or-none assay I than by the efficiency assay (Fig. 1) is to be expected if a partially active enzyme species was produced by virtue of the modifying reaction. Since inert enzyme registers as zero in both assay systems, contributions from such enzyme were the same in both cases. However, partially active enzyme retaining at least 1/200 of its original activity would register as essentially 1.0 in the all-or-none assay but as <1.0 in the efficiency assay. Hence in the presence of partially active enzyme, the percentage remaining by the all-or-none assay should at all points be greater than the percentage remaining by the efficiency assay.

An alternate explanation for the differences observed in Fig. 1 is that the all-or-none assay depended only on the reaction of eq. 1 whereas the efficiency assay depended on the reactions of both eq. 1 and 2. If the modification preferentially eliminated the capacity of the enzyme to participate in step 2 but did not alter its capacity to participate in step 1, the results of Fig. 1 might be explained without invoking a partially active, modified enzyme, since phosphoglucomutase which can participate in only one step of the over-all two-step reaction mechanism is for all intents and purposes inactive as a catalyst. To test this possibility, all-or-none assay II, using G-6-P instead of G-1-P,

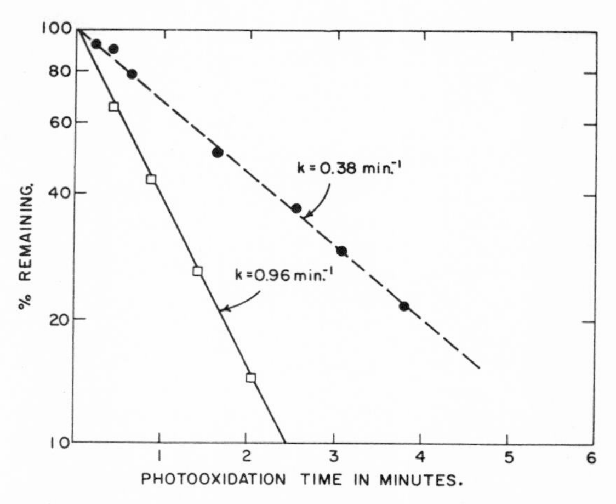

Fig. 1.—Effect of photoöxidation on the activity of phosphoglucomutase as measured by "efficiency" assay, □, and "all-or-none" assay (I), ●. The latter involves incubation of G-1-P with EP^{32}.

was run under the same conditions. In order for this test to be valid, however, it was necessary to make certain that a modified enzyme which had lost its capacity to transfer phosphate to G-6-P did not, when treated with G-6-P, transfer its phosphate to the small amount of G-1-P inevitably formed during the assay (assay II), and that a modified enzyme which had lost its capacity to transfer phosphate to G-1-P did not, when treated with G-1-P, also transfer its phosphate to the small amount of G-6-P formed in this assay (assay I). This would be impossible to demonstrate if the assay intervals employed were of sufficient duration so that equilibration of the P^{32}-label occurred between the 1- and 6-positions of glucose. A very short assay interval would have been required to preclude label shuffling with the relatively large amount of enzyme used in these assays under conditions of optimal enzymic activity, however. Assay conditions were therefore selected under which the intact enzyme was only about 1/100 as active as under optimal conditions (see Experimental), thus allowing the use of an assay interval of 5 min.

The following observations verify the essential absence of label equilibration in the glucose fraction under the conditions of these two assays: (a) Treatment of intact, P^{32}-phosphoglucomutase with a 3000-fold excess of either G-1-P or G-6-P (assays I and II) resulted in the transformation of at least 99% of the radioactivity to trichloroacetic acid soluble products. (b) Of the radioactivity in the trichloroacetic acid supernatant, 98% or more was found in the glucose phosphate region on paper electrophoresis, the remainder being found in the inorganic phosphate–glucose diphosphate region. (c) In assay I (treatment with G-1-P), 98% or more of total radioactivity in the trichloroacetic acid supernatant was acid stable, *i.e.*, was in the 6-position of glucose. (d) In assay II (treatment with G-6-P) 95% or more of the radioactivity was acid labile, *i.e.*, was in the 1-position of glucose. (e) Longer assay intervals invariably resulted in increased shuffling of label in the glucose phosphate fraction. (f) Points b, c, d and e applied equally well to samples of photoöxidized as well as intact phosphoglucomutase, the only difference being that successively smaller percentages of the total radioactivity appeared in the trichloroacetic acid supernatant with increasing photoöxidation time.

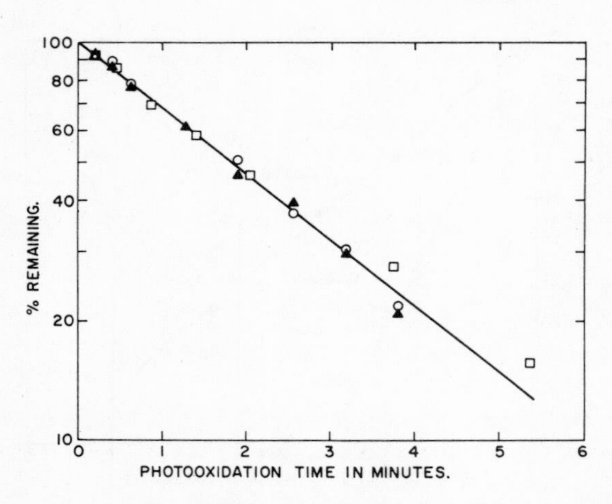

Fig. 2.—Comparison of "all-or-none" assays during photooxidation of phosphoglucomutase: assay I, EP^{32} and G-1-P, ○; assay II, EP^{32} + G-6-P, ▲; assay III, EP + E and G-1-P^{32} + G-6-P^{32} + G-1,6-P_2^{32}, □.

From these considerations it is evident that the position of the radioactive label in the trichloroacetic acid supernatant in both assays is the position in the glucose molecule to which it was first transferred. This is equivalent to saying that only a single phosphate transfer between substrate and enzyme is actually required to register in these assays.

The identity of quantities measured by assays I and II in Fig. 2 thus indicates that the photoöxidation reaction affects at an identical rate the capacity of the enzyme to participate in both step 1 and step 2 of the over-all reaction mechanism. The difference in the fractional quantities measured by the efficiency and all-or-none assays must therefore be the result of a partially active enzyme produced by the modification reaction.

Evidence that the All-or-None Assay is Measuring an Enzymatic Process.—It might be argued that part of the phosphate transfer measured by all-or-none assays I and II was non-enzymatic, and was the result of the increased tendency, resulting from chemical modification, of the phosphate group in phosphoenzyme to undergo nucleophilic displacement. To test this point, all-or-none assay III was carried out using unlabeled enzyme which was treated with a mixture of G-1-P^{32} and G-6-P^{32} (containing trace amounts of G-1,6-P_2^{32} from the method of preparation). In such a system, dephospho-enzyme could be labeled in a one-step reaction with G-1,6-P_2^{32}, but there is no reason to believe that the enzyme used in these assays contained appreciable amounts of the dephospho-form, both from the method of preparation, and from Najjar's previous work.[6] Phospho-enzyme on the other hand must engage in two phosphate transfers in order to register, the first to remove the unlabeled phosphate group and the second to replace this phosphate by reaction with G-1,6-P_2^{32}. Time for completion of both processes was allowed in the assay. The fact that assay III, which requires at least two phosphate transfer steps, measures the same quantity (Fig. 2) as assays I and II, which require a minimum of only one transfer, indicates that an enzymatic process is indeed being studied in each of the all-or-none assays and that these assays are not measuring simply a non-enzymic capacity for release of phosphate by the enzyme.

Assay III is not only more complex, theoretically, than assays I and II, but it also requires more careful control of conditions. A large molar excess of radioactive substrate relative to enzyme in this assay would eliminate isotope dilution effects, but would make measurement of covalently bound enzymic phosphate very difficult, both because of increased soluble radioactivity (due to labeled substrate) and decreased amounts of phospho-enzyme (*i.e.*, increased amounts of dephospho-enzyme) owing to the equilibria in eq. 1 and 2. If a small excess of substrate is used, however, isotope dilution and equilibrium effects must both be considered. To minimize corrections, a variable dilution technique was employed which made it possible to maintain a reasonably constant ratio of active sites to substrate. Aliquots from the photoöxidation reaction were thus diluted by trial and error in pilot runs until approximately 15% of the P^{32}-phosphate in a constant amount of substrate was incorporated by the enzyme in equal volumes of these dilutions. If ideally performed, *i.e.*, if the radioactivity incorporated by each time aliquot were exactly equal to that incorporated at $t = 0$, the relative numbers of catalytic sites undergoing phosphate exchange would then be proportional to the extent of the aliquot dilution required to achieve this situation. In practice, however, there was some deviation from this ideal situation and corrections both for equilibrium and isotope dilution must be applied. Fortunately these corrections tended to balance each other, as indicated in Table I and explained in the Appendix.

Correlation of Assays with Amino Acid Residues.—It now remains to correlate the results of Fig. 1 with the modification of individual amino acid residues by the photoöxidation reaction. That the quantity measured by the all-or-none assays decreased during photoöxidation less rapidly than the simultaneous rate of activity loss by the efficiency assay indicates, in the light of the foregoing section, that partially active species are indeed being produced by photoöxidation. The fact that the quantity measured by the all-or-none assays actually decreases with time indicates that the partially active enzyme is itself being converted to an inert form of the enzyme by the photoöxidation reaction. To identify the residues involved in these processes, *i.e.*, production of partially active and of inert enzyme, the method of kinetic analysis previously described[7] must be applied. General equations for inactivation *via* a partially active enzyme have been derived and are adequately represented by the scheme shown in eq. 3, which is consistent with the results shown in Fig. 1.

$$\begin{array}{ccccc} & \xrightarrow{k_1} & E_1\ (A=F) & \xrightarrow{k_2} & \\ E\ (A=1) & & & & E_{1,2}\ (A=0) \\ & \xrightarrow{k_2} & E_2\ (A=0) & \xrightarrow{k_1} & \end{array} \quad (3)$$

In this scheme E represents native enzyme, E_1, enzyme in which residue 1 has been modified, E_2, enzyme in which residue 2 has been modified, and $E_{1,2}$, enzyme in which both residues 1 and 2 have been modified, and constants k_1 and k_2 are the rate constants for modification of the amino acid residues, while A represents the specific activity of the various enzymic species and F is a fraction between 1.0 and 0. For completeness, modification of E_2 to yield $E_{1,2}$ is indicated in the above scheme, although such a conversion would have no influence on the activity assay. It would, however, be significant in the amino acid analyses.

(7) W. J. Ray, Jr., and D. E. Koshland, Jr., *J. Biol. Chem.*, **236**, 1973 (1961).

TABLE I

Photo-oxidation time, min.	Assay dilution	Enzyme[a] radioact., c./min.	Calcd.[b] [G-1,6-P2]	Equil.[b] corrn. factor	Isotope[b] dil. corrn. factor	Over-all corrn. factor	[EP]i	Act. sites remaining, %
0.00	10.0	6230	12,000	1.96	1.44	2.82	17,600	100.0
.44	8.2	6550	12,200	1.93	1.46	2.81	18,500	86.4
.87	6.0	7260	12,500	1.86	1.51	2.81	20,300	69.3
1.40	5.3	6970	12,400	1.88	1.48	2.78	19,500	58.5
2.04	4.0	7290	12,500	1.86	1.51	2.81	20,400	46.4
3.74	2.5	6830	12,300	1.90	1.48	2.81	19,200	27.4
5.38	1.3	7530	12,600	1.84	1.52	2.80	21,100	15.6

[a] Average of two duplicate aliquots; initial substrate radioactivity, 40,200 c./min. [b] See Appendix.

It follows from such a scheme that k_2 should be related to the slope of the all-or-none assay plot since this represents the rate at which the sum of the active and partially active enzyme ($E + E_1$) is converted to inert enzyme ($E_2 + E_{1,2}$). This rate constant is 0.38 in Fig. 1 which agrees very well with the figure 0.37 for the rate of oxidation of a single "surface" methionine residue determined from amino acid analyses under identical conditions.[3]

When partially active enzyme is produced by the chemical modification under study, the rate of activity loss (determined by the efficiency assay) will frequently be non-logarithmic (*i.e.*, non-linear on a semi-log plot, as in Fig. 3). However, the initial rate of activity loss in such cases will frequently be almost equal to the sum of the modification rates of those residues involved with the production of inert and partially active enzyme, respectively. The rate of decrease of the all-or-none assay, on the other hand, will be equal to the sum of the modification rates of only those residues producing inert enzyme. The difference between these rates should then approximate the modification constant of the residue producing partially active enzyme. The difference in rate constants in the present case (0.96–0.37, Fig. 1) is 0.6 min.$^{-1}$, which approximates the rate constant of a single surface histidine residue[3] and therefore constitutes *prima facia* evidence that the partially active enzyme is produced by modification of such a histidine residue of the enzyme.

It should be mentioned that the interpretation of the all-or-none assay is intimately related to the assumed mechanism of action of the enzyme. A change in the mechanism may or may not require a change in the conclusion derived from the all-or-none assay. For example, in the above analysis the Najjar mechanism has been assumed, but Cleland[7a] has suggested that certain kinetic observations of Bodansky[7b] are difficult to reconcile with this mechanism. It might well be asked what differences in interpretation of the all-or-none results would be required if the Najjar mechanism is modified. The answer would seem to be that the above interpretation is valid as long as the amount of phospho-enzyme produced is equal for all partially active species of enzyme. Thus, even such a drastic change in mechanism as one in which the phospho-enzyme appeared only as a side reaction and not an obligatory intermediate would require no reinterpretation of the all-or-none assay as long as this side reaction proceeded for all of the partially active species. If the phospho-enzyme is produced only as a side reaction, however, it is possible that modification of the enzyme might produce a species which could transfer phosphate only to G-1-P and not G-6-P. In this case the all-or-none assay could still provide valuable information in comparison to the efficiency assay, but the conclusions would be modified somewhat. The latter alternative does not seem likely for phosphoglucomutase, but the principle is important to bear in mind in the general application of the all-or-none assay (see below).

(7a) W. W. Cleland, *Biochim. Biophys. Acta*, **67**, 104 (1963).
(7b) O. Bodansky, *J. Biol. Chem.*, **236**, 328 (1961).

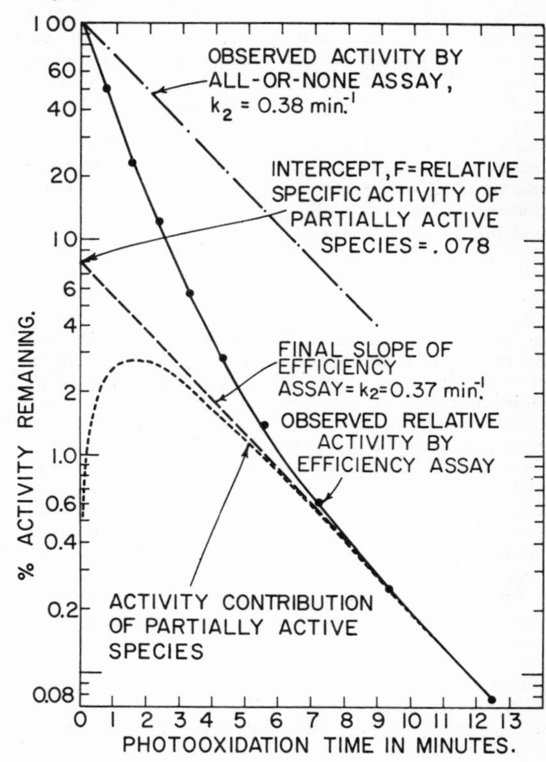

Fig. 3.—Comparison of final slope of "efficiency" assay curve with "all-or-none" assay. If the scheme of eq. 3 is valid, final slope represents rate of conversion of partially active enzyme to inert enzyme and, therefore, this slope equals $k_2/2.3$. "All-or-none" slope, shown by upper broken line, also represents k_2 as discussed in text. Lowest dashed curve represents calculated activity contribution of partially active enzyme during photooxidation assuming mechanism of eq. 3.

Comparison with Previous Results.—Figure 3 shows a plot of activity destruction during photoöxidation extended to 99.9% loss of activity[3] followed by the efficiency assay as described previously. The decrease in rate of activity loss during the middle and latter phases of the reaction, *i.e.*, the decrease in slope of the semilogarithmic plot, has previously been attributed to the accumulation of a partially active enzyme according to the scheme shown in eq. 3. With the efficiency assay alone, however, the additional possibilities could not be eliminated that the non-linear activity loss in this plot was caused by (a) residual intact enzyme resulting from a decreased photoöxidation efficiency, (b) the presence in the enzyme sample of a structurally different mutase, or (c) a different type of enzyme which was positive in the efficiency assay[8] and oxidized more slowly than the phosphoglucomutase. These explanations may now be excluded, since (a) and (b) would result in a loss of catalytic efficiency (efficiency assay) equal to destruction of catalytic sites (all-or-none assay), while (c) would result in a loss of mutase sites at a rate exceeding the rate of destruction of catalytic activity. This leaves the activity loss scheme of eq. 3 as the remaining consistent explanation for the data represented by both Fig. 1 and 3.

The validity of this interpretation is further substantiated by the good agreement of the values of k_2 (eq. 3), determined by the all-or-none assays, with the previously obtained value for the production of inert enzyme on photoöxidation, using an analysis of the data from the efficiency assay in Fig. 3. Since k_2 (0.37 min.$^{-1}$) by the latter method was obtained from the final slope of the activity destruction curve, *i.e.*, where the rate of formation of $E_{1,2}$ from E_1 was apparently being measured, whereas k_2 assessed by the all-or-none assays (0.36 to 0.38 min.$^{-1}$) was obtained by measuring the conversion of $E + E_1$ to $E_2 + E_{1,2}$, the concordance is strong support for the events depicted in eq. 3.

It is only fair to state that the identification of the individual residues of methionine and histidine with the respective steps in eq. 3 is somewhat more tenuous than the evidence for eq. 3 itself. The reason for this is that the constants for the amino acid residues were obtained by assuming certain groups of residues act as "surface" and others as "buried" residues. Until individual residue constants are obtained, these conclusions must be taken as working hypotheses. Recently, however, Schachter and Dixon[9] have measured the extent of reaction of individual residues in the photoöxidation of a different enzyme (chymotrypsin) and their values coincided quite well with the values predicted from our previous kinetic analysis of the same system based on the concept of surface and buried residues.[10]

Finally it should be mentioned that the numerical value F (0.08) of the partially active species, which was obtained by analysis of the rate of decrease in activity with photoöxidation time using the enzymic efficiency assay, is in agreement with the observations of the all-or-none assay. Thus, since an enzyme retaining only 0.5% of the efficiency of native enzyme would register as active in all-or-none assays I and II, it is obvious that a species which had retained 8% of its activity would register likewise, as observed.

General Applicability of the All-or-None Assay.—Although the type of all-or-none assays applied here is feasible only with the enzymes which form enzyme–substrate intermediates, the list of known enzymes which fall into this category is rapidly increasing. At present aldolase,[11] chymotrypsin,[12] trypsin,[12] phosphatase,[13] phosphoglyceric acid mutase,[14] transaldolase,[15] triosephosphate dehydrogenase,[16] thrombin,[17] elastase,[18] subtilisin,[19] acetoacetate decarboxylase,[20] deoxyribose phosphate aldolase[21] and 2-keto-3-deoxy-6-phosphogluconate aldolase[21] in addition to phosphoglucomutase have been shown to carry out their respective catalyses *via* such intermediates. Evaluation of the chemical modification of these enzymes using the present approach might well be feasible. This possibility has been recently substantiated by a preliminary account of the effect of dye-sensitized photoöxidation of chymotrypsin.[10] A more detailed account of this work will be the subject of a forthcoming communication.

Acknowledgments.—The authors wish to acknowledge the invaluable assistance of John J. Ruscica and H. G. Latham during the course of this work. A portion of this research was carried out at Brookhaven National Laboratory under the auspices of the U. S. Atomic Energy Commission.

Appendix

The total radioactivity incorporated into phosphoglucomutase under the conditions of all-or-none assay III is linearly related to the amount of enzyme active in the assay only over very narrow ranges of active enzyme/added substrate ratios. Corrections must therefore be made to relate the measured radioactivity incorporated to the amount of enzyme that reacts during the assay if this ratio changes appreciably. In order to minimize these corrections, the amount of enzyme used in the assay was varied by trial and error until almost the same ratio of active enzyme to added substrate was achieved as evidenced by incorporation of similar fractions of the total radioactivity of the added substrate by the enzyme, *i.e.*, about 15%. The required correction (to compensate for both equilibrium and isotope dilution effects) varied only slightly from sample to sample as indicated below.

Equilibrium Correction Factor.—Considering *only* the enzyme that was sufficiently active to exchange its label in the assay (*i.e.*, disregarding inert enzyme), $(EP)_i/(EP)_e$, the ratio of the phospho-enzyme initially present, to the phospho-enzyme remaining at equilibrium, may be calculated using Najjar's value of 3.8 for the equilibrium constant of eq. 2.[6] In these calculations, parentheses denote concentrations, brackets denote the total radioactivity of the enclosed quantity, and the subscripts i and e denote initial and equilibrium values of the indicated quantities. The assumption is also made that equivalent amounts of dephospho-enzyme and G-1,6-P_2 are always present, (*i.e.*, that $(E) = (G\text{-}1,6\text{-}P_2)$ and that $(EP)_i = (EP)_e + (G\text{-}1,6\text{-}P_2)_e$, since the enzyme is initially quite probably entirely phospho-enzyme as evidenced both from the method of preparation and from Najjar's work.[6]

In addition, the quantity of G-1-P present is neglected in all calculations. The following equations may then be written where [T] is the total added radioactivity, noting that the specific activity of the diphosphate is twice that of both EP and G-6-P.

$$[EP]_e\,[G\text{-}6\text{-}P]_e/[G\text{-}1,6\text{-}P_2]^2_e = 3.8/4 \qquad (4)$$

$$[EP]_e + [G\text{-}6\text{-}P]_e + [G\text{-}1,6\text{-}P_2]_e = [T] \qquad (5)$$

(8) Any enzyme which transformed G-1-P into an acid-stable phosphate would be positive.

(9) H. Schachter and G. H. Dixon, *Biochem. Biophys. Res. Commun.*, **9**, 132 (1962).

(10) W. J. Ray, Jr., H. G. Latham, Jr., M. Katsoulis and D. E. Koshland, Jr., *J. Am. Chem. Soc.*, **82**, 4743 (1960).

(11) E. Grazi, P. T. Rowley, T. Cheng, O. Tchola and B. L. Horecker, *Biochem. Biophys. Res. Commun.*, **9**, 38 (1962).

(12) A. K. Balls and E. F. Jansen, *Advances in Enzymol.*, **13**, 321 (1952).

(13) L. Engstrom and G. Agren, *Acta Chem. Scand.*, **12**, 357 (1958); *Arkiv Kemi*, **19**, 129 (1962); J. H. Schwartz and F. Lipmann, *Proc. Natl. Acad. Sci. U. S.*, **47**, 1996 (1961).

(14) L. I. Pizer, *J. Am. Chem. Soc.*, **80**, 4431 (1958).

(15) R. Venkataraman and E. Racker, *J. Biol. Chem.*, **236**, 1883 (1961).

(16) I. Krimsky and E. Racker, *Science*, **122**, 319 (1955).

(17) J. Gladner and K. Laki, *J. Am. Chem. Soc.*, **80**, 1263 (1958).

(18) B. S. Hartley, M. A. Naughton and F. Sanger, *Biochim. Biophys. Acta*, **34**, 243 (1959).

(19) F. Sanger, *J. Polymer Sci.*, **49**, 3 (1961).

(20) I. Fridovich and F. H. Westheimer, *J. Am. Chem. Soc.*, **84**, 3209 (1962).

(21) E. Grazi, H. Meloche, G. Martinez, W. A. Wood and B. L. Horecker, *Biochem. Biophys. Res. Commun.*, **10**, 4 (1963).

From eq. 4 and 5, the radioactivity of the G-1,6-P_2 at equilibrium, *i.e.*, [G-1,6-P_2]$_e$, may be calculated from the measured radioactivity of the added substrate and of the phospho-enzyme at equilibrium, and $(EP)_e/(EP)_i$ may then be represented as

$$(EP)_e/(EP)_i = [EP]_e/\{[EP]_e + 0.5[G\text{-}1,6\text{-}P_2]_e\} = 1/F_1 \quad (6)$$

Since $(EP)_e = [EP]_e/k_e$, the concentration of active phospho-enzyme initially present is given by eq. 7, where k_e is the specific activity of the exchangable phosphate at equilibrium and will be evaluated below.

$$(EP)_i = F_1[EP]_e/k_e \quad (7)$$

Values of F_1 calculated from eq. 6 are given in Table I as the equilibrium correction factor.

Isotope Dilution Correction Factor.—If the initial specific activity of added substrate is represented by k_i, the following equation may be written

$$k_e = k_i(G\text{-}6\text{-}P)_i/\{(EP)_i + (G\text{-}6\text{-}P)_i\} \quad (8)$$

After the appropriate manipulations this becomes

$$k_e = k_i\{[T] - [EP]_e - 0.5[G\text{-}1,6\text{-}P_2]\}/[T] = k_i/F_2 \quad (9)$$

Values of F_2 from measured values of [T] and $[EP]_e$ and calculated values of [G-1,6-P_2] are given in Table I as the isotope dilution correction factor.

Over-all Correction.—From eq. 7 and 9, $(EP)_i = F_1F_2[EP]_e/k_i$. From Table I it may be seen that the product F_1F_2 is reasonably constant, in spite of variations in both F_1 and F_2. Thus to a good approximation $(EP)_i = k[EP]_e$. It might also be observed that the F_1F_2 product will be reasonably constant no matter what value is chosen for the equilibrium of eq. 2, and hence the validity of the present procedure is dependent only on the direct measurement of $[EP]_e$, since the ratio of $(EP)_i$ at any photoöxidation time to $(EP)_i$ at a reaction time of zero is the quantity actually under consideration. Thus if $X = \{[EP]_e + {}^1/_2[G\text{-}1,6\text{-}P_2]\}$, $F_1F_2 = X[T]/\{[T] - X\}$, which means that F_1F_2 is essentially constant over narrow ranges of X.

Reprinted from THE JOURNAL OF BIOLOGICAL CHEMISTRY, Vol. 245, 1970

Kinetic Aspects of Regulation of Metabolic Processes

THE HYSTERETIC ENZYME CONCEPT*

(Received for publication, June 1, 1970)

CARL FRIEDEN

From the Department of Biological Chemistry, Washington University School of Medicine, St. Louis, Missouri 63110

SUMMARY

Hysteretic enzymes are defined as those enzymes which respond slowly (in terms of some kinetic characteristic) to a rapid change in ligand, either substrate or modifier, concentration. Such slow changes, defined in terms of their rate relative to the over-all catalytic reaction, result in a lag in the response of the enzyme to changes in the ligand level. Several mechanisms, including ligand-induced isomerization of the enzyme, displacement of tightly bound ligands by other ligands, or polymerization and depolymerization, are discussed and it can be shown that the description of the time-dependent change in enzyme activity is similar for many different cases. Examination of the literature reveals that a large number of enzymes may fall into the category termed hysteretic, that such enzymes are frequently those which are important in metabolic regulation, that the time of conversion from one kinetic form to another may vary between seconds and minutes, and that there are experimental examples of all the mechanisms which are discussed theoretically. The possible relation between those enzymes which are hysteretic and regulation of complex metabolic processes is discussed in terms of the fact that the slow response of the hysteretic enzyme to changes in ligand level will lead to a time-dependent buffering of some metabolites and that this may be important with respect to pathways which utilize common intermediates or in which there are multiple branch points. It is suggested that the question of hysteresis in enzyme systems as defined here be systematically investigated in regulatory enzymes and that this concept may be of value in discussing the regulation of complex processes *in vivo*.

Current attempts to relate the regulation of metabolic processes to the known properties of enzyme systems are primarily centered around those characteristics attributed to allosteric or "regulatory" enzymes (1–5). Thus, schemes which propose preferential use of particular metabolic pathways under certain conditions are correlated with the unique characteristics of the enzymes which are concerned with those pathways. Those characteristics of particular importance in this type of correlation are of two types (*a*) the ability of specific metabolites to influence enzymatic activity by binding to specific sites distinct from the catalytically active site, and (*b*) the unusual (*i.e.* nonhyperbolic) dependence of the initial velocity on substrate of ligand concentration.

The original and elegantly simple idea of feedback inhibition, utilizing as it does both of these characteristics, demonstrates the ability to describe and understand concepts of metabolic regulation in terms of enzymatic properties and lends considerable support to the idea that much of metabolic regulation, even in complex cases, may be discussed in these relatively simple terms.

There has been, over the few years since this simple regulatory concept was presented, considerable elaboration of the basic and underlying theme of feedback inhibition. Enzymes, much more complicated than those originally pictured, have been characterized and terms such as concerted, cooperative, or cumulative inhibition, activation, multivalent inhibition, sequential (or induced fit) mechanisms have been coined to describe some of the unique kinetic properties of such proteins. Furthermore, the ideas discussed in the relationship of structure to function and which involved only conformation changes have been expanded to include systems in which different conformational forms of an enzyme do not have the same intrinsic specific activity or enzymes which undergo reversible association-dissociation reactions in which different molecular weight forms of the enzyme have different characteristics with respect to ligand binding or activity (6, 7).

If there are any issues to be faulted with current approaches to the question of metabolic regulation *in vivo*, they are in the application of presently available equations (8–10), developed to described ligand binding, to initial velocity kinetic data and the presumption that such equations reflect the instantaneous rate of substrate disappearance in a sequence of enzymatic steps. This presumption would be valid if all of the conformational forms of an enzyme had the same intrinsic activity and if rates involving all of the different enzyme complexes were rapid prior to that in which the product is released from the enzyme. Such an assumption may not be true and the rate of certain steps in the interconversion between enzyme-containing species may be rate determining, thus making the kinetic properties of the enzyme different from those predicted by the binding equations. So far, the attempts to include this kinetic argument have centered around questions relating to the derivation of the equation to represent a particular mechanism. Thus, the use of the "steady

* This research was supported in part by United States Public Health Service Grant AM 13332 and National Science Foundation Research Grant B8-1568R.

state" assumption, rather than that assuming rapid equilibration of all species prior to product formation, has been explored. This method of approach, however, turns out to be of limited usefulness since the equations for complex systems can easily become much too cumbersome to deal with (11, 12).

In this paper, we will develop another way of approaching the question of the importance of the kinetic aspects of enzymatic reactions to metabolic processes. In particular, it is assumed that certain steps are in rapid equilibrium while others may represent slow processes relative to product formation. Such slow steps may be, for example, conformational changes in one or more of the enzyme-containing species. Thus the instantaneous rate of product formation may represent some average of the kinetic characteristics of the two (or more) conformational forms. Some of the concepts concerning slow and rapid conformational (or molecular weight) changes and their consequences were briefly considered previously (13). In this paper, the concept of the slow response to a rapid change in ligand concentration and the consequence of such slow changes with respect to enzymatic activity and metabolic regulation will be discussed in detail. The term "hysteretic enzyme" is introduced to refer to enzymes which respond (in terms of configurational or other changes) slowly to rapid changes in ligand concentration.

Development of Hysteresis Concept

Hysteretic Enzymes—In physics, hysteresis is defined as the time lag exhibited by a body in reacting to outside forces. By analogy, hysteresis in regulatory enzymes could be defined as the time lag exhibited in some kinetic or physical property of an enzyme in response to a rapid change in concentration of substances which influence those properties.[1] Hysteretic enzymes then, are defined here as those enzymes which respond slowly to rapid changes in ligand concentration. Slow and rapid, although they are relative terms, may be related for convenience to the time required to measure enzyme activity. This point will be discussed later.

That some enzymes may show this type of behavior, that is, that some enzymes undergo rather slow conformational changes, has been recognized previously. As shown later, many regulatory enzymes may exhibit such behavior, but the fact that this may have important consequences in terms of metabolic regulation has not been extensively discussed.

In general, there appear to be two mechanisms which may be primarily responsible for slow responses (*a*) isomerization processes and (*b*) displacement (under certain conditions) of a tightly bound ligand by another with a different effect on activity. A more complex process which may include either or both of these mechanisms is the polymerization or depolymerization of an enzyme.

Isomerization

In order to present the concepts relevant to the point of this paper, we shall start with simple enzyme models and develop the kinetic equations, listing along with this development the assumptions made and the reasoning for such assumptions.

Single Substrate Case—The simplest single substrate enzyme system is written as

$$E + S \rightleftharpoons ES \rightarrow E + P$$

Mechanism I

One can introduce more intermediates of the enzyme-substrate or enzyme-product type and make the reaction reversible. In all cases, and independent of the absolute value of any of the rate constants, the initial velocity shows a normal hyperbolic dependence on the substrate concentration. If the substrate level does not change as a function of time and the product concentration is low enough so that it does not inhibit the reaction (for example, if it is continuously removed, as may occur in a metabolic sequence involving a series of enzymatic steps), the velocity at any time is the initial velocity. This velocity (v_t) will also show the same substrate concentration dependence as the initial velocity (*i.e.* hyperbolic).

One property not exhibited by this system which appears to be inherent in regulatory enzymes is that the enzyme may exist in two or more conformational forms which differ in some catalytic or binding sense. Thus, the simplest extension of Mechanism I to be considered is

$$\begin{array}{ccccccc} E + S & \overset{L_1}{\rightleftharpoons} & ES & \underset{k_{-3}}{\overset{k_3}{\rightleftharpoons}} & E'S & \overset{L_2}{\rightleftharpoons} & E' + S \\ & & \downarrow k_5 & & \downarrow k_6 & & \\ & & E & \underset{k_{-2}}{\overset{k_2}{\rightleftharpoons}} & E' & & \\ & & + & & + & & \\ & & P & & P & & \end{array}$$

Mechanism II

or, written in the form in which a *solid line* represents a reversible reaction

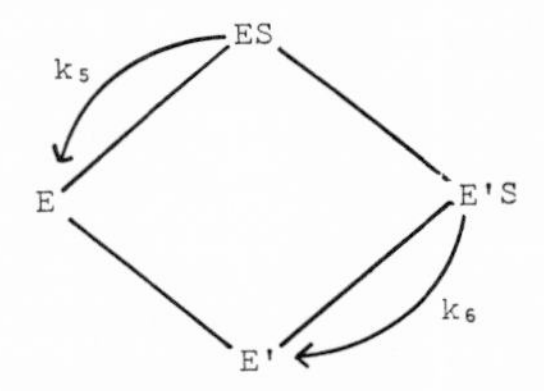

where L_1 and L_2 are equilibrium constants.

This mechanism assumes two conformational states of the enzyme, both of which can bind substrate (with different affinities). The rate constants k_5 and k_6 represent breakdown of the ES and $E'S$ complex to give product.

Derivation of the equation or equations which represent kinetic or binding data for this mechanism may be approached in several ways.

If, for example, one assumes that $k_5 = k_6 = 0$ (*i.e.* if we are concerned with ligand binding only), and if the time of measurement is long relative to adjustment of equilibrium, $\bar{Y}$, the fraction of saturation, is

$$\bar{Y} = \frac{\frac{S}{L_1} + K_2\frac{S}{L_2}}{1 + \frac{S}{L_1} + K_2\left(1 + \frac{S}{L_2}\right)} \quad (1)$$

[1] The analogy may not be a perfectly accurate one. Common examples of hysteresis in physics, as exemplified by change in magnetic induction of an iron bar on first increasing then decreasing the magnetic field strength, frequently represent irreversible energy losses. On the other hand, the definition of the Greek word appears to simply mean "to be behind, come late." The changes discussed with respect to the enzyme systems are, of course, considered to be reversible.

where $K_2 = k_2/k_{-2}$. This equation is identical to that derived by Monod, Wyman, and Changeux (8), for the case where there is only one binding site ($n = 1$).

The kinetic equation which corresponds to (has the identical form as) this binding equation may be derived by assuming that k_5 and k_6 are identical and that these steps are the rate-determining steps in the over-all reaction, *i.e.* that all steps prior to the breakdown to product are in rapid equilibrium. In this case (after rearranging)

$$\frac{v}{E_0} = \frac{k_5}{1 + \frac{L_1}{S}\left\{\frac{1 + K_2}{1 + K_2L_1/L_2}\right\}} \quad (2)$$

where again $K_2 = k_2/k_{-2}$.

If k_5 and k_6 are not identical

$$\frac{v}{E_0} = \frac{k_5(1 + K_2k_6L_1/k_5L_2)/(1 + K_2L_1/L_2)}{1 + \frac{L_1}{S}\left\{\frac{1 + K_2}{1 + K_2L_1/L_2}\right\}} \quad (3)$$

In either case, the velocity shows a hyperbolic dependence on substrate concentration and there is no kinetically convenient way to distinguish between the two cases. The complete steady state derivation $[d(EX_i)/dt = 0]$ of Mechanism II gives a more complex rate expression including S^2 terms. As a consequence, it has been suggested several times that allosteric behavior may, in some cases, simply be a kinetic, rather than binding, argument (11, 12). Although this kinetic situation may arise whenever an alternate pathway for the reaction exists, there is good reason to believe that the complete steady state derivation is not the primary explanation for the kinetic abnormalities of such enzyme systems. Thus, there are a large number of conditions, including equilibration of only some of the steps, under which the more complex rate reduces to the same form of the equation derived by rapid equilibrium methods (14, 15). For example, the assumption of equilibration between substrate and enzyme leads to such a simplification. In view of the known rate constants for the step $E + S \rightarrow ES$ ($\sim 10^7$ to 10^9 sec^{-1}), and given reasonable binding constants, such steps may indeed equilibrate very rapidly. Simple calculations show, for example, that equilibration of binding steps may occur in less than 10^{-3} sec. But such an equilibration is not the only mechanism for obtaining the simple form of the kinetic equation. Identity of certain rate constants or ratios of rate constants will also serve this purpose (15). Although it is reasonable to expect that equilibration in binding steps may be very rapid, the conversion of one conformational form to another may be very slow.

If one assumes (*a*) the steps $E + S \rightleftharpoons ES$ and $E' + S \rightleftharpoons E'S$ are in rapid equilibrium, (*b*) that the over-all reaction is irreversible, (*c*) that the substrate concentration is maintained at a constant level, and (*d*) that there is no product inhibition, the velocity obtained will be a measurable function of time provided that the isomerization rate constants (k_3, k_{-3}, k_2, and k_{-2}) are small enough.

Under these conditions, the time dependence of the velocity for Mechanism II may be shown to be

$$v_t = v_f + (v_0 - v_f)e^{-k't} \quad (4)$$

where v_t is the velocity at time t relative to the change in conditions (at $t = 0$) which initiates the conformational change (as for example addition of substrate), v_f is the velocity at $t = \infty$ (remembering the assumption that the substrate level remains

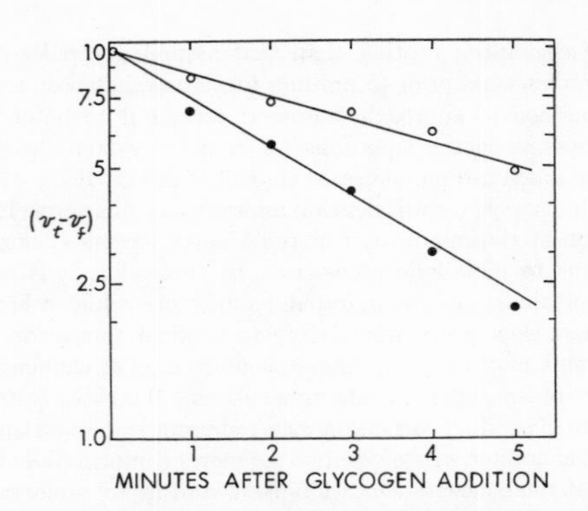

FIG. 1. Data for frog phosphorylase *a* after addition of glycogen. The reaction was run in 0.03 M β-glycerophosphate, pH 7, containing 10^{-3} M EDTA, 6×10^{-4} M 2-mercaptoethanol, and 0.2 M sodium arsenate. The enzyme concentration was 3 mg per ml and glycogen was 20 mg per ml. At the indicated times, the reaction was stopped with perchloric acid and glucose measured. The velocities were measured from the slope of a graph of glucose formed *versus* time. Data at 10°, O——O, taken from Reference 16 and at 15°; ●——●, from unpublished data of the same authors.[2]

constant), v_o is the velocity at $t = 0$ and k' is a complex rate constant which depends on the substrate concentration. For Mechanism II and, without making the assumption of microscopic reversibility (discussed below)

$$k' = \frac{k_3}{1 + \frac{L_1}{S}} + \frac{k_{-3}}{1 + \frac{L_2}{S}} + \frac{k_2}{1 + \frac{S}{L_1}} + \frac{k_{-2}}{1 + \frac{S}{L_2}} = \frac{k_3S + k_2L_1}{L_1 + S} + \frac{k_{-3}S + k_{-2}L_2}{L_2 + S} \quad (5)$$

Assuming, from the principle of microscopic reversibility, that $k_{-2}k_3L_2 = k_2k_{-3}L_1$, one of the rate constants can be eliminated. Thus

$$k' = k_2\left(1 + \frac{k_3S}{k_2L_1}\right)\left[\frac{1}{1 + \frac{S}{L_1}} + \frac{1}{K_2\left(1 + \frac{S}{L_2}\right)}\right] \quad (5a)$$

As may be seen from Equation 4, k' is easily evaluated from a plot of ln $(v_t - v_f)/(v_o - v_f)$ *versus* t and is clearly dependent on S, L_1, L_2, and the other rate constants.

An example of this type of enzyme system may be taken from data involving phosphorylase *a*. Fig. 1 shows a plot of data for the frog phosphorylase *a* measuring the increase in enzymatic activity during arsenolysis after adding glycogen at zero time. The data, obtained at both 10 and 15°, are plotted as log $(v_t - v_f)$ *versus* time and show reasonably good first order plots. The values of k' at 10 and 15° are 2.3 and 4.8×10^{-3} sec^{-1}, respectively, and correspond to half-times of 5.05 and 2.4 min at the two temperatures.

Equation 4 holds whether substrate is added to the enzyme alone or whether a new level of substrate is introduced into a system which already contains substrate, provided that the new substrate level remains unchanged during the time of the conversion from one form of the enzyme to another. Equation 4

[2] B. E. Metzger, L. Glaser, and E. Helmreich, unpublished data.

is also independent of the relative values of k_5 and k_6 (including 0) since the total concentration of $E + ES$ or $E' + E'S$ is unchanged by $ES \to E$ or $E'S \to E$ because there is (by the previous assumptions) a rapid equilibration of these species. The rate constants k_5 and k_6 do appear, of course, in the velocity terms, v_o and v_f, and it is instructive at this point to consider the substrate dependence of these velocities.

If the enzyme exists in the E form only before substrate addition, v_o is defined by

$$\frac{v_0}{E_0} = \frac{k_5}{1 + \frac{L_1}{S}} \quad (6)$$

Assuming microscopic reversibility,[3] the final velocity, v_f, is

$$\frac{v_f}{E_o} = \frac{k_5(1 + K_2k_6L_1/k_5L_2)/(1 + K_2L_1/L_2)}{1 + \frac{L_1}{S}\left\{\frac{1 + K_2}{1 + K_2L_1/L_2}\right\}}$$

as previously derived in Equation 3 for the velocity of the reaction assuming rapid equilibration of all enzyme-containing forms.

The ratio v_f/v_o is then

$$\frac{v_f}{v_0} = \frac{(S + L_1)(1 + k_6K_3/k_5)}{S(1 + K_3) + L_1(1 + K_2)} \quad (7)$$

where $K_3 = k_3/k_{-3}$ and $K_2 = k_2/k_{-2}$.

Clearly the velocity will change as a function of time but the extent of change will be dependent on the substrate level and the relative values of L_1, K_2, and K_3. The rate of change is still defined by Equation 4. If both E and E' exist at time zero or if the substrate concentration is rapidly changed from one level to another, the ratio v_f/v_o is

$$\frac{v_f}{v_0} = \left(\frac{S_2}{S_1}\right)\left\{\frac{S_1(1 + K_3) + L_1(1 + K_2)}{S_2(1 + K_3) + L_1(1 + K_2)}\right\} \quad (8)$$

where S_2 is the new level of substrate. The rate of change is again defined by Equations 4 and 5 with $S = S_2$.

As pointed out by Rabin (17), the progress curve (the product concentration as a function of time) for such a system may show a lag, but will then reach a constant value. It should be noted that the same is true of the case where $k_5 = 0$ provided that $v_f > v_o$. If one assumes substrate depletion, the curve will then level off and the over-all progress curve would be sigmoidal. If, as is the assumption made previously in the derivation of these equations, the substrate level remains constant throughout the time involved, it becomes a relatively simple matter to express the change in product concentration as a function of time. Thus, according to Equation 4

$$\frac{dP}{dt} = v_f + (v_0 - v_f)e^{-k't} = v_f\left[1 - \left(\frac{v_f - v_0}{v_f}\right)e^{-k't}\right] \quad (9)$$

where at a given substrate concentration, v_o, v_f, and k' are constants. Integration yields

$$P_t = v_f t - \frac{1}{k'}(v_f - v_0)(1 - e^{-k't}) \quad (10)$$

If $v_o = 0$ (*i.e.* $k_5 = 0$ and only one form of the enzyme is present at $t = 0$) then

$$P_t = v_f\left[t - \frac{1}{k'}(1 - e^{-k't})\right] \quad (11)$$

The extent of the lag before reaching a constant rate of product formation clearly depends on k', larger values of k' yielding shorter lag times. Product formation becomes a linear function of time when $e^{-k't} \ll 1$. Under these conditions

$$P_t = v_f t - \frac{(v_f - v_0)}{k'}$$

Extrapolation of this linear portion of a P_t *versus* t curve yields $P = -(v_f - v_o)/k'$ at $t = 0$ or $t = (v_f - v_o)/k'v_f$ at $P = 0$.

An equation similar to Equation 10 has recently been derived by Ray and Hatfield (19). The latter authors, however, used somewhat different assumptions for the mechanism as well as conditions under which the substrate level is assumed to saturating. Equation 10 as derived in this paper assumes only that the substrate level does not change during the time of the experiment. If the substrate level is saturating

$$k' = k_3 + k_{-3} = k_{-3}(1 + K_3)$$

while at low substrate levels

$$k' = k_2 + k_{-2} = k_{-2}(1 + K_2)$$

Single Substrate-Single Modifier Case—The single substrate Mechanism II can be expanded to include the modifier. The mechanism can then be represented as

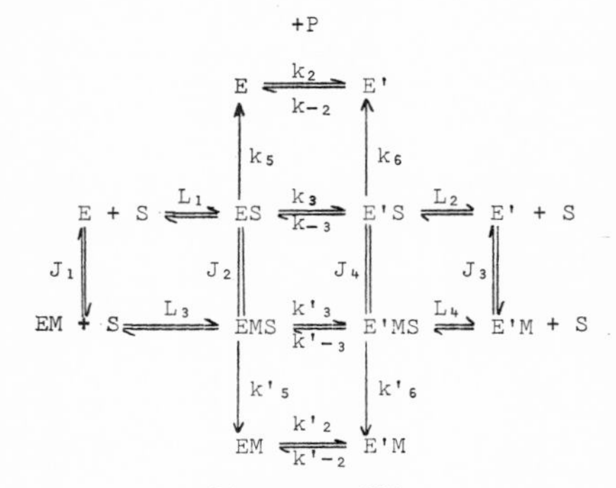

MECHANISM III

[3] If, assuming Mechanism II, one arbitrarily sets $k_{-3} = k_2 = 0$ as did Rabin when discussing this mechanism (17), the system might still be considered to be able to operate since E can be regenerated (unidirectionally) from E' and $E'S$ is formed (unidirectionally) from ES. Such a mechanism will still obey Equation 4 but now includes terms in S^2 and may therefore be claimed to be a mechanism which would show allosteric type behavior. Fundamentally this difference arises from the fact that the principle of microscopic reversibility cannot be applied to a system in which certain rate constants have been set to 0. If one were not to evoke this principle (which asserts that the product of rate constants in the forward cycle is equal to the product of rate constants in the reverse cycle) for Mechanism II, S^2 would also appear in the derived kinetic equations even without setting k_{-3} and k_2 equal to 0. There is little question that the principle must be obeyed in an equilibrium system, and presumably it would also apply in a kinetic system in which none of the rate constants is equal to 0. We will assume therefore, that the principle of microscopic reversibility (or detailed balance) must be valid in these kinetic systems and that it is therefore incorrect to arbitrarily omit some rate constants, thereby effectively equating them to 0. The problem has also been discussed by Weber (18).

or

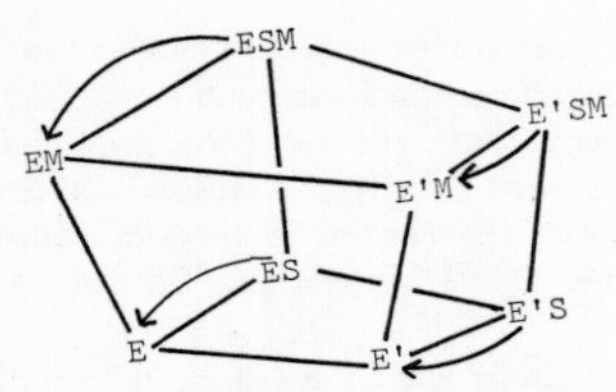

The rate constants which define steps in the presence of modifiers are primed (*i.e.* k_3 *versus* k'_3, and so on). The dissociation constants for modifier binding J_1 to J_4 are of the type $J_1 = [(E)(M)]/EM$, etc.

The derivation of the equation for this system can be made relatively easily using the same assumptions as in Mechanism II: (*a*) all steps defined by dissociation constants equilibrate rapidly, (*b*) the reaction to give product is irreversible and product does not inhibit by competing with the substrate, and (*c*) the concentration of substrate is maintained at a constant level throughout the time course of measurement.

Using these assumptions, one can obtain $v_t = v_f + (v_o - v_f)e^{-k't}$, that is, an equation identical to Equation 4. In this case, however, k' is a much more complex function. Invoking the principle of microscopic reversibility yields Equation 12 of Scheme 1. As before, the amount of product formed as a function of time is defined by Equation 10 which rests in turn on $v_f - v_o$ and on the value of k'. If M binds preferentially to E' or $E'S$, the lag in achieving a constant rate of product formation will be longer than if M bound preferentially to E and ES, since this, in essence, decreases the concentration of ES and ESM.

The values of v_o and v_f will, of course, depend upon the conditions chosen. For example, v_o might be the velocity of the running system in the absence of modifier and might have the form of Equation 3. The final velocity, assuming microscopic reversibility, is shown in Equation 13 of Scheme 1. If it is assumed that modifier does not affect substrate binding ($J_1 = J_2$; $J_3 = J_4$), then Equation 14 of Scheme 1 results. If it is further assumed that $k_5 = k_5' = k_6 = k_6' = 1$, then Equation 14a of Scheme 1, which is identical to the Monod Equation (8) for $n = 1$ in the presence of a modifier, results.

In any case, v_o and v_f could differ markedly so that the addition of modifier to an enzyme system which is already producing product may yield a hysteretic response with the apparent rate constant, k', being defined by Equation 12 of Scheme 1.

Multisite Systems—Most regulatory enzymes, of course, exist as multiple polypeptide chains, each chain usually containing an active and perhaps a modifier site. Mechanism III would therefore be greatly expanded and becomes too complex to indicate pictorially. The expansion to multiple substrate or modifier sites is not too difficult to visualize conceptually, however. Assuming as before, that all ligand binding steps are essentially equilibrium situations, that the substrate concentration is maintained constant throughout the experiment, that the reaction is irreversible, and that the product, if it accumulates, does not inhibit, the system yields the same Equation 4, $v_t = v_f + (v_o -$

$$k' = \frac{1}{1 + \frac{S}{L_1}\left\{\frac{1 + M/J_2}{1 + M/J_1}\right\}}\left[\frac{k_2 + k_3\ S/L_1}{1 + M/J_1} + \frac{k'_2 + k'_3\ S/L_3)M/J_1}{1 + M/J_1}\right]$$

$$+ \frac{1/K_2}{1 + \frac{S}{L_2}\left\{\frac{1 + M/J_4}{1 + M/J_3}\right\}}\left[\frac{k_2 + k_3\ S/L_1}{1 + M/J_3} + \frac{k'_2 + k_3\ S/L_3)M/J_3}{1 + M/J_3}\right] \quad (12)$$

$$\frac{v_f}{E_o} = \frac{\frac{S}{L_1}(k_5 + k'_5 M/J_2) + K_2\frac{S}{L_2}(k_6 + k'_6 M/J_4)}{1 + M/J_1 + \frac{S}{L_1}(1 + M/J_2) + K_2(1 + M/J_3) + \frac{S}{L_2}(1 + M/J_4)} \quad (13)$$

$$\frac{v_f}{E_o} = \frac{\frac{S}{L_1}(k_5 + k'_5\frac{M}{J_1}) + K_2\frac{S}{L_2}(k_6 + k'_6\frac{M}{J_3})}{(1 + \frac{M}{J_1})(1 + \frac{S}{L_1}) + K_2(1 + \frac{M}{J_3})(1 + \frac{S}{L_2})} \quad (14)$$

$$\frac{v_f}{E_o} = \bar{Y} = \frac{\frac{S}{L_1} + \frac{K_2 S}{L_2}\left\{\frac{1 + M/J_3}{1 + M/J_1}\right\}}{1 + \frac{S}{L_1} + K_2(1 + \frac{S}{L_2})\left\{\frac{1 + M/J_3}{1 + M/J_1}\right\}} \quad (14a)$$

$$k' = k_2\left[\frac{1 + \frac{k_3 S}{k_2 L_1}(1 + \frac{S}{L_1})^{n-1}}{(1 + \frac{S}{L_1})^n} + \frac{1 + \frac{k_{-3}S}{k_{-2}L_2}(1 + \frac{S}{L_2})^{n-1}}{K_2(1 + \frac{S}{L_2})^n}\right] \quad (15)$$

Scheme 1

$v_f)e^{-k't}$, as has been derived for all the other cases so far considered. As can be imagined, the expression for k' becomes enormously more complex for the general case. In the absence of modifier, for example, and assuming that $k_3 = k'_3$, $k_2 = k'_2$, $k_{-3} = k'_{-3}$, and $k_{-2} = k'_{-2}$ (equivalent to assuming that the rates of isomerization are independent of the number of substrate sites occupied), then Equation 15 of Scheme 1 results. At infinite substrate concentration, $k' = k_3 + k_{-3}$.

The presence of modifier introduces terms in M raised to the power n, the exact relation being determined by the assumptions made.[4] If assumptions similar to those above are made, the time dependence of the velocity may still obey that described by Equation 4. At infinite substrate concentration

$$k' = \frac{k_3 + k_3'M/J_2}{1 + M/J_2} + \frac{k_{-3} + k_{-3}'M/J_4}{1 + M/J_4} \tag{16}$$

so that at infinite substrate and modifier levels $k' = k'_3 + k'_{-3}$.

For all of the mechanisms described so far, it should be emphasized that the substrate dependence of the initial and final velocities is that which would be calculated by assuming rapid equilibration of all the enzyme species involved. If, for example, the hysteretic response appears as a lag in the rate of product formation after substrate addition, the final velocity of the enzymatic reaction is that value obtained when the product formation becomes a linear function of time. This final velocity may show typical allosteric behavior as a function of substrate concentration. Furthermore, such behavior can be discussed in terms of the models which have been proposed by Koshland *et al.* (9) or by Monod *et al.* (8), and such a treatment of kinetic data would be perfectly valid. The hysteretic response observed, therefore, is not inconsistent with these models; it only indicates that a measurable amount of time is required to convert one conformation form of the enzyme to another.

Although it may be possible to fit either the initial or final velocity with equations derived for the Koshland *et al.* or Monod *et al.* models, kinetic data obtained at intermediate times may show an abnormal substrate dependence which is not easily described. Such data might be obtained, for example, if the time of the hysteretic response is strongly dependent on substrate concentration. In cases in which readily explicable kinetic data are not obtained, one should certainly investigate this type of possibility.

Displacement Reactions

If a ligand is bound to an enzyme surface quite tightly, albeit reversibly, its rate of dissociation from the enzyme could be relatively slow. Under these conditions, the step $E + M \rightleftharpoons EM$ cannot be assumed to be rapidly equilibrating and the previous assumptions concerning the equilibration of this process are not valid. Thus if a tightly bound inhibitor is displaced by an activator, the rate of increase in enzymatic activity could be a measurable function of time depending on how rapidly the inhibitor is released from the enzyme surface. This mechanism, in its simplest form, would be represented by

$$EM_1 \underset{k_2M_1}{\overset{k_1}{\rightleftharpoons}} E \underset{k_4}{\overset{k_3M_2}{\rightleftharpoons}} EM_2$$

MECHANISM IV

Under these conditions and applying the steady state assumption that $d(E)/dt = 0$, one can derive an expression for the concentration of EM_1 or EM_2 as a function of time.

$$EM_1 = E_0 \left\{ \frac{k_2k_4M_2 + (k_1k_4 + k_1k_3M_2)e^{-k't}}{k_1k_4 + k_1k_3M_2 + k_2k_4M_1} \right\} \tag{17}$$

and

$$EM_2 = E_0 \left\{ \frac{k_1k_3M_2(1 - e^{-k''t})}{k_1k_4 + k_1k_3M_2 + k_2k_4M_1} \right\} \tag{18}$$

where

$$k' = \frac{k_1k_4 + k_2k_4M_1 + k_1k_3M_2}{k_1 + k_2M_1 + k_3M_2} \tag{19}$$

and

$$k'' = \frac{k_1k_4 + k_2k_4M_1 + k_1k_3M_2}{k_4 + k_2M_1 + k_3M_2} \tag{20}$$

If k_1 and $k_4 \ll k_2M_1 + k_3M_2$, which is likely, then

$$k' = k'' = \frac{k_1k_4 + k_2k_4M_1 + k_1k_3M_2}{k_2M_1 + k_3M_2} \tag{21}$$

Since the steady state assumption is used, the concentration of E is independent of time. If it is assumed that activity at saturating substrate levels measures relative concentration of EM_1 and EM_2 (which is equivalent to assuming that k_1 and k_4 are $\ll k_2M_1 + k_3M_2$ and that the concentration of free E is small), then it may be shown that $v_t = v_f + (v_o - v_f)e^{-k't}$. Under these conditions, it will be noted that the velocity as a function of time is identical to the expression derived previously for the isomerization case. This case may therefore be treated in an identical way to the previous case where k' may be obtained from a first order plot and where the product P may be shown to be $P_t = v_ft - (1/k')(v_f - v_o)(1 - e^{-k't})$. That is, there will be a lag in the concentration of P and the extent of the lag will depend upon k' and $v_f - v_o$.

Fig. 2 shows an example of such a displacement reaction using glutamate dehydrogenase. The enzymatic activity is measured while the tightly bound inhibitor GTP is displaced by the activator ADP. Since the dissociation constant for GTP is about 4×10^{-7} M under these conditions (20), its rate of dissociation is slow and the result is that the change in rate of product formation increases rapidly after an initial lag. Extrapolation of the linear portion of the curve to $P = 0$ yields, as indicated before, $t_{p=0} = (v_f - v_o)/k'v_f$. Assuming $v_o = 0$, k' is calculated to be 0.3 sec^{-1} and the apparent half-time is 2.3 sec for the displacement reaction under these conditions. The (——) in Fig. 2 is drawn according to Equation 11 using this value of k'.

The mechanism which should be used is, however, somewhat more complex than shown by Mechanism IV. Thus a more accurate mechanism would be

[4] It is of interest to note that the equations presented by Koshland, Nemethy, and Filmer (9), frequently used to discuss allosteric behavior, also include the assumption that particular rate constants are equal to 0 (*i.e.* $E \rightarrow E'$). Thus one would expect, in the general case, that terms in substrate raised to a power greater than the number of binding sites would appear (see Footnote 3). However, such terms are ignored in the cases discussed by Koshland *et al.* as being too small to contribute to the final equation compared with terms important in the interaction between subunits of a multichain enzyme. Thus the question of whether the principle of microscopic reversibility is valid or not does not arise.

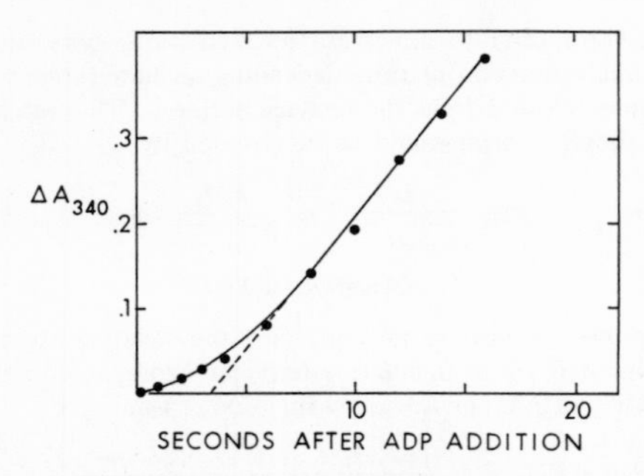

FIG. 2. Stop flow experiment in which ADP and α-ketoglutarate in one syringe are mixed with glutamate dehydrogenase, DPNH, GTP, and NH_4Cl in the other syringe. The final concentrations after mixing were enzyme 1 mg per ml; DPNH, 150 μM; NH_4Cl, 500 μM; GTP, 50 μM; α-ketoglutarate, 500 μM; ADP, 5 mM. The experiment was performed in 0.1 M Tris-acetate buffer, pH 7.4, at 10° using a Durrum stop flow spectrophotometer. ——, calculated according to Equation 11 using $k' = 0.3$ sec^{-1} and $v_o = 0$. The explanation of - - - is given in the text.

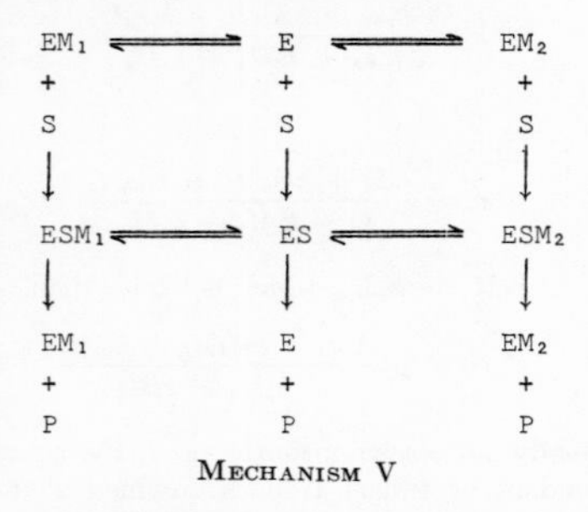

MECHANISM V

An equation identical to Equation 4 may be derived if it is assumed that substrate binding steps are equilibrated and that the conversion of ESM_1 and ESM_2 to ES is slow relative to their breakdown to give product, or if the substrate concentration is saturating so that all enzyme forms are forced into ESM_1, ES, and ESM_2.

The equation for the displacement type case quickly becomes more complex and takes a form different from Equation 4 when the enzyme involved is a multisite enzyme and the sites are not independent of each other. The treatment of such cases will not be pursued further in this paper.

Polymerizing and Depolymerizing Systems

A number of enzymes, especially those which are important in regulation of metabolic pathways, undergo reversible self-association reactions to higher molecular weight forms. Frequently, the equilibrium constant for the polymerization can be effected by ligands which are either substrates or modifiers of enzymatic activity. Previous papers have discussed the fact that anomalous (*i.e.* nonhyperbolic) binding of ligands may occur when the different molecular weight forms differ in affinity for the ligand or in the number of available binding sites (6, 7). In particular, the former case holds for the bovine liver glutamate dehydrogenase with respect to GTP binding in the presence of reduced coenzyme (20). If the different molecular weight forms have different kinetic characteristics, and if conversion between them is slow, then there should be a measurable time-dependent change in the enzymatic activity or response to modifier. The situation with respect to the description for such systems could be considerably more complicated than the isomerization or displacement mechanisms considered earlier, although under some conditions, the equations could be analogous to those already developed.

Very few studies of the rate of polymerization or depolymerization in such enzyme systems as these have been carried out, but it is clear that, at least in some cases, the rate-controlling step is quite slow and could represent a conformational change within the protein with a molecular weight change resulting as the consequence. Certainly, for studies of the rate of depolymerization, one would expect that a conformational change would precede a dissociation forced by the addition of a ligand. In such cases, the system could be treated in a manner analogous to the development already presented for the isomerization case and it might be expected that the time dependence of the velocity would follow that described by Equation 4. The situation for a polymerizing case, however, may be somewhat more complex, depending upon whether the polymerization itself is rate determining or whether it is rapid relative to conformational change which allows the enzyme to polymerize. Which process is rate limiting may depend upon the concentration of the protein. For example, at an enzyme level of 10^{-5} M or higher, the rate of polymerization *per se* may be too rapid to conveniently measure, whereas, if the enzyme concentration were as low as 10^{-7} M, polymerization of two or more monomeric forms of the enzyme could be a rather slow process. In the former case, the time dependence of the change in velocity might be described by an equation similar to Equation 4, but in the latter case, the expression would be more complex. Experimentally the two cases could be distinguished by using a range of enzyme concentrations.

If polymerization *per se* is not a rate-limiting step in the conversion between different molecular weight forms, then the system might be treated as an isomerization case and the equations developed earlier would be applicable here.

Development of the equations for the more complex system in which polymerization is rate limiting is beyond the scope of this paper, since we are primarily concerned with the hysteresis concept as such and the various mechanisms which may be involved rather than the complete kinetic description.

DISCUSSION

The rapid response of an enzyme to a change in ligand concentration is a condition which is inherent in the models currently proposed to relate the kinetic characteristics of regulatory enzyme to their observed molecular thermodynamic or initial velocity properties. The possible physiological consequences of the current mechanisms have been discussed extensively in the literature and excellent correlation may be made between the kinetic characteristics of an enzyme and its role in metabolic regulation. However, complex regulatory processes, which include systems in which different pathways utilize common intermediates or in which there are many branch points within a given pathway, may require more complex kinetic regulation of the key enzymatic steps. For such cases, the hysteretic enzyme concept may be useful. It will be apparent from the discussion below that the physiological consequences of a slow response can be quite distinct from those of a rapid response and that such responses may be important in the over-all regulation of complex metabolic processes.

Under the assumptions made in the body of this paper, and which presumably apply to an enzyme located within a sequence of events, there are several aspects of hysteretic enzyme systems which are relevant to metabolic regulation. One of these is the activity of the enzyme as a function of time at a given substrate or modifier level. If a modifier induces a slow conformational change, for example, the kinetic characteristics of two extreme cases (at $t = 0$ and $t = \infty$) may be defined by velocity studies before and well after addition of the modifier. Clearly, if the kinetic characteristics of the different forms of the enzyme differ, the activity immediately after modifier addition will be time dependent. The activity at any given time will be some weighted average of the activity of the two different forms.

Thus, under conditions in which the addition of a modifier results in the slow conversion of one form of the enzyme to another, but the activity responds immediately to changes in substrate levels, the type of substrate dependence curve obtained will be a function of time. Further, if the conversion from one form to another is slow enough, then, at any given time, the substrate dependence curve could be described by the allosteric models which assume rapid responses (8, 9).

Since a hysteretic enzyme will retain, for some period of time, some of the characteristics of the enzyme before a change in ligand concentration, it acts as a time-dependent buffer, while at the same time slowly responding to changes which will eventually alter the kinetic characteristics of the enzyme to correspond to the altered level of metabolites in the cell. The buffering capacity of such an enzyme system, which depends upon the rate of conversion of one form of the enzyme to another, serves as a mechanism to prevent immediate changes in the concentrations of other metabolites in the particular pathway, perhaps allowing the rate of other interacting pathways which use a common starting metabolite to be maintained. One would expect that such buffering systems might be common in complex metabolic regulatory systems and perhaps less frequent in pathways which could be characterized as simple biosynthetic pathways controlled by feedback inhibitors.

In order of increasing complexity, we can consider the following possible processes

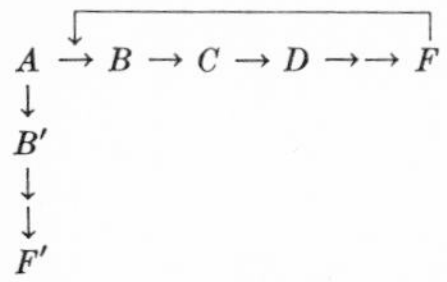

$$\begin{array}{l} A \rightarrow B \rightarrow C \rightarrow D \rightarrow\rightarrow F \quad (F \text{ feeds back to } A \rightarrow B) \\ \downarrow \\ B' \\ \downarrow \\ \downarrow \\ F' \end{array} \qquad (P1)$$

$$\begin{array}{l} A \rightarrow B \rightarrow C \rightarrow D \rightarrow\rightarrow F \quad (F \text{ feeds back to } A \rightarrow B) \\ \downarrow \leftarrow \\ B' \\ \downarrow \\ \downarrow \\ F' \quad (F' \text{ feeds back to } A \rightarrow B') \end{array} \qquad (P2)$$

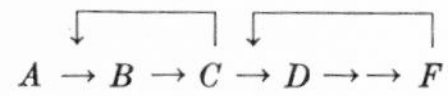

$$A \rightarrow B \rightarrow C \rightarrow D \rightarrow\rightarrow F \quad (C \text{ feeds back to } A \rightarrow B;\ F \text{ feeds back to } C \rightarrow D) \qquad (P3)$$

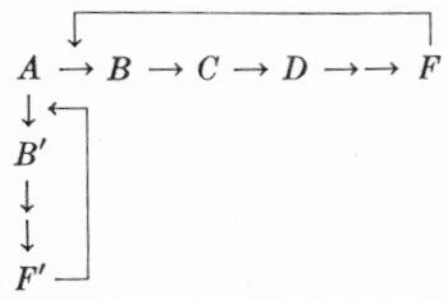

$$\begin{array}{ll} A \rightarrow B \rightarrow C \rightarrow D \rightarrow\rightarrow F & (C \text{ feeds back to } A \rightarrow B;\ F \text{ feeds back to } C \rightarrow D) \\ \downarrow \quad\ \downarrow \leftarrow & \\ B' \quad\ C'' & \\ \downarrow \quad\ \downarrow & \\ \downarrow \quad\ \downarrow & \\ F' \quad\ F'' & (F'' \text{ feeds back to } C \rightarrow C'') \end{array} \qquad (P4)$$

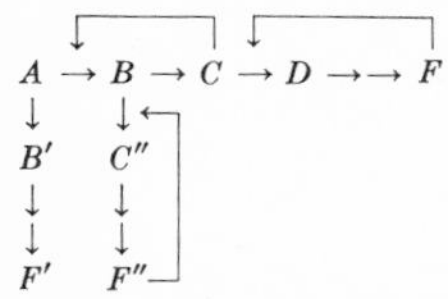

where, when indicated by an *arrow*, F, C, F', and F'' are feedback inhibitors. For purposes of discussion it is assumed (*a*) that the rate of formation of the substrate, A, is constant, (*b*) that the hysteretic response is only to the feedback inhibitor and not to the substrate, and (*c*) that the utilization of the end products, F' and F'', is constant. In the cases to be discussed we will assume that the level of F rises rapidly either because of an increase in the concentration of F itself from some other pathway (or an intermediate metabolite preceding F) or because of a decreased rate of F utilization.

Pathway $P1$, the simplest feedback system, would normally be expected to show the behavior, that, as F rises, the rate of F formation decreases and more A is available for F' production. The predicted sequence of events as a consequence of an increase in F level is shown in Table I. The symbols in the table are defined in the legend. If the step $A \rightarrow B$ does not show a hysteretic response to the feedback inhibitor, the sequence of events (shown by section labeled "None") can be read as follows: as the rate of production of B falls rapidly, $\underline{B}\downarrow$, the level of A rises rapidly, $A\uparrow$. As a consequence, the rate of formation of F decreases, the rate of formation of B' and F' rise, and the level of F' rises. All changes occur rapidly. If, on the other hand, $A \rightarrow B$ shows a hysteretic response to the inhibitor, F, then, after a rapid rise in F level the decrease in the rate of F formation will lag behind the rise in the F concentration. As shown in Table I, the rate of B formation decreases slowly, $(\underline{B}\downarrow)$, and as a consequence the rate of F formation decreases slowly (slow

TABLE I

Sequence of events for pathway P1 after rapid rise in F level

This table indicates the predicted sequence of events which may occur after a rapid rise in the level of end product F. The time scale is read from left to right. Letters placed above each other indicate that the changes occur approximately simultaneously. The conventions for rapid or slow changes in the level or rate of production of a metabolite are indicated in the footnote to this table.

Enzyme step with hysteretic response to modifier		Sequence of events after $F\uparrow$ [a] Time ⟶		
None	Rate of formation	$\underline{B}\downarrow$	$\begin{cases}\underline{F}\downarrow \\ \underline{B'}\uparrow,\end{cases}$	$\underline{F'}\uparrow$
	Level	$A\uparrow$		$F'\uparrow$
$A \rightarrow B$	Rate of formation	$(\underline{B}\uparrow)$	$\begin{cases}(\underline{F}\downarrow) \\ (\underline{B'}\uparrow),\end{cases}$	$(\underline{F'}\uparrow)$
	Level	$(A\uparrow)$		$(F'\uparrow)$

[a] Italic letter followed by ↑ or ↓ ≡ a rapid increase or decrease in level; italic letter enclosed in parentheses followed by ↑ or ↓ ≡ a slow increase or decrease in level; italic letter with underbar followed by ↑ or ↓ ≡ a rapid increase or decrease in formation; italic letter with underbar enclosed in parentheses followed by ↑ or ↓ ≡ a slow increase or decrease in formation.

TABLE II

Sequence of events for pathway P2 after rapid rise in F level

Conventions used are defined in legend and footnote to Table I.

Enzyme step with hysteretic response to modifier		Sequence of events after $F\uparrow$ Time $\longrightarrow$		
None	Rate of formation	$\underline{B}\downarrow \left\{\begin{matrix}\underline{F}\downarrow \\ \underline{B}'\uparrow\end{matrix}\right. \underline{F}'\uparrow$		$\underline{B}'\downarrow$ $\underline{F}'\downarrow$
	Level	$A\uparrow$	$F'\uparrow$	
$A \rightarrow B$	Rate of formation	$(\underline{B})\downarrow \left\{\begin{matrix}(\underline{F}\downarrow) \\ (\underline{B}'\uparrow),\end{matrix}\right. (\underline{F}'\uparrow)$		$(\underline{B}'\downarrow)$, $(\underline{F}'\downarrow)$
	Level	$(A\uparrow)$	$(F'\uparrow)$	
$A \rightarrow B'$	Rate of formation	$\underline{B}\downarrow$ $\underline{F}\downarrow$ $\underline{B}'\uparrow$ $\underline{F}'\uparrow$		$(\underline{B}\downarrow)$, $(\underline{F}'\downarrow)$
	Level	$A\uparrow$	$F'\uparrow$	

TABLE III

Sequence of events for pathway P3 after rapid rise in F level

Conventions used are defined in legend and footnote to Table I.

Enzyme step with hysteretic response to modifier		Sequence of events after $F\uparrow$ Time $\longrightarrow$		
None	Rate of formation	$\underline{D}\downarrow$	$\left\{\begin{matrix}\underline{F}\downarrow \\ \underline{B}\downarrow,\end{matrix}\right.$	$\underline{C}\downarrow$
	Level	$C\uparrow$		$A\uparrow$
$A \rightarrow B$	Rate of formation	$\underline{D}\downarrow$	$\left\{\begin{matrix}\underline{F}\downarrow \\ (\underline{B}\downarrow),\end{matrix}\right.$	$(\underline{C}\downarrow)$
	Level	$C\uparrow$		$(A\uparrow)$
$C \rightarrow D$	Rate of formation	$(\underline{D}\downarrow)$	$\left\{\begin{matrix}(\underline{F}\downarrow) \\ (\underline{B}\downarrow)\end{matrix}\right.$	$(C\downarrow)$
	Level	$(C\uparrow)$		$(A\uparrow)$

changes indicated by the parentheses). Concurrently, the level of A rises slowly and, as a consequence, the rate of F' formation and the F' level rise slowly. It should be noted that the extent of the rise in F' level in this case may depend on the rate of utilization of F', but we are assuming that the utilization rate is constant and independent of end product concentration.

The importance of a hysteretic enzyme in an isolated system like $P1$ is questionable and the other systems, $P2$ through $P4$, are of more interest. Table II shows the predicted sequence of events for $P2$ when either $A \rightarrow B$ or $A \rightarrow B'$ shows a hysteretic response to the inhibitor. Note that when $A \rightarrow B'$ is hysteretic the level of F' could rise rapidly followed by a slow decrease in the rate of F' formation. On the other hand, the rate of F formation falls rapidly in this system. If the $A \rightarrow B$ step is hysteretic, the rate of formation of both F and F' will change slowly.

Table III shows the sequence of events for pathway $P3$ after a rapid rise in the level of F. Note that if the hysteretic response is at the $A \rightarrow B$ step, an increase in F will cause a rapid rise in the concentration of C, but the rate of C formation will then decrease slowly. Thus a high level of C will be conserved. If the $C \rightarrow D$ step is hysteretic, C is also conserved but at a lower level.

Pathway $P4$ represents a more likely combination of metabolic steps than do the preceding three systems. In this case, there are multiple branch points and more than one feedback inhibitor. Table IV shows some possible predictions of the sequence of events after a rapid rise in F. Examination of Table IV shows that it is possible to buffer out rapid changes in the formation of either F and F' or F'', or all of these, depending upon which enzymatic step responds slowly to its inhibitor.

The predictions listed in the tables must be viewed with a great deal of care. A number of assumptions have been made to help define the systems and different assumptions could give rise to different predictions. Thus, differences could arise if one of the inhibitor levels was initially high enough to block that pathway, if an increased A level could overcome the F inhibitor after the rise in F concentration, if the pool sizes of intermediates were different, if the hysteretic response were either to the inhibitor or to the substrate, or both, if the utilization of the end product or production of the initial substrate was not constant, or if an end product, or some other intermediate, could activate an enzymatic step or displace an inhibitor, and so on. The essential point, however, is that it is possible to buffer out rapid changes in end product concentrations and the exact prediction would depend upon the characteristics of the system under discussion. It is also to be noted in reference to these tables that the concepts of a slow increase or decrease is relative only to the rate of formation of the enzyme induced by the inhibitor and defined by Equation 4 and relative to the rate of enzymatic reaction. The tables indicate that qualitatively the same final results would be achieved regardless of whether the pathways contain a hysteretic enzyme step or not. In actuality, this may not be the case at all because of changing rates of utilization of the end products.

The diagram $P4$ can be considered to be similar to the pathway involved in utilization of aspartate in *Escherichia coli* if $A \equiv$ aspartic semialdehyde, $F' \equiv$ lysine, $F'' \equiv$ methionine, and $F \equiv$ threonine. (For the sake of accuracy, F should go on to another metabolite, isoleucine, which is a feedback inhibitor of the threonine deaminase, and the step $A \rightarrow C'$ is controlled by F'.) Stadtman (2) has pointed out that feedback control of

TABLE IV

Sequence of events for pathway P4 after rapid rise in F level

Conventions used are defined in legend and footnote to Table I.

Enzyme step with hysteretic response to modifier		Sequence of events after $F\uparrow$ Time ⟶					
None	Rate of formation	$\underline{D}\downarrow$	$\underline{F}\downarrow$ $\underline{B}\downarrow$		$\underline{B'}\uparrow$ $\underline{C''}\downarrow$	$\underline{F'}\uparrow$ $\underline{F''}\downarrow$	
	Level	$C\uparrow$		$A\uparrow$ $B\downarrow$			$F'\uparrow$ $F''\downarrow$
$A \rightarrow B$	Rate of formation	$\underline{D}\downarrow$	$\underline{F}\downarrow$ $\underline{B}\downarrow$			$(\underline{F'}\uparrow)$ $(\underline{F''}\downarrow)$	
	Level	$C\uparrow$		$(A\uparrow)$ $(B\downarrow)$[a]			$(F'\uparrow)$ $(F''\downarrow)$
$C \rightarrow D$	Rate of formation	$(D\downarrow)$	$(\underline{F}\downarrow)$ $(\underline{B}\downarrow)$			$(\underline{F'}\uparrow)$ $(\underline{F''}\downarrow)$	
	Level	$(C\uparrow)$		$(A\uparrow)$ $(B\downarrow)$			$(F'\uparrow)$ $(F''\downarrow)$
$B \rightarrow C''$	Rate of formation	$\underline{D}\downarrow$	$\underline{F}\downarrow$ $\underline{B}\downarrow$			$\underline{F'}\uparrow$ $\underline{F''}\downarrow$	
	Level	$C\uparrow$		$A\uparrow$ $B\downarrow$			$F'\uparrow$ $F''\downarrow$

[a] Unless piling up C causes $B\uparrow$, in which case the sequence would be $B\uparrow$, $F''\uparrow$, $\underline{F}''\downarrow$, $(\underline{B}\downarrow)$, $(F''\downarrow)$.

$A \rightarrow B$ (homoserine dehydrogenase) by threonine is disadvantageous because it also blocks the production of methionine (F'') and, in fact, methionine is required for normal growth rates when the organism is grown on excess threonine (21). If, on the other hand, the step $A \rightarrow B$ were controlled by a hysteretic enzyme, the methionine level, over a short period of time at least, would be conserved. Under such conditions normal protein biosynthesis with excess threonine might occur and the system might be able to catch up with itself without significantly decreasing the level of methionine. In other words, if homoserine dehydrogenase were a hysteretic enzyme, the change in the rate of formation of methionine would be buffered against changes in threonine levels. It is of interest that the threonine inhibition of the *E. coli* homoserine dehydrogenase is in fact time dependent (22) (see Table V). On the other hand, if the threonine deaminase is hysteretic, the concentrations of all the products, F, F', and F'' (lysine, methionine, and threonine), are buffered against changes in isoleucine levels. In *Bacillus subtilis*, the threonine deaminase is a hysteretic enzyme, responding slowly to a rapid increase in the concentration of the inhibitor isoleucine (24).

These examples are presented to show the value of the hysteretic enzyme concept in control of metabolic regulation. There are undoubtedly many more possibilities and the pathways chosen above are certainly not the most complex ones. Furthermore, the tables give only what might be expected if only one enzyme of those involved shows hysteretic behavior. Better control, however, would be obtained if more than one enzyme was hysteretic.

Two examples of hysteretic enzymes are given above, but examination of the literature reveals a number of enzymes which might be classed as hysteretic enzymes. A partial list of such enzymes, and some of their characteristics, is shown in Table V. It is of interest that this table includes all of the mechanisms which have been proposed in this paper for hysteretic enzymes. For example, the increase in activity of the yeast glyceraldehyde 3-phosphate dehydrogenase after DPN addition is an example of isomerization, but a similar observation for phosphorylase *a* on glycogen addition appears to result from depolymerization of the enzyme. The slow rise in activity of phosphofructokinase on raising the pH from 6 to 7.5 or above appears to be a result of polymerization of less active molecular weight forms; on the other hand the slow increase in activity (also shown in Fig. 2) on addition of ADP to GTP-inhibited glutamate dehydrogenase appears to be due to the slow release of GTP from the enzyme surface. Several other points can be made about the enzymes listed in the table. First, most, if not all, of the enzymes involved are regulatory enzymes in the sense that they occur at a sensitive point in a metabolic pathway and frequently show allosteric (*i.e.* sigmoidal) kinetic behavior. Second, the approximate times involved for the interconversion of different forms may vary from seconds to minutes (perhaps hours) and there appears to be no way to predict *a priori* what these times will be. Third, the table does not include other possible slow interactions like those for hemocyanin (from changes in P_{O_2}) (33); for hemoglobin (after ATP addition to deoxy Hb) (34); or for the myosin-actin interactions (35). Nor does it include a rather large class of cold-labile enzymes which can be reactivated by warming (see Reference 36, for example) and for which the reactivation may be a relatively slow process. Furthermore, Table V does not include slow refolding processes which occur on renaturing enzymes from their denatured forms. These latter cases, of which there are many examples, probably fall into a very different class with respect to their importance

TABLE V

Partial list of enzymes which appear to show a hysteretic response

The following list of enzymes are those which appear to show a hysteretic response as discussed in the text. The half-times are given only in terms of seconds or minutes since they depend upon a number of factors including temperature, buffer, pH, and so on. Not all of the hysteretic responses for a given enzyme are indicated in the table nor are all of the enzymes which show such a response listed here.

Enzymes	Time dependent change		Approximate half-times	Type	Reference
	Occurs on	Appears as			
Glyceraldehyde 3-phosphate dehydrogenase (yeast)	Addition of DPN	Increase in velocity	Seconds	Isomerization	(23)
Homoserine dehydrogenase					
E. coli	Addition of threonine	Inhibition	Seconds	Isomerization	(22)
Rhodospirillum rubrum	Enzyme dilution	Loss of threonine inhibition	Minutes	Molecular weight change	(25)
Threonine deaminase					
E. coli	Addition of AMP	Inhibition	Minutes	Molecular weight change	(26)
B. subtilis	Addition of threonine to enzyme isoleucine complex	Increase in velocity	Minutes	Isomerization	(24)
	or addition of isoleucine to running reaction	Inhibition	Minutes	Isomerization	
Glutamate dehydrogenase (bovine liver)	Addition of ADP to GTP-inhibited enzyme	Increase in velocity	Seconds	Displacement	**Fig. 2**
Phosphorylase *a* (rabbit muscle)	Addition of glycogen	Increase in velocity	Minutes	Depolymerization	(27)
Phosphofructokinase (rabbit muscle)	Raising pH from 6 → 7.5	Increase in velocity	Minutes	Polymerization	(13)
Acetyl-CoA carboxylase (rat adipose tissue)	Addition of citrate	Increase in velocity	Minutes	Polymerization	(28)
DPNH oxidase (*Mycobacterium tuberculosis*)	Dissociation of AMP	Change in extent of AMP activation	Minutes	Isomerization	(29)
UDPG Hydrolase (*E. coli*)	Addition of protein fraction	Inhibition	Minutes	Isomerization of added protein?	(30)
Glutamine-PP-ribose-P amido transferase	Addition of substrate	Increase in velocity	Minutes	Isomerization	(31)
Lactic dehydrogenase	Addition of pyruvate as product	Inhibition	Seconds	Abortive complex formation	(32)

in regulatory processes. Fourth, the large number of such enzymes is a surprise since investigators have not systematically examined regulatory enzymes for this type of effect.

It would be hoped that systematic investigation might be undertaken in regard to the possibility of a hysteretic response with respect to regulatory enzyme systems. Such investigation could take the form of examination for time-dependent changes in activity, either as a lag in the product formation after substrate or effector addition or after dilution of the enzyme, or a lag in the change in activity when displacing one allosteric modifier by another. Of particular interest is the question of whether preliminary incubation of the enzyme with a substrate or effector will alter any observed lag. However, it is possible for a hysteretic enzyme that even preliminary incubation would not decrease the lag time if both the substrate and modifier are required to induce the interconversion. Such an effect could be termed concerted hysteresis. While this paper discusses the cases where the substrate concentration is not maximal, experimentally, velocity measurements at very high substrate levels may be required if one is to watch the reaction over a long period of time. If it is known that different molecular weight species have significantly different kinetic or ligand-binding characteristics, an examination of the rate of molecular weight change may be in order. One can also examine the enzyme system for time-dependent conformational changes (as indicated by spectral shifts or change in reactivity of amino acid residues) but in such cases, a clear correlation between these changes and a change in the kinetic or ligand-binding characteristics should be established.

There are, of course, pitfalls in such experiments as these, since it is always possible that the changes one observes are not metabolically significant. For activity measurements one must avoid conditions in which product accumulation or substrate depletion would alter the rate of the reaction, but attempts to do so by using coupled assay systems may present difficulties inherent in the coupling system.

It may also be helpful at this point to discuss briefly the concept of slow and rapid changes as they have been used in this paper. Arbitrarily, it is most convenient to relate such changes to the time required to measure the actual rate of enzymatic reaction as measured by product formation or substrate disappearance under the particular conditions. It will be noted that the enzymes of Table V undergo changes which in general take at least seconds if not minutes or hours. At saturating levels of substrate and modifier, the terms fast and slow could, of course, be related to the turnover number of the enzyme under those conditions. Since turnover numbers vary greatly and since it is not a requirement, under the assumptions used, to saturate the system, relating rates of conversion of one form to another to the time required to make an initial velocity measurement is a convenient definition. It should also be pointed out again that equilibration of certain steps in the mechanism, *i.e.* binding steps, is indeed probably quite rapid relative to the time chosen to define slow conversion of one form to another. Thus the kinetic treatment, as described in this paper, is probably valid and avoids some of the more complex equations which arise when no assumptions are made as to what might be a rate-limiting step in an enzymatic reaction.

The examples of the importance of hysteretic enzymes to metabolic pathways given above are discussed in terms of the assumption that the level of one of the end products or intermediates in a pathway rises rapidly. The term rapid as used here may be misleading in the sense that changes in such concentrations need only be faster than the rate of the hysteretic response. Thus, the level of an intermediate which is common to two pathways may change as a consequence of metabolic process in one of the pathways and the change could be a relatively slow one. However if the change is faster than the hysteretic response in the other pathway, it may be considered as a rapid change relative to that pathway. Table V shows that some of the responses of hysteretic enzymes may be quite slow, on the order of minutes or longer, and large changes in the flow of metabolites through a given pathway may occur in such times. In this regard, the question of fast or slow changes may only be relative to the rate of substrate flow through a given pathway under a given set of conditions. It is perhaps this latter information which will be of importance in correlating hysteretic response to complex metabolic processes.

The introduction of the concept of hysteretic enzymes clearly adds another dimension to the question of metabolic regulation *in vivo*. Although it is difficult at this time to establish any relation between the degree of hysteresis an enzyme may exhibit and the complexity of the regulatory process, it would appear that the concept is most important in relation to complex metabolic pathways. It is hoped that this concept, rather than complicating the issue of regulation in the cell, will eventually lead to some simplification of these complex systems.

REFERENCES

1. Monod, J., Changeux, J.-P., and Jacob, F., *J. Mol. Biol.*, **6**, 306 (1963).
2. Stadtman, E. R., *Advan. Enzymol.*, **28**, 41 (1966).
3. Atkinson, D. E., *Ann. Rev. Biochem.*, **35**, 85 (1966).
4. Scrutton, M. C., and Utter, M. F., *Ann. Rev. Biochem.*, **37**, 249 (1968).
5. Umbarger, H. E., *Ann. Rev. Biochem.*, **38**, 323 (1969).
6. Frieden, C., *J. Biol. Chem.*, **242**, 4045 (1967).
7. Nichol, L. W., Jackson, W. J. H., and Winzor, D. J., *Biochemistry*, **6**, 2449 (1967).
8. Monod, J., Wyman, J., and Changeux, J.-P., *J. Mol. Biol.*, **12**, 88 (1965).
9. Koshland, D. E., Jr., Nemethy, G., and Filmer, D., *Biochemistry*, **5**, 365 (1966).
10. Kirtley, M. E., and Koshland, D. E., Jr., *J. Biol. Chem.*, **242**, 4192 (1967).
11. Volkenstein, M. V., and Goldstein, B. N., *Biochim. Biophys. Acta*, **115**, 478 (1966).
12. Sweeny, J. R., and Fisher, J. R., *Biochemistry*, **7**, 561 (1967).
13. Frieden, C., in Kvamme, E. and Pihl, A. (Editors), *Regulation of enzyme activity and allosteric interactions*, Academic Press, New York, 1968, p. 59.
14. Hearnon, J. F., Bernard, S. J., Friess, S. C., Botts, D. J., and Morales, M. F., in *The enzymes, Vol. I*, Academic Press, New York, 1959, p. 49.
15. Frieden, C., *J. Biol. Chem.*, **239**, 3522 (1964).
16. Metzger, B. E., Glaser, L., and Helmreich, E., *Biochemistry*, **6**, 2021 (1968).
17. Rabin, B. R., *Biochem. J.*, **102**, 222 (1967).
18. Weber, G. in Pullman, B. and Weissbluth, M. (Editors), *Molecular biophysics*, Academic Press, New York, 1965, p. 205.
19. Ray, W. J., and Hatfield, G. W., *J. Biol. Chem.*, **245**, 1753 (1970).
20. Frieden, C., and Colman, R. F., *J. Biol. Chem.*, **242**, 1705 (1967).
21. Patte, J. C., LeBras, G., Loving, T., and Cohen, G. N., *Biochim. Biophys. Acta*, **67**, 16 (1963).
22. Barber, E. D., and Bright, H. J., *Proc. Nat. Acad. Sci. U. S. A.*, **60**, 1370 (1968).
23. Kirschner, K. in E. Kvamme and A. Pihl (Editors), *Regulation of enzyme activity and allosteric interactions*, Academic Press, New York, 1968, p. 39.
24. Hatfield, G. W., and Umbarger, H. E., *J. Biol. Chem.*, **245**, 1742 (1970).
25. Mankovitz, R., and Segal, H. L., *Biochemistry*, **8**, 3757 (1969).
26. Gerlt, J. A., and Rabinowitz, K. W., *Fed. Proc.*, **29**, 400 (1970).
27. Metzger, B., Helmreich, E., and Glaser, L., *Proc. Nat. Acad. Sci. U. S. A.*, **57**, 994 (1967).
28. Vagelos, P. R., Alberts, A. W., and Martin, D. B., *J. Biol. Chem.*, **238**, 533 (1963).
29. Worcel, A., Goldman, D. S., and Cleland, W. W., *J. Biol. Chem.*, **240**, 3399 (1965).
30. Glaser, L., Melo, A., and Paul, R., *J. Biol. Chem.*, **242**, 1944 (1967).
31. Rowe, P. R., Coleman, M. D., and Wyngaarden, J. B., *Biochemistry*, **9**, 1498 (1970).
32. Gutfreund, H., Cantwell, R., McMurray, C. H., Criddle, R. S., and Hatheway, G., *Biochem. J.*, **106**, 683 (1968).
33. DePhillips, H. A., Nickerson, K. W., and Van Holde, K. E., *Fed. Proc.*, **29**, 336 (1970).
34. Lo, H. H., and Schimmel, P. R., *J. Biol. Chem.*, **244**, 5084 (1969).
35. Finlayson, B., Lymn, R. W., and Taylor, E. W., *Biochemistry*, **8**, 811 (1969).
36. Irias, J. J., Olsmsted, M. R., and Utter, M. F., *Biochemistry*, **8**, 5136 (1969).

Reprinted from THE JOURNAL OF BIOLOGICAL CHEMISTRY, Vol. 240, 1965

The Acyl-enzyme Intermediate and the Kinetic Mechanism of the Glyceraldehyde 3-Phosphate Dehydrogenase Reaction*

CHARLES S. FURFINE† AND SIDNEY F. VELICK‡

From the Department of Biological Chemistry, Washington University School of Medicine, St. Louis 10, Missouri

(Received for publication, August 7, 1964)

The acyl-enzyme compound formed by D-glyceraldehyde 3-phosphate dehydrogenase has been studied by several workers in a variety of ways (1), primarily in experiments at high enzyme concentrations and in many instances with model substrates. In the present investigation an attempt is made to apply kinetic criteria to the mechanism of the reversible oxidative phosphorylation of the natural substrates. Although the oxidative and acyl transfer steps are clearly separable at substrate level enzyme concentrations, the initial velocity kinetics at low enzyme concentration is described by a random order of substrate addition to enzyme with the formation of a kinetically important quaternary enzyme-substrate complex. One does not obtain, in initial velocity measurements, the same kinetic signs of group transfer through an enzyme-bound intermediate that are observed, for example, with the glutamate-oxaloacetate transaminase (2). It is found that the rate-limiting step in glyceraldehyde 3-phosphate oxidation at pH 7.4 is the acyl group transfer from a site on the enzyme to an external acceptor. The properties of the partial reactions are such that in the physiological pH region the kinetics is dominated by a high steady state concentration of the intermediate. There is a large effect of the intermediate on the atypical interactions of pyridine nucleotide with enzyme, apparently mediated by protein isomerizations. Events which might not be recognized in a purely kinetic approach are clarified by an examination of isolated enzyme-substrate interactions and of the equilibria of the partial reactions in which the enzyme is a stoichiometric participant.

EXPERIMENTAL PROCEDURE

DL-Glyceraldehyde 3-Phosphate—This compound was obtained as the barium salt of the diethylacetal from Sigma Chemical Company and from California Corporation for Biochemical Research. It was converted to the free aldehyde by heating 20 mg for 3 minutes at 100° in 1.5 ml of an exhaustively washed and recycled suspension of Dowex 50, 20% by volume, in the acid form. Total concentration of the aldehyde in the chilled filtrate was determined as alkali-labile phosphate (3). The concentration of the D isomer was determined enzymatically in a solution which contained 0.1 M sodium carbonate-bicarbonate, pH 8.6, 0.01 M sodium arsenate, 1.5 mM DPN, and enzyme sufficient to catalyze complete reaction in less than 1 minute. Absorbance changes were read at 340 mμ in a Beckman DU or Zeiss spectrophotometer.

3-Phosphoglyceroyl Phosphate—This derivative was prepared by the oxidative phosphorylation of the aldehyde. Conditions for the oxidation were those of Negelein and Bromel (4) except that 0.02 M sodium pyruvate and lactic dehydrogenase, 0.06 mg per ml, were substituted for acetaldehyde and alcohol dehydrogenase to reoxidize the DPNH. The crude product, precipitated with acid-acetone as described by the above authors, was dissolved in a small volume of cold water and brought to pH 7 by the addition of solid imidazole. Diluted aliquots containing about 20 μmoles in 20 ml of 0.01 M imidazole chloride, pH 7.5, were placed on a column (0.9 × 10 cm) of DEAE-Sephadex, "fine" grade. The column was developed by a gradient in which 0.6 M sodium chloride in the imidazole buffer was fed into a mixer containing 200 ml of buffer. Fractions of approximately 4 ml were collected at a rate of 0.7 ml per minute. The product appeared in tubes 34 to 37 in about 65% yield and contained no detectable nucleotide or aldehyde. Aliquots from each tube were assayed spectrophotometrically in test solutions which contained 0.15 mM DPNH and sufficient enzyme for rapid reaction at pH 7. The acyl phosphate can be stored frozen at pH 7.0 to 7.4 but must be used promptly after melting.

Pyridine Nucleotides—The presence of nucleotide impurities in DPN, up to 4% in the best samples examined (5), was confirmed. In the present case, however, there was no sign of inhibitory action of the impurities. DPNH (Sigma) before or after chromatography on DEAE-cellulose contained no detectable inhibitors and was used directly. Concentrations of the 3-acetylpyridine analogue of DPN were determined spectrophotometrically as the cyanide complex (6) and also by enzymatic reduction at pH 9.6 in 0.1 M glycine-10 mM arsenate-1 mM glyceraldehyde-3-P and 50 μg of enzyme per ml with measurement of absorbance changes at 363 mμ. The 3-acetylpyridine analogue of DPNH was prepared with ethanol and yeast alcohol dehydrogenase (7) and purified on columns identical with those employed for 3-phosphoglyceroyl phosphate purification.

Enzyme—Enzyme was prepared from rabbit skeletal muscle (8) with 4 mM EDTA present in all solvents, and was recrystallized three or more times. Dilute stock solutions, when kept cold, lost approximately 5% of initial activity over the course of several hours, for which corrections were made. Published absorption coefficients (9) were used in determining protein concentrations spectrophotometrically. In experiments in-

* This work was supported in part by Research Grant H-02732 from the National Heart Institute, National Institutes of Health.
† Predoctoral Fellow of the National Institutes of Health. Present address, Department of Biochemistry, Albert Einstein College of Medicine, New York 61, New York.
‡ Present address, Department of Biological Chemistry, University of Utah College of Medicine, Salt Lake City, Utah.

volving AcPyDPN,[1] the bound DPN of the enzyme was first removed by treatment with charcoal.

Commercial preparations (Sigma) of triosephosphate isomerase and α-glycerophosphate dehydrogenase contained no interfering enzyme activities and could be used directly. However, crystalline preparations of phosphoglycerate kinase from yeast were heavily contaminated with the isomerase and also with the glyceraldehyde-3-P dehydrogenase. These impurities were removed by ion exchange chromatography. It may be noted that the dehydrogenase from rabbit muscle, after several recrystallizations, still contains traces of the isomerase. The contamination is not detectable at the low enzyme concentrations used in kinetic work, but may be sufficient to equilibrate the triose phosphates in experiments of long duration carried out at high enzyme concentration, for example, during ultracentrifugation or dialysis equilibrium.

Fluorometric Method—Rates of DPN reduction were measured in a Farrand model A fluorometer with a low pressure mercury arc source, primary filter Corning No. 5860 and secondary filters Corning Nos. 3385 and 4308. Reactions in a final volume of 1 ml were run in test tubes and were initiated by the addition of enzyme. The temperature of the tube holder was controlled by a water jacket and circulating water bath. An adjustable zero, adjustable range recorder (Leeds and Northrup) allowed, at highest sensitivity, a full scale deflection from a change in DPNH concentration of 0.5 μM. This sensitivity was frequently necessary to minimize product inhibition and to determine the smaller Michaelis constants. A fluorescent impurity in the imidazole buffer was reduced to a low level by filtration of stock imidazole solutions through Norit A. To minimize troublesome noise from dust particles at high instrumental sensitivities, the test solutions were also filtered through Whatman No. 50 filter paper. In order to obtain equivalent sensitivities in studying the reverse reaction fluorometrically, initial DPNH concentrations had to be kept below 5 μM to avoid phototube fatigue and drift.

Absorption Spectrophotometry—Kinetic measurements were also made in a Beckman DU spectrophotometer with cuvettes of 10-cm light path (6-ml volume) and with the same recorder that was employed with the fluorometer. At the maximal scale expansion a change of 1 μM in DPNH concentration gave a full scale recorder deflection. This instrument was used primarily with the AcPyDPN reactions because of the higher absorption coefficient and lower fluorescence quantum yield of AcPyDPNH. It was also useful in exploring higher concentration ranges of DPNH than were practical fluorometrically.

Fluorometric Titration—In these experiments the protein is excited by light of any wave length within the tryptophan absorption bands, and the right angle fluorescence is measured at 350 mμ as a function of the concentration of added DPN (10). In order to minimize undesired absorption of the excitation beam in the tail of the 260 mμ absorption band of DPN, the excitation was done at 305 mμ in a quartz cuvette of 2-mm light path. Even under these conditions absorption corrections are required, based on measured absorbances at the wave lengths of interest, when the DPN concentration approaches or exceeds 100 μM. An Aminco spectrophotofluorometer was employed with an efficient alternating current voltage regulator on the photometer unit and the recently redesigned high pressure xenon arc (Hanovia), which exhibited less than 1% fluctuation on a rectified alternating current power supply. DPN additions to enzyme were made from a microsyringe in continuous titrations.

[1] The abbreviations used are: AcPyDPN and AcPyDPNH, the 3-acetylpyridine analogue of DPN and its reduced form.

RESULTS

Kinetic Results

Limiting Conditions—The equilibrium of the oxidative phosphorylation (Equation 1) favors aldehyde and DPN in the physiological pH region.

$$R{-}CHO + DPN^+ + HPO_4^{=} \leftrightarrow R{-}CO{-}OPO_3^{=} + DPNH + H^+ \quad (1)$$

It is therefore easy to measure initial velocities of acyl phosphate reduction by DPNH, but satisfactory initial velocity measurements of aldehyde oxidation require substrate concentrations that are too high to be of primary mechanistic interest. The equilibrium restriction at pH 7.4 can be avoided by using arsenate as external acyl acceptor in place of phosphate. This adds a hydrolytic step to the reaction and leads to the irreversible formation of 3-phosphoglycerate. The alternative way of circumventing the equilibrium restriction at this pH is to take advantage of the high oxidation-reduction potential of the 3-acetylpyridine analogues of DPN and DPNH (11). Both orthophosphate ion and the natural coenzyme may be used at pH 8.6. The equilibrium at the alkaline pH is more favorable for aldehyde oxidation, and secondary substrate and product effects are also greatly diminished. Each of these conditions has been examined.

Graphical Analysis—The general procedure in the three-substrate or forward reaction direction is to measure initial velocities as a function of the concentration of one substrate at a constant concentration of a second substrate and a series of fixed concentrations of the third. Reciprocal plots of velocity against concentration from the data of such experiments yield sets of lines which may be parallel or which may intersect at a common point in a left-hand quadrant. The former event in a three-substrate reaction provides evidence for a compulsory sequence of substrate addition to enzyme, but under all adequately defined conditions that have been examined the latter relation prevails.

Aldehyde Oxidation at pH 8.6—Reciprocal plots of the data obtained at pH 8.6 are collected in Fig. 1. The experiments illustrated all involve arsenate as the external acyl group acceptor. Results of the identical form are obtained when orthophosphate is employed. The Michaelis constants of DPN and glyceraldehyde-3-P are unaltered in phosphate, and essentially the same maximal velocity is observed. Within the limits of experimental error the families of lines all intersect on the negative abscissa. The negative reciprocal of the intersection point gives a Michaelis constant directly, as conventionally defined. Over a relatively wide concentration range the Michaelis constant of each substrate is seen to be independent of the concentration of the cosubstrates.

Aldehyde Oxidation at pH 7.4—The data obtained in the physiological pH region and plotted in Fig. 2 take a form identical with the results at the alkaline pH, but the concentration ranges are different and in certain instances are much more limited. In particular, the Michaelis constant of DPN is much higher than at pH 8.6 and that of aldehyde much lower. The range of aldehyde and arsenate concentrations employed in Fig. 2 is

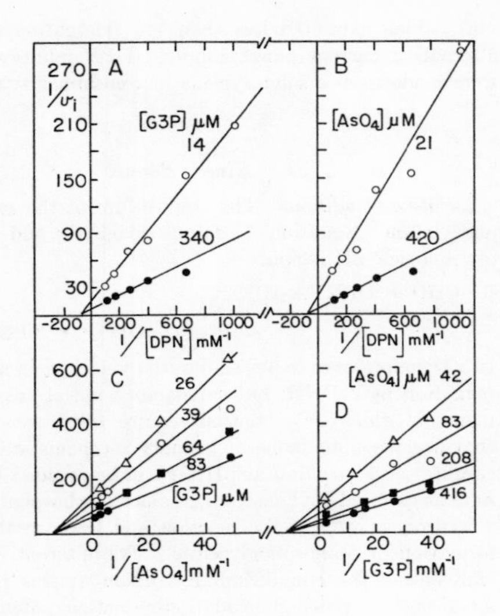

FIG. 1 (*left*). Kinetics of DPN reduction by glyceraldehyde-3-P (*G3P*) at pH 8.6 in 0.1 M sodium carbonate bicarbonate buffer at 26°. *A*, reciprocal plots of velocity against DPN concentration at 420 μM arsenate and a series of fixed aldehyde concentrations; *B*, reciprocal plots of velocity against DPN concentration at 340 μM aldehyde and a series of fixed arsenate concentrations; *C* and *D*, reciprocal plots of velocity against substrate concentrations at 114 μM DPN and other variables as indicated. The enzyme concentration in *A* and *B* was 2.9×10^{-9} M, and in *C* and *D* 6.1×10^{-10} M. The *ordinates* in this and in the following reciprocal plots of kinetic data are in units of (millimicromoles per ml per second)$^{-1}$ unless otherwise indicated.

FIG. 2 (*right*). Kinetics of DPN reduction by glyceraldehyde-3-P (*G3P*) at pH 7.4 in 0.1 M imidazole (chloride) buffer at 26°. In *A*, the arsenate concentration was held constant at 15 μM and aldehyde was varied as indicated; in *B*, the aldehyde was held at 4.2 μM at a series of fixed arsenate concentrations. Enzyme concentration in *A* and *B* was 2.8×10^{-9} M. In *C* and *D*, the enzyme concentration was 1.3×10^{-9} M and the DPN concentration was constant at 260 μM.

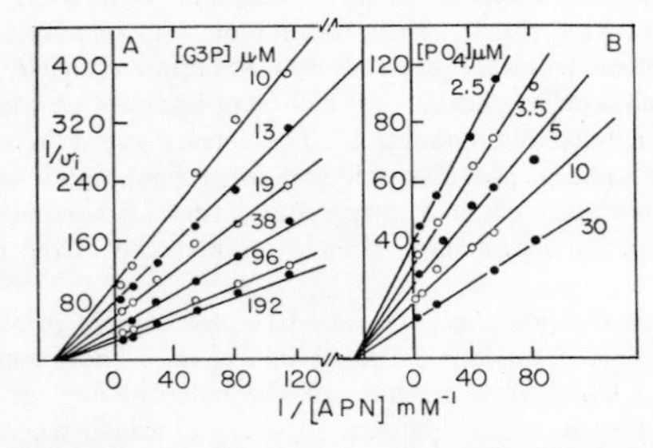

FIG. 3. Kinetics of reduction by glyceraldehyde-3-P (*G3P*) of the 3-acetylpyridine analogue of DPN (*APN*) at pH 7.4 and 26° in 0.1 M imidazole buffer. In *A*, the phosphate concentration was 30 mM and the enzyme concentration 2×10^{-8} M; in *B*, the aldehyde concentration was 164 μM and the enzyme concentration 4.8×10^{-8} M.

restricted to avoid substrate activation and inhibition effects, which are described in a following section.

Reciprocal plots of kinetic data obtained at pH 7.4 with AcPyDPN and phosphate, respectively, as the hydrogen and acyl acceptors in aldehyde oxidation are shown in Fig. 3. The Michaelis constant of AcPyDPN is independent of aldehyde and phosphate concentrations, and the kinetic parameters of the latter substrates are independent of the concentration of AcPyDPN.

3-P-Glyceroyl-P Reduction at pH 7.4—Fig. 4 illustrates the

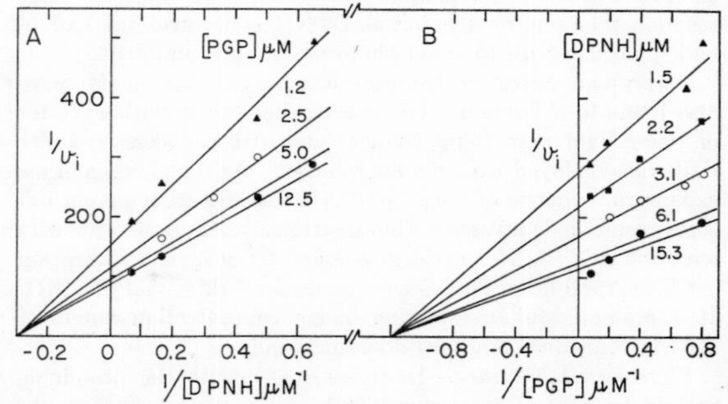

FIG. 4. Kinetics of the reduction of 3-phosphoglyceroyl phosphate (*PGP*) by DPNH at pH 7.4 in 0.1 M imidazole at 26°, at an enzyme concentration of 4.6×10^{-11} M. The Michaelis constant of each substrate is independent of the concentration of the other.

initial velocity kinetics of the reverse reaction, between the acyl phosphate and DPNH. The results take the same form as those of the forward reaction and are characterized by particularly small Michaelis constants for both substrates.

Secondary Substrate Effects

Enzyme Activation by Arsenate—The responses in initial velocity to changes in arsenate concentration at pH 7.4 differ in three separate arsenate concentration ranges and are illustrated in Fig. 5 by the effects of arsenate upon the kinetic parameters of glyceraldehyde-3-P. At arsenate concentrations below 20 μM, $K_{glyceraldehyde\text{-}3\text{-}P}$ is independent of arsenate concentration. In the arsenate concentration range of 50 to 500 μM, there is a pronounced secondary enzyme activation as shown in the *lower pair of lines* of Fig. 5A. $K_{glyceraldehyde\text{-}3\text{-}P}$ in this concentration range is dependent on arsenate concentration. When arsenate is used in the millimolar concentration range (Fig. 5B), the high maximal velocity still prevails but arsenate is also seen to be an inhibitor, competitive with aldehyde. Reciprocal plots of related data in which arsenate is the independent variable are shown in Fig. 6. It is seen in Fig. 6A that in the concentration region in which arsenate is an activator K_{AsO_4} is a function of aldehyde concentration. Fig. 6B shows a secondary plot for maximal velocity covering the normal and activating regions of arsenate concentration.

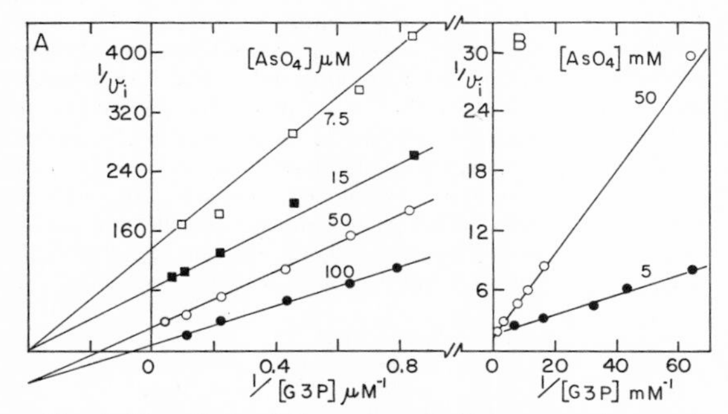

FIG. 5. The effects of three concentration ranges of arsenate on the kinetic behavior of glyceraldehyde-3-P (*G3P*) at pH 7.4 in 0.1 M imidazole buffer at 26°. In *A*, at low arsenate concentration, the lines intersect on the *abscissa*, as they do in Fig. 2*D*, and $K_{glyceraldehyde\text{-}3\text{-}P}$ is independent of arsenate concentration. At arsenate concentrations in the 50 to 100 μM range, the intersections on the *abscissa*, and hence $K_{glyceraldehyde\text{-}3\text{-}P}$, depend upon arsenate concentration and there is a secondary enzyme activation with an increase in maximal velocity. In *B*, at arsenate concentrations in the millimolar concentration range, there is secondary enzyme activation but arsenate is now an inhibitor, competitive with aldehyde. The DPN concentration in these experiments was 260 μM and the enzyme concentration 1.7×10^{-9} M.

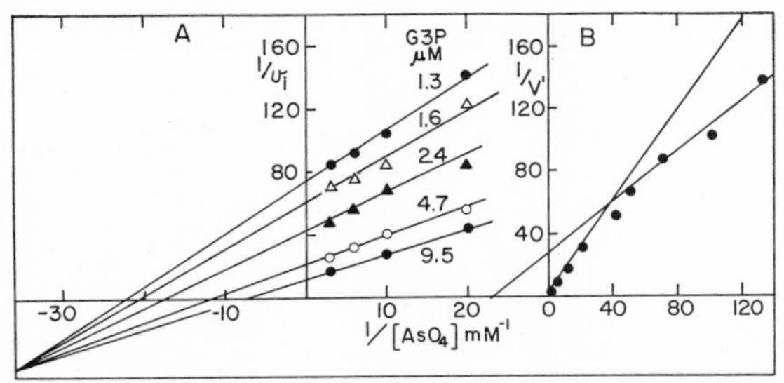

FIG. 6. Reciprocal plots of velocity against arsenate concentration in the concentration region in which enzyme is activated by arsenate. In *A*, the Michaelis constant of arsenate is seen to depend upon glyceraldehyde-3-P (*G3P*) concentration. *Curve B* is a reciprocal plot of velocities, which are maximal with respect to glyceraldehyde-3-P, against arsenate concentration. These experiments were done in 0.1 M imidazole, pH 7.4, at 26°, at a DPN concentration of 260 μM and an enzyme concentration of 1.7×10^{-9} M.

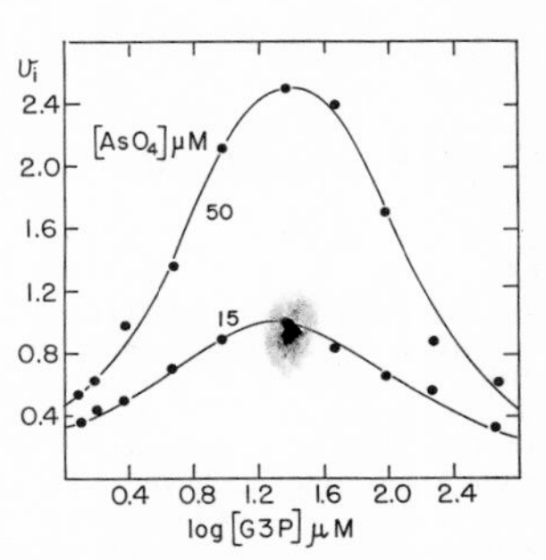

FIG. 7. Substrate inhibition by glyceraldehyde-3-P (*G3P*) at two levels of arsenate concentration in imidazole buffer, pH 7.4, at 26° and an enzyme concentration of 1.7×10^{-9} M and a DPN concentration of 230 μM. Initial velocities are plotted against log [glyceraldehyde-3-P].

Substrate Inhibition by Glyceraldehyde-3-P—In addition to the sensitivity of the behavior of glyceraldehyde-3-P to the regions of arsenate concentration, the aldehyde also exerts a substrate inhibition. Initial velocities in Fig. 7 at a constant high DPN concentration are plotted against log [aldehyde] at two levels of arsenate. The curves are approximately bell-shaped and resemble the activity-pH curves of reactions the rates of which are governed by two hydrogen dissociations (12). In the present case the apparent dissociation constants of the enzyme-glyceraldehyde-3-P complexes at the active and inhibitory sites may be obtained from the location and separation of the inflection points.

Rate Laws and Kinetic Parameters

Rate Laws—The empirical rate laws for the forward and reverse reactions under conditions in which there are no kinetically detectable secondary substrate effects are given, respectively, in Equations 2 and 3.

$$\frac{E_0}{v} = \frac{1}{V_f}\left(1 - \frac{K_A}{(A)}\right)\left(1 + \frac{K_B}{(B)}\right)\left(1 + \frac{K_C}{(C)}\right) \quad (2)$$

$$\frac{E_0}{v} = \frac{1}{V_r}\left(1 + \frac{K_D}{(D)}\right)\left(1 + \frac{K_F}{(F)}\right) \quad (3)$$

where the K values are Michaelis constants and v and V are, respectively, initial and maximal velocities. The independence of the K values of the concentrations of the cosubstrates eliminates the need for constants of the type K_{AB} or K_{ABC} dependent on two or more substrates, although constants of these types occur in the rate laws for reactions carried out at arsenate concentrations which produce secondary enzyme activation.

All of the pyridine nucleotide dehydrogenase reactions that have so far been examined kinetically (13–18) have been described by steady state mechanisms that have been simplified by the assumption of an ordered sequence of substrate addition to enzyme. When this approach is attempted for the forward or three-substrate reaction in the present case, the resulting equations lack terms that are required in Equation 2 and hence are contrary to the experimental results. The desired equation can be derived in principle by the general steady state method by assuming a random order of substrate addition to enzyme, but the results are too complicated to be useful at the present

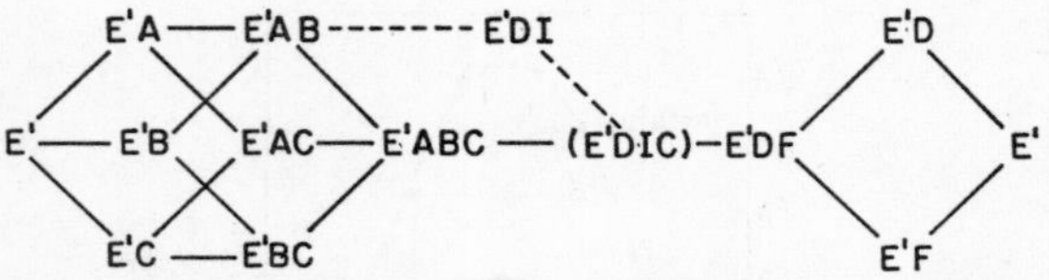

E'A — E'AB - - - - - - - E'DI
E' — E'B E'AC — E'ABC — (E'DIC) — E'DF E'D E'
E'C — E'BC E'F

FIG. 8. Pathways of intermediate complex formation considered in the rapid equilibrium kinetic mechanism. *A*, *B*, and *C* are, respectively, DPN, aldehyde, and phosphate (or arsenate). *D*, *F*, and *E'* are, respectively, DPNH, acyl phosphate, and enzyme. The complex *EDI*, acyl-enzyme-DPNH, is known to be formed but is not kinetically significant. The complex *EDIC*, in *parentheses*, is presumed to occur but need not be recognized explicitly. Enzyme is designated as *E'* instead of *E* to indicate an isomerized form which is discussed later in the text.

time. Equation 2 can be derived directly, however, if it is assumed that the substrates add to enzyme in random order to form complexes that are maintained in rapid equilibrium with respect to rate-limiting steps that occur in the breakdown of a quaternary enzyme-substrate complex. In accord with the mutual independence of the Michaelis constants, sets of equilibria of the type $E + A \leftrightarrow (EA)$, $(EB) + A \leftrightarrow (EAB)$, and $(EBC) + A \leftrightarrow (EABC)$ are described by the identical dissociation constant, K_A, defined as the dissociation constant of the enzyme-substrate complex, (EA). The possible pathways of intermediate enzyme-substrate complex formation in the rapid equilibrium mechanism are summarized in Fig. 8. The intermediate *EDI*, connected to the main scheme by *dashed lines*, corresponds to an acyl-enzyme-DPNH complex known to be formed in reactions in which an external acyl acceptor is omitted (19, 20). Such an intermediate is evidently kinetically insignificant under the conditions described. The point is treated in more detail under "Quaternary Complex" below.

The validity of the rapid equilibrium kinetic mechanism may be subjected to further kinetic test by an examination of the results of product inhibition. It may also be tested by a comparison of Michaelis and inhibition constants with the corresponding dissociation constants measured by equilibrium binding methods, since the two types of constant are assumed to be identical. It is found, in the application of these tests, that although the rapid equilibrium mechanism may be formally retained there are additional conditions affecting the mechanism which must be considered.

Product Inhibition

Inhibition by Glyceraldehyde-3-P and DPNH—Both aldehyde and DPNH are potent inhibitors of the reactions in which they are produced. These inhibitions are competitive not only with the homologous precursor substrate but also with the opposing cosubstrate. Thus, as shown in Fig. 9, the aldehyde is competitive both with acyl phosphate and with DPNH. Conversely, DPNH, in Fig. 10, is seen to be competitive with DPN and aldehyde. The rate laws which describe these inhibitions are given in Equations 4 and 5 for the forward and reverse reactions, respectively,

$$\frac{E_0}{v} = \frac{1}{V_f}\left[\left(1 + \frac{K_A}{(A)}\right)\left(1 + \frac{K_B}{(B)}\right)\left(1 + \frac{K_C}{(C)}\right) + \frac{(I)}{K_I}\left(\frac{K_AK_B}{(A)(B)} + \frac{K_AK_BK_C}{(A)(B)(C)}\right)\right] \quad (4)$$

$$\frac{E_0}{v} = \frac{1}{V_r}\left[\left(1 + \frac{K_D}{(D)}\right)\left(1 + \frac{K_F}{(F)}\right) + \frac{(I)}{K_I}\frac{K_DK_F}{(D)(F)}\right] \quad (5)$$

where I is the inhibitor and K_I the inhibition constant. Inhibitions of this type are not compatible with a simple compulsory order steady state mechanism but are readily derived for a rapid equilibrium mechanism by considering the complexes in Fig. 8 plus the inhibited complexes *EI* and *EIC*. A product inhibition analogous to that exerted by the aldehyde has been observed by Reynard, Hass, Jacobsen, and Boyer (21) in the pyruvate kinase reaction, which also exhibits rapid equilibrium kinetics.

Although the crossed competitive product inhibitions provide formal support for the rapid equilibrium kinetic mechanism, the inhibition constants calculated according to Equations 4 and 5 are extremely small and are not identical with the corresponding Michaelis constants. If the glyceraldehyde-3-P generated in the reverse reaction were as inhibitory as that added prior to initiation of the reaction, the earliest measurable initial velocities would have been subject to product inhibition. An attempt was made, therefore, to relieve any product inhibition during initial velocity measurements in the reverse reaction by adding a large excess of triosephosphate isomerase, which converts most of the aldehyde to dihydroxyacetone phosphate. No acceleration was observed. When α-glycerophosphate dehydrogenase was added together with the isomerase, the rate of DPNH oxidation was doubled and this was accounted for entirely by the

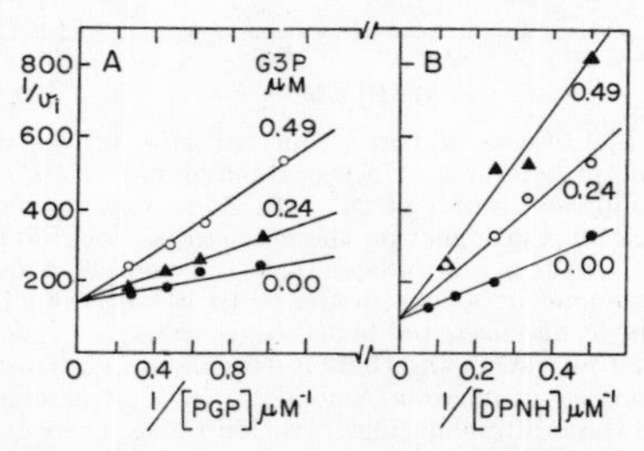

FIG. 9. Product inhibition by glyceraldehyde-3-P (*G3P*) in imidazole buffer, pH 7.4, at 26°, at an enzyme concentration of 4.7×10^{-11} M. In *A*, glyceraldehyde-3-P is seen to be competitive with 3-P-glyceroyl-P (*PGP*) at a constant DPNH concentration of 6.6 μM. In *B*, the aldehyde is seen to be competitive with DPNH at a constant 3-P-glyceroyl-P concentration of 3.9 μM.

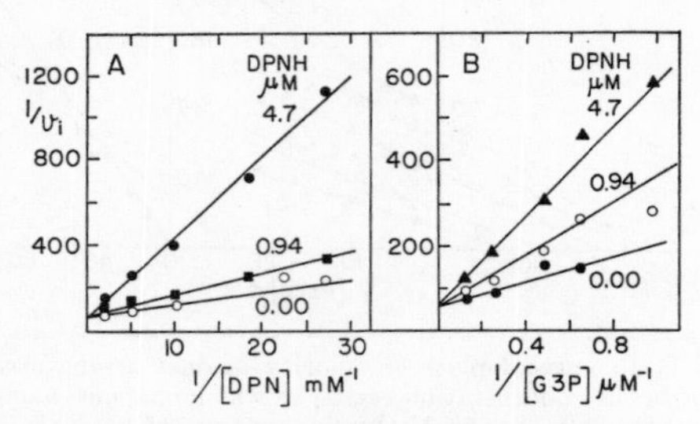

FIG. 10. Product inhibition by DPNH in imidazole buffer, pH 7.4, at 26°, at an enzyme concentration of 4.1×10^{-9} M. In *A*, DPNH is competitive with DPN at constant arsenate and glyceraldehyde-3-P concentrations of 15 and 7.7 μM, respectively. In *B*, DPNH is competitive with glyceraldehyde-3-P (*G3P*) at constant arsenate and DPN concentrations.

consumption of the extra mole of DPNH in the reduction of the ketone.

An independent test for product inhibition by DPNH in the forward reaction was provided by allowing the reaction to occur in the presence of sufficient beef heart lactic dehydrogenase to bind all DPNH formed up to concentrations of 28 μM. The lactic and glyceraldehyde-3-P dehydrogenases have approximately equal affinities for DPNH, but the 10,000-fold excess of lactic dehydrogenase allows it to compete successfully for the DPNH in a rapid binding reaction (10). Product inhibition by the DPNH generated in the irreversible glyceraldehyde-3-P-arsenate reaction is seen in Fig. 11 to be very strong at low DPN concentrations. The same reaction run in the presence of lactic dehydrogenase shows the same initial velocity but proceeds nearly to completion. Competitive binding of the inhibitory DPNH by the lactic dehydrogenase thus relieves the inhibition in extended reactions at low DPN concentration, but does not accelerate initial rate or appreciably affect K_{DPN}.

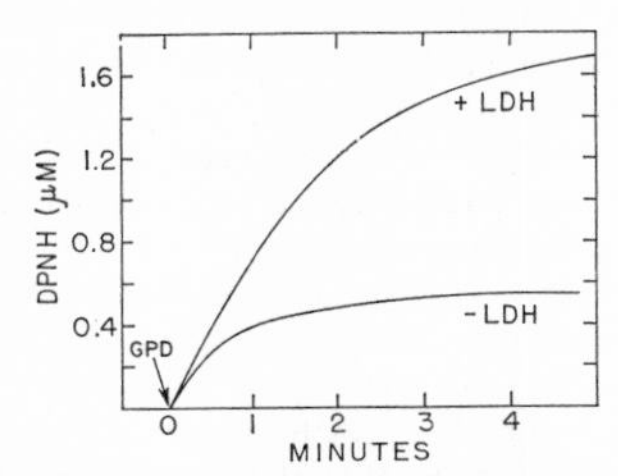

FIG. 11. Removal of product inhibition by competitive binding of product (DPNH) on substrate level concentrations of a second protein. The initial concentrations in both reactions were: DPN, 2.5 μM; glyceraldehyde-3-P, 15 μM; arsenate, 1 mM; and enzyme, 7.9×10^{-9} M, at pH 7.4 and 26°. The reaction described by the *upper curve* was performed in the presence of 1.4 mg of beef heart lactic dehydrogenase (*LDH*) per ml as a preferential DPNH binder. Reaction was initiated at the *arrow* by addition of glyceraldehyde-3-P dehydrogenase. The *curves* are redrawn from recorder tracings, and the *upper curve* has been reduced by a factor of 1/2.1 to correct for the fluorescence enhancement of DPNH bound to lactic dehydrogenase observed with the particular filter combination that was used.

Product Inhibition by DPN—The action of DPN as an inhibitory product of the reaction between DPNH and 3-P-glyceroyl-P is illustrated in Fig. 12, where $1/v$ is plotted against [DPN] at a series of fixed DPNH concentrations. In this type of plot for a two-substrate reaction, the inhibition constant is given by the projection on the *abscissa* of the intersection point at the *left of the ordinate*. DPN is found to be competitive with DPNH but is not competitive with the acyl-P in the concentration range that could be tested. The rate law for the single competitive inhibition is given in Equation 6.

$$\frac{E_0}{v} = \frac{1}{V_r}\left[\left(1 + \frac{K_D}{(D)}\right)\left(1 + \frac{K_F}{(F)}\right) + \frac{(I)}{K_I}\frac{K_D}{(D)}\right] \quad (6)$$

Although the inhibition constants of aldehyde and DPNH are both much smaller than the corresponding Michaelis constants, that of DPN is, within limits of error, equal to the Michaelis constant for DPN.

Product Inhibition by 3-P-Glyceroyl-P—The acyl phosphate is a competitive inhibitor of the oxidative phosphorylation at pH 8.6 with an inhibition constant in the range of 2 to 4 μM. At

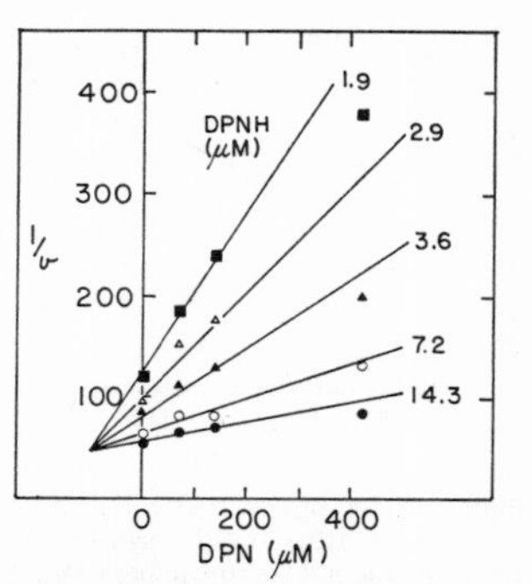

FIG. 12. Product inhibition of DPNH oxidation by DPN at pH 7.4 in 0.1 M imidazole chloride buffer at 25°. The concentration of 3-P-glyceroyl-P is fixed at 4.4 μM. DPN concentration is varied over a series of fixed DPNH concentrations as indicated.

pH 7.4, the unfavorable equilibrium in phosphate and the occurrence of arsenolysis of the acyl phosphate in arsenate hinder a kinetic study of the product inhibition. A strong product inhibition by the acyl phosphate in the physiological pH region would be anticipated in view of its very low Michaelis constant. It has so far been possible to approach this inhibition only indirectly by examining the effects of differential substrate and product binding at equilibrium.

When the simpler pyridine nucleotide-linked dehydrogenase reactions are run to equilibrium at low and at substrate levels of enzyme, it is found that at high enzyme concentration the equilibria are shifted toward increased DPNH formation (22–24). The preferential binding of DPNH over DPN becomes thermodynamically significant when the enzyme concentration is comparable to that of substrate or product. In the case of the oxidative phosphorylation equilibrium catalyzed by the dehydrogenase, the equilibrium shifts at high enzyme concentration are marginal, presumably because of compensating binding effects of substrates and products (25). One may circumvent this limitation by running the reaction at low enzyme concentration in the presence of a high concentration of lactic dehydrogenase, which binds DPNH strongly but has a low affinity for DPN and triose phosphates. Such an experiment is shown in Fig. 13, *Curves A* and *B*, which depict reactions at low enzyme concentration, respectively, in the absence and presence of lactic dehydrogenase. The difference in apparent equilibrium level of DPNH is seen to be very small. By application of Alberty's (26) general formulation to this case, Equation 7 is obtained.

$$K_{\text{app}} = \frac{[\text{DPNH}][\text{acyl-P}][\text{H}^+] f_{\text{DPN}} f_{\text{aldehyde}} f_{\text{PO}_4}}{[\text{DPN}][\text{aldehyde}][\text{PO}_4] f_{\text{DPNH}} f_{\text{acyl-P}}} \quad (7)$$

where K_{app} is the apparent equilibrium constant and the f's are the mole fractions of each reactant in the *unbound* form. From the concentrations employed and the known dissociation constants of the lactic dehydrogenase-DPNH (10, 27) and -DPN complexes (28), the f_{DPN}/f_{DPNH} ratio is in excess of 20 and the other f values are in the range of unity. A large increase in absorbance at 340 mμ in the presence of substrate level lactic dehydrogenase at equilibrium would therefore be expected. Absence of such a shift is a sign of nearly total inhibition of glyceraldehyde-3-P dehydrogenase by the low equilibrium concentration of acyl-P, about 6 μM.

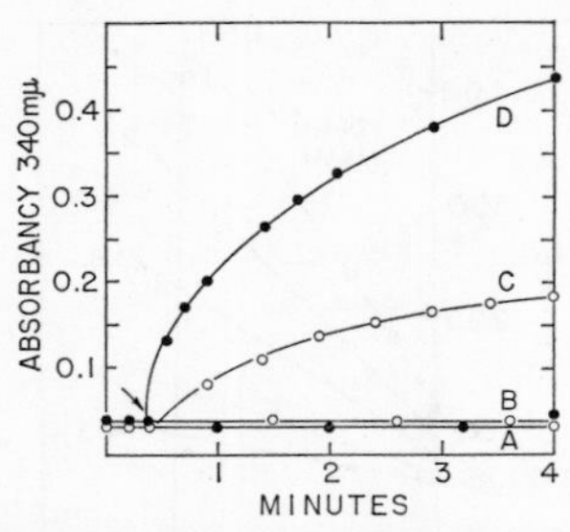

FIG. 13. Product inhibition by 3-P-glyceroyl-P. Concentrations in all of the reactions illustrated were: DPN, 250 μM; glyceraldehyde-3-P, 100 μM; and potassium phosphate, 4 mM, in 0.05 M imidazole buffer, pH 7.5, at 25°. The reaction illustrated by *Curve A* had come to a stop within 10 seconds and remained constant. *Curve B* represents an identical mixture which also contained 1.4 mg of lactic dehydrogenase per ml. *Curve C* describes a reaction similar to *A* but containing 5 μg of phosphoglycerate kinase per ml and 0.5 mM $MgCl_2$. Extended reaction was initiated at the *arrow* by addition of ADP at an initial concentration of 250 μM. The reaction of *Curve D* is identical with *C* except that the initial mixture also contained 1.4 mg of lactic dehydrogenase per ml.

That DPNH binding by lactic dehydrogenase is operative under the above conditions is shown in Fig. 13, *Curves C* and *D*. In *Curve C*, which is identical with *Curve A* but contains an incomplete 3-phosphoglycerate kinase system, the kinase reaction is initiated at the *arrow* by addition of any missing component: magnesium, ADP, or kinase. Additional DPNH production occurs when the acyl-P is converted to 3-phosphoglycerate. When the same experiment is performed in the presence of high lactic dehydrogenase, the rate and extent of the subsequent glyceraldehyde-3-P oxidation are greatly increased. Thus the weaker DPNH inhibition is detected and relieved by binding to lactic dehydrogenase when the acyl-P restriction is removed.

Product inhibition by the acyl-P in initial velocity measurements involving the coenzyme analogue AcPyDPN in phosphate at pH 7.4 are not obvious in the results shown in Fig. 3. It is likely that the inhibition is weaker in the AcPyDPN than in the DPN reaction. There are several possible reasons for the difference, but further investigation is required.

Table I contains a summary of the kinetic parameters that have so far been obtained. The low Michaelis constant of 3-P-glyceroyl-P and the relatively high $V_r/2$ make it necessary, for rapid equilibrium conditions to be maintained, that the bimolecular complex formation between 3-P-glyceroyl-P and enzyme approach a diffusion-controlled rate. There is now precedent for rate constants of this magnitude in the combination of proteins with small ligands (2, 29, 30).

Kinetic Parameters and Dissociation Constants

Contrary to the simple kinetic theory that describes the form of the results, several discrepancies have been noted between dissociation constants of enzyme-substrate complexes determined kinetically in different ways or kinetically and by equilibrium titration. The latter difference, in the case of DPN, is three orders of magnitude and raises the question whether the catalytic sites are distinct from the strong DPN-binding sites or whether the multiple binding sites are all equivalent but become modified under reaction conditions. The second explanation is the correct one. It is directly demonstrable and may be reconciled with the known properties of the enzyme and with the proposed kinetic mechanism.

Partial Reactions under Initial Velocity Conditions—At the limits of instrumental sensitivity, initial velocities of aldehyde oxidation can be measured during the formation of 0.3 μM product. Under such conditions with arsenate as external acyl acceptor and DPN in the range of 100 μM, there should be no appreciable direct product inhibition. It is pertinent to inquire, however, what the steady state concentration of acyl-enzyme intermediate might be in the initial steady state. The equilibrium constant of the oxidative step with enzyme as a stoichiometric participant in the absence of external acceptor was determined a number of years ago (19) and expressed as in Equation 8.

$$K_1 = \frac{(\text{acyl-enzyme})(\text{DPNH})(\text{H}^+)}{(\text{free enzyme})(\text{DPN})(\text{aldehyde})} = 10^{-5} \quad (8)$$

Contributions from differential substrate and product binding

TABLE I
Kinetic parameters of glyceraldehyde 3-phosphate dehydrogenase reactions

Reaction	Substrate	Michaelis constant	Inhibitory product	Inhibition constant
		μM		μM
Glyceraldehyde-3-P + DPN + arsenate; 0.1 M imidazole; pH 7.4; 26°; V_f = 1,250 min^{-1}	DPN	90	DPNH	0.3
	Glyceraldehyde-3-P	2.5	3-P-Glyceroyl-P	1
	Arsenate	28		
3-P-Glyceroyl-P + DPNH; pH 7.4; 26°; V_f = 14,900 min^{-1}	DPNH	3.3	DPN	100
	3-P-Glyceroyl-P	0.8	Glyceraldehyde-3-P	0.06
Glyceraldehyde-3-P + DPN + arsenate or phosphate; pH 8.6; 26°; V_f = 22,000 min^{-1}	DPN	13	DPNH	2
	Glyceraldehyde-3-P	90	3-P-Glyceroyl-P	2–4
	Arsenate	69		
	Phosphate	290		
Glyceraldehyde-3-P + AcPyDPN + phosphate; pH 7.4; 26°	AcPyDPN	28		
	Glyceraldehyde-3-P	37		
	Phosphate	6,800		

are not separated but are included in K_1. This equation would also apply to the initial steady state of irreversible glyceraldehyde-3-P oxidation in arsenate if the acyl transfer step were rate-limiting. In such a case one may calculate the fractional acylation of enzyme sites at a series of substrate concentrations when the DPNH concentration has reached 0.3 μM in an initial velocity measurement. Thus, if (acyl-enzyme)/(free enzyme), calculated from Equation 8, equals a, then the fractional acylation equals $a/(a + 1)$. In Table II, calculations are presented in which (DPN) is varied and (aldehyde) is fixed at 5 K_{aldehyde}, which is just below the onset of substrate inhibition. Fractional acylation of the enzyme is seen to parallel fractional apparent maximal velocity, v_i/V', which is in accord with the assumption that acyl transfer is rate-limiting. Similar results are obtained when (DPN) is fixed at 5 K_{DPN} and (aldehyde) is varied. The calculations indicate, moreover, that in all of the concentrations ranges that have been experimentally accessible, an average of one or more of the three or four acyl acceptor sites of the enzyme have been acylated during initial velocity measurements. The kinetic K_{DPN} therefore applies to acylated protein, and it is thus not necessary that it be identical with the K_{DPN} determined fluorometrically in the absence of cosubstrate.

It was observed in previous studies of the effect of p-mercuribenzoate on DPN binding that 4 mole equivalents of the mercurial promote the release of all of the bound DPN from the enzyme (9) and that 1 mole equivalent of mercurial per mole of protein weakens the affinity of the protein for DPN at all sites (10, 31). This effect was attributed to a conformational perturbation of the protein induced by the large anionic ligand in mercaptide linkage with a reactive thiol group. A similar effect may occur as a result of acylation of an acceptor site with the 3-phosphoglyceroyl group. In such a case the equilibria and relative rates of the partial reactions are such that all enzyme species are modified or isomerized during initial velocity measurements.

Correlation of Active Sites with DPN-binding Sites—If the above interpretation is correct, it should be possible to observe kinetically the strong DPN binding that is characteristic of the nonacylated enzyme-DPN when a substrate is employed which does not give a high initial steady state concentration of the acyl intermediate or which forms an intermediate that does not perturb the DPN-binding site. This condition may be realized by the use of the nonphosphorylated DL-glyceraldehyde, which has an extremely high Michaelis constant and which at substantial concentrations reacts at less than 0.001 the rate of glyceraldehyde-3-P. Such an experiment was described several years ago (9) and is repeated here with a much lower concentration of enzyme, more sensitive recording, and a wider range of DPN concentrations. With this substrate the enzyme is used in the micromolar rather than millimicromolar concentration range, and one may therefore measure directly the initial rates as a function of the concentration of bound DPN. The results are shown in Fig. 14.

Starting with charcoal-treated, DPN-free enzyme at a concentration of 3.2 μM (assumed molecular weight of 136,000), the initial rates increase linearly with added DPN, level off abruptly at 9.4 μM DPN, and remain constant up to DPN concentrations as high as 1,400 μM. If the enzyme is only 95% saturated with DPN at the apparent equivalence point, which corresponds to 2.9 binding sites per molecule, the dissociation constant of the active enzyme-DPN complex could be no larger than 10^{-7} M. This agrees with the equilibrium binding of DPN by the nonacylated enzyme.

TABLE II

Extent of acylation of glyceraldehyde 3-phosphate dehydrogenase during initial steady state of glyceraldehyde 3-phosphate oxidation in arsenate at pH 7.4

Concentration ratios of acylated sites to total sites are calculated from Equation 8 as described in the text for the conditions under which DPNH concentration is 0.3 μM, H^+ concentration is 0.04 μM, acyl transfer is rate-limiting, and all of the acyl that has been transferred has been converted irreversibly in arsenate to 3-phosphoglycerate. Glyceraldehyde-3-P concentration is fixed at 12.5 μM and v_i/V' is calculated from the velocity dependence on DPN concentration for $K_{\text{DPN}} = 90$ μM.

DPN concentration	Concentration ratio of acylated sites to total sites	V_i/V'
μM		
30	0.24	0.25
60	0.37	0.4
90	0.48	0.5
180	0.65	0.67

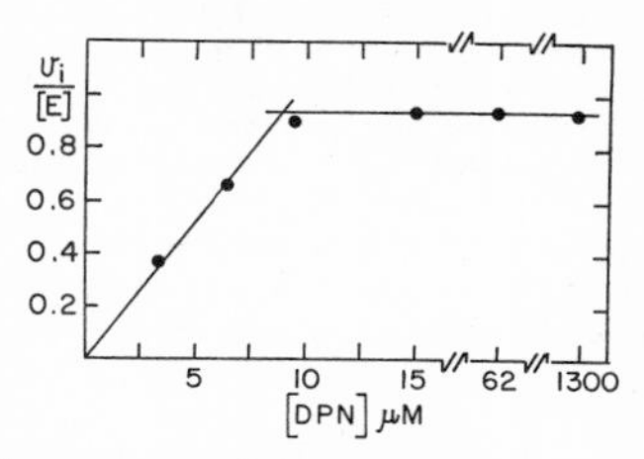

FIG. 14. The apparent dissociation constant of the enzyme-DPN complex in the reaction with glyceraldehyde and arsenate. The concentration of apoenzyme in these experiments was 3.2 μM in 0.05 M sodium arsenate, 1 mM EDTA, and 4 mM DL-glyceraldehyde, pH 7.5, at 26°. The rates were measured spectrophotometrically in a cuvette of 10-cm light path with expanded scale recording.

One can also approach the above condition kinetically with glyceraldehyde-3-P by working at pH 8.6. At this pH there is a large increase in maximal velocity associated with an acceleration of the acyl transfer step. The initial steady state concentration of the acyl-enzyme intermediate should therefore be below the maximum permitted by K_1 of Equation 8, and the kinetic K_{DPN} should become more nearly equal to the equilibrium binding value. The number observed kinetically is 13 μM, about one-seventh the value at pH 7.4. The number obtained by fluorometric quenching titration at this pH is approximately 3 μM. A similar value is obtained as described below by measuring the rates of spontaneous inactivation of the enzyme at 39° as a function of DPN concentration.

The inactivation of the apoenzyme follows first order kinetics, and the rate constants decrease as an inverse function of the DPN concentration as a result of the increased stability of the enzyme-DPN complex. The first order rate constants so obtained are plotted against log [DPN] in Fig. 15. A sigmoid curve is obtained with an inflection point corresponding to a DPN concentration of 6 μM, within a factor of 2 of the kinetic K_{DPN} at the same pH.

Effect of 3-P-Glyceroyl-P on Dissociation of Enzyme-DPN—

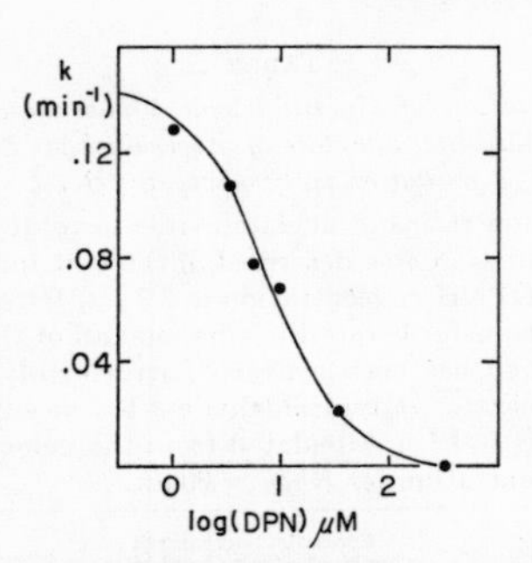

FIG. 15. The apparent dissociation constant of the enzyme-DPN complex at pH 8.6 determined by the stabilization of apoenzyme against thermal inactivation. The incubation mixtures, in 0.1 M sodium carbonate-bicarbonate buffer at 39.5°, contained charcoal-treated apoenzyme at a concentration of 8.2×10^{-9} M in increasing concentrations of added DPN. At time intervals, aliquots of each mixture were assayed for enzyme activity in the same buffer at 25° with DPN (300 μM), glyceraldehyde-3-P (800 μM), and orthophosphate (500 μM). Total enzyme concentration in the activity tests was 6.8×10^{-10} M. Loss of activity in the incubation mixtures was found in each case to follow first order kinetics. The first order rate constants are plotted against the log of the DPN concentration in the incubation mixture.

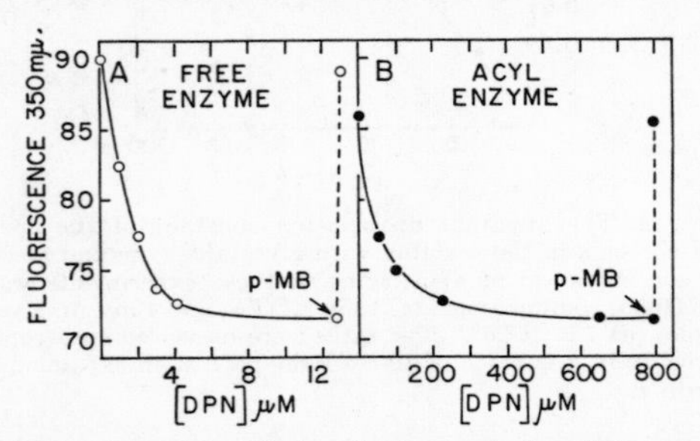

FIG. 16. The effect of 3-phosphoglyceroyl phosphate upon the dissociation constant of the enzyme-DPN complex. *A*, apoenzyme, 0.125 mg per ml in 0.02 M imidazole buffer, pH 7.4, at 25°, was titrated fluorometrically with DPN as described under "Experimental Procedure." The fluorescence quenching limit shown remained constant out to DPN concentrations of 800 μM. *B*, the same type of titration carried out in the presence of 100 μM ATP, 270 μM 3-phosphoglycerate, 100 μM $MgCl_2$, and 5 μg of phosphoglycerate kinase per ml. When the ATP or 3-phosphoglycerate or $MgCl_2$ was omitted, the titration curve resembled that in *Part A*. *p-MB*, *p*-mercuribenzoate.

The equilibrium of the acyl transfer reaction, with enzyme as a stoichiometric participant, has been formulated as in Equation 9,

$$K_2 = \frac{(\text{R—CO—OPO}_3)(\text{Enz})}{(\text{R—CO—Enz})(\text{HPO}_4)} \qquad (9)$$

and has been estimated to have an equilibrium constant, K_2, of 10^{-2} (19). In a phosphate-free solvent, the enzyme should therefore be nearly quantitatively acylated by very low concentrations of 3-P-glyceroyl-P. In the studies of DPN inhibition of the acyl-P-DPNH reaction (Fig. 12), the initial acyl-P concentrations were 4.4 μM, a level which approaches kinetic saturation. This condition can be duplicated in an equilibrium binding experiment in which apoenzyme is titrated fluorometrically with DPN in the presence and absence of 3 μM 3-P-glyceroyl-P. In these titrations, illustrated in Fig. 16, complex formation is measured by the quenching within the complex of the 350 mμ emission of the tryptophan residues of the protein as originally described for pyridine nucleotide complexes of the glyceraldehyde-3-P and lactic dehydrogenases (10) and for the combination of dinitrophenyl haptens with their specific purified antibodies (32). In the absence of 3-P-glyceroyl-P or in the presence of an incomplete 3-P-glyceroyl-P-generating system lacking 3-phosphoglycerate or ATP, the fluorescence quenching approaches a limit of about 12 μM DPN (*Curve A*). In the presence of the complete kinase system, calculated to yield an equilibrium 3-P-glyceroyl-P concentration of 3 μM, the fluorescence quenching approaches approximately the same limit at 800 μM DPN. From the magnitudes of the quenching effects, the DPN binding stoichiometrics in the two cases are very nearly the same. In both cases the original fluorescence is restored by the addition of 100 μM *p*-mercuribenzoate. The apparent dissociation constants in the absence and presence of 3-P-glyceroyl-P are, respectively, 10^{-7} and 10^{-4} M.

The results in Fig. 16 establish by a nonkinetic method that a cosubstrate interaction on the acyl-enzyme is responsible for the high inhibition constant of DPN in the reduction of the acyl phosphate by DPNH. From the value of K_2 in Equation 9, the predominant enzyme species in which the perturbation of DPN binding occurs should be the acyl-enzyme intermediate. This is in accord with the preceding calculations concerning initial steady state concentrations of the intermediate in the forward reaction. Acyl-enzyme is the only enzyme species common to the forward reaction between aldehyde and arsenate, the reverse reaction between acyl-P and DPNH, and the equilibrium titration with DPN in the presence of acyl-P. The results of Fig. 15 are in accord with the recent qualitative observation of Krimsky and Racker (33) that 3-P-glyceroyl-P promotes the release of bound DPN.

Dissociation of Enzyme-AcPyDPN Complex—For comparison with the kinetic determination of K_{AcPyDPN} in Fig. 3, the binding of AcPyDPN by apoenzyme was studied at pH 7.4 and 26° by spectrophotometric titration and at 39° by protection of the protein against thermal inactivation. The spectrophotometric method is similar in principle to one originally employed in measuring the binding of DPN (34) and is based upon the absorption band in the 360 mμ region characteristic of the enzyme complex with the oxidized form of the dinucleotide. The present results are analyzed by the method of Stockell (35). In this method the mass law is expressed as in Equation 10.

$$d/p = K/(e - p) + n \qquad (10)$$

where K is the apparent dissociation constant, n is the number of sites per molecule, d is the total concentration of AcPyDPN added, e is total enzyme concentration, and p is the concentration of occupied sites, *i.e.* the fraction of total enzyme bound times e. The plot of d/p against $1/(e - p)$ is linear with a slope of K and an intercept of n. K in Fig. 17 is found to be 28 μM, identical with the kinetic K_{AcPyDPN}. The intercept is somewhere between 2 and 4 but cannot be accurately determined. A determination of K by protection against thermal inactivation at 39° yields a sigmoid plot analogous to that in Fig. 14 and an apparent K_{AcPyDPN} of 31 μM. The coenzyme analogue appears to be bound similarly to DPN in that it gives the same type of 360 mμ absorption band in the complex, but, unlike the natural coenzyme, the dissociation constants of the complex in

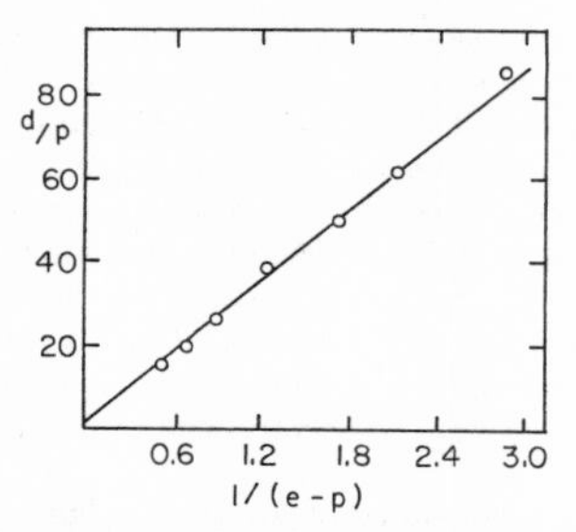

FIG. 17. The apparent dissociation constant of the enzyme-AcPyDPN complex determined by spectrophotometric titration. The apoenzyme concentration was 2.1 μM in 0.1 M imidazole buffer, pH 7.4. Absorbances at 350 mμ were measured in a cuvette of 10-cm light path as a function of the concentration of AcPyDPN. The plot is according to Equation 10.

the presence and absence of substrate are identical. It is possible that the oxidative step in the reaction with AcPyDPN is the slow one and that the initial steady state concentration of the intermediate is too low to produce a detectable effect. More detailed investigation is required.

DISCUSSION

The results that have been described are unusual in the sense that an enzyme that is particularly susceptible to modification by substrate catalyzes a reaction described by a random order, mutually independent substrate binding. The apparent paradox is a consequence of the fact that the major cosubstrate effect is produced by a nondissociable intermediate and is fully expressed under initial velocity conditions at the lowest substrate concentrations that can be handled esperimentally. For this reason all enzyme species that are dealt with kinetically in reactions with the natural substrates are modified or isomerized and are hence designated as E' rather than E in the diagram of Fig. 8.

Isomerization—Use of the term isomerization to explain the observed effects may be questioned. For example, the interference by acyl-P in DPN binding and the crossed competitive product inhibitions exerted by aldehyde and DPNH could be the result of direct steric interactions between the nonhomologous substrate-product pairs. Such direct steric interactions, in contrast with the behavior of the normal cosubstrate pairs, could arise from conformational differences in the oxidized and reduced bound forms both of the nucleotides and of the 3 carbon substrates. However, this interpretation does not account for events in the aldehyde-DPN-arsenate reaction at pH 7.4. If the acyl-enzyme intermediate interacted exclusively with the DPN at the same catalytic site, then the $1/v$ against $1/(\text{DPN})$ plots would be linear only if all sites were fully acylated during the initial steady state. This possibility is not excluded by the accuracy of K_1 (Equation 8), but it would require that all measurable initial velocities be zero order with respect to aldehyde, and this is not the case. K_{aldehyde} is quite measurable kinetically.

The acyl-enzyme-DPN interaction is therefore viewed as a special example of enzyme modifier action in which the modifier is a normal reaction intermediate. Formation of the intermediate at one site causes a protein isomerization which affects the properties of the unoccupied but otherwise equivalent catalytic sites. On a multisite enzyme of this type the cosubstrate interaction need not be a direct one. It is not necessary that the isomerization be grossly detectable, by a change, for example, in the rotatory dispersion curve of the protein. In tests that have been carried out under a variety of conditions, acylation, or removal of bound DPN with charcoal, or reaction with p-mercuribenzoate does not cause the dissociation of the protein into its subunits, although Deal (36) has found that the subunits are separable when the protein is denatured. However, removal of DPN itself reversibly affects the thermal stability and crystallizability of the protein.

Quaternary Complex—The isomerization effect is closely associated with the occurrence of a kinetically important quaternary enzyme-substrate complex. Prior indication of the need for such a complex was provided by the observation that the isolated acyl transfer step, studied by phosphate exchange or arsenolysis on acyl phosphates (37), requires or is strongly promoted by the presence of bound pyridine nucleotide. This function can be provided by bound DPN and hence is a reverse manifestation of the effect of acylation upon DPN binding. The two types of ligand are presumed to produce antagonistic or counteracting changes in the secondary structure of the protein, and each therefore promotes the release or increases the reactivity of the other. If bound DPN promotes acyl transfer to phosphate, it should also promote the acylation of the protein by acyl phosphates in the reverse reaction. Hilvers and Weenen (38) have in fact reported that under rather special conditions the presence of DPN is necessary for the catalysis of acetyl phosphate reduction by DPNH. There is no sign of such a requirement in the present studies of 3-P-glyceroyl-P reduction. It is likely that if there is a secondary nucleotide requirement it can be provided by DPNH itself. Such a function is not readily tested in the partial reaction because DPNH rapidly reduces P-glyceroyl-P.

It is clearly established that glyceraldehyde-3-P reduction leading to acyl-enzyme formation can occur in the absence of external acceptor (19, 20). Such a step is included in the rapid equilibrium reaction scheme of Fig. 7 but is designated by *broken lines* since it need not be considered in the derivation of the rate law of Equation 2. Although oxidation of aldehyde can occur under such conditions, it occurs much more rapidly when arsenate or phosphate is already present in the quaternary enzyme-substrate complex. In view of the general anion-binding tendencies of glyceraldehyde-3-P dehydrogenase and the strong secondary enzyme activation produced by intermediate ranges of arsenate concentration (Fig. 4), this behavior is not surprising. Some years ago Koeppe, Boyer, and Stulberg (39) compared initial rates of glyceraldehyde oxidation at high enzyme concentration in the presence and absence of phosphate and observed no stimulation by phosphate. However, the conditions of the experiments and the kinetic behavior of glyceraldehyde and glyceraldehyde-3-P are sufficiently different that the two sets of results are not necessarily comparable.

In the absence of other information, the kinetic evidence for a quaternary complex would be interpreted mechanistically as a sign that under conditions of optimal catalysis the oxidative phosphorylation occurs by a modified nucleophilic displacement or S_N2 mechanism, in which the oxygen of a phosphate ion attacks the carbonyl carbon of the aldehyde at the same time that the hydrogen atom is transferred to the pyridine C-4 of bound DPN. Such a mechanism is in fact a limiting case of the two-step oxidation and acyl transfer mechanism as the time interval between the two steps approaches zero. The rate-limiting step at pH 7.4 would then not be acyl transfer but dissociation of acyl phosphate (or arsenate) from the protein. In terms of all of the evidence at present available, however, the acyl acceptor function at the catalytic site still appears to be real and not virtual even though it occurs within a quaternary enzyme-substrate complex.

Other Approaches—The present results describe only the gross features of the glyceraldehyde-3-P dehydrogenase reaction kinetics but provide the basis for a more extended attack on the mechanism both by standard and by newer methods. In several laboratories, applications are being made of fast reaction techniques which may clarify the events occurring in the ternary and quaternary complexes. From tryptic digests of glyceraldehyde-3-P dehydrogenase, Harris, Meriweather, and Park (40) and Perham and Harris (41) have isolated and determined the structure of an octadecapeptide which in the native protein carries the thiol groups that act as acyl acceptors with model substrates. By limited proteolysis with chymotrypsin, Krimsky and Racker (33) have destroyed phosphorylating activity without a commensurate loss in the activity of the oxidizing step and are exploring the possibility of intramolecular acyl migrations which on other grounds have so far been contraindicated.

Conditions in Muscle—In considering the performance of glyceraldehyde-3-P dehydrogenase in skeletal muscle, a chief point of interest is the fact that the enzyme concentration, expressed as active sites without correction for excluded volumes, is of the order of 70 μM or higher, whereas the steady state concentrations of the aldehyde and acyl phosphate have been reported to be in the range of 5 and less than 1 μM, respectively (42). The enzyme concentration thus normally exceeds those of its 3 carbon substrates but is considerably less than the concentrations of DPN and orthophosphate, which are in the millimolar range. Any restriction in the activity of the 3-phosphoglycerate kinase, such as might occur during rest by a limitation of the magnesium ion or ADP supply, would lead to a nearly total inhibition of glyceraldehyde-3-P dehydrogenase in the acylated form. Free 3-P-glyceroyl-P as such would not accumulate to any appreciable extent under any circumstances. Some of the DPNH produced can be removed by pyruvate reduction. To the extent that pyruvate is oxidized in mitochondria, there is an excess of free DPNH which must be removed by another pathway. DPNH production is self-limiting in the sense that it can take over product inhibition of the dehydrogenase when the restraint by acyl-P is removed. To what extent this may occur under physiological conditions is unknown.

SUMMARY

1. The oxidative and acyl group transfer steps previously shown to be catalyzed by glyceraldehyde 3-phosphate dehydrogenase under a variety of restricted conditions have been considered under conditions of rapid enzyme turnover in the reversible oxidative phosphorylation of the natural substrates.

2. Initial velocity kinetics has been measured in both reaction directions. The kinetic mechanism at pH 8.6 is described by a random order of substrate addition to enzyme and rate-limiting steps that occur in the breakdown of a kinetically important quaternary enzyme-substrate complex.

3. In the physiological pH region, similar kinetics is observed at low substrate concentrations, but at higher substrate concentrations a variety of secondary activations and inhibitions occur.

4. By kinetic criteria, the dissociation constant of the enzyme-diphosphopyridine nucleotide complex at pH 7.4, tested with diphosphopyridine nucleotide both as a substrate and as an inhibitory product, is several orders of magnitude larger than the value obtained by equilibrium titration in the absence of the phosphorylated 3 carbon substrates. It is identical, however, with the K_{DPN} determined by fluorescence quenching titration in the presence of very low concentrations of 3-phosphoglyceroyl phosphate.

5. This result is accounted for by the equilibria of the partial reactions and the fact that in the physiological pH region the acyl transfer to external acceptor is rate-limiting. The kinetic K_{DPN} under such conditions describes the dissociation of DPN from acyl-enzyme rather than from free enzyme. At pH 8.6 the acyl group transfer is greatly accelerated and the kinetic K_{DPN} approaches the equilibrium value determined in the absence of substrate. Perturbation of DPN binding at pH 7.4 disappears when the nonphosphorylated D-glyceraldehyde is employed as substrate. There is also no perturbation of pyridine nucleotide binding by the acyl intermediate at pH 7.4 with phosphorylated substrate when the 3-acetylpyridine analogue is used as hydrogen acceptor in place of DPN.

6. The interaction between DPN and the acyl intermediate is reciprocal. Thus the binding of DPN promotes acyl group transfer to external acceptor. A reciprocal interaction also occurs between glyceraldehyde 3-phosphate and DPNH, each of which, as an inhibitory product, competes with the other as well as with its homologous precursor.

7. The effect of acyl intermediate on K_{DPN} is analogous to the weakening of DPN binding at three sites, promoted by 1 equivalent of *p*-mercuribenzoate per mole of enzyme. It is therefore believed that a protein isomerization occurs. The presumed isomerization does not involve the dissociation of the protein subunits but may involve their mutual reorientation.

REFERENCES

1. Velick, S. F., and Furfine, C., in P. D. Boyer, H. Lardy, and K. Myrbäck (Editors), *The enzymes, Vol. VII*, Academic Press, Inc., New York, 1963, p. 243.
2. Velick, S. F., and Vavra, J., *J. Biol. Chem.*, **237**, 2109 (1962).
3. Lowry, O. H., and Lopez, J. A., *J. Biol. Chem.*, **162**, 421 (1946).
4. Negelein, E., and Bromel, H., *Biochem. Z.*, **303**, 132 (1939).
5. Dalziel, K., *J. Biol. Chem.*, **238**, 1538 (1963).
6. *Pabst Laboratories, Circular OR-18*, Pabst Brewing Company, Milwaukee, 1961.
7. Rafter, G. W., and Colowick, S. P., in S. P. Colowick and N. O. Kaplan (Editors), *Methods in enzymology, Vol. III*, Academic Press, Inc., New York, 1957, p. 887.
8. Cori, G. T., Slein, M. W., and Cori, C. F., *J. Biol. Chem.*, **173**, 605 (1948).
9. Velick, S. F., Hayes, J. E., Jr., and Harting, J., *J. Biol. Chem.*, **203**, 527 (1953).
10. Velick, S. F., *J. Biol. Chem.*, **233**, 1455 (1958).
11. Kaplan, N. O., Ciotti, M. M., and Stolzenback, F. E., *J. Biol. Chem.*, **221**, 833 (1956).
12. Alberty, R. A., and Massey, V., *Biochim. et Biophys. Acta*, **13**, 347 (1954).
13. Theorell, H., in F. F. Nord (Editor), *Advances in enzymology, Vol. 20*, Interscience Publishers, Inc., New York, 1958, p. 31.
14. Dalziel, K., *Biochem. J.*, **84**, 244 (1962).
15. Schwert, G. W., *Ann. N. Y. Acad. Sci.*, **75**, 311 (1958).
16. Raval, D. N., and Wolfe, R. C., *Biochemistry*, **1**, 64 (1962).

17. Fromm, H. J., and Nelson, D. R., *J. Biol. Chem.*, **237**, 215 (1962).
18. Frieden, C., *J. Biol. Chem.*, **234**, 2891 (1959).
19. Velick, S. F., and Hayes, J. E., Jr., *J. Biol. Chem.*, **203**, 545 (1953).
20. Segal, H. L., and Boyer, P. D., *J. Biol. Chem.*, **204**, 265 (1953).
21. Reynard, A. M., Hass, L. F., Jacobsen, D. D., and Boyer, P. D., *J. Biol. Chem.*, **236**, 2277 (1961).
22. Theorell, H., and Bonnichsen, R., *Acta Chem. Scand.*, **5**, 1105 (1951).
23. Hayes, J. E., Jr., and Velick, S. F., *J. Biol. Chem.*, **207**, 225 (1954).
24. Nielands, J., *J. Biol. Chem.*, **199**, 373 (1952).
25. Cori, C. F., Velick, S. F., and Cori, G. T., *Biochim. et Biophys. Acta*, **4**, 160 (1950).
26. Alberty, R. A., *J. Am. Chem. Soc.*, **75**, 1925 (1953).
27. Winer, A. D., Schwert, G. W., and Millar, D. B. S., *J. Biol. Chem.*, **234**, 1149 (1959).
28. Takenaka, Y., and Schwert, G. W., *J. Biol. Chem.*, **223**, 157 (1956).
29. Hammes, G. G., and Fasella, P., *J. Am. Chem. Soc.*, **84**, 4644 (1962).
30. Day, L. A., Sturtevant, J. M., and Singer, S. J., *J. Am. Chem. Soc.*, **54**, 3768 (1962).
31. Velick, S. F., in R. Benesch et al. (Editors), *Sulfur in proteins*, Academic Press, Inc., New York, 1959, p. 267.
32. Velick, S. F., Parker, C. W., and Eisen, H. N., *Proc. Natl. Acad. Sci. U. S.*, **46**, 1470 (1960).
33. Krimsky, I., and Racker, E., *Biochemistry*, **2**, 512 (1963).
34. Velick, S. F., *J. Biol. Chem.*, **203**, 563 (1953).
35. Stockell, A., *J. Biol. Chem.*, **234**, 1286 (1959).
36. Deal, W. C., Jr., *Federation Proc.*, **22**, 290 (1963).
37. Harting, J., and Velick, S. F., *J. Biol. Chem.*, **207**, 867 (1954).
38. Hilvers, A. G., and Weenen, J. H. M., *Biochim. et Biophys. Acta*, **58**, 380 (1962).
39. Koeppe, O. J., Boyer, P. D., and Stulberg, M., *J. Biol. Chem.*, **219**, 509 (1956).
40. Harris, J. I., Meriweather, B. P., and Park, J. H., *Nature*, **198**, 154 (1963).
41. Perham, R. N., and Harris, J. I., *J. Molecular Biol.*, **7**, 316 (1963).
42. Hohorst, H. J., Reim, M., and Bartels, H., *Biochem. and Biophys. Research Communs.*, **7**, 137 (1962).

Reprinted from PROCEEDINGS OF THE NATIONAL ACADEMY OF SCIENCES, Vol. 47, 1961

ON THE NATURE OF THE TRANSALDOLASE-DIHYDROXYACETONE COMPLEX

BY B. L. HORECKER, S. PONTREMOLI,* C. RICCI,† AND T. CHENG

DEPARTMENT OF MICROBIOLOGY, NEW YORK UNIVERSITY SCHOOL OF MEDICINE

Communicated October 23, 1961

During recent years there has been considerable progress in the elucidation of the mechanism of action of the aldolases, particularly with respect to the activation of the substrate. (For a review of this subject, see ref. 1.) However, little information has been gained relating to the nature of the active site on the enzyme, apart from the work of Swenson and Boyer[2] with sulfhydryl reagents.

The enzyme transaldolase is distinguished from other aldolases by the fact that it catalyzes not merely the cleavage of an aldol linkage to form a pair of products, but rather a transfer of a dihydroxyacetone group from one molecule to another: dihydroxyacetone itself is not a substrate for the enzyme. It has been proposed[3–5] that a stable intermediate is formed with the dihydroxyacetone group firmly bound to the enzyme, from which it is removed only when a suitable acceptor is present or when the enzyme is denatured. This enzyme-dihydroxyacetone complex has recently been described by Venkataraman and Racker.[5]

A study of the formation and properties of the transaldolase-dihydroxyacetone intermediate has been in progress in our laboratory. We have identified the bound group as dihydroxyacetone; when the enzyme complex is treated with borohydride the dihydroxyacetone group is reduced and becomes firmly and irreversibly fixed to the enzyme protein. The complex can then be hydrolyzed with acid to yield a small molecule containing the bound group. In this way it has been established that the dihydroxyacetone group must be linked through the β-carbon atom. Some of the properties of the reduced fragment are described in this report.

Experimental Procedures.—Materials: Most of the experiments reported were carried out with a preparation of transaldolase[3] which is about 30–60 per cent pure, compared with the crystalline preparation.[4]

D-Erythrose 4-phosphate was kindly provided by Drs. Sprinson and Srinivasan of the College of Physicians and Surgeons of Columbia University. Commercial dihydroxyacetone was recrystallized from hot ethanol. This was essential to remove glyceraldehyde which was present in significant amounts. Other enzymes and substrates were as previously reported.[4]

Methods: Transaldolase was measured spectrophotometrically as described in the preceding paper.[4] Other procedures were as previously described.[3,4]

Results.—Formation and stability of the enzyme-dihydroxyacetone complex: The conditions for formation and precipitation of the radioactive complex are described in the preceding report.[4] With the noncrystalline enzyme preparations similar conditions were employed, except that hexose phosphate isomerase was present as a contaminant and addition of this enzyme was unnecessary. With the less pure enzyme preparations the extent of labeling was less than has been observed with the crystalline enzyme and with prolonged incubation the amount of complex decreased (Table 1).

TABLE 1

CONDITIONS FOR FORMATION OF THE TRANSALDOLASE-DIHYDROXYACETONE COMPLEX

Time of incubation, minutes	Radioactivity, cpm/mg protein
1	1,460
3	2,500
5	1,900

The incubation mixture (0.075 ml) contained 0.6 mg of transaldolase (specific activity = 16.1), 0.1 *M* triethanolamine buffer, pH 7.4, and 10^{-3} *M* glucose 6-phosphate-1-C^{14} (1×10^6 cpm per μmole). After incubation at 34° for the period indicated, 0.15-ml samples were mixed at 0° with 1 ml of 85 per cent saturated ammonium sulfate solution, centrifuged, the precipitate resuspended in 0.2 ml of 85 per cent saturated ammonium sulfate, centrifuged, and the residue dissolved in 0.2 ml of triethanolamine buffer. An aliquot was plated for counting and another taken for the determination of protein.

The enzyme complex is stable to repeated precipitation with ammonium sulfate (Table 2). The first precipitate is contaminated with substrate from the incubation mixture, but following the second precipitation at 0° there is no further loss in activity. With each successive precipitation at 37° about 5–10 per cent of the label is lost. At 80° the complex is completely dissociated and the precipitated protein shows only traces of radioactivity.

Characterization of the enzyme-dihydroxyacetone complex: When glucose 6-phosphate-1-C^{14} is replaced by glucose 6-phosphate-P^{32}, there is little labeling of the protein (Table 3). Therefore it is not the intact substrate molecule which is involved in complex formation. When the labeled enzyme is incubated with a suitable acceptor, such as erythrose 4-phosphate or glyceraldehyde, all of the radioactivity is removed from the protein (Table 3).

TABLE 2

STABILITY OF TRANSALDOLASE-DIHYDROXYACETONE COMPLEX

	Precipitate dissolved at— 0°		37°	
Precipitation	Total counts, cpm	Specific activity, cpm/mg protein	Total counts, cpm	Specific activity, cpm/mg protein
First	63,300	9,500	60,800	9,100
Second	39,700	7,400	36,500	6,500
Third	38,900	7,600	32,800	6,200
Fourth	36,500	7,900	25,200	5,500
Fifth (80°)	3,500	100	7,700	300

Each incubation mixture (0.25 ml) contained 0.01 M triethanolamine buffer, pH 7.4, 0.25×10^{-3} M glucose 6-phosphate-1-C^{14} (3×10^6 cpm/μmole) and 1.8 mg of transaldolase (specific activity = 16.4). After 3 min at 34° the solutions were precipitated with 10 volumes of 85 per cent saturated ammonium sulfate solution, centrifuged, and the precipitate dissolved in 0.2 ml of triethanolamine buffer. One solution was kept at 0° for 3 min and the other at 37°. Each was then precipitated as before and the process repeated. Aliquots of each solution were taken for determination of radioactivity and protein content. The last solutions, following the fourth precipitation, were kept for 3 min at 80° and then treated with ammonium sulfate. The over-all recovery of protein was about 60 per cent.

TABLE 3

DISSOCIATION OF THE COMPLEX BY TRANSALDOLASE ACCEPTORS

First incubation addition	Second incubation addition	Total counts, cpm	Specific activity, μmoles C^{14} or P^{32} per mg protein $\times 10^3$
Glucose 6-phosphate-1-C^{14}	None	3,200	0.68
" " " " "	D-erythrose 4-phosphate	20	0.004
" " " " "	D-glyceraldehyde	0	0.0
Glucose 6-phosphate-P^{32}	...	3	0.05

The first incubation mixtures contained 1.2 mg of transaldolase (specific activity = 16.2 units/mg), 0.01 M triethanolamine buffer, pH 7.6, and 0.11 μmole of glucose 6-phosphate-1-C^{14} (1×10^6 cpm/μmole) or 0.11 μmole of glucose 6-phosphate-P^{32} (5.4×10^4 cpm/μmole). After incubation for 3 min at 34° the protein was precipitated with 5 ml of 85 per cent saturated ammonium sulfate, centrifuged, and washed with 1 ml of 85 per cent saturated ammonium sulfate. The washed precipitate was dissolved in 0.1 ml of 0.04 M triethanolamine buffer, pH 7.6. Aliquots were removed for counting and for protein determination. The remainder was incubated for 3 min at 34° with no other addition or with 0.12 μmole of D-erythrose 4-phosphate or 0.38 μmole of D-glyceraldehyde. The enzyme was then precipitated as before, and the precipitate washed and dissolved in 0.1 ml of triethanolamine buffer and counted.

The product formed from the labeled protein and D-erythrose 4-phosphate was identified as sedoheptulose 7-phosphate by co-chromatography with an authentic sample (Fig. 1). The autoradiogram obtained on exposure of the paper to x-ray film showed a perfect fingerprint reproduction of the spot developed by spraying with aniline phthalate reagent.[6] The enzyme complex therefore has the properties of a transaldolase-dihydroxyacetone complex.

This was confirmed by identification of dihydroxyacetone itself following decomposition of the complex at 80°. The isolated product, co-chromatographed with dihydroxyacetone in two different solvents, showed identical spots on autoradiograms and the paper developed with aniline phthalate reagent.[6] One of these chromatograms is shown in Figure 2.

Fixation of the labeled group by reduction with borohydride: While the enzyme-dihydroxyacetone complex is relatively stable under ordinary conditions, it would not survive hydrolysis of the protein, which is essential to a determination of the nature of the linkage. It was found, however, that the labeled group could be fixed to the protein by reduction with borohydride, and indeed without changing the solubility properties of the enzyme. Reduction was found to be most effective at pH 6.0. Following borohydride treatment at this pH, none of the counts were removed from the protein by incubation with D-glyceraldehyde. At pH 7.5, reduction was far less effective, and almost three-fourths of the counts remained transferable to D-glyceraldehyde. Without borohydride treatment, even at pH

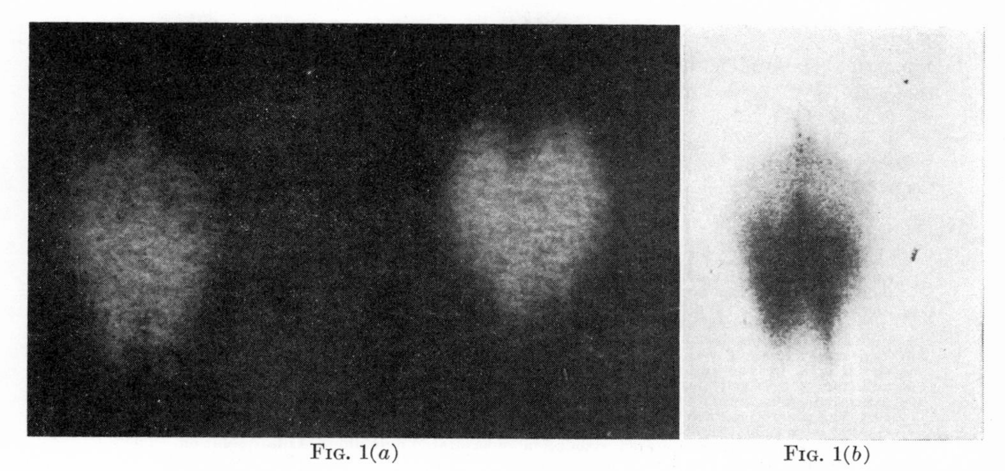

FIG. 1(a) FIG. 1(b)

FIG. 1.—(a) Chromatogram (developed with aniline phthalate spray) with authentic sample (right) and (b) autoradiogram of the product formed from D-erythrose 4-phosphate and labeled transaldolase. In this experiment 4.5 mg. of transaldolase (specific activity = 14.5) was incubated with 1.3 μmoles of glucose 6-phosphate-1-C^{14} (3.0×10^6 cpm per μmole) in 0.07 ml of 0.01 *M* triethanolamine buffer, pH 7.6. After 3 min at 34° the reaction mixture was precipitated with 4.0 ml of saturated ammonium sulfate solution, centrifuged, and the precipitate dissolved in 0.5 ml of 0.04 *M* triethanolamine buffer, pH 7.6. The protein was again precipitated as before and dissolved in 0.2 ml of 0.04 *M* triethanolamine buffer and treated with 0.05 ml of D-erythrose 4-phosphate solution containing 0.35 μmole. After 3 min at 34°, 3 μmoles of carrier sedoheptulose 7-phosphate was added amd the solution deproteinized with 0.02 ml of 100 per cent trichloroacetic acid solution. The supernatant solution contained 80 per cent of the counts in the original enzyme complex (1.11×10^4 cpm, total). The trichloroacetic acid solution was thoroughly extracted with ether and the phosphate ester precipitated with barium acetate and 4 volumes of ethanol. The precipitate was dissolved in 0.5 ml of 0.01 *M* acetic acid. Barium ion was removed with a slight excess of sodium sulfate and the supernatant solution concentrated to a small volume, placed on Whatman No. 1 paper, and chromatographed with butanol:pyridine:water (6:4:3).

FIG. 2(a) FIG. 2(b)

FIG. 2.—(a) Chromatogram developed with aniline phthalate spray and (b) autoradiogram of dihydroxyacetone released by warming the transaldolase complex to 80°. The formation of the labeled enzyme complex was as in Fig. 1, with 4.0 mg of transaldolase (specific activity = 24.6). The protein was precipitated twice with ammonium sulfate, dissolved in 0.5 ml of 0.1 *N* acetic acid to maintain an acid pH, and warmed to 80° for 10 min. The solution was cooled and 0.04 ml of 0.1 *M* dihydroxyacetone was added as carrier. The solution was then deproteinized with 2.5 ml ethanol and 0.002 ml of mercaptoethanol added to prevent oxidation of dihydroxyacetone. Ethanol was removed by evaporation under a stream of N_2 and the residue dissolved in water containing 10^{-3} *M* mercaptoethanol. The solution was deionized by passage through a 1-ml bed of Amberlite MB-3 resin. The effluent was concentrated to a small volume and placed on Whatman No. 1 paper for ascending chromatography in water-saturated phenol. The over-all yield of counts from the labeled enzyme was 35 per cent. A similar chromatogram was obtained by chromatography in 80 per cent acetone-water solvent.

In Fig. 2(a) the left-hand spot is a dihydroxyacetone standard.

TABLE 4

REDUCTION OF THE TRANSALDOLASE-DIHYDROXYACETONE COMPLEX WITH BOROHYDRIDE

Reduced at pH	Total counts	Fixed counts*	Fixed, %
7.5	3,600	960	27
6.5	6,200	3,200	52
6.0	3,300	3,500	100
Control at pH			
6.5	3,200	390	12
6.0	5,000	0	0

* Remaining after incubation with D-glyceraldehyde. The enzyme-transaldolase complex was prepared as in Fig. 1 with 2.7 mg of transaldolase (specific activity = 9.7) and 0.05 μmole of glucose 6-phosphate-1-C^{14}. The labeled enzyme was twice precipitated with ammonium sulfate, and finally dissolved in 0.5 ml of phosphate buffer (0.05 *M*) at the pH indicated. Reduction with borohydride was carried out at 0° by the addition of 0.01 ml quantities of fresh borohydride solution containing 40 mg of sodium borohydride per ml. Ten such additions were made over a period of 30 min; the pH was maintained at the initial level by the addition of 0.01-ml portions of 1 *N* acetic acid. Following the last addition of borohydride, the mixture was kept at 0° for 30 min, treated with 4.0 ml of saturated ammonium sulfate solution and centrifuged. To test for completeness of reduction, the precipitate was dissolved in 0.4 ml of 0.02 *M* triethanolamine buffer, pH 7.4, and treated with 0.005 ml of 1.8 *M* D-glyceraldehyde. This solution was warmed to 34° for 3 min and the protein precipitated with 2.5 ml of saturated ammonium sulfate solution. The last precipitate was dissolved in 0.2 ml of 0.04 *M* triethanolamine buffer, pH 7.4, and an aliquot counted.

The control experiments were identical except that borohydride was not added.

6.0, the enzyme-complex still retained full activity, and the dihydroxyacetone group was quantitatively transferred to D-glyceraldehyde, leaving the protein completely nonradioactive (Table 4).

Reduction of the enzyme-dihydroxyacetone complex with borohydride permitted a determination of the extent of formation of the complex. The loss of enzymatic activity corresponds to the degree of blocking of the active site by the reduced group. The reduced complex, as has been shown above, is without enzymatic activity. Following addition of the dihydroxyacetone group and reduction of the complex with borohydride, 48 per cent of the enzyme activity was lost (Table 5),

TABLE 5

LOSS OF TRANSALDOLASE ACTIVITY FOLLOWING REDUCTION OF THE DIHYDROXYACETONE COMPLEX WITH BOROHYDRIDE

	Complex, units/mg	Control, units/mg
Before reduction	32.0	29.6
After reduction	17.1	29.2

For formation of the enzyme-dihydroxyacetone complex 3.2 mg of transaldolase was incubated at 25° with 20 μmoles of fructose 6-phosphate in 1 ml of 0.04 *M* triethanolamine buffer, pH 7.4. After 15 min the enzyme was precipitated with 5 volumes of saturated ammonium sulfate and reduced at pH 6.0 with borohydride as in Table 4. The enzyme was removed by precipitation with ammonium sulfate, dissolved in 0.04 *M* triethanolamine buffer and assayed. The over-all recovery of protein was 50 per cent. The control was treated in identical fashion except that fructose 6-phosphate was omitted.

which corresponds almost exactly to the per cent of enzyme protein labeled in a similar experiment with fructose 6-phosphate 1-C^{14}.[4]

Stability to dialysis: The C^{14}-labeled transaldolase-dihydroxyacetone complex was slowly dissociated during dialysis against distilled water; in 16 hr approximately one-half of the radioactivity of the C^{14}-labeled complex was lost. Under the same conditions of dialysis the reduced complex lost no radioactivity.

Hydrolysis and periodate oxidation: Some preliminary experiments have been carried out to determine which carbon atom of dihydroxyacetone is involved in complex formation. For this purpose we have used the borohydride-reduced complex, in which the three-carbon group is fixed by a stable covalent linkage. Two possibilities must be considered. The complex may be either a derivative involving the carbonyl group at position 2 of the original fructose 6-phosphate molecule or a derivative formed through carbon atom 3. Following reduction with borohydride the former would yield a β-glycerol derivative, while the latter would be converted to an α-glycerol derivative. These can be distinguished by their susceptibility to oxidation with periodate since the α-glycerol derivative would yield formaldehyde derived from the C-1 of fructose 6-phosphate, while the β-glycerol derivative would not be attacked. For this purpose, the borohydride reduced complex, derived from glucose 6-phosphate 1-C^{14}, was dialyzed overnight against water and hydrolyzed in a sealed tube with 6 N HCl at 110° for 22 hr. HCl was removed by repeated evaporation *in vacuo* over KOH and $CaCl_2$ and an aliquot of the hydrolysate, containing 1,500 cpm, was treated with excess periodate at pH 4.5. After 1 hr at room temperature glycerol was added to remove excess periodate and the formaldehyde formed collected as the dimedon derivative. This was found to be completely nonradioactive and all the counts were recovered in the supernatant solution. A similar experiment was carried out with the reduced enzyme complex, without acid hydrolysis, and again no activity was obtained in the dimedon derivative. These results suggest that dihydroxyacetone is linked to the enzyme through carbon atom 2, rather than through the hydroxymethylene group at carbon 3.

Properties of the active site: With respect to the possible nature of the amino acid involved in formation of the complex some information is available. Following hydrolysis in 6 N HCl all of the radioactivity is absorbed by Dowex 50-H^+ resin, but not by Dowex 1-formate or Dowex 1-OH^-. The reduced complex is therefore not a strong anionic compound. On the other hand, following exposure to nitrous acid all of the counts passed through a Dowex 50-H^+ column, which indicates that the reduced complex possesses a free amino group.

Treatment with HBr or HI failed to yield glycerol, suggesting that the reduced 3-carbon residue is not attached by an 0-ether linkage to serine or threonine. It is pertinent to note that transaldolase is not inhibited by diisopropylfluorophosphate (DFP).

The possibility remains that dihydroxyacetone is attached to a protein SH-group as a thioacetal derivative. However, transaldolase is not significantly inhibited by iodoacetic acid nor by incubation with oxidized glutathione.

In two-dimensional paper chromatography, with methanol, pyridine, water (60:20:20) as the first solvent and *t*-butanol, water, diethylamine (85:15:4) as the second solvent, the radioactive fragment formed after borohydride reduction and acid hydrolysis is found to be located between aspartic acid and serine. It has thus been converted to a molecule of low molecular weight, not identical with glycerol.

Discussion.—We have described a stable C^{14}-labeled transaldolase complex formed when the enzyme is incubated with fructose 6-phosphate-1-C^{14}. The labeled group can be transferred to erythrose 4-phosphate to form sedoheptulose 7-phosphate, or liberated as free dihydroxyacetone when the enzyme is warmed to 80°. This confirms the work of Venekataraman and Racker[5] and proves that the intermediate formed in the enzymic action of transaldolase is, as originally postulated,[3] an enzyme-dihydroxyacetone complex.

Some of the properties of this complex have been determined. It can be reduced with sodium borohydride to form a stable derivative, presumably a glycerol derivative. This derivative does not yield labeled formaldehyde when treated with periodate, and we have therefore concluded that the dihydroxyacetone group is attached through the 2-position, possibly as a ketal derivative, rather than through the terminal hydroxymethylene group.

The borohydride reduced protein complex has been hydrolyzed with 6*N* HCl at 110°. This procedure yields a small molecule which is strongly adsorbed by cationic but not by anionic resins. The cationic group is probably a free amino group, since the molecule is no longer adsorbed following treatment with nitrous acid.

It is unlikely that either serine or threonine is involved in formation of the complex since the reduced product is not susceptible to cleavage by hydrobromic or hydroiodic acids, as would be expected for an 0-ether linkage.

Swenson and Boyer[2] have obtained evidence which indicates that the active group in aldolase may be a sulfhydryl group. This does not appear to be the case for transaldolase. However, in view of the similarity in the action of this enzyme and transaldolase it is reasonable to expect that they will possess similar groups at the active site. In the case of transaldolase a definitive answer to this question may now be possible since we can obtain the active group firmly linked as a C^{14}-labeled derivative, presumably a derivative of glycerol. The availability of large quantities of crystalline enzyme[4] should facilitate an approach to this problem.

This work was supported by grants in aid from the National Science Foundation and the National Institutes of Health.

* Present address: Institute of Biological Chemistry of the University of Genoa, Genoa, Italy.

† Fulbright Exchange Fellow. Present address: Institute of Biological Chemistry of the niversity of Siena, Siena, Italy.

[1] Horecker, B. L., *J. Cell and Comp. Physiol.*, **54**, Sup. 1, 89 (1959).

[2] Swenson, A. D., and P. D. Boyer, *J. Am. Chem. Soc.*, **79**, 2174 (1957).

[3] Horecker, B. L., and P. Z. Smyrniotis, *J. Biol. Chem.*, **212**, 881 (1955).

[4] Pontremoli, S., B. D. Prandini, A. Bonsignore, and B. L. Horecker, these PROCEEDINGS (preceding paper), **47**, 1942 (1961).

[5] Venkataraman, R., and E. Racker, *J. Biol. Chem.*, **236**, 1883 (1961).

[6] Partridge, S. N., *Nature*, **164**, 443 (1949).

Reprinted from THE JOURNAL OF BIOLOGICAL CHEMISTRY, Vol. 241, 1966

The Nature of the Hydrogen Transfer in the Dimethylbenzimidazolylcobamide Coenzyme-catalyzed Conversion of 1,2-Propanediol to Propionaldehyde*

(Received for publication, December 20, 1965)

ROBERT H. ABELES‡ AND B. ZAGALAK§

From the Graduate Department of Biochemistry, Brandeis University, Waltham, Massachusetts 02154

SUMMARY

This paper shows that the hydrogen which is removed from C-1 of a particular diol during the conversion to the corresponding aldehyde is not necessarily the same hydrogen which replaces the hydroxyl group of the diol and that therefore the reaction is not necessarily intramolecular.

* This work was supported by Grant GM-012633 from the National Institutes of Health, United States Public Health Service. Contribution No. 411 from the Graduate Department of Biochemistry, Brandeis University, Waltham, Massachusetts.

‡ To whom inquiries should be addressed.

§ Present address, Department of Biochemistry, College of Agriculture, Poznan, Poland.

TABLE I

Intermolecular hydrogen transfer in reaction catalyzed by dioldehydrase

The reaction mixture contained 118 units of dioldehydrase (2) (specific activity 45 to 55), 0.067 μmole of dimethylbenzimidazolylcobamide coenzyme, 0.05 M potassium phosphate buffer, pH 8.0, and 1,2-propanediol as indicated, 1.4 × 10^6 dpm. Total volume was 10 ml and the reaction time was 35 min at 37°. In Experiment 4, 10 μmoles of acetaldehyde were added at the beginning of the reaction and 70 μmoles after 35 min. Products were isolated as follows. For the isolation of aldehydes, the reaction mixture was added directly to 50 ml of 0.1% 2,4-dinitrophenylhydrazine in 2 N HCl. The 2,4-dinitrophenylhydrazones were isolated, dissolved in benzene, and passed through a column (2 × 5 cm) of Fisher Alumina. The 2,4-dinitrophenylhydrazones were separated and purified by chromatography on Whatman No. 1 paper impregnated with dimethylformamide. The solvent was cyclohexane. For the isolation of sodium acetate and sodium propionate, 100 mg of nonisotopic propionaldehyde and acetaldehyde were added at the end of the reaction. The pH was adjusted to 5.0 and the aldehydes distilled from the reaction mixture at atmospheric pressure. The aldehydes were then oxidized to the corresponding acid at pH 7.5 at 5° with $KMnO_4$ (B. Zagalak, P. A. Frey, G. L. Karabatsos, and R. H. Abeles, unpublished observations). The acids were purified by chromatography on silicic acid (2). DL-1,2-Propanediol-1-T was prepared by the catalytic reduction of DL-lactaldehyde (reduction carried out by New England Nuclear Corporation); 18% of the tritium content was at C-2 and 82% at C-1.

Additions to reaction	Amount	Products isolated	Total counts recovered
	μmoles		*dpm*
Experiment 1			
Ethylene glycol	80	Acetaldehyde	5.4 × 10^5
1,2-Propanediol-1-T	20	Propionaldehyde	8.7 × 10^5
Experiment 2			
Ethylene glycol	160	Sodium acetate	5.0 × 10^5
1,2-Propanediol-1-T	20	Sodium propionate	2.0 × 10^5
Experiment 3			
Ethylene glycol	40	Acetaldehyde	2.5 × 10^5
1,2-Propanediol-1-T	72	Propionaldehyde	11 × 10^5
Experiment 4			
1,2-Propanediol-1-T	20	Propionaldehyde	1.4 × 10^6
Acetaldehyde	80	Acetaldehyde	0

When a mixture of DL-1,2-propanediol containing tritium at C-1 and unlabeled ethylene glycol is converted to propionaldehyde and acetaldehyde in the presence of dioldehydrase (1) and dimethylbenzimidazolylcobamide coenzyme, acetaldehyde is obtained containing tritium in the α position. These results show that the hydrogen which is removed from C-1 of a particular diol during the conversion to the corresponding aldehyde is not necessarily the same hydrogen which replaces the hydroxyl group of the diol and that therefore the reaction is not necessarily intramolecular.

Several experiments which establish the transfer of hydrogen from propanediol to acetaldehyde are summarized in Table I. Experiments 1, 2, and 3 establish that tritium from C-1 of propanediol is transferred to the α position of the acetaldehyde derived from unlabeled ethylene glycol. Furthermore, the amount of tritium transferred to acetaldehyde depends upon the relative ratio of ethylene glycol to 1,2-propanediol (Experiments 1 and 3). In Experiments 1 and 2, in which the ratio of ethylene glycol to propanediol is 4:1 and 8:1, respectively, essentially all of the tritium that can potentially be transferred is found in acetaldehyde and sodium acetate. The propanediol added to the reaction contained 1.1 × 10^6 dpm at C-1 and 0.28 × 10^6 dpm at C-2. Since the hydrogen at C-1 was not introduced stereospecifically, only 50%, 0.55 × 10^6 dpm, is subject to transfer (3) and is the maximal amount which can be expected in acetaldehyde. It is, in fact, the amount found in acetaldehyde (Experiment 1) and sodium acetate (Experiment 2). When the ratio of ethylene glycol to 1,2-propanediol is reduced, an appreciably smaller percentage of tritium is transferred (Experiment 3). The presence of tritium in the α position of acetaldehyde is established by the fact that essentially no tritium is lost when acetaldehyde is converted to sodium acetate (Experiments 1 and 2). Experiment 4 shows that the addition of acetaldehyde prior to or after the conversion of 1,2-propanediol to propionaldehyde does not lead to tritium incorporation into acetaldehyde.

The conversion of propanediol or ethylene glycol to the corresponding aldehyde involves the abstraction of a hydrogen from C-1 and the replacement of an hydroxyl group at C-2 by hydrogen. The results presented here show that the hydrogen which is removed from C-1 is not necessarily the hydrogen that is introduced at C-2. The enzymatic reaction, therefore, does not involve a direct 1,2-hydride shift. Two possible mechanisms can be proposed to account for the observed results. (*a*) Two molecules of substrate participate in the reaction and interact in such manner that a hydrogen from C-1 of one substrate molecule is transferred to C-2 of the other and *vice versa*. This mechanism appears improbable to us on chemical grounds. (*b*) The enzyme-coenzyme complex functions as a hydrogen transfer agent in the conversion of propanediol to propionaldehyde. In the course of the reaction, hydrogen, as a proton, hydride ion, or hydrogen atom, is transferred from the substrate to the enzyme-coenzyme complex and another hydrogen from the complex to the substrate. If this is the case, the enzyme-coenzyme complex must contain two or more equivalent hydrogen atoms since the hydrogen which is donated by the complex is not necessarily the same as that removed from the substrate.

Final proof for the participation of the enzyme-coenzyme complex as a hydrogen transfer agent can only be obtained if the presence of hydrogen, derived from the substrate, can be shown in the enzyme-coenzyme complex. Experiments designed to establish this are now in progress.

REFERENCES

1. LEE, H. A., JR., AND ABELES, R. H., *J. Biol. Chem.*, **238**, 2367 (1963)
2. VARNER, J. E., *Methods Enzymol.*, **3**, 397 (1955).
3. FREY, P. A., KARABATSOS, G. J., AND ABELES, R H *Biochem. Biophys. Res. Commun.*, **18**, 551 (1965).

Reprinted from THE JOURNAL OF BIOLOGICAL CHEMISTRY, Vol. 241, 1966

The Role of the B_{12} Coenzyme in the Conversion of 1,2-Propanediol to Propionaldehyde*

(Received for publication, April 4, 1966)

PERRY A. FREY‡ AND ROBERT H. ABELES
From the Graduate Department of Biochemistry, Brandeis University, Waltham, Massachusetts 02154

SUMMARY

When 1,2-propanediol-1-^{3}H is converted to propionaldehyde in the presence of dioldehydrase and 5,6-dimethylbenzimidazolylcobamide 5′-deoxyadenosine, 2 hydrogen atoms per mole of the coenzyme are replaced by tritium. The tritiated coenzyme so obtained transfers tritium to the product when it is added to unlabeled 1,2-propanediol and dioldehydrase. The coenzyme containing tritium at C-5′ of the adenosyl moiety was prepared synthetically; when it was added to dioldehydrase and unlabeled 1,2-propanediol, all of the tritium initially present in the coenzyme was found in the propionaldehyde. From these results we concluded that the hydrogen transfer that occurs in the conversion of 1,2-propanediol to propionaldehyde involves two steps: transfer of hydrogen from C-1 to the carbon bonded to the cobalt of the coenzyme and, from this position, to C-2 of the product.

In a previous communication (1) we reported that the hydrogen transfer which occurs in the dioldehydrase reaction is not intramolecular, and we proposed that, in the course of the reaction, hydrogen is transferred from the substrate to the coenzyme and from the coenzyme to the product. This hypothesis has now been verified.

Transfer of Tritium from 1,2-Propanediol-1-^{3}H to Coenzyme—The reaction mixture consisted of 580 units of dioldehydrase, 0.044 μmole of 5,6-dimethylbenzimidazolylcobamide 5′-deoxyadenosine, 12.5 μmoles of 1,2-propanediol-1-^{3}H (1.1×10^5 cpm per μatom of hydrogen), and 17 μmoles of K_2HPO_4 in a total volume of 3.3 ml. The reaction was allowed to proceed for 6 min at 25° and then was stopped by the addition of 0.2 ml of 40% trichloracetic acid. DBC coenzyme[1] was isolated by paper electrophoresis in 0.5 M NH_4OH after removal of the protein that had been precipitated. DBC coenzyme so obtained was purified by paper chromatography in H_2O-saturated 2-butanol and 1-butanol-H_2O-acetic acid-2-propanol (100:99:1:70). After chromatography in these two solvents, the coenzyme had a specific activity of 2.4×10^5 cpm per μmole. To confirm radiopurity, the coenzyme was again chromatographed on paper with 2-butanol-H_2O-NH_4OH (100:36:14). After elution, the specific activity of the DBC coenzyme was unchanged. The specific activity of the coenzyme is twice that of the substrate. This may be due to the isotope effect or more probably to the fact that 2 hydrogen atoms of the coenzyme can exchange with the substrate.

Transfer of Tritium from DBC Coenzyme to Product—DBC coenzyme containing tritium was prepared and purified as described above. It was allowed to react under the following conditions: 4.9×10^{-3} μmole of DBC coenzyme (2.1×10^4 cpm), 94 units of dioldehydrase, 240 μmoles of 1,2-propanediol, and 80 μmoles of K_2HPO_4 in a total volume of 3.5 ml. The reaction was allowed to proceed for 20 min at 25°, and propionaldehyde was then isolated by distillation. Propionaldehyde was converted to propionic acid, which was then isolated (2). A total of 1.1×10^4 cpm was recovered in the propionic acid. This represents approximately 50% of the radioactivity originally present in the DBC coenzyme.

These experiments establish that tritium that has been initially introduced into DBC coenzyme by the substrate can be transferred to the product. To eliminate the possibility that the source of the tritium found in the substrate was a contaminant carried along through the purification, the following control experiment was carried out. Tritiated DBC coenzyme was inactivated by exposure to light and then added to the reaction mixture described above (containing unlabeled DBC coenzyme). Propionic acid derived from propionaldehyde was isolated, and it contained no tritium.

* Publication 432 from the Graduate Department of Biochemistry, Brandeis University. Supported in part by Research Grant GM-012,633 from the National Institute of General Medical Science.
‡ Public Health Service Predoctoral Fellow (National Institute of General Medical Science Fellowship 7-F1-GM-20,226-02)

[1] The abbreviation used is: DBC coenzyme, 5,6-dimethylbenzimidazolylcobamide 5′-deoxyadenosine.

Transfer of Tritium from Chemically Synthesized DBC Coenzyme to Product—DBC coenzyme containing tritium at C-5′ of the adenosyl moiety was chemically synthesized by the following sequence of reactions.

$$\text{2}',\text{3}'\text{-Isopropylidene adenosine-5}'\text{-aldehyde}^{2} \xrightarrow{NaBH_4(^3H)} \text{2}',\text{3}'\text{-isopropylidene adenosine-5}'\text{-}^3H \xrightarrow{TsCl} \text{2}',\text{3}'\text{ isopropylidene 5}'\text{-tosyladenosine-5}'\text{-}^3H$$

DBC coenzyme was then prepared by a modification of the procedures of Johnson *et al.* (4) from 2′,3′-isopropylidene 5′-tosyladenosine-5′-^3H and was purified by carboxymethyl cellulose chromatography, paper electrophoresis (0.5 M NH_4OH), and paper chromatography in two solvent systems. The yield, based on isopropylidene adenosinealdehyde, was low (5%). However, no attempt was made to maximize the yield. Although the synthetic method used establishes the location of the tritium in the DBC coenzyme, the following two experiments were done to verify the location of the tritium. (*a*) The coenzyme was photolyzed under anaerobic conditions (5). All of the radioactivity was recovered in the resulting cyclic nucleoside. This establishes the presence of tritium in the adenosyl moiety of the coenzyme. (*b*) The coenzyme was then photolyzed aerobically (6). Aerobic photolysis led to the liberation of 50% of the tritium into the solvent water. The resulting adenosine-5′-aldehyde was isolated by paper chromatography and contained tritium. Loss of tritium during aerobic photolysis confirms the presence of tritium at C-5′ of the adenosyl moiety.

Transfer of tritium from the synthetically prepared DBC coenzyme to the reaction product was established by the following experiment: 9.3×10^{-3} μmole of tritiated DBC coenzyme (7.2×10^4 cpm) was added to 192 units of enzyme and 10 μmoles of K_2HPO_4 in a total volume of 0.80 ml. The reaction mixture was cooled in an ice bath, and 0.2 ml of 0.036 M 1,2-propanediol was added with rapid mixing. The reaction mixture was then brought to 37°. After 5 min, the reaction was stopped by the addition of 2 ml of 2% 2,4-dinitrophenylhydrazine in 3 N H_2SO_4. Propionaldehyde 2,4-dinitrophenylhydrazone was isolated and purified as described (1); it contained 7.6×10^4 cpm. Therefore, quantitative transfer of tritium occurs from C-5′ of the adenosyl moiety of DBC coenzyme to the reaction product.

In a control experiment, tritium-containing DBC coenzyme was incubated with propionaldehyde without enzyme under the conditions of the enzymatic reaction. Propionaldehyde 2,4-dinitrophenylhydrazone was isolated and contained 0.35×10^4 cpm. These results suggest that the hydrogen transfer that occurs during the conversion of 1,2-propanediol to propionaldehyde is a two-step process. One step involves the transfer of hydrogen from C-1 of the substrate to C-5′ of the adenosyl moiety of DBC coenzyme, and the second involves transfer of hydrogen from the same position in the coenzyme to C-2 of the substrate. Either one of these processes could occur first, or both could occur simultaneously. During the course of the over-all reaction, the carbon-cobalt bond of the coenzyme must be so modified that either the 2 hydrogen atoms attached to the carbon, which is bonded to cobalt, are equivalent or can become equivalent. A possible way in which the 2 hydrogens can become equivalent is a transient dissociation of the carbon-cobalt bond.[3]

[2] This compound was prepared by a modification of the method of Pfitzner and Moffatt (3). We are grateful to Dr. Moffatt for advice during the preparation of this compound.

Acknowledgment—We acknowledge the competent technical assistance of Mr. T. Smith.

REFERENCES

1. Abeles, R. H., and Zagalak, B., *J. Biol. Chem.*, **241**, 1245 (1966).
2. Zagalak, B., Frey, P. A., Karabatsos, G. L., and Abeles, R. H., *J. Biol. Chem.*, in press.
3. Pfitzner, K. E., and Moffatt, J. G., *J. Am. Chem. Soc.*, **87**, 5661 (1965).
4. Johnson, A. W., Merwyn, L., Shaw, N., and Smith, L. E., *J. Chem. Soc.*, 4146 (1963).
5. Hogenkamp, H. P. C., *J. Biol. Chem.*, **238**, 477 (1963).
6. Hogenkamp, H. P. C., Ladd, J. N., and Barker, H. A., *J. Biol. Chem.*, **237**, 1950 (1962).

[3] Another possibility is that the hydrogen transfer from coenzyme to substrate is stereospecific. One of the methylene hydrogens of the adenosyl moiety participates when L-1,2-propanediol is the substrate, and the other participates when D-1,2-propanediol is used. In all of these experiments DL-1,2-propanediol was used, and, therefore, both hydrogens participate in the reaction.

Reprinted from THE JOURNAL OF THE AMERICAN CHEMICAL SOCIETY, Vol. 79, 1957

COMMUNICATIONS TO THE EDITOR

RAPID DEUTERIUM EXCHANGE IN THIAZOLIUM SALTS[1]

Sir:

In the course of work on a model system for thiamine action we have been led to investigate the stability of an anion at C-2 of thiazolium salts (I). Such a species is formally analogous to a cyanide ion, in that both are anions on a carbon atom which is multiply bonded to nitrogen,[2] and (I) might be expected to catalyze benzoin condensation and similar reactions in the same fashion as does cyanide ion. This is of special interest since thiazolium salts are known to be catalysts for the benzoin condensation,[3] and since thiamine, a thiazolium salt, is involved in biological catalysis of reactions formally analogous to the benzoin condensation.

It was decided that deuterium exchange would be the best way to demonstrate whether such an anion is indeed stable, since deuterium exchange at C-2 could only occur, under mild conditions, through the formation of (I), electrophilic attack by a deuteron on the positively charged thiazolium ring being of course quite improbable. It has been found that thiazolium salts do indeed exchange at C-2 with deuterium oxide very readily, and that *this occurs in the absence of any basic catalyst.*

3,4-Dimethylthiazolium bromide (II) incorporates one atom of deuterium (found,[4] 1.1 atoms) on standing in D_2O at room temperature for 20 hours, and then vacuum drying. Similarly 3-benzyl-4-methylthiazolium bromide incorporates 1.2 atoms of deuterium under these conditions. Both compounds show a strong C–D stretching band[5] in the infrared (KBr pellet) at 4.5μ. The high intensity compared to that in a synthetic sample of 3-benzyl(α-d_2)-4-methylthiazolium bromide suggests that the deuterium is not located on a saturated carbon, and this is supported by the

(1) This is part II of the series "The Mechanism of Thiamine Action"; for part I see R. Breslow, *Chemistry and Industry*, R 28 (1956).

(2) A related species apparently is formed during the decarboxylation of pyridine-2-carboxylic acid (B. R. Brown and D. Ll. Hammick, *J. Chem. Soc.*, 659 (1949)).

(3) T. Ugai, S. Tanaka and S. Dokawa, *J. Pharm. Soc. Japan*, **63**, 269 (1943).

(4) The author wishes to acknowledge the assistance of Miss Laura Ponticorvo with these analyses, which were done by combustion and mass-spectral analysis after conversion of water to hydrogen gas.

(5) See, for instance, R. N. Jones and C. Sandorfy in A. Weissberger "Techniques of Organic Chemistry," Vol. IX, Interscience Publishers, New York, N. Y., 1956, Chap. IV.

loss of a band at 11.0μ which can be assigned to the out-of-plane bending of the H at C-2.[5] In addition, both substances are free of D_2O (no band at 4.0μ). Thiamine chloride hydrochloride itself incorporates 5.2 atoms of deuterium (including the OH and NH_3^+ groups), and also shows the new band at 4.5μ and the loss of a band at 11μ.

Confirmation of the conclusion that it is the hydrogen at C-2 which exchanges is found in nuclear magnetic resonance studies[6] on (II) in D_2O as solvent. The compound initially shows two small equal peaks, at −108 and −47 cycles/sec., and a larger pair at +66 and +114 cycles/sec. (Varian V-4012A magnet, 7050 gauss field, 30 megacycles/sec. probe; frequencies referred to benzene capillary and increasing field). These are assigned to the groups at C-2, C-5, N-3, and C-4, respectively. On standing, the peak at −108 cycles/sec. diminishes and disappears because of exchange with the solvent. *The half time for this disappearance is of the order of 20 minutes at 28°;* more accurate studies are in progress.

Thus the hydrogen at C-2 of thiazolium salts exchanges with D_2O more rapidly than almost any other "active" carbon-bound hydrogen so far reported,[7] and this is especially striking since its activity apparently is derived merely from attachment to a doubly-bonded carbon and proximity to two electronegative atoms. Such factors are, of

I R_3—$\overset{+}{N}$ R_4 (−) S R_5

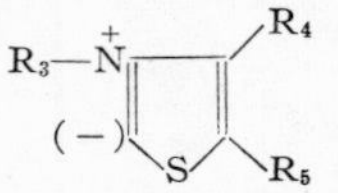

course, like the ones which stabilize cyanide ion, with its triple bond, but it is apparent that in suitable cases even a double bond is sufficient.[8]

(6) Performed by P. Corio and A. Okaya of this department.

(7) Most such exchanges require base or acid in order to proceed at an observable rate. As one example, it is reported that acetylene does not exchange with neutral D_2O after 36 hours (L. H. Reyerson and S. Yuster, THIS JOURNAL, **56**, 1426 (1934)).

(8) It seems that geometrical considerations rule out the possibility that sulfur assists ionization by valence expansion, as this would require a bent allenic carbon at C-2.

DEPARTMENT OF CHEMISTRY
COLUMBIA UNIVERSITY
NEW YORK, N. Y.
RONALD BRESLOW

RECEIVED JANUARY 30, 1957

Reprinted from THE JOURNAL OF BIOLOGICAL CHEMISTRY, Vol. 244, 1969

Initial and Equilibrium ^{18}O, ^{14}C, ^{3}H, and ^{2}H Exchange Rates as Probes of the Fumarase Reaction Mechanism*

(Received for publication, June 9, 1969)

J. N. HANSEN,‡ E. C. DINOVO,‡ AND P. D. BOYER

From the Molecular Biology Institute and the Department of Chemistry, UCLA, Los Angeles, California 90024

SUMMARY

Net conversion of malate to fumarate as measured by the A_{260} of fumarate is accompanied by a nearly equivalent loss of ^{3}H or ^{2}H from (2S,3R)-$3^{2 \text{ or } 3}$H-malate to water but by an appreciable incorporation of ^{18}O from water into unreacted malate. At equilibrium with high substrate concentrations, exchange of the hydroxyl oxygen of malate with water is faster but interchange of the methylene hydrogen is slower than the carbon skeleton interchange between malate and fumarate. Relative values for the $^{18}O_{(malate)} \rightleftarrows O_{(water)}$, $^{14}C_{(malate)} \rightleftarrows C_{(fumarate)}$, $^{3}H_{(malate)} \rightleftarrows H_{(water)}$, and the $^{2}H_{(malate)} \rightleftarrows H_{(water)}$ exchange rates are 4.0, 2.5, 1.0, and 1.0, respectively. Exchange rate differences and ^{18}O incorporation into malate as compared to net reaction are accentuated in a glycerol-water medium. The ^{14}C:^{3}H exchange ratio approaches a limiting value of below 1 at low substrate concentrations and a limiting value of between 2 and 3 at high substrate concentrations. Addition of 0.3 M ammonium sulfate accelerates the ^{3}H exchange, presumably by facilitating proton dissociation from the enzyme. The exchange patterns are consistent with a carbonium ion mechanism. Theoretical equations and a reaction scheme in accord with the results are given. These allow calculation of distribution patterns of enzyme-bound intermediates and ratios for a number of velocity constants.

Consideration of present information about the fumarase (fumarate hydratase) reaction (Equation 1) and the information obtainable from equilibrium reaction rate measurements suggested that considerable additional insight into the mechanism of this simple enzymic hydration might be obtained by measure-

$$\text{Fumarate} + \text{HOH} \rightleftarrows \text{L-malate} \quad (1)$$

ments of rates of exchange of the carbon skeleton of malate with fumarate and of the methylene hydrogen and the hydroxyl group with water. The only study reported on comparative rates of hydrogen and oxygen exchange in enzymic hydration reactions appears to be the valuable contribution of Rose and O'Connell with aconitase (1) demonstrating intramolecular transfer of hydrogen but not oxygen.

The results of our studies of equilibrium reaction rates and of velocity and isotope exchange with fumarase allow deductions about substrate and water binding and release steps, about distribution patterns of enzyme-bound intermediates, about ratios of various rate constants, and about carbonium ion participation in the catalysis. The findings appear to illustrate the value of measurement of several isotopic exchange rates in a simple enzymic catalysis and add to the value of equilibrium reaction rate measurements as probes of enzymic mechanism.

* This study was supported in part by Contract AT(11-1)34, Project 102 of the United States Atomic Energy Commission. Computing assistance was obtained from the Health Sciences Computing Facility, UCLA, sponsored by National Institutes of Health Grant FR-3.

‡ Supported by training Grant GM-00463 of the Institute of General Medical Sciences, United States Public Health Service.

EXPERIMENTAL PROCEDURE

Special Reagents—Heart muscle fumarase was obtained from Calbiochem or Boehringer-Mannheim as a crystalline suspension in ammonium sulfate solution, ^{14}C-fumarate, carboxyl-labeled, was obtained from Calbiochem (3.6 mCi per mmole) and from Nuclear-Chicago (20 mCi per mmole); tritiated water (100 mCi per ml) from New England Nuclear; D_2O (99.7%) from General Dynamics; $H^{18}O$ (1.5 and 10 atom %) from Miles Laboratories; 2,5-diphenyloxazole (PPO) and *p*-bis[2-(5-phenyloxazolyl)]benzene (POPOP) from New England Nuclear.

A water-miscible scintillation fluid prepared according to the method of Bray (2) contained 60 g of naphthalene, 100 ml of methanol, 20 ml of ethylene glycol, 4.0 g of 2,5-diphenyloxazole, and 0.2 g of *p*-bis[2-(5-phenyloxazolyl)]benzene. This was diluted to 1.0 liter with *p*-dioxane and protected from light.

Sodium fumarate was prepared by recrystallizing fumaric acid (Eastman) from hot water and titrating to pH 7.3 with saturated NaOH. Sodium malate was prepared by recrystallizing malic acid (Eastman) from a mixture of ethyl acetate and petroleum ether according to the method of Frieden, Wolfe, and Alberty (3) followed by washing with diethyl ether. The acid, washed with ether, was titrated to pH 7.3 with saturated NaOH. Alternatively, malic acid was also recrystallized from diethyl ether.

Optical Assays—Absorbance measurements were carried out at 260 or 270 mμ in a quartz cell, 1 cm in width, usually provided with quartz spacers to reduce the light path to 1.0 mm because most measurements were made at high concentrations of fumarate.

Sampling Procedure—Aliquots (usually 0.2 ml), removed at appropriate intervals, were heated for 1 min at 95° in distillation vessels or test tubes to inactivate the fumarase when exchange of ^{3}H or ^{2}H was measured. Portions of this solution were then used for various analyses. When zero time controls were necessary, an aliquot was heated prior to addition of enzyme. In experiments for which only separation of malate and fumarate was required, 0.1-ml aliquots were quenched by addition to 0.5 ml of 0.6 N HCl.

Separation of Reaction Components—The separation of malate and fumarate is based on the favorable distribution coefficients of malic and fumaric acids between water and ether. Fumaric acid is extracted into ether 100 times more readily than is malic acid (4). A series of three extractions (5 volumes of water-saturated ether used for each) of an acidified mixture of malic and fumaric acids sufficed for removal of more than 98% of the fumaric acid. Under the standard conditions used, replicate samples lost the same fraction (about 20%) of the malic acid to the ether layer. The radioactivity remaining as malate in the aqueous layer was then counted; duplicates agreed within 2%. Samples incubated with and without enzyme were compared directly to give the fraction of counts appearing in or disappearing from the malate fraction. In some cases, the radioactivity in the ether layer was counted, but the presence of malic acid reduced the sensitivity of the measurements.

Deuterated or tritiated water was separated from the other reaction components by distillation in a small, evacuated U-tube apparatus fitted with sample and receiver flasks. After an aliquot of the reaction mixture in the sample flask was frozen in liquid N_2, the apparatus was evacuated and then closed off, and the sample flask was warmed in warm water while the receiver flask was cooled in liquid N_2. An aliquot of the distillate was used for determination of ^2H or ^3H.

Preparation of Isotopically Labeled Compounds—^{14}C-Malate was prepared from ^{14}C-fumarate by equilibrating with fumarase, quenching with HCl, and removing ^{14}C-fumarate by the ether extraction procedure.

(2S,3R)-3-^3H-Malate was prepared by equilibrating a 1 M solution of fumarate in tritiated water (100 mCi per ml) with fumarase, acidifying with HCl, removing the precipitated protein, and recovering the tritiated water by distillation. Fumaric acid was removed by ether extraction, and the sample was dried under vacuum to leave a residue of malic acid.

(2S,3R)-3-^2H-Malate was prepared in the same manner as tritiated malate except that the equilibration was carried out in D_2O.

^{18}O-Malate, specifically labeled in the hydroxyl of malate, was prepared by equilibration of fumarate with 1.5 atom % of ^{18}O in the presence of fumarase. After acidification with HCl, fumarate was removed by ether extraction, and the ^{18}O-malate was precipitated as the barium salt by addition of 1.5 volumes of 95% ethanol and enough 50% ethanol saturated with $BaCl_2$ to give nearly complete precipitation. The precipitate, collected by vacuum filtration, was dried at 60° under vacuum. Barium malate prepared in this manner was used as a standard for evaluation of the ^{18}O analyses.

Determinations of Radioactivity—The ^{14}C and ^3H were counted with a Packard Tri-Carb scintillation spectrometer with the use of Bray's scintillation fluid (2). Experiments were designed so that all samples from the same experiment had the same chemical contents; thus, quench corrections were unnecessary.

^{18}O Determinations—The ^{18}O in the hydroxyl of malate was analyzed as CO_2 obtained by pyrolysis of the barium salt of malate at 270–290° in the presence of guanidine hydrochloride. Guanidine hydrochloride was found to react nonselectively with oxygens of malate to give an 80 to 90% yield of CO_2 based on all 5 oxygens of malate (5). The barium malate for ^{18}O analyses needed to be free of contaminating fumarate, carbonate, or other oxygen-containing compounds. A mixture of malate and fumarate containing 10 to 20 μmoles of malate in 1.0 ml was acidified with 1.0 N HCl to a red color of thymol blue (pH <1.5). The solution was extracted three times with 5-ml portions of water-saturated ether to remove fumaric acid. Then 0.5 ml of 50 volume % ethanol saturated with $BaCl_2$ was added, followed by 2.0 of absolute ethanol. Any acid-insoluble barium salts were removed by centrifugation and discarded. The supernatant solution was titrated to a blue-green color of thymol blue (pH about 8.5 to 9) by careful addition of saturated sodium hydroxide. The fluffy white precipitate of barium malate was collected by centrifugation. Carbon dioxide, which is absorbed rapidly under these conditions, is precipitated along with the malate. To remove carbonate, the precipitate was dissolved in 1.0 ml of water, and the solution was centrifuged to remove any insoluble barium salts. The barium malate was then reprecipitated by addition of 2.0 ml of absolute ethanol. After centrifuging, the supernatant was discarded, the barium malate was redissolved in 1.0 ml of water, the sample centrifuged once again to remove insoluble barium salts, and again precipitated with 2.0 ml of ethanol. This repeated precipitation was found to be essential for high accuracy. The remaining barium malate, resuspended in 0.5 ml of ethanol, was transferred to a breakseal tube (7 mm × 10 cm) with a Pasteur pipette. The breakseal tube was centrifuged, and the ethanol was discarded.

The breakseal tube was dried in a 60° vacuum oven for 1 hour, and then 20 to 30 mg of recrystallized guanidine hydrochloride were added. A slight excess was not harmful, but a large excess gave low yields of carbon dioxide. After further drying in the 60° vacuum oven overnight, the breakseal tubes were constricted, evacuated, and sealed off by heating the constriction with a soft gas flame. The sealed tubes were then heated overnight in a muffle furnace at 270–290°. The CO_2 was collected as described by Boyer *et al.* (6). The isotope ratio was measured with a mass spectrometer equipped with a specially designed, low volume, glass inlet system to enable determinations of isotope ratios to be made on samples as small as 1 μmole. The ^{18}O analyses by this procedure were sufficiently sensitive and reproducible to allow detection of differences of 0.002 atom % excess of ^{18}O (5) with samples containing 0 to 1.0 atom % excess.

Deuterium Determinations—The deuterium content of the water was determined by conversion of the water to H_2 and HD and measuring the isotopic ratio. The hydrogens of water were quantitatively converted to H_2 in the presence of granulated zinc at 405–415° according to the procedure of Graff and Rittenberg (7). For water analyses, a considerable simplification of their method could be made.

Five microliters of water were added to 250 mg of reagent zinc dust in a breakseal tube (7 mm × 10 cm), and the tube was constricted in a flame. The tube and zinc had been previously dried overnight at 40–60° in a vacuum oven. The breakseal tube was partially evacuated by brief exposure to vacuum such that little of the water was removed, then the constriction was closed with a soft flame.

An aluminum block, 5 × 5 × 10 cm, with 15 holes 9 mm in diameter and 4 cm deep drilled in one side, was used as a furnace. The block was heated with a burner to a temperature of 405–415° as measured by a thermocouple in a glass tube located in the middle of the block. The block was oriented so that the breakseal tubes were horizontal. Each breakseal tube was tapped gently to distribute the zinc evenly along its side to allow maximal surface area. After the tube was heated in the block for 10 min, removed, and cooled, the zinc was redistributed in the tube. An additional 10-min heating was employed to ensure complete reaction. Careful handling was required to prevent accidental explosion of the hot tube.

An aliquot of hydrogen was collected by breaking the breakseal tube with the aid of a magnet in an appropriate, evacuated vessel with a collection bulb attached, and the isotope ratios were determined with a suitable mass spectrometer. The procedure was sufficiently sensitive and reproducible to allow detection of a difference of 0.02 atom % excess with samples containing 0 to 1.0 atom % excess of deuterium (5).

Calculations—The rates of exchange between malate and fumarate for the reaction, fumarate + HOH ⇄ malate, with all substrates at equilibrium were calculated from the relation

$$R = -\frac{(\text{fumarate})(\text{malate})}{(\text{fumarate}) + (\text{malate})} \times \frac{\ln(1-F)}{t}$$

where R is the rate of exchange and F is the fraction of isotopic equilibrium attained in time t. A typical experiment is shown in Fig. 1. The rate of exchange, R, for a given reaction was evaluated by averaging the rates calculated for different times. Alternatively, R was calculated with the aid of an IBM 360 Fortran(H) program based on best least squares fit of a set of data points consisting of counts exchanged *versus* time (5). This approach often proved advantageous, because the removal of ^{14}C-malate counts during the ether extraction procedure made prediction of isotopic equilibrium difficult. The computer program gives extrapolation to infinite time and evaluation of F and R from the best fit for the data.

When exchange with a component of water is measured, the high molarity of water leaves only the malate concentration term in the equation, and $R = -(\text{malate}) \ln(1 - F)/t$.

RESULTS

Initial Reaction Rates of Malate as Measured by A_{260} or with ^{14}C, ^{3}H, or ^{2}H—Four of the five initial reaction rates measured with malate as substrate gave identical or nearly identical rates, namely, net conversion to fumarate as measured by increase in absorbance at 260 mμ, net conversion to fumarate as measured with ^{14}C, and loss of ^{3}H or ^{2}H from malate. As noted below, the initial reaction rate of the hydroxyl group of malate as measured with ^{18}O was considerably faster than other rates.

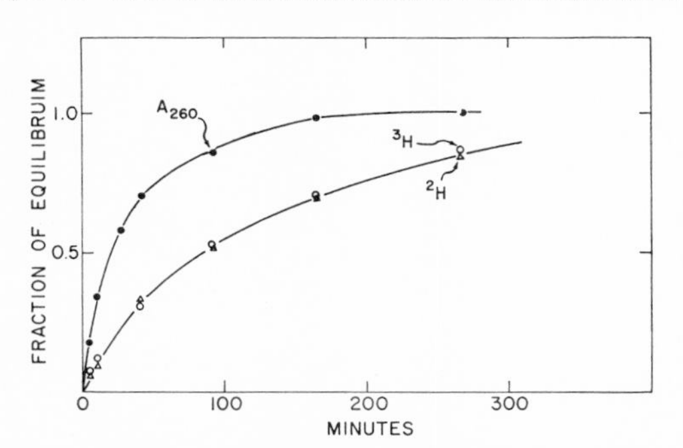

FIG. 1. Relative rates of reactions of malate catalyzed by fumarase as measured by absorbance change or appearance of ^{2}H or ^{3}H from malate into water. The 2-ml reaction mixture at pH 7.3 and 25° contained 0.105 M 3-(R)^{2}H-, 3-(R)^{3}H-malate (99 atom % excess of ^{2}H, 0.1 μCi of ^{3}H per μmole), 0.1 M Tris acetate, and 5 μl of fumarase (2 mg per ml in 2.0 M $(NH_4)_2SO_4$). Analyses were performed as described under "Experimental Procedure."

To illustrate a typical set of measurements as made to determine rates in these and subsequent studies, Fig. 1 gives the observed time course for increase in A_{260} and for loss of ^{3}H or ^{2}H from 3-(R)^{3}H- or 3-(R)^{2}H-malate in an aqueous medium. As noted in Fig. 1, the approach to equilibrium of the A_{260} value is considerably more rapid than that for the ^{2}H and ^{3}H measurements. Only about one-fifth of the malate needs to be converted to fumarate for chemical equilibrium, but, because of the high water molarity, essentially all of the malate will have reacted when ^{2}H or ^{3}H equilibrium is reached. Actual initial reaction rates calculated from the appropriate relationships are given in Table I. Within experimental error, the rates are identical.

The equality of the rate of loss of ^{2}H and ^{3}H and equality of both these rates with the net rate measured by A_{260} is consistent with the earlier observations of Alberty, Miller, and Fisher (8) and Fisher *et al.* (9) showing apparent lack of a primary deuterium isotope effect in the fumarase reaction.

As noted below, differences between initial and equilibrium reaction rates of the malate hydroxyl group and other rates are accentuated when the fumarase reaction is conducted in a medium of glycerol and water, 1:1 (v/v). Initial rates of malate reaction as measured by A_{260}, ^{14}C-malate conversion to fumarate, or loss of ^{3}H or ^{2}H from malate are nearly identical in the glycerol-water medium, as shown by the data in Table II. The slightly higher observed rate of loss of ^{3}H or ^{2}H as compared with the other measurements might be greater than the experimental error, but the possibility of a small difference was not

TABLE I

Comparison of initial rates catalyzed by fumarase in aqueous medium

Rates are calculated from the data given in Fig. 1.

Measurement	Rate
	mM/min
Malate → fumarate (A_{260})	0.98
$^{3}H_{(malate)} \rightarrow {}^{3}H_{(water)}$	1.0
$^{2}H_{(malate)} \rightarrow {}^{2}H_{(water)}$	1.0

TABLE II

Comparison of initial rates catalyzed by fumarase in glycerol-water medium

Reaction conditions were as given in Fig. 1 except 10 μM ^{14}C-malate was present (3.6 μCi per μmole), the reaction volume was 3 ml, the fumarase solution contained 10 mg per ml, and the medium consisted of glycerol and water, 1:1 (v/v).

Measurement	Rate
	mM/min
Malate → fumarate (A_{270})	0.74
$^{14}C_{(malate)} \rightarrow {}^{14}C_{(fumarate)}$	0.74
$^{3}H_{(malate)} \rightarrow {}^{3}H_{(water)}$	0.80
$^{2}H_{(malate)} \rightarrow {}^{2}H_{(water)}$	0.78

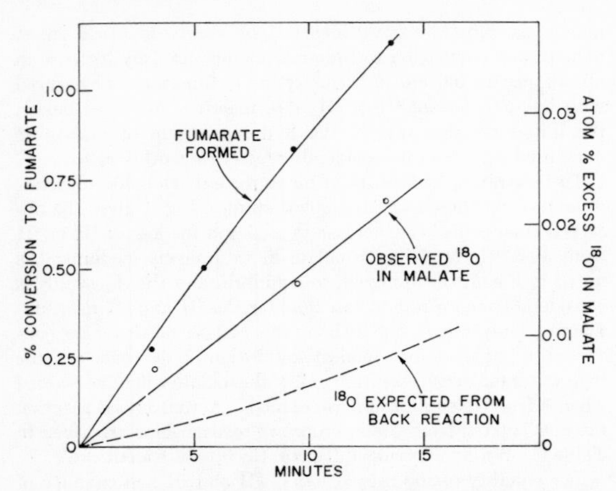

FIG. 2. ^{18}O incorporation into malate during the initial enzyme dehydration in an aqueous medium. A reaction mixture was prepared by diluting 114 μg of ^{14}C-fumarate (3.5 μCi per μmole), 26.8 mg of malic acid, and 55.6 mg of Tris base to 2.0 ml with $H^{18}OH$ (10 atom %) to give 0.1 M malate, 0.0005 M ^{14}C-fumarate, and 0.23 M Tris at pH 7.3 and 25°. Reaction was initiated by addition of 5 μl of 0.1 M Tris acetate, pH 7.3, containing 0.5 μg of fumarase, and analyses were made as described under "Experimental Procedure." The line for ^{18}O expected by back reaction was estimated from ^{14}C appearance in malate with consideration of the changing specific activity and concentration of the fumarate.

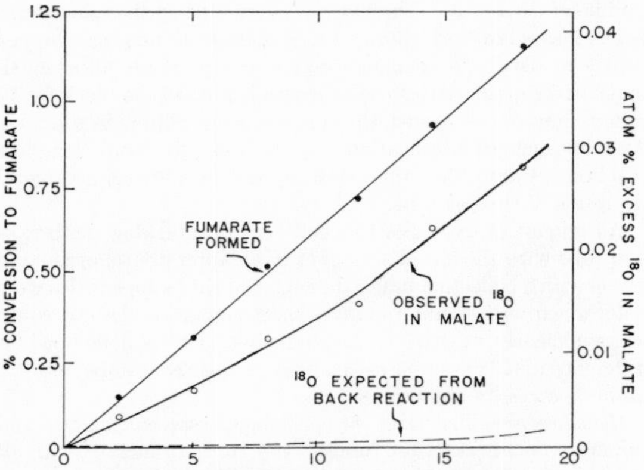

FIG. 3. ^{18}O incorporation with malate during the initial enzyme dehydration in a glycerol-water medium. Reaction conditions were as given with Fig. 2 except the medium consisted of water-glycerol (1:1, v/v) in a 3-ml final volume, pH 7.1, and the fumarase was 5 times as concentrated.

checked further. The principal point drawn from the data of Table II is that the four rates are at least close to identical.

Initial Reaction Rates of Malate as Measured by A_{260} and ^{18}O—Equilibrium rate measurements showed that the rate of the $O_{(malate)} \rightleftarrows O_{(water)}$ reaction was greater than the $C_{(malate)} \rightleftarrows C_{(fumarate)}$ reaction (5, 10). This was unexpected because of the reports of Miller (11) and Alberty (12) that the rate of ^{18}O loss from ^{18}O-malate was the same as the net rate of fumarate formation. The initial rate of the reaction of the malate hydroxyl group was thus measured with the use of a sensitive ^{18}O technique to detect any ^{18}O incorporation from $H^{18}OH$ into unreacted malate. Incorporation was found in excess of that expected for the occurrence of the over-all back reaction of fumarate hydration. To make a definitive assessment, the rate of fumarate hydration to give malate was determined by inclusion of a small amount of ^{14}C-fumarate in the reaction mixture. Results of measurements of the rate of ^{18}O incorporation as compared with the rate of net fumarate formation and ^{18}O incorporation by net fumarase hydration are given in Figs. 2 and 3. Fig. 2 gives results in an aqueous medium; Fig. 3 for glycerol-water medium, 1:1 (v/v). Both experiments were conducted with 0.1 M malate, and the ^{18}O measurements were made during conversion of the first 1% of the malate to fumarate. Hydration of fumarate to give malate was thus minimized.

The data of Figs. 2 and 3 establish that fumarase catalyzes an incorporation of water oxygens into the hydroxyl group of malate in considerable excess over that accounted for by the small net hydration of fumarate to malate. The relative rate of water-oxygen incorporation compared to net conversion to fumarate is greater in the glycerol-water medium. The ratio of the initial rates of the observed water-oxygen incorporation into malate compared to that expected from the fumarate back reaction in the aqueous medium was 4.5 and in the glycerol-water medium the ratio was 25. The initial rate of appearance of water-oxygen into malate was approximately equal to the rate of fumarate formation for both the water and the glycerol-water media.

Relative Equilibrium Reaction Rates—Results of comparative rate measurements of isotope exchanges at equilibrium using ^{18}O, ^{14}C, ^{3}H, and ^{2}H are given in Tables III and IV for a water and a glycerol-water medium, respectively. For experimental convenience, comparisons were made with only two different isotopes in a given experiment, with enzyme concentrations and sampling time as convenient for each isotope. In these experiments, malate and fumarate are present at concentrations much greater than their Michaelis constants. Equilibrium rates are calculated on the assumption that no isotope mass effects on the rate occur. This is experimentally justified for the hydrogen exchange by the near equality of the rates measured with ^{3}H and ^{2}H. Any isotope effects of ^{14}C and ^{18}O are undoubtedly small compared with the differences in equilibrium rates observed. In concurrence with the initial velocity measurements, the equilibrium rates of the ^{3}H and ^{2}H exchanges between the methylene hydrogen and water are equal. Also, the rates of ^{18}O interchange between the malate hydroxyl group and water are considerably greater than the rates of carbon skeleton interchange as measured with ^{14}C. A particularly interesting feature of the equilibrium rates is the finding that both the ^{18}O and the ^{14}C exchanges are considerably faster than the ^{3}H or ^{2}H exchanges. The differences in relative rates are accentuated in the glycerol-water medium in which the $O_{(malate)} \rightleftarrows O_{(water)}$ rate is 9.2 times greater than the $H_{(malate)} \rightleftarrows H_{(water)}$ rate.

The actual equilibrium rates in the glycerol-water medium were considerably slower than those in the aqueous medium for a given amount of fumarase. For example, the observed rate of the $H_{(malate)} \rightleftarrows H_{(water)}$ reaction in the aqueous medium (Table III) was 1.21 mM per min, and in the glycerol-water medium it was 1.0 mM per min. However, the enzyme was 11 times more concentrated in the glycerol-water medium, thus giving a relative rate for the same amount of enzyme about 13 times as great as that in the aqueous medium.

In our experiments, Tris acetate buffer, with or without addi-

TABLE III

Relative exchange rates of ^{14}C, ^{18}O, ^{2}H, and ^{3}H at equilibrium in aqueous medium

The relative rates were calculated from experiments with 0.1 M malate and 0.023 M fumarate in 0.1 M Tris-acetate at pH 7.3 and 25° in water for the following comparisons: (*a*) ^{14}C *versus* ^{18}O, 0.5 μCi of ^{14}C-fumarate per ml, $H^{18}OH$ with 1.37 atom % excess of ^{18}O, and approximately 33 μg of fumarase per ml; (*b*) ^{14}C *versus* ^{3}H, 0.5 μCi of ^{14}C-fumarate per ml, 0.2 μCi of 3-(R)^{3}H-malate per ml, and approximately 10 μg of fumarase per ml; (*c*) ^{3}H *versus* ^{2}H, 3-(R)^{2}H-malate with 99 atom % excess of ^{2}H, 0.2 μCi of 3-(R)^{3}H-malate per ml, and approximately 167 μg of fumarase per ml.

Isotope measurement	Relative rate
$^{3}H_{(malate)} \rightarrow {}^{3}H_{(water)}$	1.0
$^{2}H_{(malate)} \rightarrow {}^{2}H_{(water)}$	1.0
$^{14}C_{(fumarate)} \rightarrow {}^{14}C_{(malate)}$	2.5
$^{18}O_{(water)} \rightarrow {}^{18}O_{(malate)}$	4.0

TABLE IV

Relative exchange rates of ^{14}C, ^{18}O, ^{2}H, and ^{3}H at equilibrium in water-glycerol medium

The relative rates were calculated from experiments with 0.1 M malate and 0.048 M fumarate in 0.07 M Tris-acetate at pH 7.3 and in glycerol-HOH, 1:1 (v/v), for the following comparisons: (*a*) ^{14}C *versus* ^{18}O, 0.5 μCi of ^{14}C-fumarate per ml, $H^{18}OH$ with 0.4 atom % excess of ^{18}O, and approximately 17 μg of fumarase per ml; (*b*) ^{14}C *versus* ^{3}H, 0.5 μCi of ^{14}C-fumarate per ml, 0.2 μCi of ^{3}H-malate per ml, and approximately 110 μg of fumarase per ml; (*c*) ^{3}H *versus* ^{2}H, 3-(R)^{2}H-malate with 99 atom % excess of ^{2}H, 0.2 μCi of 3-(R)^{3}H-malate per ml, and approximately 83 μg of fumarase per ml.

Isotope measurement	Relative rate
$^{3}H_{(malate)} \rightarrow {}^{3}H_{(water)}$	1.0
$^{2}H_{(malate)} \rightarrow {}^{2}H_{(water)}$	1.0
$^{14}C_{(fumarate)} \rightarrow {}^{14}C_{(malate)}$	2.3
$^{18}O_{(water)} \rightarrow {}^{18}O_{(malate)}$	9.2

tions of NaCl, was used. Tests showed that the equilibrium constant was not changed by variation in acetate concentration. Acetate thus behaves like chloride but unlike phosphate; with phosphate, K_{eq} decreases with decreasing ionic strength (13).

Substrate Concentrations and Equilibrium Reaction Rates—The preceding experiments were performed with substrate concentrations far in excess of their Michaelis constant values and probably in excess of dissociation constant values. Inhibition of equilibrium exchanges at high substrate concentrations is particularly diagnostic of compulsory order of addition or release of substrates (14, 15). Thus, the continuation of the exchange of malate oxygen and malate hydrogen with water at high fumarate concentrations is a result of considerable mechanistic importance.

Further probes of substrate concentration effects were made with measurements of the rate of exchange of ^{14}C and ^{3}H. Data in Fig. 4 show the effect of increasing concentrations of malate and fumarate at equilibrium on the ^{14}C exchange rate. Substrate concentrations in this experiment were considerably lower than in the previous experiments. The results show a simple saturation dependence of exchange rate on the concentration. From the data of Fig. 4, the value of the "exchange constant" (14) for malate is 29 μM, and for fumarate it is 1.8 μM (12). These may

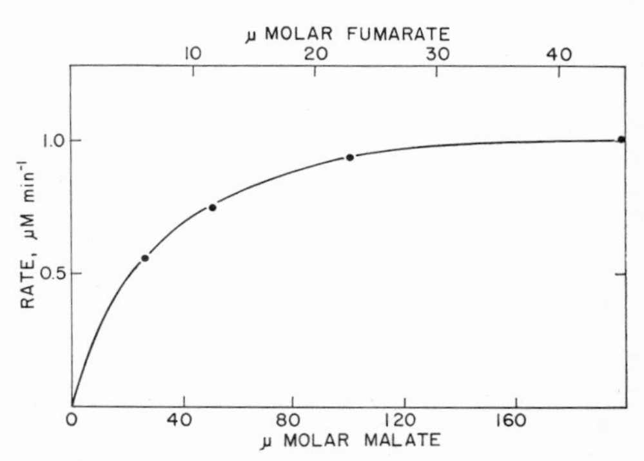

FIG. 4. Effect of substrate concentrations on the ^{14}C exchange at equilibrium. A reaction mixture containing 2.08×10^{-4} M malate and 4.73×10^{-5} M ^{14}C-fumarate (10 μCi per μmole) in 0.001 M phosphate buffer at pH 7.3 was diluted with the same buffer to give equilibrium concentrations of malate and fumarate as noted. Enzymic reaction was initiated by adding fumarase to give a final concentration of 0.013 μg per ml in the 10-ml total reaction volume at 25°. The reaction was quenched by addition of 1 ml of 1 N HCl. The fraction of approach to isotopic equilibrium, *F*, was calculated for each substrate concentration by comparison with samples containing 1 μg of fumarase per ml, sufficient for attainment of isotopic equilibrium under the conditions used.

be compared to the Michaelis constants under similar conditions of 7.5 μM for malate and of 1.8 μM for fumarate (12). The exchange constants are thus somewhat higher than the respective Michaelis constants.

As noted in the "Appendix" and considered below under "Discussion," simpler relationships for exchange rates are obtained at high substrate concentrations. Measurement of the effect of substrate concentration on exchange rates and on the ratios of these rates can allow deductions about the relative value of rate constants and about partitioning of intermediates among various reaction paths. Accordingly, the effect of a wide variation in substrate concentration on the absolute and relative rates of the ^{14}C and ^{3}H exchanges was measured. Control of ionic strength and of ionic milieu over such a wide range of substrate concentrations presents some problems. Thus, experiments were performed with and without addition of sodium chloride to maintain ionic strength invariant over part or most of the range tested. The experiments necessitated the use of widely different enzyme concentrations and incubation times at the extremes of the concentration ranges. Appropriate tests showed stability of the enzyme during the time periods used and linearity of the rates with time. To assess the validity of the assays, rates at an intermediate substrate concentration were determined with both the high and low enzyme concentrations used and with appropriate assay times. Satisfactory agreement of rates was obtained.

Observed ratios of the ^{14}C and ^{3}H exchange rates for substrate concentration change over four orders of magnitude are reported in Fig. 5. The ratio of the ^{14}C:^{3}H exchanges increased from a value of definitively less than 1 at low equilibrium concentrations of malate and fumarate to values of 1.9 (with added NaCl) or 2.5 with essentially only malate and fumarate as anions present. With added NaCl to help maintain ionic strength approximately constant, the ratios of the ^{14}C:^{3}H exchange were decreased somewhat throughout.

Absolute values of rates are not reported in Fig. 5. Both the ^{14}C and ^{3}H exchange rates showed similar patterns of an initial increase in rate with increase in substrate concentration at all ionic strengths. Under the conditions used, the rates did not show simple saturation behavior but continued to increase slightly with increase in substrate concentration up to about 0.05 M malate. Above this concentration, some inhibition of both the ^{14}C and ^{3}H exchange rates was noted, with over 50% inhibition of the ^{14}C exchange rate at the highest concentration tested (0.5 M malate). Preferential inhibition of the ^{14}C exchange resulted in the observed decrease in the ratio of the ^{14}C:^{3}H rates.

Effect of Other Conditions on Equilibrium Exchange Rates—An interesting observation is that the presence of 0.3 M ammonium sulfate considerably accelerates the ^{3}H exchange with relatively small increases in the ^{18}O and ^{14}C exchanges as shown by the data in Table V. As a result, the ^{3}H and ^{14}C exchange rates become close to equal. In some manner, the increase in salt nearly eliminates the factor or factors causing a slow $H_{(malate)} \rightleftarrows H_{(water)}$ exchange.

Penner and Cohen (16) have reported that ATP is a strongly negative effector and that AMP is a weakly positive effector for the pig heart fumarase reaction. In our studies, addition of up to 0.1 mM ATP was observed to inhibit the equilibrium exchange of the ^{14}C-fumarate with malate by about 40%. This inhibition was not overcome by increase in the AMP concentration up to 0.12 mM. When the glycerol-water medium (1:1, v/v) was used, addition of 10 mM ATP caused about the same inhibition of both the ^{14}C and ^{3}H exchanges.

An increase in hydroxyl ion concentration could accelerate the ^{3}H exchange at equilibrium if OH^- were the acceptor of a proton from a group on the enzyme at the catalytic site. Increase in pH from 7.3 to 8.4 did eliminate the difference in the ^{3}H and ^{14}C exchange rates. However, both rates were decreased, the ^{14}C being decreased nearly 5-fold. Thus, it was not feasible to determine whether increased OH^- concentration assisted proton removal from the catalytic site.

Buffer components present could likewise be acceptors of a proton at the catalytic site. Change in buffer from Tris acetate to imidazole acetate at pH 7.3 had no effect on the relative rates of the ^{14}C and ^{3}H exchanges.

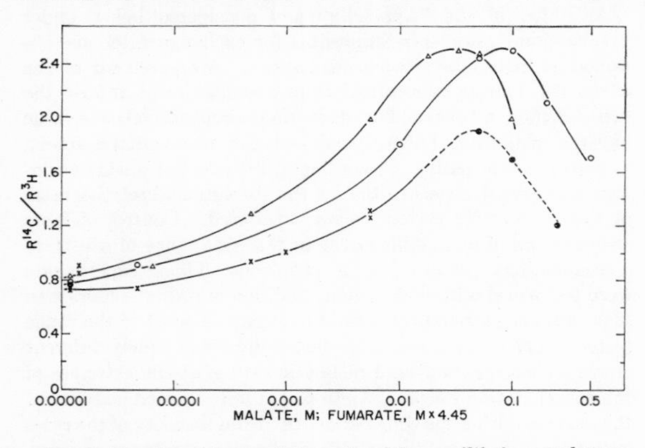

FIG. 5. The dependence of the ratio of the equilibrium exchange rates ($R^{14}C$:$R^{3}H$) on the concentration of substrates, ionic strength, and anion type. Substrates were increased in concentrations as indicated in the following solutions. △——△, 0.01 M Tris acetate, pH 7.3; ○——○, 0.1 M Tris acetate, pH 7.3; 0.1 M Tris acetate, pH 7.3, with sufficient NaCl added to keep total ionic strength (including substrates) constant at 0.25 (×——×) or at higher substrate concentration with 0.05 M NaCl added (⊗- - -⊗). The reaction mixtures contained the equivalent of approximately 10 μg of fumarase per ml for concentrations of 5 mM malate and above, and one-fiftieth of this amount for concentrations of <5 mM malate. Times of incubation ranged from 15 sec to 10 min for rate determinations and up to 24 hours for measurement of the distribution of ^{3}H and ^{14}C at isotopic equilibrium.

DISCUSSION

A minimal reaction scheme for the fumarase reaction, consistent with the results presented without $(NH_4)_2SO_4$ addition, is given in Fig. 6. This scheme includes only the requisite addition and release steps for malate, fumarate, OH^-, and H^+ that affect rates measured in our experiments. The intermediate $E \cdot X$ could represent a bound carbonium ion formed by loss of OH^- from malate or a bound fumarate and proton derived from malate ($E \cdot H \cdot F$) or an equilibrium mixture of these components.

The fastest reaction catalyzed by fumarase is the removal and addition of the hydroxyl group of malate, as shown both by the incorporation of water oxygen into unreacted malate during

TABLE V

Effect of ammonium sulfate on ^{18}O, ^{14}C, and ^{3}H exchanges at equilibrium in glycerol-water medium

A reaction mixture contained 0.104 M 3-(R)^{3}H-malate, 0.050 M ^{14}C-fumarate (1.5 μCi), 0.03 M Tris-acetate, and 110 μg of fumarase per ml in 7.2 ml of glycerol and water, 1:1 (v/v), at pH 7.3 and 25°. Rates were calculated from five measurements for each exchange at appropriate time intervals ranging from 6 to 300 min.

$(NH_4)_2SO_4$	Exchange	Relative rate
M		
0	$^{3}H_{(malate)} \rightleftarrows {}^{3}H_{(water)}$	1.0
0	$^{14}C_{(malate)} \rightleftarrows {}^{14}C_{(fumarate)}$	2.3
0	$^{18}O_{(malate)} \rightleftarrows {}^{18}O_{(water)}$	9.2
0.3	$^{3}H_{(malate)} \rightleftarrows {}^{3}H_{(water)}$	2.3
0.3	$^{14}C_{(malate)} \rightleftarrows {}^{14}C_{(fumarate)}$	2.7
0.3	$^{18}O_{(malate)} \rightleftarrows {}^{18}O_{(water)}$	11.6

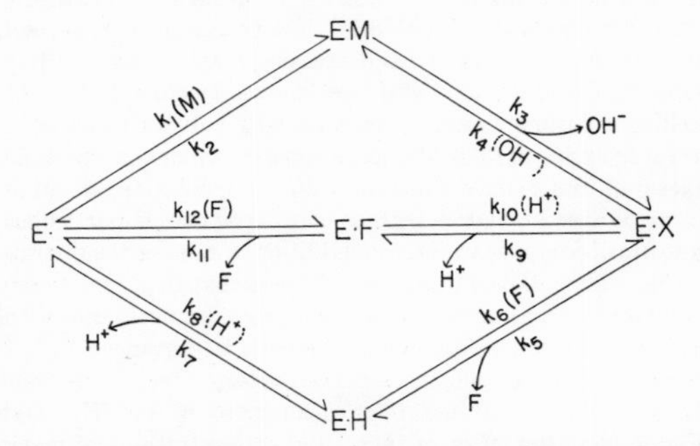

FIG. 6. A reaction pattern for fumarase. Participation of malate (M) and fumarate (F) in rate expressions are indicated in parentheses by the appropriate constants. Participation of (H^+) and (OH^-) are indicated in *dashed* parentheses by the appropriate rate constants. In derivations of equations, (H^+) and (OH^-) are regarded as constant and incorporated into values of the appropriate rate constants.

initial net fumarate formation and by the relative equilibrium exchange rates. This rapid reaction occurs when essentially only malate is present or in the presence of both substrates. Such a reaction is most logically explained by a binding of malate, conversion of the bound malate to bound fumarate (or to a carbonium ion precursor to fumarate) with release of OH^- to the medium, and reversal of these steps prior to release to the medium of enzyme-bound fumarate or of the hydrogen withdrawn from the $—CH_2—$ group of the malate. This is depicted in Fig. 6 by the malate binding and release steps governed by k_3 and k_4. The possibility exists that the hydroxyl group released from malate might be transitorily bound to the enzyme, but nothing in the present data requires such a binding. Hence Fig. 6 depicts direct release of the OH^- to water.

In the design of our experiments, the possibility was also considered that the OH^- released from malate might be bound to the enzyme in such a manner that it could form water only after the fumarate had dissociated. This arrangement of substrates at the active site is made impossible by the observation that the $O_{(malate)} \rightleftarrows O_{(water)}$ exchange rate exceeds other exchange rates at equilibrium, even in the presence of quite high concentrations of malate and fumarate. For example, the highest concentration of fumarate used, 90 mM, is about 50,000 times the Michaelis constant for fumarate (12) and 10,000 times the "exchange constant" (see Fig. 4) for the exchange of ^{14}C-fumarate with malate. An ordered sequence, in which fumarate formed from malate must be released before the OH^-, would require that the $C_{(malate)} \rightleftarrows C_{(fumarate)}$ exchange be at least equal to the $O_{(malate)} \rightleftarrows O_{(water)}$ exchange. In addition, such an ordered sequence would make likely an inhibition of the oxygen exchange by high fumarate. Neither of these effects was observed.

The observation from the equilibrium exchange measurements at high substrate concentrations that the carbon skeleton of fumarate can exchange with malate more rapidly than the methylene hydrogen of malate exchanges with water necessitates inclusion in the reaction scheme of an enzyme form with a bound proton but without bound fumarate. This is depicted by the formation of $E \cdot H$ from $E \cdot X$ as governed by k_5 and k_6 in Fig. 6. Formation of free E from $E \cdot H$ as governed by k_7 and k_8 completes the over-all reaction by this pathway. The inability of increase in the fumarate concentration at equilibrium to inhibit preferentially the ^3H exchange (Fig. 5) and the occurrence of ^3H exchange greater than ^{14}C exchange mean that the proton must be able to escape from the $E \cdot X$ complex without fumarate release. Thus, it is necessary to include the formation of $E \cdot F$ from $E \cdot X$ as governed by k_9 and k_{10}. Formation of E from $E \cdot F$ completes the over-all reaction by this pathway.

The "Appendix" gives derivations of relationships governing equilibrium reaction rates and isotope exchanges for the reaction scheme given in Fig. 6. Particularly instructive results may be obtained by comparison of the ratios of various exchange reactions and the effect of varying substrate concentration (while maintaining equilibrium) on these ratios. For both the ^{18}O and ^{14}C exchanges, increase in rate with increase in substrate concentration depends only on increase in the amount of enzyme with bound substrates (Equations 4 and 5, "Appendix"). Thus, the relationships predict that the exchanges will increase to a maximum as substrate concentration is increased analogous to dependence of initial rate on substrate in a simple Michaelis-Menten relationship. Such a behavior is shown by the ^{14}C exchange (Fig. 4); analogous tests with the ^{18}O exchange have not been made, in part because of the difficulty of ^{18}O measurement at the low substrate concentrations required.

In contrast to the ^{18}O and ^{14}C exchange rates, the ^3H exchange shows an additional dependence on the concentration of fumarate (see "Appendix," Equation 3). This arises because, with increase in the fumarate concentration, loss of enzyme-bound proton via the k_7 step can be prevented by the reaction of $E \cdot H$ with fumarate by the k_6 step to give $E \cdot X$. As a consequence, the ^{14}C:^3H exchange ratio (see "Appendix," Equations 9 and 10) can approach a maximum value greater than 1 at high substrate concentrations. Perhaps more important, the ratio must approach a limiting value of less than 1 at low substrate concentrations. The measurements of the concentration dependence of the ^{14}C:^3H exchange ratio (Fig. 5) are in accord with these predictions and give values for the ^{14}C:^3H exchange ratio of 1.9 to 2.5 at high substrate concentration, depending upon the nature and concentration of anion present, and of 0.75 to 0.90 at low substrate concentration.

The exchange ratio measurements allow calculation of the relative rates of formation of the $E \cdot X$ complex from substrates by the three alternative paths depicted in Fig. 6. These values, calculated as indicated in the "Appendix," are given in Tables VI and VII. The ranges given in the tables correspond to values for R_4 of 0.94 to 1.0; this is the range that gives positive values for various constants according to the reaction scheme used and evaluations as in the "Appendix." Experimental values for R_4 are close to this range. As noted in the tables, change in R_4 has only a small effect on the distribution patterns reported in Tables VI and VII. Table VI gives the relative rates of flow of proton from malate or water into $E \cdot X$ as fumarate and malate concentrations approach infinity or zero. Table VII gives the relative rates of incorporation of carbon into $E \cdot X$ by the three pathways. These relative rates are independent of substrate concentration. One important feature of these data is that no one pathway predominates for interchange of $E \cdot X$ with substrates and thus no single step can be regarded as rate-limiting in the catalysis.

The tabulations given in Tables VI and VII and the values for the ratios of various constants that may be calculated as indicated in the "Appendix" can at this stage be taken as approximations only. The number of experimental observations are somewhat limited. Perhaps more important, the effects of salts on the catalytic characteristics are somewhat complicated and are hard to control. The possibility clearly exists that as substrate concentrations are increased, either with or without maintenance of constant ionic strength, the actual values of some rate constants change. This would, of course, cause deviations from the relationships given in the "Appendix."

As noted in the "Appendix," it is theoretically possible to obtain values for the ratio of k_{10}:k_{11} from the present data. However, because the actual equilibrium flow of ^{14}C or ^3H through the k_{10} pathway is small, as characteristic of the system, quite small differences in experimental values considerably change the apparent ratio of k_{10}:k_{11}. Thus estimates are not reported herein.

The retention of a proton abstracted from substrate on the enzyme for sufficient time for a substrate to again bind and react, as noted by Rose and O'Connell with aconitase (1) and in the present studies with fumarase, is an interesting phenomenon. A possibility that must be considered is that the proton is on an imidazole residue of fumarase and that the dissociation of the

TABLE VI
Source of hydrogen in $E \cdot X$ at equilibrium

	Fraction from M by $k_1(M)$ k_3 pathway	Fraction from H_2O by k_{10} pathway	Fraction from H_2O by k_8 $k_6(F)$ pathway
As $(F) \rightarrow \infty$	0.75	0.25	0
As $(F) \rightarrow 0$	0.34–0.38[a]	0.11–0.12[a]	0.50–0.55[a]

[a] Range of values for permissible extremes of R_4.

TABLE VII
Source of carbon in $E \cdot X$ at equilibrium

Fraction from M by $k_1(M)$ k_3 pathway (a)[b]	Fraction from F by $k_{12}(F)$ k_{10} pathway (b)	Fraction from F by k_8 $k_6(F)$ pathway (c)
0.36–0.38[a]	0–0.12[a]	0.50–0.64[a]

[a] Range of values for permissible extremes of R_4.
[b] See "Appendix" for definition of a, b, and c.

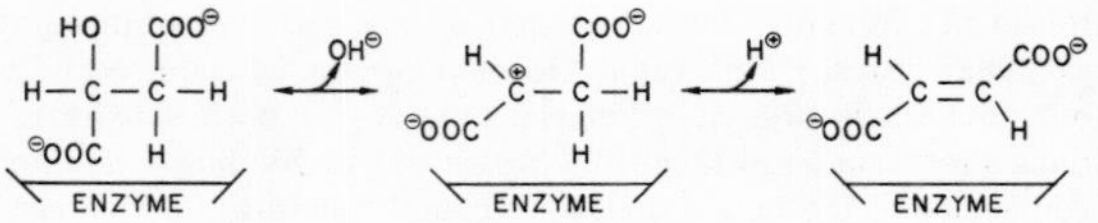

imidazolium proton ($k = 1.5 \times 10^3$ sec^{-1} according to Eigen and Hammes (17)) may be rate-limiting. The V_{max} for the initial rate of fumarate formation, however, corresponds to a k of about 1.7×10^2 sec^{-1}, assuming 4 catalytic sites per molecule. The pH-velocity data of several workers (18–20) are consistent with participation of 1 or 2 imidiazole residues in the catalysis, and Wagner, Chai, and Charlson (21) have demonstrated alkylation of a histidine residue by α-bromo-succinate. That the addition of ammonium sulfate nearly abolishes proton release as the slow step (Table V) is important in this regard. Such an effect of salt is not consistent with the dissociation of a proton from an imidazolium group as being rate-limiting, or for the approach of an OH^- or another anion as a proton acceptor. The ammonium sulfate effect could reflect a change in protein conformation that increases accessibility of water to the protonated site.

Occurrence of the fumarase reaction by a carbonium ion mechanism as suggested by previous researches (8, 22) is favored by the recent data of Nigh and Richards on the rate of reaction of mono- and difluorofumarates (23) and by the findings of Schmidt *et al.* on secondary isotope effects (24). That the exchange with water occurs more rapidly with the hydroxyl group than with methylene hydrogen of malate is clearly consistent with carbonium ion formation as depicted by Scheme 1. However, the diagnostic value of this test is obviated under most conditions as used herein by the observation that the exchange of methylene hydrogens with water is slower than the $^{14}C_{(malate)} \rightleftarrows {}^{14}C_{(fumarate)}$ exchange. Such retention of protons by the enzyme could mask the occurrence of the reaction by a concerted or a carbanion mechanism. Because proton dissociation from the enzyme in the presence of 0.3 M $(NH_4)_2SO_4$ is nearly as rapid as fumarate dissociation, the possibility exists that nearly all of the protons abstracted from the methylene hydrogen are released to the media in preference to reformation of the —CH_2— group. Were this the case, the more rapid oxygen exchange would argue strongly for a carbonium ion mechanism. Additional probes of salt and concentration effects might give additional clarification.

The preceding interpretations do not take into account the inhibition of the exchange rates and the decrease in the ratio of

SCHEME 1

the ^{14}C:^3H rates noted at high substrate concentrations (Fig. 5). These inhibitory effects arise at substrate concentrations far above the K_m values, and they probably represent decreases in the rate of one or more steps in the catalytic process. For example, decrease in the rate of malate binding and release or decrease in the rate of interconversion of $E \cdot M$ and $E \cdot X$ (Fig. 6) would cause decrease in and trend toward equality of the ^{14}C and ^3H exchange rates. Because the ^{14}C exchange rate was originally greater, it would decrease faster than would the ^3H exchange rate, to give the drop in ratio as observed.

Although the results of Table II indicate a possible incorporation of ^2H or ^3H from water into unreacted malate during the initial dehydration in the absence of fumarate, the amount of any such incorporation is much less than that of the water oxygen incorporation. This warrants further consideration because some incorporation of ^3H or ^2H into unreacted malate by the k_9 k_4 k_2 pathway without net conversion from fumarate would be a logical result of the scheme depicted in Fig. 6. Lack of a greater observed incorporation may represent a balance of two effects; the proton and substrate associations and conversions that favor incorporation and a secondary isotope effect that decreases incorporation (24).

The valuable data of Teipel and Hill (25) show 4 binding sites per mole of fumarase. The simplest interpretation is that the four sites have equivalent catalytic reactivity. If site or subunit interaction effects are brought about by change in salt or substrate concentrations and if these occur by changes in rate constants or by numbers of catalytically active sites, some of the considerations of substrate concentration effects reported herein will need to be modified accordingly.

APPENDIX

Relations for equilibrium reaction rates for the scheme depicted in Fig. 6 may be attained by adaptations of approaches previously used (14, 15). An approach similar in some conceptual aspects to that developed for these studies was recently reported by Yagil and Hoberman (26).

For example, to derive expressions for incorporation of ^3H from water into malate, use of the following definitions is convenient:

a = fraction of $E \cdot X$ containing hydrogen from malate of the reaction medium derived by the $K_1(M)$-k_3 pathway;

b = fraction of $E \cdot X$ containing hydrogen from water derived by the k_{10} pathway;

c = fraction of $E \cdot X$ containing hydrogen from water derived by the k_8-k_6 (F) pathway.

The sum of $a + b + c$ will equal 1 because all of the hydrogen in $E \cdot X$ must arise from either water or malate by one of the three routes. The fraction of $E \cdot X$ containing hydrogen from water is obviously $(a + b)/(a + b + c)$.

Similarly:

d = fraction of $E \cdot M$ derived most recently from $E \cdot X$;

e = fraction of $E \cdot M$ derived most recently from $E + M$.

The product $(b + c)d$ will equal the fraction of $E \cdot M$ containing hydrogen derived from water. Thus, the rate of ^{3}H from water into malate, assuming no isotope rate effects, will be equal to this product times the total $(E\cdot M)$ times k_2, the rate constant for release of M from $E\cdot M$, or

$$\frac{d^3\text{H}}{dt} = (b + c)dk_2(E\cdot M) \tag{1}$$

The fraction d is given by the ratio of the rate of formation of $E\cdot M$ from $E\cdot X$ to the total rate of formation of $E\cdot M$, or

$$d = \frac{k_4(E\cdot X)}{k_1(M)(E) + k_4(E\cdot X)}$$

Because the system is at equilibrium, $k_1(M)(E) = k_2(E\cdot M)$ and $k_4(E\cdot X) = k_3(E\cdot M)$, therefore,

$$d = \frac{k_3(E\cdot M)}{k_2(E\cdot M) + k_3(E\cdot M)} = \frac{k_3}{k_2 + k_3} \tag{2}$$

And, similarly,

$$e = \frac{k_2}{k_2 + k_3}$$

The rate of incorporation of hydrogen originating in medium malate into $E\cdot X$, designated r_a, is given by

$$r_a = ek_3(E\cdot M) = ek_4(E\cdot X) = \frac{k_2k_4(E\cdot X)}{k_2 + k_3}$$

By a similar approach

$$r_b = k_9(E\cdot X)$$

and

$$r_c = \frac{k_5k_7(E\cdot X)}{k_6(F) + k_7}$$

The fraction c is given by

$$c = \frac{r_c}{r_a + r_b + r_c}$$

or

$$c = \frac{\dfrac{k_5k_7}{k_6(F) + k_7}}{\dfrac{k_2k_4}{k_2 + k_3} + \dfrac{k_5k_7}{k_6(F) + k_7} + k_9}$$

Substitution into Equation 1 of this relation for c, an analogous relation for b, and the relation for d given by Equation 2 gives

$$\frac{d^3H}{dt} = \frac{k_9 + \dfrac{k_5k_7}{k_6(F) + k_7}}{\dfrac{k_2k_4}{k_2 + k_3} + k_9 + \dfrac{k_5k_7}{k_6(F) + k_7}} \times \frac{k_2k_3(E\cdot M)}{k_2 + k_3} \tag{3}$$

By a similar approach, incorporation of ^{14}C-fumarate into malate at equilibrium is given by

$$\frac{d^{14}C}{dt} = \frac{k_5 + \dfrac{k_9k_{11}}{k_{10} + k_{11}}}{\dfrac{k_2k_4}{k_2 + k_3} + k_5 + \dfrac{k_9k_{11}}{k_{10} + k_{11}}} \times \frac{k_2k_3(E\cdot M)}{k_2 + k_3} \tag{4}$$

The rate of ^{18}O incorporation from water into malate at equilibrium is given by the fraction of $E\cdot M$ derived from $E\cdot X$ times the concentration of $E\cdot M$ and its rate of formation of $E + M$, or

$$\frac{d^{18}O}{dt} = \frac{k_2k_3(E\cdot M)}{k_2 + k_3} \tag{5}$$

Equations 3, 4, and 5 will suffice for most considerations in this paper. If desired, the $(E\cdot M)$ term may be equated to substrate concentrations and E_t by use of the enzyme conservation equation and respective equilibria involved to give

$$(E\cdot M) = \frac{E_t}{1 + \dfrac{k_2}{k_1(M)} + \dfrac{k_3}{k_4}\left(1 + \dfrac{k_5}{k_6(F)} + \dfrac{k_9}{k_{10}}\right)} \tag{6}$$

Substitution of this relation for $(E\cdot M)$ into Equations 3, 4, or 5 gives relations for exchanges in terms of E_t, substrate concentrations, and rate constants.

Ratios of exchange rates lead to useful expressions. Ratios that may be estimated from experimental data include

$$R_1 = \frac{d^{18}\text{O}}{d^3\text{H}} \text{ as } (F) \rightarrow \infty$$

$$R_2 = \frac{d^{18}\text{O}}{d^3\text{H}} \text{ as } (F) \rightarrow \text{zero}$$

$$R_3 = \frac{d^{14}\text{C}}{d^3\text{H}} \text{ as } (F) \rightarrow \infty$$

$$R_4 = \frac{d^{14}\text{C}}{d^3\text{H}} \text{ as } (F) \rightarrow \text{zero}$$

$$R_5 = \frac{d^{18}\text{O}}{d^{14}\text{C}}$$

Relationships for these ratios as derived from preceding equations, and with additional designation

$$f = \frac{k_{11}}{k_{10} + k_{11}}$$

the fraction of $E\cdot F$ derived most recently from $E + F$, are as follows:

$$R_1 = 1 + \frac{k_4e}{k_9} \tag{7}$$

$$R_2 = 1 + \frac{k_4e}{k_5 + k_9} \tag{8}$$

$$R_3 = \frac{1}{1 + \dfrac{k_4e}{k_5 + k_9f}} \times \left(1 + \frac{k_4e}{k_9}\right) \tag{9}$$

$$R_4 = \frac{1}{1 + \dfrac{k_4e}{k_5 + k_9f}} \times \left(1 + \frac{k_4e}{k_5 + k_9}\right) \tag{10}$$

$$R_5 = \frac{R_1}{R_3} = 1 + \frac{k_4e}{k_5 + k_9f} \tag{11}$$

and also that

$$\frac{R_3}{R_4} = \frac{1 + \frac{k_5}{k_9}}{1 + \frac{k_5}{k_9 + k_4e}} \quad (12)$$

From Equations 7 and 10 and measured R values, one can obtain values for k_4e/k_9, k_5/k_9, and k_5/k_4e. From Equation 9 it is apparent that R_3 can be >1 if $k_5 + k_9f > k_9$. Also, R_4 must be <1 because f must be <1 and thus $k_5 + k_9f$ must be $<k_5 + k_9$.

Equation 12 gives the minimum possible value for k_5/k_9 as equal to R_3/R_4 and the maximum possible value of k_4e/k_9 as equal to $(R_3/R_4) - 1$. From the ratios of k_4e/k_9 and k_5/k_9, Equation 11 allows evaluation of f, and because $f = (k_{11})/(k_{10} + k_{11})$, this allows evaluation of k_{11}/k_{10}.

The fractions a, b, and c for hydrogen incorporation into $E \cdot X$ as defined previously and as given in Table VI are obtained simply from the above. For example, as $(F) \rightarrow$ zero

$$a = \frac{k_4e}{k_4e + k_5 + k_9}$$

$$b = \frac{k_9}{k_4e + k_5 + k_9}$$

$$c = \frac{k_5}{k_4e + k_5 + k_9}$$

Similar evaluations are obtainable for incorporation of H into $E \cdot X$ as $(F) \rightarrow \infty$ and for incorporation of carbon into $E \cdot X$.

REFERENCES

1. Rose, I. A., and O'Connell, E. L., *J. Biol. Chem.*, **242,** 1870 (1967).
2. Bray, G. A., *Anal. Biochem.*, **1,** 279 (1960).
3. Frieden, C., Wolfe, R. G., Jr., and Alberty, R. A., *J. Amer. Chem. Soc.*, **79,** 1523 (1957).
4. Collander, R., *Acta Chem. Scand.*, **3,** 717 (1949).
5. Hansen, J. N., Ph.D. thesis, University of California, Los Angeles, 1968.
6. Boyer, P. D., Graves, D. J., Suelter, C. H., and Dempsey, M. E., *Anal. Chem.*, **33,** 1906 (1961).
7. Graff, J., and Rittenberg, D., *Anal. Chem.*, **24,** 878 (1952).
8. Alberty, R. A., Miller, W. G., and Fisher, H. F., *J. Amer. Chem. Soc.*, **79,** 3973 (1957).
9. Fisher, H. F., Frieden, C., McKinley-McKee, J. S., and Alberty, R. A., *J. Amer. Chem. Soc.*, **77,** 4436 (1955).
10. Hansen, J. N., *Fed. Proc.*, **27,** 389 (1968).
11. Miller, W. G., Ph.D. thesis, University of Wisconsin, 1958 (University Microfilms 58-5363).
12. Alberty, R. A., in P. D. Boyer, H. A. Lardy, and K. Myrback (Editors), *The enzymes, Vol. 5*, Ed. 2, Academic Press, New York, 1961, p. 531.
13. Bock, R. M., and Alberty, R. A., *J. Amer. Chem. Soc.*, **75,** 1921 (1953).
14. Boyer, P. D., *Arch. Biochem. Biophys.*, **82,** 387 (1959).
15. Boyer, P. D., and Silverstein, E., *Acta Chem. Scand.*, **17,** Suppl. 1, 195 (1963).
16. Penner, P. E., and Cohen, L. H., *J. Biol. Chem.*, **244,** 1070 (1969).
17. Eigen, M., and Hammes, G. G., *Advan. Enzymol.*, **25,** 1 (1963).
18. Hammes, G. G., and Alberty, R. A., *J. Phys. Chem.*, **63,** 274 (1959).
19. Frieden, C., and Alberty, R. A., *J. Biol. Chem.*, **212,** 859 (1955).
20. Wigler, P. W., and Alberty, R. A., *J. Amer. Chem. Soc.*, **82,** 5482 (1960).
21. Wagner, T. E., Chai, H. G., and Charlson, M. E., *Biochem. Biophys. Res. Commun.*, **28,** 1019 (1967).
22. Teipel, J. W., Hass, G. M., and Hill, R. L., *J. Biol. Chem.*, **243,** 5684 (1968).
23. Nigh, W. G., and Richards, J. H., *J. Amer. Chem. Soc.*, **91,** 5847 (1969).
24. Schmidt, D. E., Jr., Nigh, W. G., Tanzer, C., and Richards, J. H., *J. Amer. Chem. Soc.*, **91,** 5849 (1969).
25. Teipel, J. W., and Hill, R. L., *J. Biol. Chem.*, **243,** 5679 (1968).
26. Yagil, G., and Hoberman, H. D., *Biochemistry*, **8,** 352 (1969).

Reprinted from THE JOURNAL OF THE AMERICAN CHEMICAL SOCIETY, Vol. 92, 1970

Nuclear Magnetic Resonance Assignment of the Vinyl Hydrogens of Phosphoenolpyruvate. Stereochemistry of the Enolase Reaction[1]

M. Cohn,[2a,b] J. E. Pearson,[2b] E. L. O'Connell,[2c] and I. A. Rose[2c]

Contribution from the Johnson Research Foundation, University of Pennsylvania School of Medicine, Philadelphia, Pennsylvania 19104, and Institute for Cancer Research, Fox Chase, Philadelphia, Pennsylvania 19111.
Received October 24, 1969

Abstract: From an analysis of the nmr spectra of phosphoenolpyruvate and phosphoenolpyruvate-1-^{13}C it is concluded that the downfield proton represents the position *trans* to the phosphate group. The phosphoenolpyruvate-3-*d* formed from (2*R*,3*R*)-2-phosphoglycerate-3-*d* by reaction with enolase is found to contain deuterium only in this position. Thus the elimination of water is specifically *anti*.

Enolase (EC: 4.2.1.11, 2-phospho-D-glycerate hydrolyase) catalyzes the reversible interconversion of 2PGA[3] and PEP[3] (reaction 1).

$$CO_2^- - HCOPO_3^{2-} - HCH(OH) \rightleftharpoons {}^-O_2C(OPO_3^{2-})C{=}C(H_A)(H_B) + H_2O \quad (1)$$

The problem of the stereochemistry of this reaction would be resolved if 3-monodeuterio-2-PGA of known stereochemistry were converted to PEP in which the position of the deuterium were established.

(3*R*)-2PGA-3-*d* was readily obtained by use of glycolytic enzymes; deuterium was introduced from D_2O in the conversion of G6P[3] to (1*R*)-fructose-6-P-1-*d* with phosphoglucose isomerase.[4,5]

Determination of the stereochemically appropriate position of deuterium in PEP-3-*d* formed by enolase from (3*R*)-2PGA-3-*d* was accomplished by comparing its nmr spectrum with that of unlabeled PEP, the H_A, H_B compound above, and that of PEP-1-^{13}C.

A preliminary assignment of the two proton peaks, H_A and H_B, in the nmr spectrum of PEP was made from a consideration of the difference of the chemical shifts $\nu_A - \nu_B$ and from the relative magnitudes of J_{AP} and J_{BP}, the coupling constants of the H_A and H_B proton, respectively, with phosphorus. Unequivocal confirmation of the assignment was made from the relative magnitudes of the ^{13}CH coupling constants J_{AC} and J_{BC} in PEP-1-^{13}C. The nmr spectra were recorded with a Varian HA-60 spectrometer except for the 1-^{13}C compound for which the proton spectrum was recorded with a Varian HA-100.

Preparation of Deuterated PEP. (1) G6P to (1*R*)-FDP-1-*d*.[3] G6P (2000 μmol of the K_2 salt, 50 μmol of $MgCl_2$, 10 μmol of adenosine diphosphate, 200 μmol of Tris[3]-HCl (pH 8.0) in 10 ml of D_2O (99%) and ^3H_2O (1540 cpm/μatom of H) was incubated with 1 mg of P-glucose isomerase (390 units) for 3 hr at 25° and 10 hr at 3°. Complete equilibration was shown by the specific activity of G6P + F6P (1800 μmol), 1540 cpm/μmol. To this solution were added the enzymes, P-glucose isomerase (1 mg, 390 units), pyruvate kinase (2 mg, 250 units), and P-fructokinase (2 mg, 200 units). K-PEP (2.4 *M*) in D_2O was added stepwise to the solution at 25° over a period of about 2 hr. By using the pyruvate kinase reaction to generate ATP, it was possible to maintain ATP, a strong inhibitor of P-fructokinase, at low concentration. FDP was determined[6] on acidified samples to be 1600 μmol. Water was removed by freeze-drying and the residue treated with 50 m*M* $HgCl_2$ (10 ml) to destroy excess PEP and inactivate the enzymes. The FDP was purified by ion exchange on Dowex-1-Cl (2 × 11 cm) by elution with 0.1 *N* HCl and was precipitated as the Ba salt, 1430 μmol, 1300 cpm/μmol.

(2) FDP to (2*R*,3*R*)-Glyceraldehyde-3-P-3-*d*. FDP was converted to the triose phosphate mixture in the presence of hydrazine to displace the reaction toward products:[7] FDP (1400 μmol), bovine serum albumin (30 mg), EDTA[3] (300 μmol), hydrazine sulfate (20 mmol), $NaHCO_3$ buffer (2.5 mmol), and 30 mg of muscle aldolase (420 units) purified to be free of triose-P isomerase were incubated at pH 8.5, 25°, in 25 ml for 60 min, at which time the increase in alkali labile phosphate[8] due to triose phosphates was complete. Protein was precipitated with $HClO_4$ (20 ml of 70%) and the solution neutralized in the cold with KOH and treated with five 20-ml portions of benzaldehyde to remove the hydrazine[7] and then with ether to remove benzaldehyde. The neutral mixture of triose phosphates was diluted to 1 l. and poured on a Dowex-1-Cl

(1) This work was supported in part by a National Science Foundation grant, GB-7114, and by U. S. Public Health Service grants, GM-12446 and CA-07818; in addition partial support came from institutional grants from the U. S. Public Health Service, CA-06927 and FR-05539, and the Commonwealth of Pennsylvania to the Institute for Cancer Research.

(2) (a) Career Investigator of the American Heart Association; (b) University of Pennsylvania School of Medicine; (c) Institute for Cancer Research.

(3) The following abbreviations will be used in this paper: 2PGA, (2*R*)-2-phosphoglycerate; PEP, phosphoenolpyruvate; G6P, glucose 6-phosphate; F6P, fructose 6-phosphate; FDP, fructose 1,6-diphosphate; Tris, tris(hydroxymethyl)aminomethane; P-, phospho; EDTA, ethylenediaminetetraacetate; 3PGA, (2*R*)-3-phosphoglycerate; DSS, Na salt of 2,2-dimethyl-2-silapentane-5-sulfonic acid.

(4) I. A. Rose, *Biochim. Biophys. Acta*, **42**, 159 (1960).

(5) C. K. Johnson, E. J. Gabe, M. R. Taylor, and I. A. Rose, *J. Amer. Chem. Soc.*, **87**, 1802 ('965).

(6) T. Bücher and H. J. Hohorst, "Methods of Enzymatic Analysis," H. U. Bergmeyer, Ed., Academic Press, New York, N. Y., 1965, p 246.

(7) P. Oesper and O. Meyerhof, *Arch. Biochem.*, **27**, 223 (1950).

(8) K. Lohmann and O. Meyerhof, *Biochem. Z.*, **273**, 60 (1943).

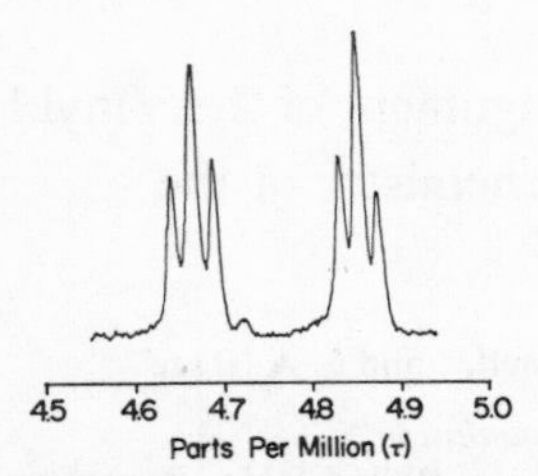

Figure 1. 60-MHz spectrum of 0.3 *M* PEP in D_2O at approximately pH 7.5 and 28°.

column (2 × 15 cm); the mixture of dihydroxyacetone-P (800 μmol) and D-glyceraldehyde-3-P (670 μmol) was eluted with 0.025 *N* HCl in 400 ml. This solution was treated with excess Br_2 at pH 5 (0.025 *M* Na acetate buffer) for 12 hr in the cold in order to oxidize the glyceraldehyde-P. The Br_2 was extracted with ether, the neutralized solution was poured on Dowex-1-Cl column (2 × 15 cm), and dihydroxyacetone-P (570 μmol) was eluted with 0.02 *N* HCl. It preceded the appearance of 3PGA by 100 ml.

(3) Conversion of Dihydroxyacetone-P to (3*R*)-3PGA-3-*d*. To the concentrated solution of dihydroxyacetone-P in 20 ml with 2 *M* Tris-HCl, pH 8.5, was added 2 mg of triose-P isomerase and after 30 min at 25° the solution, now largely the Tris adduct of glyceraldehyde-P, was put through Dowex-50-H^+ to acidify the product and remove the Tris. The combined column eluate and wash were treated with Br_2 as above and 380 μmol of 3PGA (1490 cpm/μmol) was recovered from a Dowex-1-Cl column with 0.02 *N* HCl.

(4) Conversion of 3PGA to PEP-3-*d*. The combined phosphoglycerate mutase and enolase equilibria were established at 55° to achieve optimal conversion, 32% to PEP as follows: 15 ml contained at pH 8, $MgCl_2$ (150 μmol), EDTA (30 μmol), 2,3-diphosphoglycerate (0.3 μmol), 15 mg of bovine serum albumin, enolase (1 mg, 27 units), and mutase (1 mg, 18 units). After 1 hr the reaction was terminated with acid and the 3PGA and PEP were separated on Dowex-1-Cl by elution with 0.02 *N* HCl and 0.04 *N* HCl, respectively. The 3PGA was treated with enzymes as above to increase the yield of PEP. The combined PEP (180 μmol, 1250 cpm/μmol), which had been neutralized immediately after elution, was adsorbed on Dowex-1-Cl and eluted as before, and the HCl removed by lyophiliza-The K^+ salt was used for nmr studies.

All enzymes used for this preparation, including the rabbit muscle enolase, were from Boehringer and Sons.

Preparation of PEP-1-^{13}C. Pyruvate-1-^{13}C was prepared by condensation of cuprous cyanide with acetyl bromide.[9] The CuCN was prepared by reaction with $CuSO_4$ and $K^{13}CN$ (61.0 atom % from Capintec Nuclear Co.) scaled down to use about 1 g of KCN.[10] The pyruvate obtained by hydrolysis of the amide in HCl was purified by ion exchange chromatography, Dowex-1-Cl, rather than distillation.

Phosphoenolypyruvate-1-^{13}C was prepared by phosphorylation of pyruvate with ATP in the PEP synthase reaction as follows: pyruvate-1-^{13}C (200 μmol), adenosine triphosphate (1 mmol), $MgCl_2$ (1 mmol), Tris-HCl (2 mmol), pH 8.0, and 2 units of *E. coli* PEP synthase in 30 ml were incubated at 25° for 3 hr leading to the complete conversion to PEP. The PEP was isolated on Dowex-1-Cl and HCl was removed by freeze drying.

Results

Pmr Spectra of Phosphoenolpyruvate. The 60-MHz proton spectrum of PEP in D_2O consists of two groups of peaks, one centered at 4.67 ppm and the other at higher field, 4.85 ppm (DSS[3] is used as reference at +10 ppm) as shown in Figure 1 at pH 7.5. One group corresponds to H_A, *trans* to the phosphate, the other to H_B, *cis* to the phosphate (*cf.* eq 1). The assignments are based on the difference in chemical shifts $\nu_A - \nu_B$ and on the fact that the coupling constant $J_{AP} > J_{BP}$. The chemical shift argument rests on the data of an extensive study by Brugel, *et al.*,[11] of the spectra of vinyl compounds of the general form

$$\begin{array}{ccc} H_A & & H_C \\ & C{=}C & \\ H_B & & R \end{array}$$

When R = diethylphosphate, $\nu_A - \nu_B = +0.31$ ppm; when R = carboxyl, $\nu_A - \nu_B = +0.49$ ppm. In PEP the contribution of the *trans*-phosphate group would be +0.31 ppm for $\nu_A - \nu_B$ and the contribution of the *cis* carboxyl group should be −0.49 ppm and the anticipated value for $\nu_A - \nu_B$ would be −0.18 ppm, in quantitative agreement with the observed value if the H_A resonance (*trans* to phosphate) is assigned to the lower field peak at τ 4.67 and H_B to the resonance at 4.85.

The line for each proton in PEP is split into two overlapping doublets due to couplings with the *gem* proton and with phosphorus corresponding to J_{AB}, J_{AP} for the proton H_A, *trans* to the phosphate group, and J_{AB}, J_{BP} for the proton H_B, *cis* to the phosphate group. J_{AB}, J_{AP}, and J_{BP} were evaluated from the relationships that the separation of the outermost peaks of the apparent H_A triplet, $J_{AB} + J_{AP} = 3.0$ cps, and of the corresponding H_B triplet, $J_{AP} + J_{BP} = 2.7$ cps. The ^{31}P spectrum of PEP was recorded and also consists of two overlapping doublets and the separation of the outermost lines, $J_{AP} + J_{BP} = 2.6$ cps. Thus $J_{AB} = 1.55$ cps, $J_{AP} = 1.45$ cps, and $J_{BP} = 1.15$ cps. Since the *trans*-coupling constant in vinyl phosphate compounds has been shown to be greater than the *cis*[11] and $J_{AP} > J_{BP}$, the assignment is again consistent with the assignment of H_A for the peak at τ 4.67. It is of interest to note that the coupling constants of the *gem* protons J_{AB} in vinyl compounds is linearly related to the electronegativity of the R group[12] and the value found for PEP; $J_{AB} = 1.55$ cps is the average of $J_{AB} = 2.3$ for R = diethylphosphate and $J_{AB} = 0.8$ cps for R = COOH. The chemical shifts of H_A and H_B and the coupling constants J_{AP} and J_{BP} are given as a function of pH in Table I.

Pmr Spectrum of Phosphoenolpyruvate-1-^{13}C. The small difference in the magnitude of the coupling constants J_{AP} and J_{BP} and, more important, the exception to the generalization that the *trans*-coupling constant is

(9) M. Calvin, C. Heidelberger, J. C. Reid, B. M. Tolbert, and P. F. Yankwich, "Isotopic Carbon," John Wiley & Sons, Inc., New York, N. Y., 1949, p 210.

(10) "Organic Syntheses," Coll. Vol. I, H. Gilman and A. H. Blatt, Ed., John Wiley & Sons, Inc., New York, N. Y., 1948, p 46.

(11) W. Brugel, Th. Ankel, and F. Kruckeberg, *Z. Elektrochem.*, **64**, 1121 (1960).

(12) T. Schaeffer, *Can. J. Chem.*, **40**, 1 (1962).

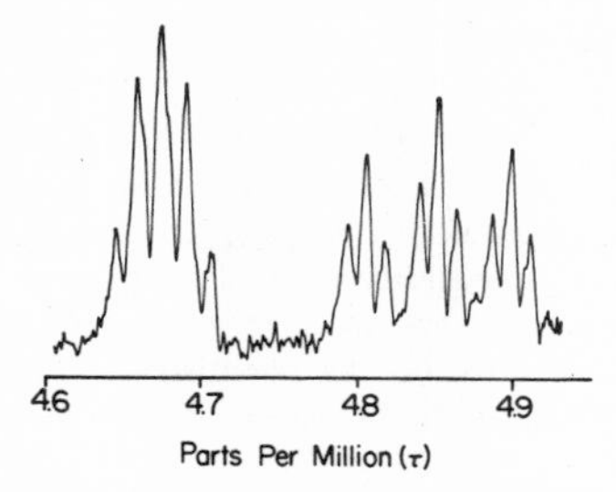

Figure 2. 100-MHz spectrum of 0.28 M PEP 60% 1-^{13}C, 40% 1-^{12}C in D_2O at approximately pH 7.6 and 31°.

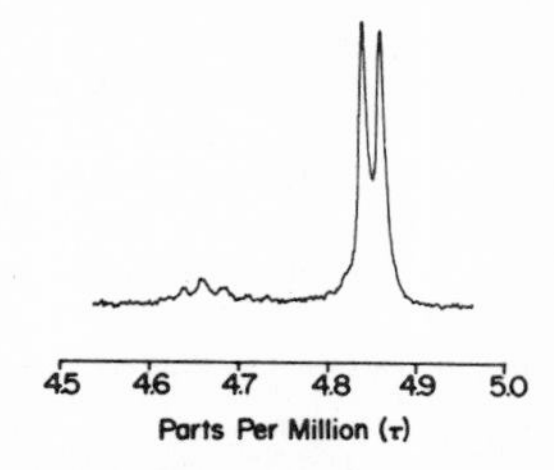

Figure 3. 60-MHz spectrum of 0.3 M deuterated PEP in D_2O at approximately pH 7.5 and 28°.

greater than the *cis* assigned to the vinyl phosphate compound Phosdrin[13] led us to seek confirmation of our tentative assignments in PEP by measuring ^{13}CH coupling constants. The *trans*-^{13}CH coupling constant in acrylic acid has been shown to be larger than the *cis*.[14]

Table I. Chemical Shifts and Proton–Phosphorus Coupling Constants as a Function of pH (0.1 M Phosphoenolpyruvate)

pH[a]	H_A	H_B
Chemical Shift Data, ppm (DSS[3] = +10 ppm)		
7.55	4.672 ± 0.007	4.850 ± 0.003
7.11	4.665 ± 0.013	4.848 ± 0.013
6.61	4.637 ± 0.007	4.835 ± 0.010
6.43	4.617 ± 0.013	4.825 ± 0.015
6.12	4.597 ± 0.008	4.815 ± 0.013
5.90	4.573 ± 0.010	4.810 ± 0.010
5.63	4.552 ± 0.008	4.800 ± 0.012
4.90	4.485 ± 0.008	4.765 ± 0.012
3.90	4.390 ± 0.010	4.692 ± 0.012
3.20	4.272 ± 0.003	4.592 ± 0.008
2.40	4.140 ± 0.008	4.483 ± 0.012
1.0	4.012 ± 0.023	4.383 ± 0.023
pH	J_{AP}	J_{BP}
Coupling Constant Data, cps		
7.55	2.62 ± 0.21	2.36 ± 0.12
7.11	2.66 ± 0.16	2.32 ± 0.32
6.61	3.01 ± 0.09	2.64 ± 0.05
6.43	3.00 ± 0.26	2.73 ± 0.06
6.12	3.24 ± 0.19	2.91 ± 0.09
5.90	3.26 ± 0.22	2.75 ± 0.10
5.63	3.40 ± 0.11	3.08 ± 0.08
4.90	3.97 ± 0.15	3.62 ± 0.05
3.90	4.03 ± 0.20	3.64 ± 0.15
3.20	4.30 ± 0.13	4.08 ± 0.12

[a] The pH values are directly observed readings on a pH meter with glass electrode of solutions in D_2O.

The spectrum of PEP-1-^{13}C synthesized as described and containing about 60% ^{13}C is shown in Figure 2. The usual unresolved triplet spectrum of the upfield peak at 4.85 ppm due to the ^{12}C component (~40%) is flanked on either side by two triplets due to the ^{13}C splitting. The ^{13}CH coupling constant is 9.2 cps. The ^{13}CH coupling constant for the low-field peak is 3.1 cps calculated by measuring the separation of the outermost peaks on the low-field multiplet (6.1 cps) and subtracting from it the separation of the outermost peaks of the ^{12}C PEP low-field triplet (3.0 cps). The PEP-1-^{13}C gives a larger difference between the coupling constants J_{AC} and J_{BC} than J_{AP} and J_{BP} as might have been anticipated from the fact that the ^{13}C is one bond closer to the vinyl protons than the ^{31}P. The ^{13}CH coupling constant data confirm the assignment of the downfield peak as H_A, the proton *trans* to the phosphate group.

Pmr Spectrum of Deuterated Phosphoenolpyruvate. The nmr spectrum of deuterated PEP formed from (3*R*)-PGA-3-*d* by the enolase reaction was determined in D_2O and was compared with a solution of PEP at the same pH and concentration. The PEP-3-*d* spectrum shown in Figure 3 has only a doublet due to ^{31}P coupling centered at τ 4.86 compared with the two sets of overlapping doublets centered for the nondeuterated compound at τ 4.67 and at 4.85. Thus the deuterium is in the position *trans* to the phosphate group and the PEP is designated (*E*)-PEP-3-*d*.[15] A small contamination of the protonated PEP with the usual unresolved triplet is seen at 4.67 ppm.

Discussion

The present study establishes that the elimination of the elements of water from (2*R*)-2-phosphoglycerate by enolase is stereospecific. The abstraction of –OH from the C-3 position of 2PGA is completely stereospecific within the error of the experiment since the low-field region of the deuterated species in Figure 3 shows a small contaminant of the fully protonated form but no evidence of a downfield doublet that could be attributed to a PEP species containing one deuterium at H_B. That the abstraction of hydroxide occurs *anti* to the C-2 proton that is eliminated is seen from the formulation

H; D; H; ^-O_2C; OPO_3^{2-}; OH — (3*R*)-2PGA-3-*d* ⇌ (enolase) ^-O_2C; H; D; OPO_3^{2-} — PEP-3-*d* (2)

The stereochemistry of the C-3 of the 2PGA-3-*d* which is established in the reaction of P-glucose isomerase to form fructose-6-P in D_2O is ultimately based[4] on the comparison of phosphoglycolate-2-*d* derived from C-1 and C-2 of this F6P with specifically deuterated ^{6}Li glycolate. The absolute configuration of the latter was determined by neutron diffraction.[5] The finding that deuterium in this PEP is only *trans* to the phosphate group leads to the conclusion that hydroxide addition occurs specifically from the *re* face[16] of the plane at C-3.

(13) A. R. Stiles, C. A. Reilly, G. R. Pollard, C. H. Tieman, L. F. Ward, D. D. Phillips, S. B. Soloway, and S. R. Whetstone, *J. Org. Chem.*, **26**, 3960 (1961).
(14) Personal communication from Dr. A. A. Bothner-By.
(15) J. E. Blackwood, C. L. Gladys, K. L. Loening, A. E. Petrarca, and J. E. Rush, *J. Amer. Chem. Soc.*, **90**, 509 (1968).
(16) The designation of the faces is according to the system of K. R. Hanson, *ibid.*, **88**, 2731 (1966).

Since (2*R*)-2-phosphoglycerate is the specific product, proton addition at C-2 must be from the *si* face[16] of the plane of PEP.

Having established the stereochemistry of specifically deuterated PEP, the way is open to the steric analysis of additions to C-3 that occur in many enzymatic reactions.[17,18]

(17) D. K. Onderka and H. G. Floss, *J. Amer. Chem. Soc.*, **91**, 5894 (1969).

(18) I. A. Rose, E L. O'Connell, P. Noce, M. F. Utter, H. G. Wood, J. M. Willard, T. G. Cooper, and M. Benziman, *J. Biol. Chem.*, **244**, 6130 (1969).

Acknowledgments. The authors wish to acknowledge the preparation of the pyruvate-1-^{13}C by Mr. J. R. Trabin and of PEP synthase by Dr. K. M. Berman, and to thank Dr. A. A. Bothner-By for suggesting the ^{13}C experiment and making the values of the long-range ^{13}CH coupling constants available to us prior to publication. We also wish to acknowledge our thanks to Miss Carol Hardy of the chemistry department for recording the 100-MHz spectrum shown in Figure 2.

Facilitated Proton Transfer in Enzyme Catalysis

It may have a crucial role in determining the efficiency and specificity of enzymes.

Jui H. Wang

A question frequently asked concerning enzyme action is, When a substrate molecule is bound at the active site of an enzyme, is the susceptible bond already distorted or under strain so that it is rendered more reactive? Recent x-ray data on the lysozyme-tri-*N*-acetylglucosamine complex suggest that when a larger substrate is bound to this enzyme there may be considerable distortion in the susceptible section of the substrate molecule (*1*). On the other hand, infrared studies show that the CO_2 molecule bound at the active site of carbonic anhydrase is definitely not distorted or under appreciable strain (*2*). These observations show that, although the "strain theory" (*3*) might be applicable in some cases, it cannot be the general explanation of enzyme catalysis. There is also the alternative theory of enzyme action based on the activation entropy effect, according to which enzyme catalysis is only a special case of general acid-base catalysis, having the particular advantage that the activation step does not involve a large decrease in entropy since the responsible acid and base groups are already nearby. While this activation entropy effect is undoubtedly an important factor, it is generally believed that enzymes must have additional characteristics which enable them to carry out their remarkable function. In this article I suggest that facilitated proton transfer along rigidly held hydrogen bonds (*4*) may play a crucial role in determining the efficiency and specificity of many enzymes. For clarity, let us examine these possibilities by considering selected examples.

The author is Eugene Higgins Professor of Chemistry and Molecular Biophysics, Yale University, New Haven, Connecticut. This article is based on a lecture delivered 4 April 1968 at the Symposium on Fundamental Aspects of Catalysis of the 155th American Chemical Society National Meeting, San Francisco.

Carbonic Anhydrase

Carbonic anhydrase contains one firmly bound zinc ion per enzyme molecule. It catalyzes the hydration of CO_2 to HCO_3^- and the reverse dehydration of HCO_3^-. Accurate difference infrared spectrometry shows that CO_2 bound at the active site of carbonic anhydrase exhibits an infrared absorption peak at wave number 2341 cm^{-1}, due to the asymmetric stretching of this linear molecule. Since this wave number is very close to the corresponding values for dissolved CO_2 (2343.5 cm^{-1} for CO_2 dissolved in water, 2340 cm^{-1} for CO_2 dissolved in methanol, and 2336 cm^{-1} for CO_2 dissolved in benzene) it has been concluded that the CO_2 at the active site is neither coordinated to the Zn(II) nor appreciably distorted, but is loosely bound to a hydrophobic surface or cavity of the protein, as in clathrate compounds (*2*). The infrared studies also show that the inhibitor azide ion is coordinated to the Zn(II) of the enzyme, and that the binding of a single azide ion at this Zn(II) prevents the binding of CO_2 at the specific CO_2 site mentioned above. Since the binding of inhibitors has not been observed to lead to gross conformational change in this enzyme (*5*), it was concluded that the specific CO_2 site must adjoin the Zn(II) so that the ligand azide can protrude at least partly into that site to interfere sterically with the binding of CO_2.

Nitrate and bicarbonate were found, in these infrared studies, to displace both the azide from the Zn(II) and the CO_2 from its specific binding site. But since the specific CO_2 site is merely a hydrophobic surface or cavity which loosely binds nonpolar molecules as CO_2 or N_2O in preference to polar water molecules or ions, we must conclude that the HCO_3^- is coordinated to the Zn(II) through its negatively charged oxygen atom in such a way that its relatively neutral oxygen atom and OH group are placed at the specific CO_2 site, as illustrated by structure I in Fig. 1 (*2*).

These results show that, in the dehydration reaction represented by the lower arrow in Fig. 1, proton transfer must accompany the breaking of the C–O bond, since we already know that only CO_2 is to be left in the hydrophobic binding site. Conversely, because of microscopic reversibility, it must be the OH^- on the Zn(II) which attacks the bound CO_2 and converts the latter to HCO_3^- in the reverse hydration reaction represented by the upper arrow in Fig. 1.

Although the foregoing conclusions are consistent with earlier suggestions (*6*), with kinetic data (*7*), and with the results of recent titration (*8*) and fluorescence (*9*) studies, two important riddles regarding the catalytic mechanism remain unsolved.

1) The value of first-order rate constant k_2 for the enzyme-catalyzed hydration step in Fig. 1 is 4×10^5 sec^{-1} at *p*H 7 and 25°C (*7*); this is 10^7 times the rate of hydration of CO_2 in the absence of a catalyst. The observed bimolecular rate constant for the reaction of OH^- and CO_2 at 25°C is $\sim 8 \times 10^3$ sec^{-1} M^{-1} (*7*). Using this latter value, we may estimate the pseudo-first-order rate constant for a hypothetical system in which a given OH^- ion is placed next to a CO_2 molecule. The estimated pseudo-first-order rate constant is $\sim 8 \times 10^3$ $(10^3/10)$ $(2/4) = 4 \times 10^5$ sec^{-1}. Because of the numerical uncertainties in this estimate, the exact value may be debatable, but its order of magnitude is significant. That this estimated value is of the same order of magnitude as the observed k_2 for carbonic anhydrase is surprising indeed, because the free OH^-, with $K_a = 10^{-15.7}$ as the acid dissociation constant of its conjugate acid, is a much stronger base than the OH^- coordinated at the Zn(II) of carbonic anhydrase (*6*), with $K_a = 10^{-7.1}$. Although the ratio of the nucleophilic reactivities of free and coordinated OH^- need not be equal to the ratio of their K_a values, for very similar reacting groups one would not expect these ratios to differ very much in order of magnitude. In other words, the OH^- coordinated to the Zn(II) of carbonic anhydrase reacts faster by several orders of magnitude than one would expect on the basis of a pure activation entropy effect. Therefore, the enzyme must have additional means of expediting the reaction. But what are the additional means?

2) The foregoing conclusion that proton transfer must accompany C–O bond breaking or bond formation is a very vague one. Specifically, we want to know whether the proton transfer precedes, is concerted with, or immediately follows the breaking or formation of the C–O bond. Are all three processes of proton transfer—namely, pretransition-state proton transfer, concerted proton transfer, and posttransition-state proton transfer—important enough to be considered, and, if so, what is the path of transfer in each process?

Unfortunately, in spite of the in-

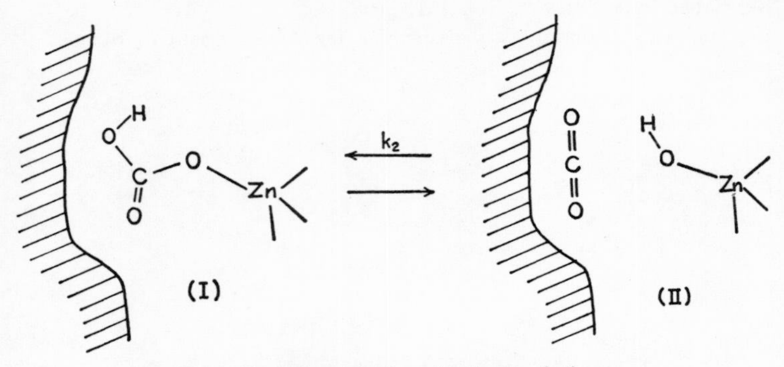

Fig. 1. Catalytic mechanism of carbonic anhydrase.

Fig. 2. Mechanism of the acylation step in the chymotrypsin-catalyzed hydrolysis of anilides or peptides.

teresting kinetic studies that have been made of carbonic anhydrase with other substrates (*10*), the experimental information seems still insufficient to allow us to reach a definite conclusion on these problems. On the other hand, the catalytic hydrolysis of a large number of substrates by α-chymotrypsin has been systematically and thoroughly investigated in many laboratories, under a variety of conditions. A careful examination of the existing data on chymotrypsin might, one would hope, throw some light on the nature of proton transfer in enzyme catalysis.

Chymotrypsin

α-Chymotrypsin catalyzes the hydrolysis of a large number of compounds according to the following scheme:

$$E + S \rightleftarrows ES \rightleftarrows \cdots \rightleftarrows E{-}\overset{\overset{\displaystyle O}{\|}}{C}{-}R + P_1 \rightleftarrows E + P_1 + P_2 \quad (1)$$

where E is the enzyme, S is the substrate, and ES is the enzyme-substrate complex, which transforms, through one or more steps, first to the acyl enzyme

$$E{-}\overset{\overset{\displaystyle O}{\|}}{C}{-}R$$

and splits off the product P_1, then splits off the second product, P_2, and regenerates the enzyme (*11*). Studies of amino acid sequences (*12*) show that in the acyl enzyme the acyl residue is attached to the OH group of serine-195 of α-chymotrypsin. Fast kinetic measurements suggest that this active serine may be hydrogen-bonded to a basic imidazole group (*13*). Recent x-ray data indicate probable hydrogen bonding between the active serine-195 and the basic imidazole group of histidine-57 (*14*).

In the hydrolysis of *p*-nitrophenyl esters and mixed acid anhydrides, the deacylation step—that is, the splitting off of P_2 and regeneration of the active enzyme—is rate-limiting. But in the hydrolysis of proteins, peptides, amides, anilides, and normal esters, the acylation step is rate-limiting. Let us first consider the chymotrypsin-catalyzed hydrolysis of the C–N bond, since presumably such catalysis is the principal function of chymotrypsin in nature.

Careful studies of the *p*H-dependence of the chymotrypsin-catalyzed hydrolysis of amides show that only one basic imidazole group is involved in the catalysis (*15*). On the other hand, the catalysis definitely involves protonation of the substrate, since the measured values of log k_2 bear a linear relationship to the values of pK_a of the protonated anilides, as predicted from the Hammett relationship $\log (K_a)_i/(K_a)_o = \rho\,\sigma$, where the parameter σ is characteristic only of the substituent and the parameter ρ is characteristic of the type of reaction under consideration (*16*). The only apparent way to reconcile these two sets of observations is to assume that the proton which is added to the nitrogen atom of the substrate came from the OH group of serine-195 via the nitrogen atom of histidine-57. A plausible path for this essential proton transfer is discussed below.

Let us assume that a good substrate RCONHR′ is bound to the enzyme in such a stereospecific way that the susceptible C–N bond of the substrate is placed in juxtaposition to the serine-histidine hydrogen bridge of the enzyme, and that the plane of the imidazole group of histidine-57 is roughly perpendicular to the plane determined by the basic nitrogen atom of the imidazole, the oxygen atom of serine-195, and the nitrogen atom of the substrate (*17*), as illustrated by structure I in Fig. 2. In view of the distribution of the outer σ- and π-electrons around the basic nitrogen atom of this imidazole, one expects structures I, II, and III in Fig. 2 to be in relatively fast protonation equilibrium via the electronic charge of this nitrogen atom. Complex III in Fig. 2 is expected to be highly reactive, since it contains an alkoxide group in juxtaposition to the carboxyl carbon atom of an already protonated anilide. Let us consider the rate of hydrolysis along the path

$$I \rightleftarrows II \rightleftarrows III \rightleftarrows IV \rightarrow V$$

in solutions where the concentration of the product $R'NH_2$ is negligible. To simplify the notation, let us define the first-order rate constants k_f, k_b, and k_a by the following equivalent reaction scheme

$$I \underset{k_b}{\overset{k_f}{\rightleftarrows}} III \overset{k_a}{\rightarrow} V \quad (2)$$

According to Eigen and his co-workers (*18*), the second-order rate constant for the recombination of H^+ and OH^- in ice at $-10°C$ is 0.86×10^{13} sec^{-1} M^{-1}; this is 70 times as fast as the corresponding process in water at 25°C. From this value we estimate the pseudo-first-order rate constant for the transfer of an excess proton of a hydronium ion in ice to a given water molecule among its four nearest neighbors to be (0.86×10^{13}) (55.5) (0.9) (1/4) (3/4), or $\approx 10^{14}$ sec^{-1}. This unusually high value is, according to Eigen and DeMaeyer (*4*), due to facilitated proton transfer along rigidly held H bonds in ice. In fact, for a hydronium ion held next to a hydroxide ion in an ice lattice, we may, in view of the absence of any restoring force, expect the rate of proton transfer from the former to the latter to be even faster. Since the thermodynamically favorable transition

$$III \overset{k_b}{\rightarrow} I$$

also represents proton transfers along preformed H bonds, we may infer that k_b is of the same order of magnitude. Making use of the relationship

$$\frac{k_f}{k_b} = \frac{(K_a)_{Ser}}{(K_a)_{SH^+}} \quad (3)$$

where $(K_a)_{Ser}$ and $(K_a)_{SH^+}$ represent the acid dissociation constants of serine-195 and the protonated substrate, respectively, we obtain

$$k_f \approx k_b\,(K_a)_{Ser}/(K_a)_{SH^+} \approx 10^{14}\,(10^{-13})/(10^{0.5}) \approx 10^{0.5} > 1\ sec^{-1}$$

if we choose a typical anilide with $pK_a \approx -0.5$ as the substrate. Since this value of k_f is larger than the measured values of k_2 for most anilides (*16, 17*), the proton-transfer

$$I \overset{k_f}{\rightarrow} III$$

cannot be the rate-limiting step in the above reaction scheme. Consequently it is justifiable to treat the catalytic hydrolysis of these substrates approximately as a pretransition-state protonation problem and to calculate the relative rates of hydrolysis of structurally similar anilides as follows (*17*):

$$\frac{[III]}{[I]} = \frac{k_f}{k_b + k_a} \approx \frac{k_f}{k_b} \quad (4)$$

$$\text{Rate} = k_2\,[I] = k_a\,[III] \approx k_a\,(k_f/k_b)\,[I]$$

or

$$k_2 \approx k_a\,(K_a)_{Ser}/(K_a)_{SH^+}, \quad (5)$$

For the enzyme discussed above, $(K_a)_{Ser}$ is a constant, and for a series of structurally very similar substrates, k_a may be assumed to be approximately constant. Therefore Eq. 5 gives

$$\log (k_2)_i - \log (k_2)_o = (pK_a)_i - (pK_a)_o = \rho(\sigma_i - \sigma_o) \quad (6)$$

for two substrates represented by the subscripts *i* and *o*, respectively. This result is entirely consistent with experimental data on the chymotrypsin-catalyzed hydrolysis of acetyl-L-tyrosyl anilides (*16, 17, 19*).

In addition, by combining the kinetic data for the hydrolysis of a series of acetyl-L-tyrosyl anilides in H_2O mixtures with the titration data for the corresponding acetanilides in protonated and deuterated mixtures, respectively, it is possible, from the treatment presented in the foregoing paragraph, to predict the k_2 values for the corresponding acetyl-L-tyrosyl anilides in D_2O mixtures. The predicted values are, within the limits of experimental uncertainty, in agreement with the directly measured values (*19*). These results suggest that for most anilides, and also for amides and peptides, which are even stronger bases, treatment as essentially a pretransition-state protonation problem provides an adequate description of the catalytic process, and that consequently for these substrates the transformation path

$$\mathrm{I} \rightleftarrows \mathrm{II} \rightleftarrows \mathrm{IV} \rightarrow \mathrm{V}$$

in Fig. 2 may be neglected in this treatment.

This approximation breaks down in the case of benzoyl-L-tyrosyl-nitroanilides, where not only is the substrate a much weaker base but the strong electron-withdrawing property of the p-NO_2 group weakens the susceptible C–N bond by decreasing its partial double bond character; this results in a much higher value of observed k_2 (*20*), so that Eqs. 4 and 5 are no longer valid.

For the chymotrypsin-catalyzed hydrolysis of esters the situation becomes even more unfavorable to treatment as pretransition-state protonation, both because esters are weaker bases and because the susceptible C–O bonds are also more labile than the corresponding C–N bonds. Therefore, for the hydrolysis of alkyl esters, the direct transformation of II to IV through the concerted and posttransition-state protonation mechanisms illustrated in Fig. 2 becomes more important.

The general catalytic scheme for α-chymotrypsin is summarized in Fig. 3, where the general substrate is represented by RCOXR′. For the hydrolysis of *p*-nitrophenyl esters, $XR' = OC_6H_4NO_2$, no protonation of substrate is necessary, since at $p\mathrm{H} > 7$ the resulting *p*-nitrophenylate ion, p-$NO_2C_6H_4O^-$,

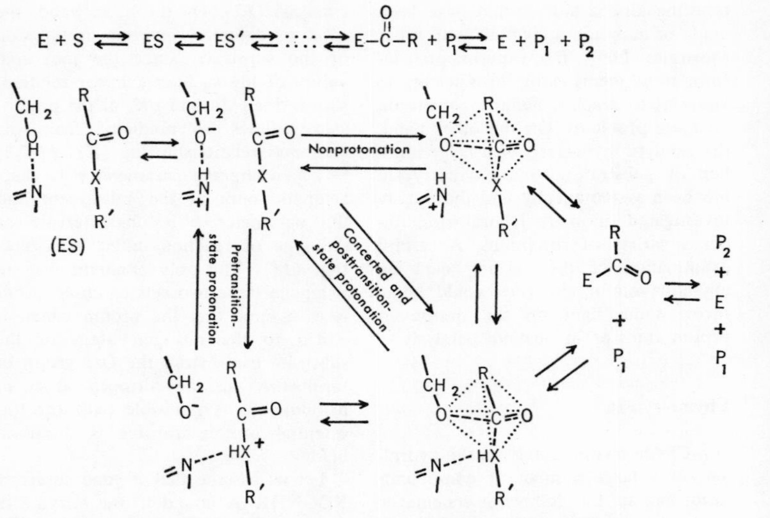

$$E + S \rightleftarrows ES \rightleftarrows ES' \rightleftarrows :::: \rightleftarrows E\text{–}\overset{O}{\overset{\|}{C}}\text{–}R + P_1 \rightleftarrows E + P_1 + P_2$$

Fig. 3. General reaction scheme for chymotrypsin catalysis.

might not even be protonated in solution. Consequently the hydrolysis can take place rapidly through the upper, nonprotonation path. In fact, for this case the acylation step becomes so rapid that it is no longer rate-limiting (*11*). Nevertheless, in all three of the reaction paths of Fig. 3, facilitated proton transfer along the rigidly held H bonds always occurs during the transformation of the enzyme-substrate complex I to the active-form complex II. From there on the paths differ according to the nature of the substrate.

Specificity

The proposed mechanism also suggests an attractive interpretation of the unsettled question of substrate specificity of enzymes (*21*). In the case of α-chymotrypsin, the substrates which are most efficient are presumably those in which the specificity group R in Fig. 2—for example, L-phenylalanyl, L-tyrosyl, or L-tryptophanyl residue—interacts with the enzyme and stabilizes the H bonds crucial for facilitated proton transfer. By analogy, R in Fig. 2 may also be taken to represent the L-arginyl or L-lysyl residue of the efficient substrate in tryptic hydrolysis, presumably via the same catalytic mechanism (*22*).

Recently two groups of workers have independently succeeded in replacing the active serine residue of subtilopeptidase A (subtilisin) by an SH group without either configurational or conformational change (*23*). The "chemical mutant" so obtained has been christened thiol-subtilisin. It is one-third as active as subtilisin in catalyzing the hydrolysis of *p*-nitrophenyl acetate (NPA), almost inert toward acetyl-L-tyrosine ethyl ester, and completely inert toward natural proteins. By contrast, subtilisin hydrolyzes proteins 10^2 to 10^3 times as fast as it hydrolyzes NPA, and hydrolyzes acetyl-L-tyrosine ethyl ester 3 to 4 times as fast as NPA (*23*).

This surprising observation cannot be reconciled with the usual activation entropy theory of general acid-base or "push-pull" catalysis. While the concentration of –SH and –OH groups at the respective active sites of these two proteins, subtilisin and thiol-subtilisin, cannot be very different, the reactivities should greatly favor thiol-subtilisin. This should be the case since not only does the mercaptide ion have a nucleophilic action 10^2 to 10^3 times as fast as that of the alkoxide ion but, at $p\mathrm{H}$ 7 to 8, the concentration of RS^- is higher than that of RO^- by a factor of $(10^{-9})/(10^{-13})$, or $\sim 10^4$. Thus, on the basis of the activation entropy theory we would expect thiol-subtilisin to be 10^6 to 10^7 times as active as subtilisin—an expectation in complete disagreement with the facts!

On the other hand, if we assume that subtilisin catalyzes through a facilitated proton transfer mechanism similar to that depicted in Fig. 2 for chymotrypsin, we would expect the replacement of an oxygen atom (van der Waals diameter, 2.80 angstroms; covalent diameter, 1.32 angstroms; single bond angle, 105 degrees) by a sulfur atom (van der Waals diameter, 3.70 angstroms; covalent diameter, 2.08 angstroms; single bond angle, 92 degrees) at the active center to disrupt the rigidly and accurately held H bonds in the natural enzyme-substrate complex which are required for facilitated proton transfer from the active OH group to the susceptible nitrogen atom of the substrate. For a substrate such as NPA, which does not need protonation, the effect of replacing the active OH group in subtilisin by an SH group may not be very pronounced, since at *p*H 7 to 8 the equilibrium concentration of the very reactive mercaptide group may be high enough to catalyze the hydrolysis of the substrate at a moderate rate. But for normal alkyl esters, and particularly for proteins where protonation of the leaving moiety of the substrate is essential, such a disruption of the crucial H bonds could completely wipe out facilitated proton transfer, with the consequence that the enzyme-substrate complex I (see Fig. 2) cannot reach its transition state IV fast enough, even though the free energy of IV in the modified system may not be very different from its free energy in the natural system.

Ribonuclease

Recent x-ray results on ribonuclease A (*24*) and ribonuclease S (*25*), in combination with the data on amino acid sequence (*26*), generated a three-dimensional structure for ribonuclease which is consistent with most of the chemically deduced structural information (*27*). In particular, the structure shows that histidine-12 and histidine-119 are both very close to the active site. With this information, it is possible to construct a catalytic mechanism based similarly on facilitated proton transfer, as illustrated in Fig. 4.

The mechanism for the cyclization step in ribonuclease catalysis, as illustrated in Fig. 4, is quite similar to the mechanisms previously suggested by workers in several laboratories (*28*), except with respect to the following:

1) The basic histidine residue and the 2′-OH group in Fig. 4 are assumed to play roles in ribonuclease catalysis similar to those of histidine-57 and serine-195, respectively, in chymotrypsin catalysis, and facilitated proton transfer can take place as illustrated in the transformations I $\rightleftarrows$ II, II $\rightleftarrows$ III, III $\rightleftarrows$ IV, and possibly IV $\rightleftarrows$ V.

2) The initially protonated nitrogen atom of another histidine or lysine side chain is assumed to be hydrogen-bonded to a negatively charged oxygen atom to form a salt linkage, not to the neutral oxygen atom attached to the next nucleotide R. This assumption is supported by the observations that substrates stabilize the enzyme against denaturation, whereas triesters of phosphoric acid are not affected by the enzyme (*27*).

3) The intermediates III and IV are included in Fig. 4 to make the catalysis consistent with the elegant results obtained by Westheimer's group on the hydrolysis of small phosphates and phostonates (*29*). Protonation is assumed to assist the recipient oxygen atom to reach an apical position in the trigonal-bipyramidal configuration III, and deprotonation returns it to the equatorial position, in IV.

Figure 4 is otherwise self-explanatory. To find the mechanism of the decyclization step, all one need do is replace the HOR in configuration V by H_2O and follow the reaction backward.

Ribonuclease is known to catalyze specifically the formation and breakdown of the 2′ O–P bond of ribose residues with a pyrimidine group (*Py* in Fig. 4) attached to it. In chymotrypsin catalysis, the essential specificity group (the phenyl, 4′-hydroxyphenyl, or 3′-indolyl group of the susceptible L-amino acid residue) is three atoms away from the newly formed C–O bond in the acyl-enzyme intermediate. Similarly, for ribonuclease, the pyrimidine group is three atoms away from the newly formed P–O bond in the cyclic intermediate. If the function of the specificity group in chymotrypsin catalysis is to stabilize the crucial H bonds by secondary interactions with the enzyme, one may wonder if the pyrimidine group in ribonuclease catalysis plays a similar crucial supporting role without itself participating directly in the chemical reaction.

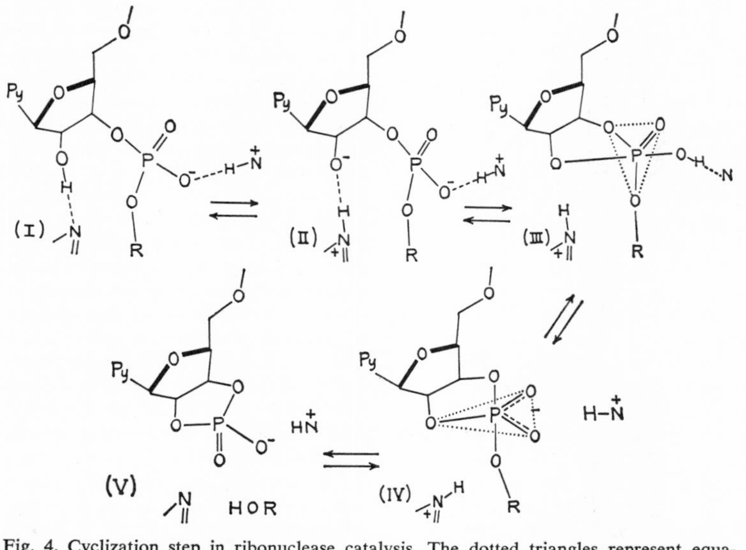

Fig. 4. Cyclization step in ribonuclease catalysis. The dotted triangles represent equatorial planes in the trigonal-bipyramidal conformation. *Py* represents a pyrimidine group. The tertiary nitrogen atom represents the basic nitrogen atom of the imidazole group of a histidine residue. The other nitrogen atom belongs either to another histidine residue or to a lysine residue.

Alcohol Dehydrogenase

Alcohol dehydrogenases of both yeast and liver are zinc enzymes which catalyze the following reaction (*30*) in a stereospecific manner (*31*):

$$CH_3CHO + NADH + H^+ \rightleftarrows CH_3CH_2OH + NAD^+ \quad (7)$$

The competition of imidazole with substrates for binding by the enzyme suggests that the substrates may be bound to the Zn(II). The displacement of bound NAD+ by p-mercuribenzoate and the protection of the enzyme from iodoacetamide by NAD+ suggest the presence of a sulfhydryl group in close proximity to the bound NAD+ (30).

Since, in media of low dielectric constant, 1-methyl-nicotinamide and donor anions are known to form stable ion pairs with characteristic charge-transfer spectra (32), one would expect that the binding of NAD+ by alcohol dehydrogenase might result in the formation of a similar ion pair involving the positively charged nitrogen atom of NAD+ and a negatively charged sulfur atom of the adjacent mercaptide group of the enzyme. If one assumes that the bound ethanol is coordinated to the Zn(II), with one of the hydrogen atoms of its C-1 atom oriented for stereospecific transfer to the C-4 atom in the nicotinamide ring of the coenzyme (31) and with its methyl group still free to rotate (33), one arrives at the arrangement illustrated by structure I in Fig. 5.

In structure I of Fig. 5, the OH group of the bound ethanol is shown hydrogen-bonded to the mercaptide group of the protein. Facilitated proton transfer along this H bond brings structure I into rapid equilibrium with structure II. Structure II should be thermodynamically very unstable, because of its net positive charge immersed in a medium of relatively low dielectric constant, and hence tends to decrease, through inductive effect, the electron density at the C-4 atom, as indicated by the arrows. Meanwhile the activated oxygen atom of the alkoxide ion in II tends to form an additional bond with its own carbon atom, and thereby to weaken the adjacent C–H bond. These two effects in II could effectively cooperate to assist the hydride ion transfer which changes II to III. Conversely, in the reverse hydrogenation reaction, III could change to II by hydride ion transfer and then transform to I by facilitated proton transfer.

Since the hydride and proton transfers illustrated in Fig. 5 are linked processes (34), there is a reciprocal relation between them. In other words, neither process has to precede the other. Consequently, the reaction path I ⇄ IV ⇄ III shown in Fig. 5 should be just as satisfactory as the path I ⇄ II ⇄ III.

A similar mechanism may be suggested for L-glutamate dehydrogenase, in which the NH_2 group of the substrate replaces the OH group of ethanol in Fig. 5. In addition, the observed esterase activity of D-glyceraldehyde-3-phosphate dehydrogenase after the removal of NAD+ (35) and the identification of the active SH group in this enzyme by means of radioactive labels (36) are also consistent with the present interpretation. These observations suggest that dehydrogenase may catalyze through a mechanism which has much in common with the mechanisms for hydrases and hydrolytic enzymes.

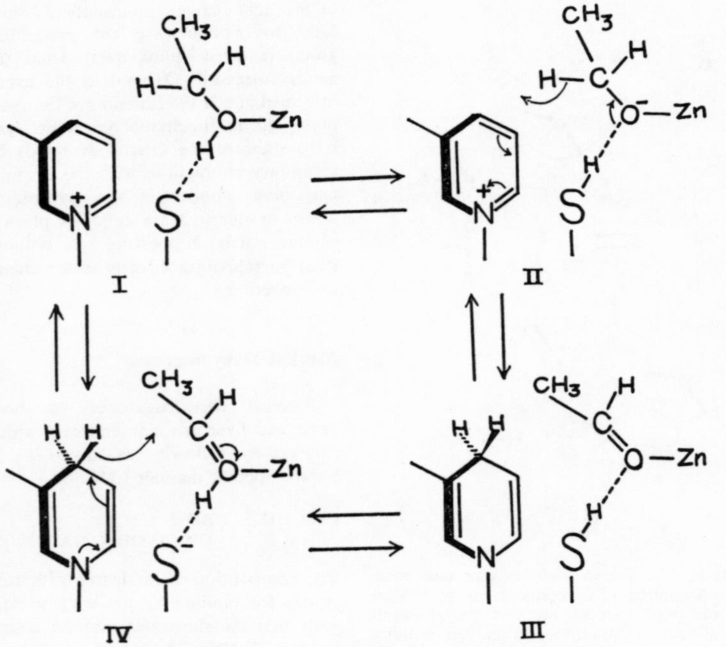

Fig. 5. Possible catalytic mechanism for alcohol dehydrogenase.

Conclusion

Ever since Fischer first developed his famous "lock and key" analogy (37), the secret of enzyme action has posed an intellectual challenge to scientists in several fields. The crux of this problem may not be resolved for some time, but any attempt to seek a broad catalytic principle which transcends the idiosyncracies of individual enzymes is likely to accelerate our rate of progress toward the truth. Two of the most fruitful examples of such attempts are the activation entropy theory inspired by the work of Swain and Brown (38) and the strain theory of Quastel (3), which has recently been revived by the exciting work of Phillips' group (1).

As an additional attempt, the present article proposes that facilitated proton transfer along rigidly and accurately held H bonds in the enzyme-substrate complexes, similar to the transfer mechanism in ice, discovered by Eigen and his co-workers (4, 18), may play a crucial role in enzyme catalysis. In other words, enzymes catalyze not only by lowering the free energy of the transition state (or states), as assumed in all previous theories, but also by enabling the system to reach the transition-state (or states) faster through facilitated proton transfer along strategically fixed H bonds. The available evidence concerning carbonic anhydrase, α-chymotrypsin, trypsin, thiolsubtilisin, ribonuclease, and alcohol dehydrogenases seems to support the proposed mechanism, although decisive information is still wanting. If the present interpretation proves to be correct, it will give us a new dimension for understanding the efficiency and specificity of enzyme action and even more appreciation of the advantage of using proteins to make enzymes.

References and Notes

1. C. C. F. Blake, L. N. Johnson, G. A. Mair, A. C. T. North, D. C. Phillips, V. R. Sarma, *Proc. Roy. Soc. London Ser. B* **1967**, 378 (1967).
2. M. E. Riepe and J. H. Wang, *J. Amer. Chem. Soc.* **89**, 4229 (1967); *J. Biol. Chem.*, **243**, 2779 (1968).
3. J. H. Quastel, *Biochem. J.* **20**, 166 (1926); R. Lumry, *Enzymes* **1**, 157 (1959).

4. M. Eigen and L. De Maeyer, *Proc. Roy. Soc. London Ser. A* **247**, 505 (1958).
5. K. Fridborg, K. K. Kannan, A. Liljas, B. Lundin, B. Strandberg, S. Strandberg, B. Tilander, G. Wiren, *J. Mol. Biol.* **25**, 505 (1967).
6. E. L. Smith, *Fed. Proc.* **8**, 581 (1949); R. P. Davis. *J. Amer. Chem. Soc.* **80**, 5209 (1958).
7. J. C. Kernohan, *Biochem. Biophys. Acta* **81**, 346 (1964); ———, *ibid.* **96**, 304 (1965); B. H. Gibbons and J. T. Edsall, *J. Biol. Chem.* **238**, 3501 (1963); C. Ho and J. M. Sturtevant, *ibid.* **238**, 3499 (1963); S. Lindskog, *Biochemistry* **5**, 2641 (1966).
8. J. E. Coleman, *J. Biol. Chem.* **242**, 5212 (1967).
9. R. F. Chen and J. C. Kernohan, *ibid.*, p. 5813; L. Stryer, private communication (1967).
10. F. Schneider and M. Liefländer, *Z. Physiol. Chem.* **334**, 279 (1963); Y. Pocker and J. T. Stone, *J. Amer. Chem. Soc.* **87**, 5497 (1965); Y. Pocker and J. E. Meany, *ibid.*, p. 1809; ———, *Biochemistry* **4**, 2535 (1965); ———, *ibid.* **6**, 239 (1967); Y. Pocker and D. R. Storm, *ibid.* **7**, 1202 (1968).
11. M. L. Bender, *J. Amer. Chem. Soc.* **84**, 2580 (1962); ——— and F. Kezdy, *ibid.* **86**, 3704 (1964). The idea of an acyl-enzyme intermediate was first suggested by H. Gutfreund and J. M. Sturtevant [*Proc. Nat. Acad. Sci. U.S.* **42**, 719 (1956).
12. N. K. Schaffer, L. Simet, S. Harshman, R. R. Engle, R. W. Drisko, *J. Biol. Chem.* **225**, 197 (1957); R. A. Oosterbaan and M. E. Adrichem, *Biochim. Biophys. Acta* **27**, 423 (1958); B. S. Hartley, *Nature* **201**, 1284 (1964).
13. H. Gutfreund and J. M. Sturtevant, *Proc. Nat. Acad. Sci. U.S.* **42**, 719 (1956).
14. B. W. Matthews, P. B. Sigler, R. Henderson, D. M. Blow, *Nature* **214**, 652 (1967).
15. A. Himoe and G. P. Hess, *Biochem. Biophys. Res. Commun.* **23**, 234 (1966); M. L. Bender, M. J. Gibbian, D. J. Whelan, *Proc. Nat. Acad. Sci. U.S.* **56**, 833 (1966).
16. T. Inagami, S. S. York, A. Patchornik, *J. Amer. Chem. Soc.* **87**, 126 (1965).
17. J. H. Wang and L. Parker, *Proc. Nat. Acad. Sci. U.S.* **58**, 2451 (1967).
18. M. Eigen, L. De Maeyer, H. C. Spatz, *Ber. Bunsengesellschaft* **68**, 19 (1964).
19. L. Parker and J. H. Wang, *J. Biol. Chem.*, **243**, 3729 (1968).
20. H. F. Bundy and C. L. Moore, *Biochemistry* **5**, 808 (1966).
21. M. Bergmann and J. S. Fruton, *Advan. Enzymol.* **1**, 63 (1941).
22. H. Gutfreund, *Trans. Faraday Soc.* **51**, 441 (1955).
23. L. Polgar and M. L. Bender, *J. Amer. Chem. Soc* **88**, 3153 (1966); K. E. Neet and D. E. Koshland, Jr., *Proc. Nat. Acad. Sci. U.S.* **56**, 1606 (1966).
24. G. Kartha, J. Bello, D. Harker, *Nature* **213**, 862 (1967).
25. H. W. Wyckoff, K. D. Hardman, N. M. Allewell, T. Inagami, D. Tsernoglu, L. N. Johnson, F. M. Richards, *J. Biol. Chem.* **242**, 3984 (1967). I thank Dr. H. W. Wyckoff for showing me the intricate structure of their ribonuclease model.
26. C. H. W. Hirs, S. Moore, W. H. Stein, *J. Biol. Chem.* **235**, 633 (1960); D. G. Smyth, W. H. Stein, S. Moore, *ibid.* **238**, 227 (1963); J. T. Potts, D. M. Young, C. B. Anfinsen, *ibid.*, p. 2593.
27. H. A. Scheraga and J. A. Rupley, *Advan. Enzymol.* **24**, 161 (1962); A. M. Crestfield, W. H. Stein, S. Moore, *J. Biol. Chem.* **238**, 2421 (1963).
28. A. Deavin, A. P. Mathias, B. R. Rabin, *Nature* **211**, 252 (1966); E. N. Ramsden and K. J. Laidler, *Can. J. Chem.* **44**, 2597 (1966); G. Gassen and H. Witzel, *European J. Biochem.* **1**, 36 (1967); D. G. Oakenfull, D. I. Richardson, Jr., D. A. Usher, *J. Amer. Chem. Soc.* **89**, 5491 (1967).
29. E. A. Dennis and F. H. Westheimer, *J. Amer. Chem. Soc.* **88**, 3431 (1966); ———, *ibid.*, p. 3432; R. Kluger, F. Kerst, D. G. Lee, E. A. Dennis, F. H. Westheimer, *ibid.* **89**, 3918 (1967). I thank Dr. D. A. Usher of Cornell and Dr. M. Caplow of Yale for calling my attention to this work.
30. H. Sund and H. Theorell, *Enzymes* **7**, 25 (1963); H. Sund, H. Diekmann, K. Wallenfels, *Advan. Enzymol.* **26**, 115 (1964); J. S. McKinley-McKee, *Progr. Biophys.* **14**, 223 (1964).
31. B. Vennesland and F. H. Westheimer, in *The Mechanism of Enzyme Action*, W. D. McElroy and B. Glass, Eds. (Johns Hopkins Press, Baltimore, 1954), p. 357.
32. E. M. Kosower, in *The Enzymes* (Academic Press, New York, 1960), vol. 3, pp. 171–194.
33. D. P. Hollis, *Biochemistry* **6**, 2080 (1967).
34. J. Wyman, *Advan. Protein Chem.* **4**, 436 (1948).
35. J. H. Park. B. P. Meriwether, P. Clodfelder, L. W. Cunningham, *J. Biol. Chem.* **236**, 136 (1961).
36. R. N. Perham and J. I. Harris, *J. Mol. Biol.* **7**, 316 (1963); J. I. Harris, B. P. Meriwether, J. H. Park, *Nature* **198**, 154 (1963); L. Cunningham and A. M. Schepman, *Biochim. Biophys. Acta* **73**, 406 (1963).
37. E. Fischer, *Ber. Deut. Chem. Ges.* **27**, 2985 (1894).
38. C. G. Swain and J. F. Brown, Jr., *J. Amer. Chem. Soc.* **74**, 2538 (1952).
39. Some of the experimental work mentioned in this article has been supported in part by a research grant (GM 04483 from the U.S. Public Health Service.

Reprinted from THE JOURNAL OF BIOLOGICAL CHEMISTRY, Vol. 246, 1971

On the Equilibrium of the Adenylate Cyclase Reaction*

(Received for publication, July 19, 1971)

OSAMU HAYAISHI,‡ PAUL GREENGARD,§ AND SIDNEY P. COLOWICK

From the Department of Microbiology, Vanderbilt University School of Medicine, Nashville, Tennessee 37203

SUMMARY

With a partially purified enzyme from *Brevibacterium liquefaciens*, the adenylate cyclase reaction, ATP → cyclic AMP + PP_i, was demonstrated to be readily reversible. Pyruvate, which was known to activate the reaction in the forward direction, had a similar stimulatory effect on the reverse reaction. The equilibrium constant at pH 7.3 and 25° was approximately 0.065 M at the magnesium concentrations used. This result indicates a high free energy of hydrolysis (ΔG^0_{obs}) for the 3′ ester bond of cyclic adenosine 3′,5′-monophosphate (cyclic AMP), estimated at −11.9 kcal mole^{-1} under these conditions.

Since the reaction involves one reactant and two products, the equilibrium is concentration-dependent, *i.e.* high concentrations of cyclic AMP and PP_i are required to show a high percentage of conversion to ATP; high enzyme concentrations and long incubation periods are therefore needed. The recent report by others of failure to confirm the reversibility of the reaction with the same enzyme system is attributed to failure to observe these requirements.

The availability of a partially purified adenylate cyclase from *Brevibacterium liquefaciens* (1) made feasible the study of the reversibility and the estimation of the equilibrium constant of the reaction:

$$\text{ATP} \rightarrow \text{cyclic AMP}^1 + \text{pyrophosphate}$$

* This work was supported by Grant CA-04910 and 1-S04-06067 from the United States Public Health Service. This work was carried out during tenure as Visiting Professor (O. H. and P. G.) in the Spring of 1968 at Vanderbilt University School of Medicine.

‡ Present address, Department of Medical Chemistry, Kyoto University Faculty of Medicine, Kyoto, Japan, and Department of Physiological Chemistry and Nutrition, The University of Tokyo, Faculty of Medicine, Tokyo, Japan.

§ Present address, Department of Pharmacology, Yale University School of Medicine, New Haven, Connecticut.

[1] The abbreviation used is: cyclic AMP, cyclic adenosine 3′,5′-monophosphate.

An abstract of these findings has been published previously (2). We present here details of our evidence for reversibility, based on ATP measurements by firefly luciferase, and the equilibrium data from which the free energy of hydrolysis was calculated. The accompanying paper (3) contains more recent data with a more highly purified enzyme and confirms the formation of ATP by isotopic and fluorimetric procedures.

The adenylate cyclase was prepared from *Brevibacterium liquefaciens* by the method of Hirata and Hayaishi (1). The final fraction was dialyzed against 0.1 M Tris·Cl, pH 7.3. The protein content was 0.6 to 1.7 mg per ml in different preparations, and the adenylate cyclase activity for the forward reaction under the standard incubation conditions of Hirata and Hayaishi (1) was approximately 120 nmoles per mg of protein per min.

The standard test system used to study the reversibility of the adenylate cyclase reaction contained, in a volume of 0.2 ml, the following components: 50 mM Tris·Cl buffer (pH 7.3), 10 mM tetrasodium pyrophosphate adjusted to pH 7.3 with HCl, 10 mM cyclic AMP (sodium salt), 10 mM $MgSO_4$, 5 mM sodium pyruvate, 3 mM NaF, and adenylate cyclase at a protein concentration of 0.3 to 0.85 mg per ml. When the concentration of reactants and products present in the initial reaction mixture was varied, the $MgSO_4$ was varied accordingly to maintain a 1:1 molar ratio of magnesium to ATP plus pyrophosphate. The reaction was started by the addition of $MgSO_4$ in order to minimize precipitation of $Mg_2P_2O_7$. After incubation for various times at room temperature (25°), aliquots of 10 or 20 μl were removed and the reaction was stopped by heating for 2 min at 100° in 10 mM Tris·Cl, pH 7, containing 10 mM $MgSO_4$. The samples were incubated in the above buffer for 15 min at 25° with 0.02 volume of inorganic pyrophosphatase solution (Schwarz BioResearch, 600 units per ml) to avoid possible interference by pyrophosphate in the luciferase assay (4). The boiling procedure converted most of the pyrophosphate to a form which was not susceptible to pyrophosphatase and which caused no serious interference with the luciferase assay. After pyrophosphatase treatment, 0.1-ml samples were injected into 1 ml of crude luciferase solution and the maximum intensity of light emission was compared with that of standard ATP solutions, essentially as described by Strehler and McElroy (4). It was found that the recovery of added ATP from the reaction mixtures was 90%. Correction has been made for this recovery in the experiments reported.

In some experiments on reversibility of the reaction, cyclic AMP and pyrophosphate were also measured. In such cases the reaction was stopped with 0.08 N HCl instead of by boiling. After neutralization, the pyrophosphate was completely hydrolyzed by the pyrophosphatase within 5 min under these conditions. Cyclic AMP was measured by the phosphorylase method (5) through the courtesy of Dr. G. Alan Robison. Pyrophosphate was measured by colorimetric phosphate determination (6) before and after pyrophosphatase treatment.

With the complete system for reversal of the adenylate cyclase reaction, the rate of ATP formation was about 2 nmoles min^{-1} mg^{-1} and showed an absolute dependence on the addition of cyclic AMP, pyrophosphate, Mg^{++}, and enzyme. Pyruvate caused an activation of about 15-fold in the formation of ATP, in agreement with the known pyruvate activation of the forward reaction (1); it therefore seems very likely that the same enzyme is responsible for both the ATP-forming reaction and the adenylate cyclase reaction as ordinarily measured in the forward direction. Fluoride, which is an activator of mammalian adenylate cyclases, had no appreciable effect on the activity of this bacterial adenylate cyclase measured either in the reverse or forward direction.

The firefly luciferase assay method is highly specific for ATP (4). Further evidence that the reaction product was ATP was obtained by treating it with crystalline yeast hexokinase in the presence or absence of glucose, prior to the assay with firefly luciferase. This treatment abolished completely the luminescence reaction only in the presence of glucose.

The time course of the reaction starting with various mixtures of the reactants and products was studied in order to establish the position of equilibrium. Fig. 1*A* shows the time course of ATP formation starting with 10 mM cyclic AMP and 10 mM pyrophosphate, or with mixtures of cyclic AMP, pyrophosphate, and ATP simulating 2.5, 5, 7.5, 10, and 20% conversion to ATP. The results show that the reaction proceeded toward ATP formation under all conditions except the one simulating 20% conversion, in which case the reaction proceeded in the direction of ATP breakdown. In order to ascertain that the decrease in ATP was indeed due to the forward reaction of adenylate cyclase, it was shown in a similar experiment that this decrease was pyruvate-dependent. Furthermore, analyses for cyclic AMP in such an experiment showed close to the expected increase in cyclic AMP, while samples forming ATP showed the expected decrease in cyclic AMP. Analysis for pyrophosphate also showed approximately the expected changes and established that there was no serious loss of pyrophosphate through side reactions. It therefore seemed justifiable to estimate the equilibrium constant from these experiments.

From the data shown in *Panel A* of Fig. 1, it was estimated that, starting with 10 mM cyclic AMP and 10 mM pyrophosphate, 12% conversion to ATP occurred at equilibrium (see *dashed line*). The value of K_{obs} for these conditions, given by the equation,

$$K_{obs} = \frac{[\text{cyclic AMP}]\cdot[\text{PP}_i]}{[\text{ATP}]}$$

is thus calculated to be 0.065 M.

The validity of this constant is supported by the data in Fig. 1, *B* and *C*, which shows that the expected equilibrium is approached at two quite different concentrations of total nucleotide. The results are essentially as predicted for an equilibrium of this type, in which the number of reactants and products are unequal, so that in this case dilution favors a lower percentage of ATP at equilibrium. Thus, at 10 mM, 3 mM, and 1 mM starting concentrations of cyclic AMP and pyrophosphate, the values for ATP approached at equilibrium approximate the conversions of 12%, 4.2%, and 1.5%, respectively, theoretically expected with an equilibrium constant of 0.065 M. Furthermore, when these equilibria were approached from the forward direction, the expected extent of reaction was also found.

From the equilibrium constant of 0.065 M for the adenylate cyclase reaction, we can calculate the standard free energy change at pH 7.3, pMg 3,[2] and 25°.

$$\Delta G^0_{obs} = -RT \ln K = +1.6 \text{ kcal mole}^{-1}$$

[2] The concentration of free Mg^{++} in the system is not known accurately, but we have estimated from the known dissociation constants (9) of magnesium complexes with pyrophosphate and ATP that pMg = 3.0, 3.2, and 3.5 for the reaction mixtures containing 10, 3, and 1 mM total nucleotide.

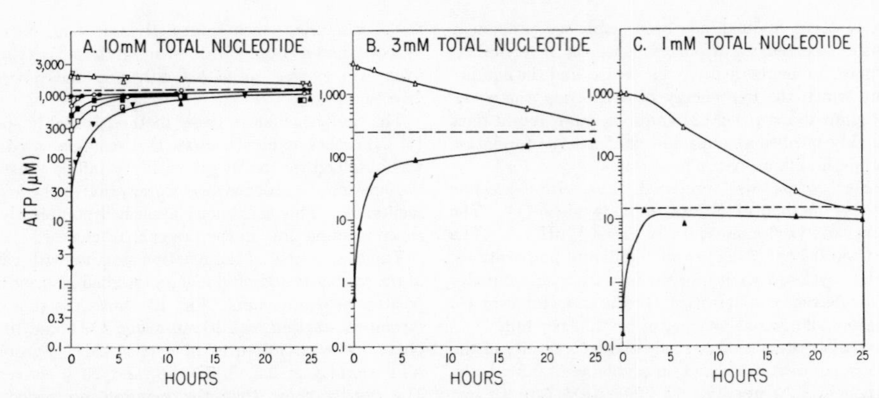

FIG. 1. Time course of approach to equilibrium of adenylate cyclase reaction. The standard test system for reversal of the reaction was used, except that the starting concentrations (mM) of the reactants were varied as indicated. Enzyme concentration was 0.85 mg per ml.

For *Panel A* (10 mM total nucleotide)

Symbol	Cyclic AMP (mM)	Pyrophosphate (mM)	ATP (mM)
▲ or ▼	10	10	0
□	9.75	9.75	0.25
■	9.5	9.5	0.50
○	9.25	9.25	0.75
●	9.0	9.0	1.0
△ or ▽	8.0	8.0	2.0

For *Panel B* (3 mM total nucleotide)

Symbol	Cyclic AMP (mM)	Pyrophosphate (mM)	ATP (mM)
▲	3	3	0
△	0	0	3

For *Panel C* (1 mM total nucleotide)

Symbol	Cyclic AMP (mM)	Pyrophosphate (mM)	ATP (mM)
▲	1	1	0
△	0	0	1

The *dashed horizontal lines* show the level of ATP predicted at equilibrium for an equilibrium constant of 0.065. Where two symbols are used, two separate experiments were done under the same conditions.

Our results indicate that the cyclase equilibrium would actually favor ATP synthesis under standard conditions, *i.e.* at concentrations of 1 M for reactants and products. Thus the standard free energy of hydrolysis of the 3′ bond of cyclic AMP must be greater than that of the α, β-pyrophosphate bond of ATP by 1.6 kcal $mole^{-1}$. The free energy of hydrolysis of ATP at the α, β bond to yield AMP and pyrophosphate is known from previous data (7, 8). The value at pH 7.3 can be estimated from the contour maps of Alberty (9) to be -10.3 kcal $mole^{-1}$ and independent of Mg^{++} over the pMg range of 3 to 3.5 used here.[2] This Mg^{++} independence is in agreement with the observed Mg^{++} independence of the adenylate cyclase equilibrium. The standard free energy of hydrolysis of cyclic AMP under these conditions must therefore be close to -11.9 kcal $mole^{-1}$. These relationships are summarized below.

	ΔG^0_{obs} kcal $mole^{-1}$
cyclic AMP + $PP_i \rightleftarrows$ ATP	-1.6
ATP + $H_2O \rightleftarrows$ 5′-AMP + PP_i	-10.3
sum: cyclic AMP + $H_2O \rightleftarrows$ 5′-AMP	-11.9

It is of interest to compare the value for hydrolysis of cyclic AMP with that for the hydrolysis of the β, γ bond of ATP to yield ADP and P_i. For pH 7.3, pMg 3, and $T = 25°$, Alberty's contour map (9) gives a value of -8.9 kcal $mole^{-1}$. Thus the 3′ bond of cyclic AMP is 3 kcal $mole^{-1}$ more "energy-rich" than the β, γ bond of ATP under these conditions. In support of these findings, the enthalpy of hydrolysis of cyclic AMP to 5′-AMP was found to be extraordinarily high, -14.1 kcal $mole^{-1}$ (10).

It is necessary to keep in mind that the adenylate cyclase reaction becomes measurably reversible only at high nucleotide concentrations, owing to the dilution effects described above. Thus, high concentrations of cyclic AMP and pyrophosphate and therefore a high concentration of enzyme should be used in order to convert a reasonable fraction of the substrate to ATP. The failure of some workers (11–13) to observe reversibility of the adenylate cyclase reaction was probably due to failure to meet these requirements. The negative results of Cheung and Chiang (11) need particular comment since they used the same enzyme source and same reaction mixtures as were used here and drew the conclusion that the reaction was not reversible. In this case, we calculate that the concentration of enzyme used and the incubation time were such as to convert only about 0.1% of the ^{3}H-labeled cyclic AMP to ATP (Table III of Reference 11). Since the starting isotope concentration was low (46,000 total cpm), the labeled ATP could have escaped detection in the aliquots taken for counting. The fact noted by Cheung and Chiang (11) that ATP at low concentrations (*e.g.* 2 μM) is converted quantitatively to cyclic AMP by the same enzyme is not at all surprising; it simply confirms the expected effect of dilution on the percentage of conversion at equilibrium. It is important to emphasize that, from the physiological point of view, it is very unlikely that the adenylate cyclase reaction would result in net ATP synthesis *in vivo* at the prevailing concentrations of reactants.

We have suggested previously (2) that cyclic AMP might serve as an adenylylating agent for proteins. Dissociation of the complex of cyclic AMP with protein kinase by various means has resulted in recovery of the original cyclic AMP (14), but the possibility that a 5′-adenylyl-protein linkage underwent cyclization prior to dissociation cannot be ruled out. A model system has been described (15) in which an aromatic 5-membered cyclic phosphate diester, catechol cyclic phosphate, reacts reversibly with chymotrypsin to form a covalently bonded catechol-phosphoryl enzyme.

Acknowledgments—We are grateful to Miss Kate Welch for aid in preparation of the adenylate cyclase and to Mrs. Maryda Colowick for assistance in the assay of ATP by the luciferase method.

Addendum—A reversal of the adenylate cyclase reaction with a partially purified enzyme from *Streptococcus salivarius* has recently been reported (16).

REFERENCES

1. Hirata, M., and Hayaishi, O., *Biochim. Biophys. Acta.* **149,** 1 (1967).
2. Greengard, P., Hayaishi, O., and Colowick, S. P., *Fed. Proc.*, **28,** 467 (1969).
3. Takai, K., Kurashina, Y., Suzuki, C., Okamoto, H., Ueki, A., and Hayaishi, O., *J. Biol. Chem.*, **246,** 5843 (1971).
4. Strehler, B. L., and McElroy, W. D., in S. P. Colowick and N. O. Kaplan (Editors), *Methods in enzymology, Vol. III*, Academic Press, New York, 1957, p. 871.
5. Rall, T. W., and Sutherland, E. W., *J. Biol. Chem.*, **232,** 1065 (1958).
6. Fiske, C. H., and SubbaRow, Y., *J. Biol. Chem.*, **66,** 375 (1925).
7. Schuegraf, A., Ratner, S., and Warner, R. W., *J. Biol. Chem.*, **235,** 3597 (1960).
8. Wood, H. G., Davis, J. J., and Lochmüller, H., *J. Biol. Chem.*, **241,** 5692 (1966).
9. Alberty, R. A., *J. Biol. Chem.*, **244,** 3290 (1969).
10. Greengard, P., Rudolph, S. A., and Sturtevant, J. M., *J. Biol. Chem.*, **244,** 4798 (1969).
11. Cheung, W. Y., and Chiang, M., *Biochem. Biophys. Res. Commun.*, **43,** 868 (1971).
12. Tao, M., and Lipmann, F., *Proc. Nat. Acad. Sci. U. S. A.*, **63,** 86 (1969).
13. Rosen, O. M., and Rosen, S. M., *Arch. Biochem. Biophys.*, **131,** 449 (1969).
14. Reimann, E. M., Walsh, D. A., and Krebs, E. G., *J. Biol. Chem.*, **246,** 1986 (1971).
15. Kaiser, E. T., Lee, T. W. S., and Boer, F. P., *J. Amer. Chem. Soc.* **93,** 2351 (1971).
16. Khandelwal, R. L., and Hamilton, I. R., *J. Biol. Chem.*, **246,** 3297 (1971).

Reprinted from BIOCHEMISTRY, Vol. 4, 1965

Stereospecific Synthesis of α-Methyl-L-glutamine by Glutamine Synthetase*

Herbert M. Kagan, Lois R. Manning, and Alton Meister

ABSTRACT: Although glutamine synthetase is active with both isomers of glutamic acid, the enzyme acts stereospecifically toward α-methylglutamic acid producing only a single α-methylglutamine isomer. α-Methyl-L-glutamine and both optical isomers of α-methylglutamic acid have been prepared.

Designation of the configuration of α-methyl-L-glutamine is supported by studies with *Escherichia coli* L-glutaminase, D-glutamic acid cyclotransferase, and *Azotobacter agilis* amidase. The last-mentioned enzyme catalyzes the hydrolysis of both L- and D-glutamine, but acts only on the L isomer of α-methylglutamine. Models of the substrates have been constructed and an explanation is proposed for the marked increase in stereospecificity associated with introduction of an α-methyl group.

Glutamine synthetase, which catalyzes the synthesis of glutamine from glutamic acid and ammonia according to the equation L-(or D-)glutamic acid + NH_3 + ATP $\rightleftharpoons$ L-(or D-)glutamine + ADP + P_i, also acts on a number of glutamate analogs and substituted glutamate derivatives to yield the corresponding amides and hydroxamates (Meister, 1962).[1] It is of particular interest that the enzyme is active with D-glutamic acid as well as L-glutamic acid; enzymatically synthesized D-glutamine has been isolated and characterized (Levintow and Meister, 1953, 1954). Although it has been reported that α-methyl-DL-glutamic acid is a substrate for glutamine synthetase (Braganca *et al.*, 1952; Lichtenstein *et al.*, 1953), the product formed was not isolated nor were studies with the individual optical isomers of this amino acid carried out. The present studies were stimulated by the observation that only about 50% of racemic α-methylglutamic acid was utilized by glutamine synthetase suggesting that, in contrast to its action on glutamic acid, the enzyme acts stereospecifically on α-methylglutamic acid to yield a single α-methylglutamine isomer. This suggestion has been substantiated by experiments in which one of the isomers of α-methylglutamine and both optical isomers of α-methylglutamic acid have been obtained. Evidence derived from studies with glutamine synthetase, D-glutamic acid cyclotransferase, and L-glutaminase indicates that the α-methylglutamine synthesized by glutamine synthetase is of the L-configuration.

Experimental

Materials. The nucleotides and sodium phosphoenolpyruvate were obtained from the Sigma Chemical Co. Pyruvate kinase and pronase were obtained from Calbiochem. α-Methyl-DL-glutamine was prepared as previously described (Meister, 1954). Glutamine synthetase from sheep brain (Pamiljans *et al.*, 1962), *Escherichia coli* L-glutaminase (Meister *et al.*, 1955), *Azotobacter agilis* amidase (Ehrenfeld *et al.*, 1963), rat kidney D-glutamic acid cyclotransferase (Meister *et al.*, 1963), and renal acylases I and II (Greenstein and Winitz, 1961) were prepared as described. Beef pancreas carboxypeptidase was obtained from the Worthington Biochemical Corp.

Methods. Glutamine synthetase activity was followed by hydroxamate and phosphate determinations as previously described (Pamiljans *et al.*, 1962). Ascending paper chromatography was carried out on Whatman No. 1 paper in solvents consisting of (a) 88% aqueous phenol–concd NH_4OH (99:1), and (b) 1-butanol–acetic acid–water (4:1:1). The R_F values for α-methylglutamic acid, α-methylglutamine, glutamic acid, and glutamine were, for (a), 0.34, 0.72, 0.20, and 0.58, and, for (b), 0.33, 0.26, 0.20, and 0.14, respectively.

α-Methylglutamic acid was determined by the ninhydrin method of Rosen (1957). Under the conditions described, α-methylglutamic acid gave about 25% of the color given by glutamic acid. Color development was considerably slower with α-methylglutamic acid than with glutamic acid, and it was found that maximum color was attained only after 30 minutes at 100°. Wilson and Snell (1962) have previously noted relatively slow development of color on treating α-methylserine with ninhydrin. In the present studies a heating time of 35 minutes was employed. The maximal color value for α-methylglutamic acid was 0.53 that for glutamic acid.

Results

Enzymatic Synthesis of α-Methylglutamine. When glutamine synthetase was incubated with α-methyl-

* From the Department of Biochemistry, Tufts University School of Medicine, Boston, Mass. *Received March 5, 1965.* Supported in part by the National Institutes of Health, U.S. Public Health Service, and the National Science Foundation.

[1] Abbreviations used in this work: ATP, adenosine 5′-triphosphate; ADP, adenosine 5′-diphosphate.

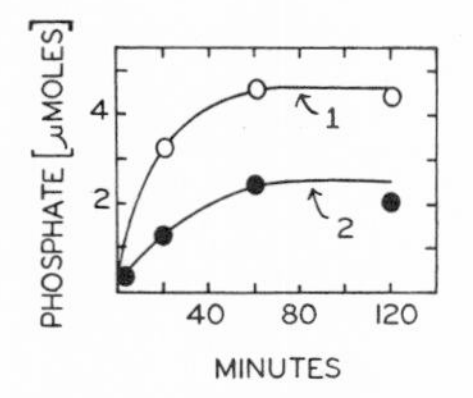

FIGURE 1: Course of amide synthesis from L-glutamic acid and α-methyl-DL-glutamic acid. The reaction mixtures contained imidazole-HCl buffer (pH 7.2, 50 μmoles), $MgCl_2$ (20 μmoles), 2-mercaptoethanol (25 μmoles), ATP (10 μmoles), NH_4Cl (100 μmoles), glutamine synthetase (44 units), and L-glutamic acid (curve 1) or α-methyl-DL-glutamic acid (curve 2) (5 μmoles) in a final volume of 1.0 ml; 37°.

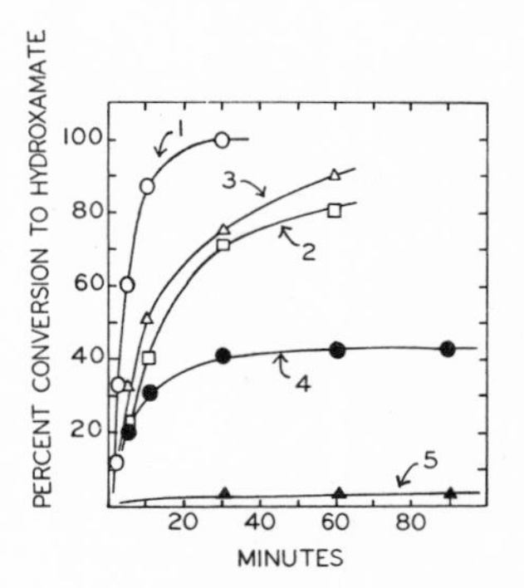

FIGURE 2: Synthesis of hydroxamates by glutamine synthetase. The reaction mixtures contained imidazole-HCl buffer (pH 7.2, 50 μmoles), $MgCl_2$ (20 μmoles), ATP (10 μmoles), 2-mercaptoethanol (25 μmoles), hydroxylamine hydrochloride (adjusted to pH 7.2 with sodium hydroxide; 100 μmoles), amino acid (1.0 μmole), and enzyme (43.2 units) in a final volume of 1.0 ml.; 37°. Curve 1, L-glutamic acid; curve 2, D-glutamic acid; curve 3, α-methyl-L-glutamic acid; curve 4, α-methyl-DL-glutamic acid; curve 5, α-methyl-D-glutamic acid.

DL-glutamic acid, ATP, Mg^{2+}, and NH_4Cl, the formation of phosphate proceeded to a value that approached 50% of that expected for complete reaction of the added amino acid (Figure 1). Experiments in which additional amounts of enzyme were added failed to increase the formation of inorganic phosphate beyond 50% of the theoretical. Under the same conditions, but with L-glutamic acid, more than 90% of the theoretical amount of inorganic phosphate was formed. Paper chromatographic study of reaction mixtures containing α-methyl-DL-glutamic acid after 60 minutes of incubation under the conditions described in Figure 1 revealed the disappearance of approximately half the substrate and formation of a new compound, which exhibited an R_F value identical to that of authentic α-methyl-DL-glutamine.

A large-scale experiment was carried out to facilitate isolation of both the enzymatically synthesized α-methylglutamine and the unsusceptible α-methylglutamic acid. In order to achieve as complete a reaction of the susceptible substrate as possible, the large-scale experiment was carried out in the presence of pyruvate kinase and phosphoenolpyruvate, and less than stoichiometric amounts of ATP.[2] The relatively small amount of nucleotide present also facilitated subsequent isolation of the amino acids. The reaction mixture contained α-methyl-DL-glutamic acid (2.254 g; 14 mmoles), ATP (1 mmole), $MgCl_2$ (2 mmoles), phosphoenolpyruvate (7.7 mmoles), NH_4Cl (10 mmoles), pyruvate kinase (7 mg; 917 units), 2-mercaptoethanol (150 μmoles), and glutamine synthetase (12 mg, 816 units) in a final volume of 31 ml. The reaction mixture was adjusted to pH 7.2 and incubated at 37° with gentle shaking. The formation of inorganic phosphate reached a value that was within experimental error of completion (i.e., 7 mmoles) after 7.5 hours; after a total of 8 hours of incubation, cold ethanol (100 ml) was added and the precipitated protein was removed by centrifugation and washed by centrifugation with 30 ml of 50% ethanol. The supernatant solutions were combined and concentrated at 25° in a flash evaporator to a thick oil weighing 2 g. This was dissolved in 10 ml of water and applied to the top of a column of Dowex 50 (H^+; 2 × 7 cm) at 4°. The column was washed with 140 ml of water and then eluted with 120 ml of 2 M NH_4OH. The water effluent contained virtually all of the phosphate, nucleotide, chloride, pyruvate, and 2-mercaptoethanol, while the amino acids were found in the NH_4OH effluent. The NH_4OH effluent was concentrated at 25° in a flash evaporator until a thick oil was obtained; this contained minimal amounts of ammonia as determined with Nessler's reagent. The oil was dissolved in 10 ml of water and chromatographed in two equal batches on a column of IRC-50 (H^+; 200-400 mesh; 90 × 2.4 cm); elution was carried out with water at 4°, and 175 fractions of 3.8 ml each were obtained. These were analyzed for α-methylglutamic acid and α-methylglutamine by paper chromatography. α-Methylglutamic acid was found in fractions 82–101; α-methylglutamine was found in fractions 104–150. Some nucleotide was present in fractions 58–74, but no chloride or ammonia was found in any of the fractions. Evaporation of the pooled amino acid and amino acid amide fractions followed by lyophilization gave 475 and 450 mg, respectively, of α-methylglutamic acid and α-methylglutamine.

Anal. (α-methylglutamic acid, mp 169–172°) Calcd for $C_6H_{11}O_4N$: N, 8.7. Found: N, 8.5. (α-methylglutamine, mp 166–167°) Calcd for $C_6H_{12}O_3N_2$: N, 17.5. Found: N, 17.2.

[2] A similar method was subsequently used for the isolation of enzymatically synthesized β-glutamine (Khedouri *et al.*, 1964).

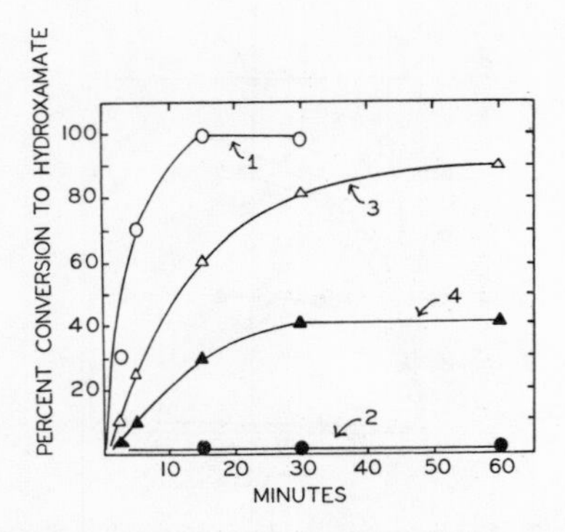

FIGURE 3: γ-Glutamyl transfer reactions. The reaction mixtures contained imidazole-HCl buffer (pH 7.2, 50 μmoles), $MnCl_2$ (5 μmoles), ADP (0.1 μmole), K_2HAsO_4 (pH 7.2, 25 μmoles), hydroxylamine hydrochloride (adjusted to pH 7.2 with sodium hydroxide, 100 μmoles), amino acid (2 μmoles), and enzyme (43.2 units), in a final volume of 1.0 ml. Curve 1, L-glutamine; curve 2, D-glutamine; curve 3, α-methyl-L-glutamine; curve 4, α-methyl-DL-glutamine.

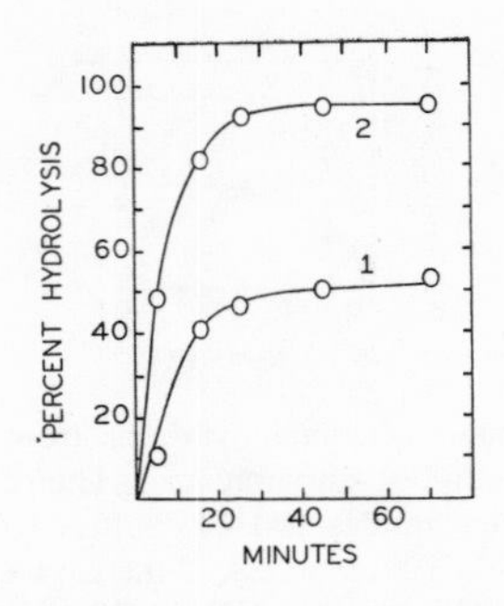

FIGURE 4: Hydrolysis of α-methyl-DL-glutamine and DL-glutamine by *A. agilis* amidase. The reaction mixtures contained imidazole-HCl buffer (25 μmoles, pH 7.2), enzyme (6.4 units), and α-methyl-DL-glutamine (3.6 μmoles) or DL-glutamine (3.3 μmoles) in a final volume of 0.5 ml. Curve 1, α-methyl-DL-glutamine; curve 2, DL-glutamine.

The products were homogeneous on paper chromatography. Because of the small amounts of material available, no attempt was made to crystallize the products. Acid hydrolysis of α-methylglutamine (3 N HCl, 120°, 30 minutes) gave stoichiometric amounts of ammonia and (after removal of ammonia by alkalinization and aeration) α-methylglutamic acid. The enzymatically synthesized α-methylglutamine and the α-methylglutamic acid derived from it are considered to be of the L configuration (*vide infra*).

Studies with Glutamine Synthetase and the Isomers of α-Methylglutamic Acid. When α-methyl-L-glutamic acid and α-methyl-D-glutamic acid, obtained as described, were incubated with glutamine synthetase, ATP, Mg^{2+}, and hydroxylamine, only α-methyl-L-glutamic acid (Figure 2, curve 3) was significantly active. Studies with L-glutamic acid (curve 1), D-glutamic acid (curve 2), and α-methyl-DL-glutamic acid (curve 4) were carried out for comparative purposes. Little, if any, hydroxamate was formed when α-methyl-D-glutamic acid was incubated under these conditions (curve 5). The very low activity observed with α-methyl-D-glutamic acid (1–2% in various experiments) probably may be ascribed to contamination with α-methyl-L-glutamic acid. Such contamination could be due to failure of the preparative enzymatic reaction to go to completion (cf. Levintow and Meister, 1954), or possibly to nonenzymatic hydrolysis of enzymatically synthesized α-methyl-L-glutamine.

Since it is known that the relative activities with L-glutamic acid and D-glutamic acid may vary depending upon pH and the nature and concentration of divalent metal ion, the synthesis of hydroxamate from α-methylglutamic acid was determined under various conditions. Studies were carried out with α-methyl-D-glutamic acid at pH values of 5.5, 7.2, and 7.8. Experiments at each of these values of pH were carried out with concentrations of Mg^{2+}, Mn^{2+}, and Co^{2+} varying from 0.005 to 0.1 M. In none of these experiments was more than 1.7% of the amino acid converted to the hydroxamate. In a similar series of studies with α-methyl-DL-glutamic acid no more than 48.3% of the added amino acid was converted to the hydroxamate. In other experiments carried out using conditions similar to those described in Figure 2, the effect of added α-methyl-D-glutamic acid on the formation of α-methyl-L-γ-glutamylhydroxamate from α-methyl-L-glutamic acid was determined. With concentrations of α-methyl-L-glutamic acid and α-methyl-D-glutamic acid of 0.0046 and 0.01 M, respectively, only 9% inhibition was observed.

α-Methyl-DL-glutamine and α-methyl-L-glutamine were examined as substrates for the γ-glutamyl transfer reaction. As indicated in Figure 3, under the conditions employed, about 90% of the added α-methyl-L-glutamine (curve 3) was utilized and about half of this amount of α-methyl-DL-glutamine (curve 4) reacted. In agreement with previous findings (Meister, 1962), the reaction with L-glutamine was rapid (curve 1) while little if any transfer was observed with D-glutamine (curve 2). The findings are consistent with the conclusion that α-methyl-D-glutamine is also not a very active substrate for the transfer reaction.

Other Enzymatic Studies. The action of *E. coli* L-glutaminase (Meister *et al.*, 1955) and *A. agilis* amidase (Ehrenfeld *et al.*, 1963) on α-methyl-L-glutamine and α-methyl-DL-glutamine was determined, using the conditions previously described. Both enzymes catalyzed the hydrolysis of α-methyl-L-glutamine virtually to completion and, within experimental error, 50% of α-methyl-DL-glutamine. In confirmation of previous findings *E. coli* glutaminase did not catalyze appreciable hydrolysis of D-glutamine, while the *A. agilis* enzyme

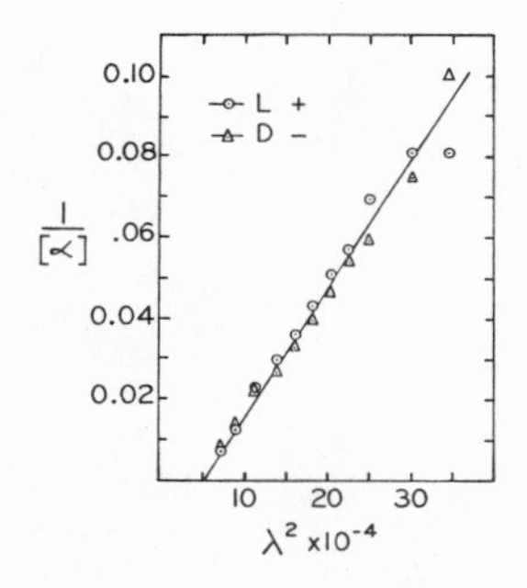

FIGURE 5: Optical rotatory dispersion curves of α-methyl-L-glutamic acid and α-methyl-D-glutamic acid. Concentrations: L-, 1.375%; D-, 2.01%; in 3 N HCl.

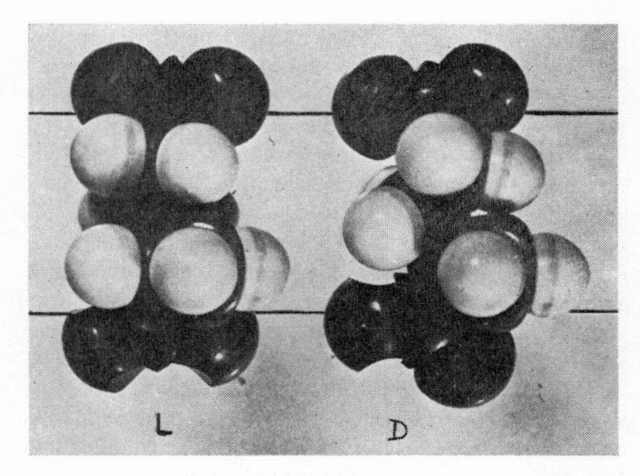

FIGURE 6: Models of L-glutamic acid and D-glutamic acid (see text).

hydrolyzed D-glutamine at about 60% of the rate observed for hydrolysis of L-glutamine. Under conditions in which the latter enzyme hydrolyzed DL-glutamine completely (Figure 4), the reaction appeared to reach a plateau when about half of the amide nitrogen of α-methyl-DL-glutamine was released as ammonia. The data suggest that the *A. agilis* enzyme does not hydrolyze α-methyl-D-glutamine at an appreciable rate.

Neither isomer of α-methylglutamic acid was found to be susceptible to L-glutamic acid decarboxylase prepared and studied as described (Meister *et al.*, 1951). *N*-Chloroacetyl-α-methyl-DL-glutamic acid was prepared (Greenstein and Winitz, 1961) and incubated with acylase I, acylase II, pronase, and carboxypeptidase in separate experiments; no hydrolysis of the *N*-chloroacetyl compound was observed as determined by the ninhydrin method (Rosen, 1957). Previous studies showed that both α-methyl-DL-glutamate and α-methyl-D-glutamate were substrates for D-glutamic acid cyclotransferase (Meister *et al.*, 1963). These observations have been confirmed and it was also shown that D-glutamic acid cyclotransferase does not act on α-methyl-L-glutamic acid.[3]

Optical Rotatory Dispersion of the Isomers of α-Methylglutamic Acid. These determinations were carried out with a Cary Model 60 recording spectropolarimeter at 25° in 3 *N* HCl with a 1-cm light path. A plot of the data based on a one-term Drude equation is given in Figure 5. Within experimental error (due largely to instrumental variability in the range of 500–600 mμ) the data indicate equal and opposite optical rotations over the spectral range examined. Specific optical rotations of +18.2°, +19.8°, +28.1°, and +43.6° were obtained, respectively, at 500, 450, 400, and 350 mμ for α-methyl-L-glutamic acid; the corresponding values for α-methyl-D-glutamic acid were −17.0°, −21.8°, −29.9°, and −43.4°.

Discussion

The data indicate that the action of glutamine synthetase on α-methylglutamic acid is specific for one optical isomer, and that such stereoselectivity can be utilized to effect the enzymatic resolution of α-methyl-DL-glutamic acid. The strict stereospecificity of the enzyme toward α-methylglutamic acid thus contrasts with its action on glutamic acid, the natural substrate. D-Glutamic acid is amidated less rapidly than L-glutamic acid, but there is evidence that both isomers are activated at substantially similar rates (Krishnaswamy *et al.*, 1962). In considering the substrate specificity of glutamine synthetase, it seems more remarkable that both optical isomers of glutamic acid are substrates than it does that only one isomer of α-methylglutamic acid is active. Although detailed information concerning the structure of the active site of glutamine synthetase is not yet at hand, a possible explanation for the observed specificity may be tentatively considered. It appears reasonable to assume that the enzyme has specific binding sites for the amino and carboxyl groups of L-glutamic acid, and that the respective functional groups of the other substrates combine with the same enzyme sites. Second, it may logically be postulated that the glutamic acid carbon chain is oriented on the enzyme surface in the fully (or almost fully) extended form so as to permit the maximum (or almost maximum) distance between the carboxyl groups. This postulate follows from the observation that aspartic acid is neither a substrate nor an inhibitor of the enzyme. If one constructs models of L- and D-glutamic acid in which the carbon chains are fully extended, it can be seen that it is possible to bring the respective amino and carboxyl groups to almost the same hypothetical enzyme binding sites (Figure 6). To effect such contacts with the enzyme would require that the β- and γ-carbon atoms of the isomers of glutamic acid occupy somewhat different

[3] We wish to thank Mr. Peter Polgar for assistance in these experiments. The formation of α-methylpyrrolidonecarboxylic acid (5-oxo-2-methylpyrrolidine-2-carboxylic acid) was determined by gas-liquid chromatography; the details of this procedure will be presented in a subsequent publication (P. Polgar and A. Meister, 1965, unpublished data).

positions; however, it is known that the enzyme acts on glutamic acid derivatives possessing β- and γ-substitutions (Meister, 1962), implying that an exact fit of this portion of the substrate to the enzyme is not necessary. Assuming this interpretation to be correct, one can arrive at a plausible explanation for the strict stereospecificity of the enzyme toward α-methylglutamic acid. Examination of the models shows that when the carboxyl and amino groups of L- and D-glutamic acids are brought as close as possible to the same respective hypothetical enzyme-binding sites, the α-hydrogen atoms of the glutamic acid isomers are oriented in markedly different ways. Thus replacement of the α-hydrogen atom of D-glutamic acid by a methyl group might be expected to exert a significantly different effect than such substitution of L-glutamic acid. Accordingly, the methyl group of α-methyl-D-glutamic acid would offer much greater steric hindrance to binding than that of its enantiomorph. For example, if the model of L-glutamic acid shown in Figure 6 is assumed to be on the enzyme surface, introduction of an α-methyl group would not be expected to interfere greatly with binding; on the other hand, replacement of the α-hydrogen atom of D-glutamic acid (on the under surface of the model shown in Figure 6) by a methyl group would displace the positions of the amino and carboxyl groups, and therefore interfere with attachment to the enzyme. Figure 7 shows models of α-methyl-L-glutamic acid and α-methyl-D-glutamic acid in which the respective methyl groups can be seen to be on opposite sides of the molecules; these models were constructed by replacing the α-hydrogen atoms of the models shown in Figure 6 with methyl groups. (The models have been turned about 75° to the left from the positions shown in Figure 6.)

Although the models constructed here suggest a way in which substitution of the α-hydrogen atom of D-glutamic acid could prevent its binding to the enzyme, while similar substitution of L-glutamic acid would not, other explanations for the experimental findings cannot be excluded and more evidence for the present hypothesis is needed. Any explanation proposed to account for the experimental findings described here must also take into account the interesting finding that β-glutamic acid is a substrate for the enzyme (Khedouri *et al.*, 1964) and that it is enzymatically converted to the D-isomer of β-glutamine. Extension of the considerations discussed above to β-glutamic acid and β-glutamine has led to conclusions which are consistent with the present ones (Khedouri and Meister, 1965).

The *A. agilis* amidase resembles glutamine synthetase in its specificity. Thus it hydrolyzes both isomers of glutamine, but cleaves only one isomer of α-methylglutamine (Figure 4). An explanation of the type suggested above for glutamine synthetase may also apply to the amidase.

Designation of the configuration of the isomer of α-methylglutamic acid that is amidated by glutamine synthetase as L is based entirely on the available enzymatic data. Thus glutamine synthetase amidates L-glutamic acid more rapidly than D-glutamic acid. *E. coli* glutaminase is highly specific for L-glutamine, and

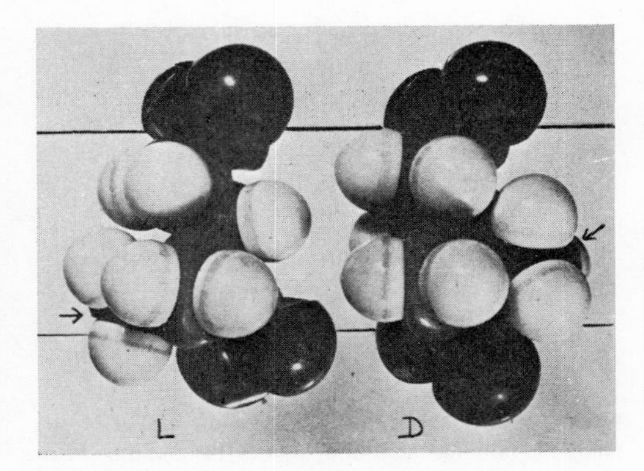

FIGURE 7: Models of α-methyl-L-glutamic acid and α-methyl-D-glutamic acid (see text). Arrows indicate the methyl carbon atoms.

this enzyme as well as the amidase from *A. Agilis* (which exhibits greater activity toward L-glutamine than D-glutamine) specifically hydrolyze the isomer of α-methylglutamine that is synthesized by glutamine synthetase. In addition, D-glutamic acid cyclotransferase, whose D-specificity has been demonstrated in studies on the isomers of glutamic acid and related substrates, acts only on the isomer of α-methylglutamic acid that is not amidated by glutamine synthetase. The designation of configurations of the α-methylglutamic acid isomers is based on the assumptions that the α-methyl group of enzymatically synthesized α-methylglutamine occupies the same position as the α-hydrogen atom of L-glutamic acid, and that the several enzymes studied here recognize α-methylglutamic acid as a derivative of glutamic acid. The isomer designated here as α-methyl-L-glutamic acid can also be considered as a derivative of D-alanine, and it is conceivable that an L-specific enzyme exists which would recognize this substrate as an L-alanine derivative and would therefore act on the enantiomorph of α-methyl-L-glutamic acid, i.e., α-(2-carboxyethyl)-L-alanine.

References

Braganca, B. M., Quastel, J. H., and Schucher, R. (1952), *Arch. Biochem. Biophys. 41*, 478.

Ehrenfeld, E., Marble, S. J., and Meister, A. (1963), *J. Biol. Chem. 238*, 3711.

Greenstein, J. P., and Winitz, M. (1961), Chemistry of the Amino Acids, New York, Wiley, p. 822.

Khedouri, E., and Meister, A. (1965), *J. Biol. Chem. 240* (in press).

Khedouri, E., Wellner, V. P., and Meister, A. (1964), *Biochemistry 3*, 824.

Krishnaswamy, P. R., Pamiljans, V., and Meister, A. (1962), *J. Biol. Chem. 237*, 2932.

Levintow, L., and Meister, A. (1953), *J. Am. Chem. Soc. 75*, 3039.

Levintow, L., and Meister, A. (1954), *J. Biol. Chem. 209*, 265.

Lichtenstein, N., Ross, H. E., and Cohen, P, P. (1953), *J. Biol. Chem. 201*, 117.

Meister, A. (1954), *J. Biol. Chem. 210*, 17.

Meister, A. (1962), *Enzymes 6*, 443.

Meister, A., Bukenberger, M. W., and Strassburger, M. (1963), *Biochem. Z. 338*, 217.

Meister, A., Levintow, L., Greenfield, R. E., and Abendschein, P. A. (1955), *J. Biol. Chem. 215*, 441.

Meister, A., Sober, H. A., and Tice, S. V. (1951), *J. Biol. Chem. 189*, 577, 591.

Pamiljans, V., Krishnaswamy, P. R., Dumville, G., and Meister, A. (1962), *Biochemistry 1*, 153.

Rosen, H. (1957), *Arch. Biochem. Biophys. 67*, 10.

Wilson, E. M., and Snell, E. E. (1962), *J. Biol. Chem. 237*, 3171.

Reprinted from THE JOURNAL OF BIOLOGICAL CHEMISTRY, Vol. 244, 1969

Interaction of a Spin-labeled Analogue of Nicotinamide Adenine Dinucleotide with Alcohol Dehydrogenase

III. THERMODYNAMIC, KINETIC, AND STRUCTURAL PROPERTIES OF TERNARY COMPLEXES AS DETERMINED BY NUCLEAR MAGNETIC RESONANCE*

(Received for publication, November 27, 1968)

ALBERT S. MILDVAN‡

From the Johnson Research Foundation, University of Pennsylvania, Philadelphia, Pennsylvania 19104

H. WEINER

From the Department of Biochemistry, Purdue University, Lafayette, Indiana 47907

SUMMARY

ADP-$C_5H_4(CH_3)_4N$—O (ADP-R·) is a paramagnetic analogue of NAD in which the unpaired electron is located in a region which corresponds to the pyridine *N*-ribose C_1 bond of the coenzyme. Studies of the effect of ADP-R· on the longitudinal nuclear relaxation rate ($1/T_1$) of the protons of water have previously shown this analogue to form binary complexes with liver alcohol dehydrogenase and to form a ternary enzyme-ADP-R·-ethanol complex. By the same technique, ternary complexes are shown here with acetaldehyde and isobutyramide. The dissociation constants of all ternary complexes are in reasonable agreement with those reported for the respective kinetically active species. All of the ternary complexes are less effective in relaxing water protons than is the binary *E*-ADP-R· complex, suggesting that ethanol, acetaldehyde, and isobutyramide displace water molecules which are hydrogen-bonded to the paramagnetic nitroxide group of enzyme-bound ADP-R·. This is directly shown by the effect of *E*-ADP-R· on the transverse relaxation rates ($1/T_2$) of the protons of ethanol, acetaldehyde, and isobutyramide. In the absence of enzyme, high concentrations of ADP-R· (5 mM) increase $1/T_2$ of the protons of ethanol. Its effect on the methylene protons is >3 times its effect on the methyl protons. The complex of ADP-R· with liver alcohol dehydrogenase at much lower concentrations (~0.2 mM) relaxes the protons of ethanol and the effect on the methyl protons is 8 to 23 times greater than the effect on the methylene protons, suggesting that the enzyme reorients the ethanol with reference to the unpaired electron of ADP-R· in the ternary complex. Depending on conditions, the enzyme enhances the effect of ADP-R· on $1/T_2$ of the methylene protons by a factor of 1.5 to 13.4 and on the methyl protons by a factor of 16 to 148. The *E*-ADP-R· complex is comparably effective in relaxing the protons of acetaldehyde and of isobutyramide, indicating that the unpaired electron is very near the bound substrates and inhibitor in the respective ternary complexes. All of these effects are reversed by the addition of excess NADH. The temperature dependence of $1/T_2$ of the methyl protons of ethanol in the presence of the enzyme-ADP-R·-ethanol complex can be fit by two rate processes: $1/\tau_M$, the rate of chemical exchange of ethanol molecules into the ternary complex, and $1/T_{2M}$, the relaxation rate of ethanol molecules which are in the ternary complex. From $1/\tau_M$, the exchange rate of ethanol into the ternary complex is 610 sec^{-1} at 25° which is two orders of magnitude faster than the maximum turnover number of the alcohol dehydrogenase reaction. Lower limits of 770 sec^{-1} and 180 sec^{-1} are set on the exchange rates of acetaldehyde and isobutyramide into their ternary complexes. From the dipolar contributions to $1/T_{2M}$, the average distances between the unpaired electron of enzyme-bound ADP-R· and the protons of the bound substrates and inhibitors, calculated from the Solomon-Bloembergen equation, are: ethanol (methyl), 3.6 A; ethanol (methylene), 4.1 A; acetaldehyde (aldehyde), 3.1 A; isobutyramide (methyl), ≤3.9 A. It is concluded that substrates bind to liver alcohol dehydrogenase on the "water side" of the bound coenzymes and directly overlie the ribosidic bond to pyridine. Structures consistent with the calculated distances permit molecular contact between the hydrogen-donating and hydrogen-accepting positions of the substrates and the coenzymes and, therefore,

* Journal paper 3524 from the Purdue University Agricultural Research Station. The work was supported in part by United States Public Health Service Grants AM-09760, GM-12446, and 5-SO1-FRO-5415 and by National Science Foundation Grant NSF-GB-7263. Preliminary reports of this work have been published (*Abstracts of the Biophysical Society Meeting, Los Angeles, March 1969*, and Reference 1).

‡ This work was done during the tenure of an Established Investigatorship from the American Heart Association. Present address, The Institute for Cancer Research, Fox Chase, Philadelphia, Pennsylvania 19111.

support the direct transfer of hydrogen between them. Although hydrogen transfer by way of stacked pyridine and indole rings can be excluded, the participation of tryptophan in an edge to edge structure with the pyridine ring cannot be excluded.

ADP-4(2,2,6,6-tetramethylpiperidine-1-oxyl) a paramagnetic analogue of NAD, has been prepared (2) and has been shown to bind to liver alcohol dehydrogenase in a manner which is indistinguishable from the binding of the competitive inhibitor adenosine diphosphoribose (2, 3). This was shown by kinetics (2) and by binding studies with electron paramagnetic resonance line broadening (2, 3) and enhancement of the proton relaxation rate ($1/T_1$) of water (3). The unpaired electron in the analogue is localized in a region which corresponds to the ribotide bond between the pyridine *N* and ribose C_1 of NAD (3). Binding of ADP-R·[1] to liver alcohol dehydrogenase enhanced the effect of the unpaired electron on the longitudinal relaxation rate ($1/T_1$) of the protons of water by a factor of 13, which was decreased to 6 upon addition of ethanol, indicating the formation of a ternary *E*-(ADP-R·)-ethanol complex. The dissociation constant of this ternary complex agreed with the reported dissociation constant of *E*-NAD-ethanol (4). The decrease in the enhancement factor in the ternary complex and the temperature dependence of $1/T_1$ suggested that ethanol combined with *E*-ADP-R· by displacing water molecules from the solvent side of the bound analogue and was very near to the unpaired electron, hence near to the ribosidic bond to pyridine in the *E*-NAD-ethanol complex.

To test this hypothesis directly, we have examined the effect of the bound analogue on the transverse relaxation rate ($1/T_2$) of the protons of ethanol. Ternary complexes containing acetaldehyde or isobutyramide are detected by studying the longitudinal relaxation rate ($1/T_1$) of the protons of water. The structural properties of these ternary complexes are examined by proton relaxation rate studies of the substrates and inhibitors themselves.

EXPERIMENTAL PROCEDURE

Materials—Alcohol dehydrogenase from horse liver was prepared by a method similar to that of Taniguchi, Theorell, and Åkeson (5). From its specific activity (4000 μmoles per min·mg) and its behavior on starch gel, acrylamide gel, and cellulose acetate electrophoresis, the purity of the enzyme was estimated to be greater than 95% (2).

The paramagnetic analogue ADP-R· and metal-free buffers were prepared and used as previously described (2). Isobutyramide was recrystallized from water prior to use, and all other compounds used were of the highest purity commercially available.

Magnetic Resonance Techniques—The concentration of unbound ADP-R· in a mixture of free and enzyme-bound forms was monitored by electron spin resonance as previously described, with a Varian E-3 electron paramagnetic resonance spectrometer (2).

The effect of the bound analogue on the longitudinal relaxation rate ($1/T_1$) of water protons was determined by the pulsed method at 24.3 MHz as previously described (3, 6).

The transverse relaxation rates ($1/T_2$) of the protons of the substrates and inhibitors were determined in D_2O solutions by measurement of the line width of the proton NMR spectra at power levels at least 5 decibels below saturation as previously described (7) with the use of the Jeolco C-60-H or the Varian HA-60 NMR spectrometers. Solutions of liver alcohol dehydrogenase (EC 1.1.1.1) in D_2O were prepared by precipitating the enzyme at 75% saturated ammonium sulfate, dissolving the pellet in 0.03 M phosphate buffer in D_2O, pD = 6.45, followed by extensive dialysis against the same buffer.

THEORY

The theory of relaxation enhancement for coordination complexes of paramagnetic ions (8, 9) and for the interactions of organic radicals with solvent protons (3) has been discussed elsewhere. We now extend our previous treatment to the interaction of enzyme-bound radicals with magnetic nuclei of substrates as well as solvent. The present treatment is similar to that for ternary enzyme-metal-substrate complexes with the major difference being that chemical bonding between the nitroxide radical and the nuclei which are relaxed by this radical is assumed to be very weak; *i.e.* the hyperfine coupling constant is assumed to be vanishingly small. Hence the dipolar interaction between the unpaired electron and the relaxing nucleus is the only one which is considered operative.

Experimentally, by continuous wave NMR we observe no effect of the radical on the chemical shifts of the protons of water or substrates under conditions in which paramagnetic broadening is seen. Hence, from Swift and Connick (10) and Luz and Meiboom (11) we may write:

$$1/T_{1p} = \frac{pq}{T_{1M} + \tau_M} \quad (1)$$

$$1/T_{2p} = \frac{pq}{T_{2M} + \tau_M} \quad (2)$$

where $1/T_{1p}$ and $1/T_{2p}$ are the paramagnetic contributions to the longitudinal and transverse relaxation rates, p is the ratio of the concentration of the radical to that of the species which is being relaxed, q is the "coordination number," the number of solvent or substrate molecules which can simultaneously interact with the radical, T_{1M} and T_{2M} are the longitudinal and transverse relaxation times of ligand molecules which are directly interacting with the radical, and τ_M is the residence time of the ligand molecule in the region of direct interaction ("coordination sphere") of the radical.

As previously shown (7, 10, 11) the most convenient way to determine the relative contributions of τ_M, T_{1M}, and T_{2M} to the experimentally determined T_{1p} and T_{2p} is to study the temperature dependence of these relaxation times. For example, if

[1] The abbreviations used are: ADP-R·, ADP-4 (1-oxyl-2,2,6,6-tetramethylpiperidine), the structure of which is:

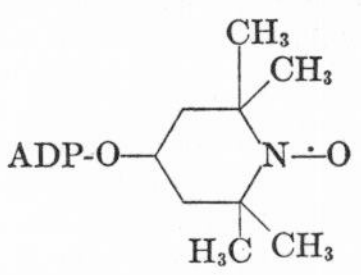

ADP-O—(piperidine ring with CH_3, CH_3, H_3C, CH_3 substituents)—N—O

NMR, nuclear magnetic resonance. All symbols related to enhancement are defined in the text and in Reference 3.

$1/T_{1p}$ increases exponentially with temperature this behavior is consistent with dominance of Equation 1 by τ_M; *i.e.* $1/T_{1p} = pq/\tau_M$. An exponential decrease of $1/T_{1p}$ with increasing temperature suggests the $1/T_{1p} = pq/T_{1M}$. Similar arguments apply to Equation 2.

Assuming that the radical binds very weakly to the solvent and to the substrates (*i.e.* neglecting the hyperfine interaction), as described above, from Bloembergen and Solomon we may write the following equations relating T_{1M} and T_{2M} of the relaxing nucleus to the spatial distance, r, between the unpaired electron and the nucleus (12, 13).

$$1/T_{1M} = \frac{2}{15}\frac{\mu_s^2\mu_I^2}{r^6}\left[3\tau_c + \frac{7\tau_c}{1 + \omega_s^2\tau_c^2}\right] \quad (3)$$

$$1/T_{2M} = \frac{2}{15}\frac{\mu_s^2\mu_I^2}{r^6}\left[3.5\tau_c + \frac{6.5\tau_c}{1 + \omega_s^2\tau_c^2}\right] \quad (4)$$

In these equations, μ_s and μ_I are the magnetic moments of the unpaired electron and the nucleus respectively, ω_s is the electron resonance frequency, and τ_c is the correlation time for the dipolar interaction.

For an organic radical with spin = 1/2 relaxing a proton, on substituting the proper values for μ_s, μ_I, and ω_s into Equations 3 and 4, we obtain the following equations at 24.3 Mc per sec:

$$r = 544\left[T_{1M}\left(3\tau_c + \frac{7\tau_c}{1 + 10^{22}\tau_c^2}\right)\right]^{\frac{1}{6}} = 544\left[T_{2M}\left(3.5\tau_c + \frac{6.5\tau_c}{1 + 10^{22}\tau_c^2}\right)\right]^{\frac{1}{6}} \quad (5)$$

and at 60.0 Mc per sec:

$$r = 544\left[T_{1M}\left(3\tau_c + \frac{7\tau_c}{1 + 6.15(10^{22})\tau_c^2}\right)\right]^{\frac{1}{6}} = 544\left[T_{2M}\left(3.5\tau_c + \frac{6.5\tau_c}{1 + 6.15(10^{22})\tau_c^2}\right)\right]^{\frac{1}{6}} \quad (6)$$

Thus if one can determine either T_{1M} or T_{2M} and τ_c, one can calculate r, the distance between the unpaired electron and the magnetic nucleus which is being relaxed.

The analogue, ADP-R·, possesses a well localized spin (3); hence r is a measure of the distance between the protons of the substrates and a fixed reference point on the analogue. The electron spin relaxation times of nitroxide radicals are long (10^{-9} to 10^{-8} sec) (3, 13, 14) compared to rotation times (10^{-13} to 10^{-11} sec) (15, 16). Hence the correlation time for the dipolar interaction, τ_c, is probably determined by the rotational correlation time, τ_r (3). Relaxation enhancements of $1/T_{1M}$ and $1/T_{2M}$ can, therefore, occur by the formation of complexes with macromolecules in which the relative rotational motion of the radical and the relaxing solvent or substrate molecule is hindered, *i.e.* in which τ_r is increased. Similar effects are well known in complexes of the manganous ion (6, 8, 9) and have previously been described for the interaction of water protons with free and enzyme-bound ADP-R· (3). By analogy with previous definitions of enhancement (ϵ^*) of the longitudinal and transverse relaxation rates (7–9, 16) we define:

$$\epsilon_1^* = (1/pT_{1p}^*)/(1/pT_{1p}) \quad (7)$$

$$\epsilon_2^* = (1/pT_{2p}^*)/(1/pT_{2p}) \quad (8)$$

where the asterisk denotes the paramagnetic contribution to the relaxation rate in the presence of a macromolecule which can bind the radical. The denominators refer to the paramagnetic contribution to the relaxation rate in the presence of the unbound radical. The normalization factor, p, is as defined above, the ratio of the total concentration of the radical to that of the species which is being relaxed.

RESULTS

Effect of ADP-R· and Alcohol Dehydrogenase-ADP-R· on Transverse Relaxation Rates of Protons of Ethanol—The NMR spectrum of 50 mM ethanol in D_2O containing 0.03 M phosphate buffer, pD = 6.45, consists of a quartet due to the methylene group 4.32 ± 0.02 ppm[2] downfield from tetramethyl silane (an external standard), and a triplet due to the methyl group 1.84 ± 0.01 ppm[2] downfield from tetramethyl silane. The hyperfine coupling constant J = 7.13 ± 0.06 cps.

The addition of 5.34 mM ADP-R· (Fig. 1) caused no significant chemical shift of the resonances or change in J but broadened both resonances. The broadening of the methylene quartet is greater than the broadening of the methyl triplet of ethanol. The paramagnetic contributions to the transverse relaxation rate of the protons of ethanol ($1/T_{2p}$) calculated from line broadenings at several concentrations of ADP-R· are summarized in Table I. In this table, the relaxation rates are also normalized by the ratio p = [ADP-R·]/[ethanol] (11). The effect of ADP-R· on the relaxation rate of the protons of ethanol is greater by a factor of ≥3 on the methylene protons as compared to the methyl protons.[3] Such a difference is consistent with a hydrogen-bonded structure (I),

$$R_2\overset{+}{N}\text{—}\ddot{\underset{-}{O}}\text{:}\cdots H\text{—}O\text{—}CH_2\text{—}CH_3$$

I

in which the methylene protons are nearer the unpaired electron than are the methyl protons. An intermolecular hydrogen bond from O—H to N—O has been detected in the crystalline six-membered nitroxide ring compound, HO—$C_5H_4(CH_3)_4$N—O (17).

The presence of liver alcohol dehydrogenase, which binds 2 molecules of ADP-R· tightly per enzyme molecule (2) and which forms a ternary E(ADP-R·)$_2$(ethanol)$_2$ complex (3), qualitatively and quantitatively alters the effects of ADP-R· on the protons of ethanol as shown in Fig. 2. A low level of ADP-R· (165 μM) produces no detectable broadening of the resonances of ethanol as expected from the titration data presented in Table I. The addition of 61 μM alcohol dehydrogenase which binds 64% of the ADP-R· results in a broadening of both resonances. Contrary to the results with ADP-R· alone, the presence of the enzyme causes a greater broadening of the methyl resonance than of the methylene resonance of ethanol, indicating that in

[2] As in previous publications (7, 16) these chemical shifts are not corrected for differences in the magnetic susceptibility of the solvent and the external standard. Such a correction to the chemical shifts (~0.12 ppm) would not alter the interpretation of the results.

[3] The effect of ADP-R· on $1/T_2$ of the methylene protons of ethanol is smaller than its effect on $1/T_1$ of water protons (3) by a factor of 3.4×10^3. Since both effects are dominated by a dipole-dipole mechanism, this difference probably reflects a greater distance and a shorter correlation time for the interaction of ADP-R· with the protons of ethanol than with the protons of water.

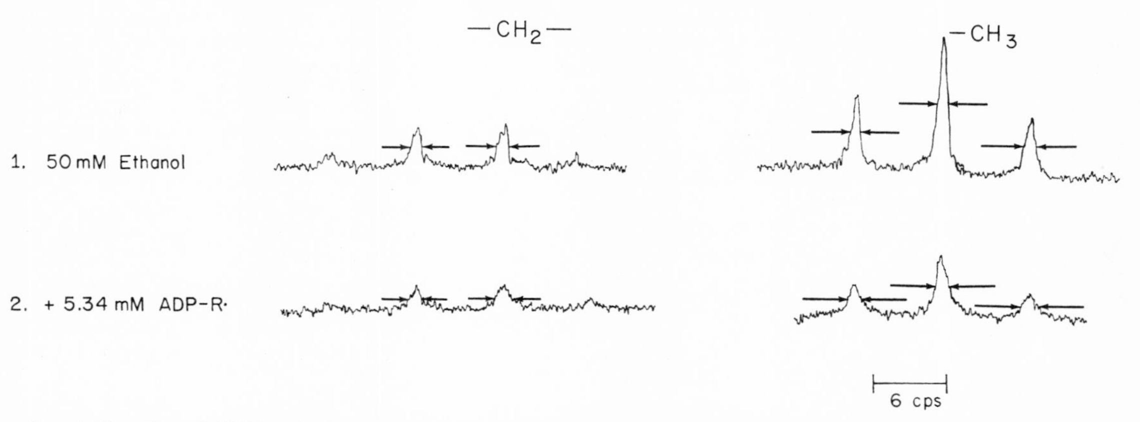

FIG. 1. The effect of ADP-R· on the proton magnetic resonance spectrum of ethanol at 60 MHz. The solutions contained 0.03 M sodium phosphate buffer (pD = 6.45) in 66% D_2O. Temperature, 27°. The *arrows* indicate the width of the resonance lines at half-height.

TABLE I
Effect of ADP-R· and alcohol dehydrogenase-ADP-R· on transverse relaxation rate of protons of ethanol

Temperature	[Ethanol]	[ADP-R·]	[E]	Methylene protons			Methyl protons		
				$1/T_{2p}$	$1/pT_{2p}$	$\epsilon_2^{*,a}$	$1/T_{2p}$	$1/pT_{2p}$	$\epsilon_2^{*,b}$
	mM	μM	μM	sec⁻¹			sec⁻¹		
26°	48.5	165		0	0		0	0	
27	50.0	1375		0.41	14.9		0	0	
27	95.0	5100		0.92	17.1		0.12	2.2	
27	50.0	5340		1.25	11.7		0.44	4.1	
26	48.5	107	61.0	0.43	195.	13.4	1.03	466.	148
26	48.5	165	61.0	0.37	107.	7.3	0.62	182.	58
25	35.0	234	88.5				1.41	211.	67
1	37.0	204	93.5	1.00	181.	12.4	1.91	346.	110
13	37.0	204	93.5	0.47	85.	5.8	1.65	299.	95
25	37.0	204	93.5	0.29	53.	3.6	1.46	265.	84
34	37.0	204	93.5	0.12	22.	1.5	0.40	72.5	16

[a] $\epsilon_2^* = 1/pT_{2p}$ (with E)/$(1/pT_{2p})$ (without E). The denominator used was the average value (14.6 sec^{-1}).
[b] The denominator used was the average value (3.15 sec^{-1}).

the ternary complex the relative orientation of the bound ADP-R· and the bound ethanol has changed from Structure I to II:

$$\begin{array}{c} CH_3{-}CH_2 \\ \quad\quad \diagdown O{-}H \\ R_2\overset{+}{N}{-}\ddot{\underset{\cdot\cdot}{O}}: \end{array}$$

II

The enzyme alone or in the presence of excess NADH (Fig. 2) causes no significant broadening of the resonances of ethanol.

The paramagnetic contributions to the transverse relaxation rates of the protons of ethanol in the presence of alcohol dehydrogenase and ADP-R· under a variety of conditions are summarized in Table I. It may be seen that the presence of the enzyme enhances the effect of ADP-R· on the methylene protons of ethanol by a factor of 1.5 to 13.4 and on the methyl protons by a factor of 16 to 148, depending on the experimental conditions. The enhancement by the enzyme of the paramagnetic effect of ADP-R· on the protons of ethanol indicates that the enzyme-bound ethanol is very near the unpaired electron in the ternary complex. The effect on the methyl protons of ethanol ranges from 7.9 to 23.4 times the effect on the methylene protons, suggesting that in the ternary complex the methyl group of ethanol is nearer to the unpaired electron than is the methylene group.

Effect of NADH on Interaction of E-ADP-R· with Ethanol—The broadenings of both the methylene and the methyl resonances of ethanol by $E(ADP\text{-}R\cdot)_2$ are removed by the addition of 2.3 mM NADH (Fig. 2) which converts the ternary ADP-R· complex to a binary or ternary NADH complex. Abortive ternary complexes of the type $E(NADH)_2(ethanol)_2$ were detected kinetically by Theorell and McKinley-McKee (18). The absence of broadening in the presence of $E(NADH)_2(ethanol)_2$ (Fig. 2) indicates that the broadening observed in the presence of $E(ADP\text{-}R\cdot)_2(ethanol)_2$ is due to a specific paramagnetic effect of the unpaired electron rather than to a nonspecific effect of binding and immobilization of ethanol by the protein.

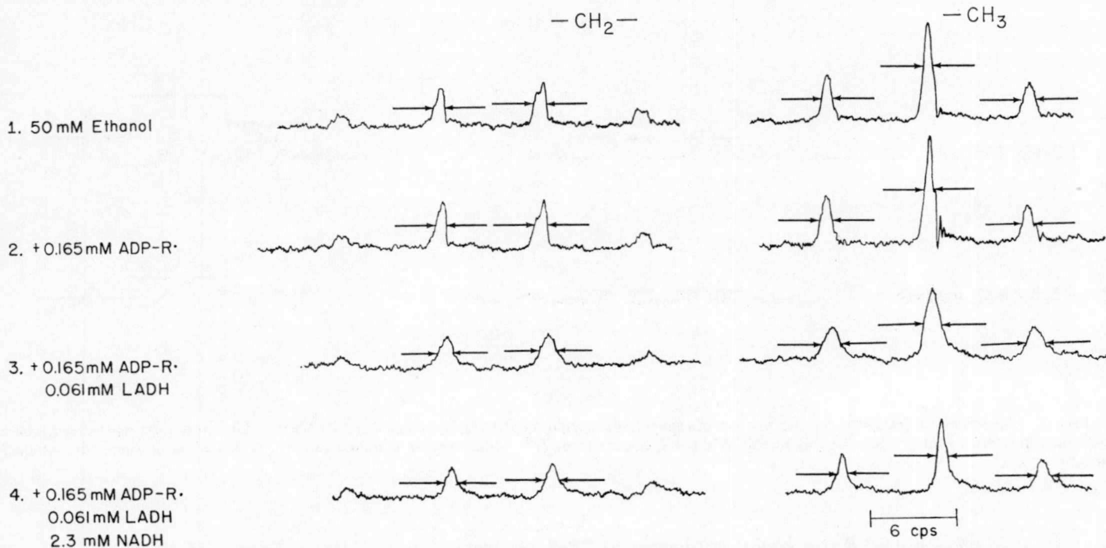

FIG. 2. The effect of ADP-R· and liver alcohol dehydrogenase (*LADH*) on the proton magnetic resonance spectrum of ethanol. The solutions contained 0.03 M sodium phosphate buffer (pD = 6.45) in ≥81% D_2O. Temperature, 25°.

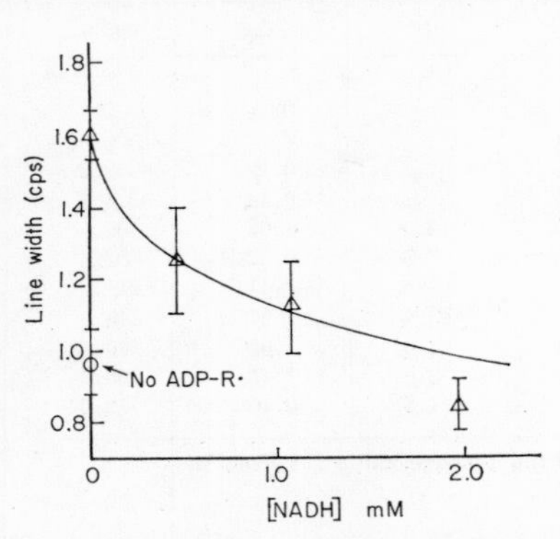

FIG. 3. The effect of NADH on the line width of the methyl resonance of ethanol in the presence of liver alcohol dehydrogenase and ADP-R·. The solution contained ethanol (48.5 mM), alcohol dehydrogenase (66 μM) and ADP-R· (107 μM). Conditions are otherwise as described in Fig. 2. The smooth curve was constructed by assuming competition between ADP-R· (K_D = 16 μM) and NADH (K_D = 45 μM), as described in the text.

Solutions containing alcohol dehydrogenase (66 to 89 μM), ADP-R· (107 to 234 μM), and ethanol (35 to 49 mM) were titrated with NADH measuring the line widths of the protons of ethanol. An example of such a titration is shown in Fig. 3. Assuming competition between NADH and ADP-R· (2) and that the dissociation constant of ADP-R· from E(ADP-R·)$_2$-(ethanol)$_2$ is 16 μM as measured for the binary E(ADP-R·)$_2$ complex (3), the dissociation constant for NADH from its enzyme complex is calculated from three titrations to be 36 ± 9 μM at 25°. A single titration at 1° reveals a K_D = 36 ± 10 μM. While these values are much greater than that obtained by fluorescence enhancement (0.31 μM) (19), they agree with K_D

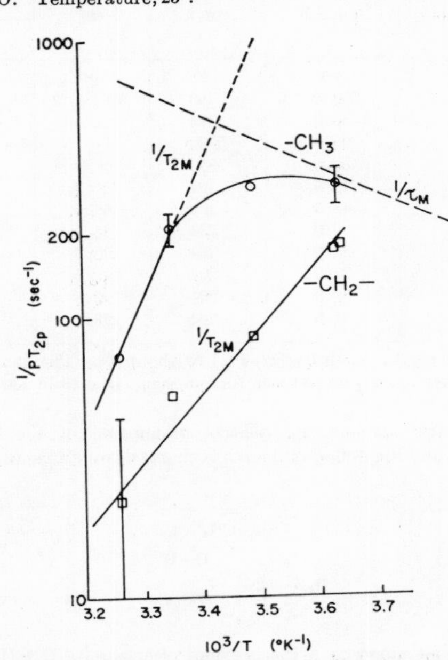

FIG. 4. Arrhenius plot of the effect of temperature on the transverse relaxation rates of the methyl and methylene protons of ethanol in the presence of liver alcohol dehydrogenase and ADP-R· (Table I). The solutions contained 37 mM ethanol, 93.5 μM alcohol dehydrogenase, 204 μM ADP-R·, and other components as indicated in Fig. 2. The *broken lines* indicate the values of $1/\tau_M$ and $1/T_{2M}$ used in Equation 2 to generate the *solid curve* which is fitted to the observed relaxation rates of the methyl protons. The energies of activation in kilocalories per mole are as follows: $1/T_{2M}$ (methyl), 21.2 ± 4.4; $1/\tau_M$ (methyl), 4.4 ± 0.9; $1/T_{2M}$ (methylene), 10.7 ± 2.0.

TABLE II

Kinetic parameters for exchange of ethanol, acetaldehyde, and isobutyramide into their ternary complexes with alcohol dehydrogenase and ADP-R·

Parameter	Units	Ethanol	Acetaldehyde	Isobutyramide
$1/\tau_M$ (at 25°)	sec^{-1}	610 ± 125	≥770 ± 360	≥180 ± 30
E_a	kcal/mole	4.4 ± 0.9[a]		
$\Delta H‡$	kcal/mole	3.8 ± 0.9		
$\Delta S‡$	cal/deg·mole	−20.3 ± 4.2		
$-T\Delta S‡$	kcal/mole	6.1 ± 1.2		

[a] Used to fit the data of Fig. 4.

(E(NADH)$_2$) = 50 ± 10 μM obtained by competition experiments with ADP-R· in the absence of ethanol, with electron paramagnetic resonance (2). The reason for the difference is not clear. It may be due to the 100-fold greater concentration of enzyme required for the present experiments or that the conformation of alcohol dehydrogenase in the presence of ADP-R· is unfavorable for the tight binding of NADH.

Effect of Temperature on Interaction of Alcohol Dehydrogenase ADP-R· with Ethanol—In order to determine the mechanism of the broadening of the NMR spectrum of ethanol by E-ADP-R·, the effect of temperature on this interaction was studied (Table I). An Arrhenius plot of the transverse relaxation rate ($1/pT_{2p}$) of the methyl and methylene protons of ethanol in the presence of E-ADP-R· is given in Fig. 4. Following Swift and Connick (10), we assume that the relaxation rate of the methyl protons can be fit by the summation of two processes (Equation 2): the chemical exchange rate of ethanol molecules into the environment of the unpaired electron ($1/\tau_M$) and the relaxation rate of the methyl protons of ethanol ($1/T_{2M}$) in the ternary E(ADP-R·)$_2$(ethanol)$_2$ complex.

The relaxation rate of the methylene protons is slower than the chemical exchange of ethanol (*i.e.* $\tau_M < T_{2M}$). Therefore, the relaxation rate of the methylene protons is dominated by $1/T_{2M}$ (Equation 2) over the entire temperature range studied. Hence the relaxation rate of the methyl protons permits an estimation of the kinetic parameters of ethanol exchange into E(ADP-R·)$_2$(ethanol)$_2$ and the distance between the methyl protons and the unpaired electron in the ternary complex (Equation 6). The relaxation rate of the methylene protons permits only a calculation of the distance between these protons and the unpaired electron in the ternary complex.

Table II summarizes the kinetic parameters for ethanol exchange into the ternary complex which were used to fit the data of Fig. 4. Assuming two processes, the exchange rate ($1/\tau_M$) extrapolated to 25° is 610 sec^{-1}. The measured relaxation rate at 25° ($1/T_{2p}$ = 200 sec^{-1}) sets a lower limit on this exchange rate. Both rates are faster than the turnover number of alcohol dehydrogenase under similar conditions (3.12 sec^{-1}) which is equal to the rate of dissociation of NADH from the enzyme (18, 20).

Determination of Average Distance (r) between Protons of Ethanol and Unpaired Electron in Ternary E(ADP-R·) (ethanol) Complex—From Equation 6 it is seen that, in order to calculate r, the distance between the unpaired electron and the methyl and methylene protons of ethanol in the ternary complex, from

TABLE III

Calculated distances between unpaired electron of ADP-R· and protons of ethanol in ternary complex with alcohol dehydrogenase

Assumed τ_c	Source	Distance (N—O···H)	
		—CH_2—	—CH_3
$sec \times 10^{12}$		A	
0.41	τ_c(ADP-R·(H_2O)$_4$)	3.1	2.7
8.2	τ_c(pyruvate carboxylase-Mn-pyruvate)[a]	4.7	4.1
28.0	τ_c(alcohol dehydrogenase-ADP-R·(H_2O)$_2$)	5.2	4.6
Range		4.1 ± 1.1	3.6 ± 1.0

[a] From Reference 16.

the values of T_{2M}, one must assign a value to τ_c, the correlation time for the interaction. This approach has previously been used to calculate distances in binary and ternary complexes of enzymes, paramagnetic metal ions, and substrates (7, 16), and, in the one case in which a comparison could be made, has given the same value for Mn-F in solution as that obtained by x-ray crystallography (7). A reasonable value for τ_c of E-ADP-R·-ethanol is 8.2 × 10^{-12} sec as estimated for the interaction of the methyl protons of pyruvate and the methyl and methylene protons of α-ketobutyrate with manganese in the pyruvate carboxylase-Mn-substrate bridge complexes (16). This value for τ_c which probably measures the short intramolecular rotation time of alkyl groups yields a value of r = 4.1 for the distance between the unpaired electron of ADP-R· and the methyl protons of ethanol in the ternary complex (Table III). Limiting values for τ_c may be obtained from Equation 5 with the use of the values of $1/T_{1M}$ for the interaction of H_2O with ADP-R· and alcohol dehydrogenase-ADP-R·. A lower limit for τ_c = 4.1 × 10^{-13} sec is obtained from $1/T_{1M}$(H_2O) = 1.23 × 10^3 sec^{-1} in the binary (ADP-R·) (H_2O)$_4$ complex[4] (3) assuming hydrogen-bonding distances between H_2O and ADP-R·. This value for τ_c results in a value for r(ADP-R· to methyl) in the ternary complex = 2.7 A which is approximately equal to the sum of the van der Waals radii of hydrogen and oxygen (2.6 A) and hydrogen and nitrogen (2.7 A) and is consistent with molecular contact between the methyl protons and the nitroxide group.

An upper limit[3] for τ_c = 2.8 × 10^{-11} sec is obtained from $1/T_{1M}$ of water in the E(ADP-R·)$_2$(H_2O)$_4$ complex[4] (3.22 × 10^4 sec^{-1}) (3), again assuming hydrogen bond distances between ADP-R· and H_2O. Higher values for τ_c are unlikely if the alkyl groups of the bound substrate remain free to undergo rotation or tortional

[4] The "nitroxide radical" is here assumed to "coordinate" 4 water molecules in aqueous solution (*i.e.* q = 4) for the following reasons. A consideration of the electronic structure of dialkyl nitroxide radicals indicates that the N—O group can accept four hydrogen bonds.

$$(R)_2\overset{+}{\dot{N}}—\overset{\cdot}{\underset{-}{O}}: \leftrightarrow (R)_2\dot{N}—\dot{O}:$$

Space-filling models of ADP-R· indicate that, at most, 4 water molecules can pack around the N—O group. Because of the immobilization of the radical by alcohol dehydrogenase (2, 3), it is assumed that enzyme-bound ADP-R· coordinates 2 water molecules. Inaccuracies in these estimates by a factor of 2 would produce errors in the estimated values of r (±11.2%) within the range of those reported in Table III.

oscillation since these processes are very rapid (16). This value for τ_c results in an upper limit for $r = 4.6$ A. Hence the distance between the bound radical and the methyl protons of ethanol in the ternary complex is in the range 3.6 ± 1.0 A (Table III).

Since the change in orientation of ethanol with respect to ADP-R· on the enzyme (Figs. 1 and 2; Table I) indicates the existence of a site and a specific stoichiometry for ethanol in the ternary complex, the Solomon-Bloembergen equation (Equation 6) is appropriate for calculating distances from correlation times and T_{2M} in the ternary E(ADP-R·) (substrate) complexes. There is no direct evidence for the existence of sites for the interaction of water molecules with ADP-R· and E-ADP-R·, as was assumed for calculating the limiting values of τ_c.[4] An alternative method for calculating τ_c from $1/T_{1M}(H_2O)$ with the outer sphere equation of Luz and Meiboom (11), which assumes a continuum of solvent up to ADP-R· and no fixed coordination number, yields values of τ_c (6.6×10^{-13} sec $\leq \tau_c \leq 8.6 \times 10^{-12}$ sec) within the range of those calculated with the above mentioned assumptions (Table III).

Ternary Complexes with Acetaldehyde—A titration of the binary enzyme-ADP-R· complex with acetaldehyde, measuring the enhancement (ϵ^*) of the longitudinal relaxation rate ($1/T_1$) of the protons of water, is shown in Fig. 5. Acetaldehyde reduces ϵ^* from a value of 15.9 ± 1.5 in the binary complex to a value of 8.4 ± 0.2 without significantly altering the amount of ADP-R· which is bound, suggesting the formation of a ternary complex in which water molecules have been displaced from the environment of bound ADP-R·. At higher concentrations of acetaldehyde (>0.6 mM) the enhancement appears to increase to $\epsilon^* = 11.2 \pm 0.6$, suggesting the formation of another complex. The dissociation constants (K_3) and enhancement factors (ϵ_T) of these ternary complexes are summarized in Table IV. The presence of two ternary complexes with acetaldehyde may be due to the presence of significant concentrations of the hydrated form of the aldehyde.

The NMR spectrum of dilute solutions of acetaldehyde in D_2O is known to consist of two spectra, that of the aldehyde and that of the hydrate (22, 23). Under our experimental conditions the

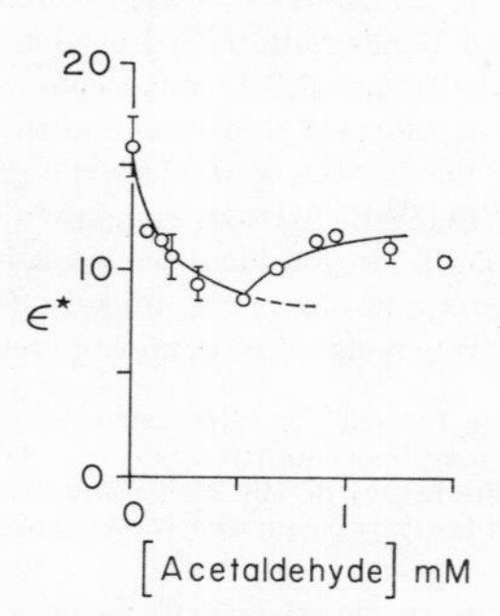

FIG. 5. The effect of acetaldehyde on the enhancement (ϵ^*) of the longitudinal proton relaxation rate of water in the presence of alcohol dehydrogenase-ADP-R·. A solution containing enzyme (50.5 μM) and ADP-R· (114 μM) was titrated with acetaldehyde, measuring ϵ^*. As determined by electron paramagnetic resonance (2), 18.2% of the ADP-R· was free at the beginning of the titration and 19.8% was free at the end of the titration. The solution contained 0.03 M sodium phosphate buffer in H_2O, pH 6.05. Temperature, 23°.

TABLE IV

Enhancements of water proton relaxation rate and dissociation constants of ternary complexes E-ADP-R·-A

A	ϵ_T/ϵ_b [a]	K_3 [b]	K_3 (E-coenzyme A)
		μM	μM
Ethanol	0.33[c]	1100 ± 200[c]	1000–2000[d]
Acetaldehyde	0.45[e]	60 ± 30[e]	84–295[d]
	0.73[f]	620 ± 40[f]	
Isobutyramide	0.66[g]	90 ± 20[g]	180[h]
	0.10[i]	330 ± 40[i]	

[a] ϵ_T/ϵ_b = ratio of the enhancement of the solvent proton relaxation rate in solutions of the ternary complex (ϵ_T) to that in solutions of the binary complex (ϵ_b). $\epsilon_b = 13.1 \pm 1.5$ when the binding sites of alcohol dehydrogenase are saturated with ADP-R· (3).

[b] K_3 = ((E-ADP-R·)(A))/(E-ADP-R·A).

[c] From Reference 3.

[d] From Reference 4.

[e] From [acetaldehyde] ≤0.5 mM (Fig. 5).

[f] From [acetaldehyde] >0.5 mM (Fig. 5).

[g] [ADP-R·]/2[E] = 1.04 (Fig. 7).

[h] From Reference 21.

[i] [ADP-R·]/2[E] = 0.52 (Fig. 7).

183 mM Acetaldehyde
+0.083 mM LADH

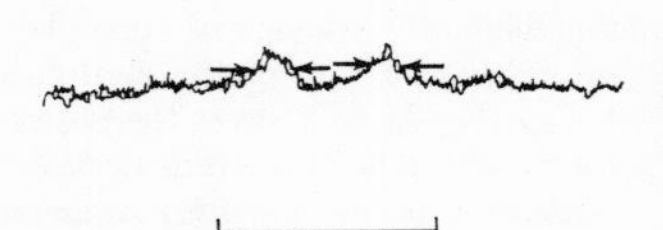

+0.078 mM LADH
+0.206 mM ADP-R·

+0.075 mM LADH
+0.200 mM ADP-R·
+0.364 mM NADH

6 cps

FIG. 6. The effect of liver alcohol dehydrogenase (*LADH*) and ADP-R· on the resonance of the aldehyde proton of acetaldehyde. Conditions are as described in Fig. 2.

areas under the methyl doublets of the aldehyde and of the hydrate suggested approximately equal amounts of each form, which agrees with the known equilibrium constant for hydration (24). The effect of alcohol dehydrogenase-ADP-R· on the downfield quartet due to the aldehyde proton (23) was studied (Fig. 6). The addition of 206 μM ADP-R· to a solution containing 0.17 M acetaldehyde and 155 μM alcohol dehydrogenase markedly broadened the downfield quartet but did not shift them or change the coupling constant (J = 3.05 ± 0.05 cps).

Table V which summarizes the paramagnetic contribution to the transverse relaxation rate reveals that E-ADP-R· strongly relaxes the aldehyde proton of acetaldehyde. The line broadening was partially reversed by NAD (Fig. 6; Table V) which was generated from added NADH. The oxidized coenzyme, NAD,

TABLE V

Effect of alcohol dehydrogenase-ADP-R· on transverse relaxation rate of aldehyde proton of acetaldehyde

Temperature, 25°, 0.03 M sodium phosphate buffer, pD = 6.45.

[Acetaldehyde]	[ADP-R·]	[E]	[NAD][a]	$1/T_{2p}$	$1/pT_{2p}$
mM	*μM*	*μM*	*μM*	*sec⁻¹*	*sec⁻¹*
183	0	84	0	0	0
169	206	78	0	0.94	770
164	200	75	364	0.71	580
157	191	73	870	0.56	460

[a] Added as NADH for reason given in text.

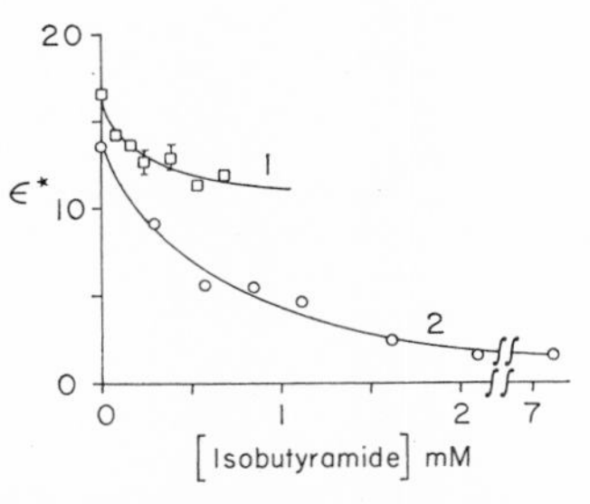

FIG. 7. The effect of isobutyramide on the enhancement (ϵ^*) of the longitudinal proton relaxation rate of water in the presence of alcohol dehydrogenase ADP-R·. *Curve 1*, titration of a solution containing 50.5 μM alcohol dehydrogenase and 114 μM ADP-R· with isobutyramide measuring ϵ^*. *Curve 2*, titration of a solution containing 50.5 μM alcohol dehydrogenase and 57.0 μM ADP-R·. As determined by electron paramagnetic resonance (2), 16.0% of the ADP-R· was free at the beginning of this titration and 13.7% was free at the end of the titration.

was not used for this experiment because our commercial preparations of NAD contained paramagnetic contaminants.

A titration of the ternary complex with NADH measuring the line width of the aldehydic proton reveals competition between ADP-R· and NAD and yields a K_D value for NAD of 20 μM from the ternary *E*-NAD-acetaldehyde complex. Such an abortive complex was proposed by Dalziel and Dickinson (25). The dissociation constant is one-sixth as great as the dissociation constant for *E*-NAD (21), suggesting that acetaldehyde raises the affinity of liver alcohol dehydrogenase for NAD. The upfield resonances of acetaldehyde were not investigated.

From the kinetic studies of Theorell and McKinley-McKee, acetaldehyde would be expected to exchange into the ternary complex at least as rapidly as ethanol (18). Hence, as with ethanol, the line broadening of acetaldehyde ($1/pT_{2p}$ = 770 sec⁻¹) is probably equal to the relaxation rate of acetaldehyde which is bound in the ternary complex ($1/T_{2M}$) and sets a lower limit for the aldehyde exchange rate into the ternary complex ($1/\tau_M >$ 770 sec⁻¹). The distance between the unpaired electron of enzyme-bound ADP-R· and the aldehyde proton of enzyme-bound acetaldehyde calculated from $1/T_{2M}$ with the use of Equation 6 and assuming that τ_c (acetaldehyde) = τ_c (ethanol), is 3.1 A. Hence the aldehyde proton of acetaldehyde is closer to the ribosidic bond to pyridine than is the methyl or methylene group of ethanol (Table VII).

Ternary Complexes with Isobutyramide—An NMR study was carried out on isobutyramide which competes with the aldehyde for the active site of the enzyme (21).

Titrations of E-ADP-R· and E(ADP-R·)$_2$ with isobutyramide, measuring the relaxation rate of water, reveal ternary complexes as shown in Fig. 7, and the results are summarized in Table IV. When the concentration of tight binding sites of alcohol dehydrogenase is in excess of [ADP-R·], a weaker binding of isobutyramide (K_3 = 330 ± 40 μM) is observed than when the tight binding sites are saturated (K_3 = 90 ± 2 μM). Hence, ADP-R· raises the affinity of alcohol dehydrogenase for isobutyramide possibly because of site-site interaction between the two tight binding sites as previously described (3). These dissociation constants are in reasonable agreement with the kinetically determined dissociation constants for *E*-isobutyramide (270 μM) and *E*-NADH-isobutyramide (180 μM) at pH 6.0 (21), suggesting that isobutyramide is binding at the active site (Table IV). At saturating [ADP-R·], the reduction in the enhancement on forming the ternary complex is less (ϵ_T/ϵ_b = 0.66) than the reduction observed at lower levels of ADP-R· (ϵ_T/ϵ_b = 0.10), suggesting that the structure of the ternary complex differs at different ratios of ADP-R· and alcohol dehydrogenase.

The NMR spectrum of isobutyramide consists of a doublet resulting from the methyl groups 1.81 ± 0.02 ppm downfield

17.7mM Isobutyramide
+0.09mM LADH

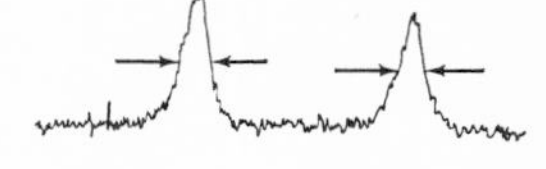

+0.082mM LADH
+0.219mM ADP-R·

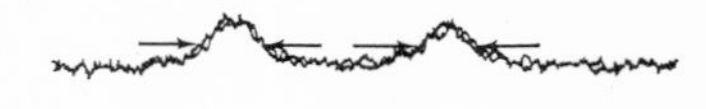

+0.075mM LADH
+0.202mM ADP-R·
+0.924mM NADH

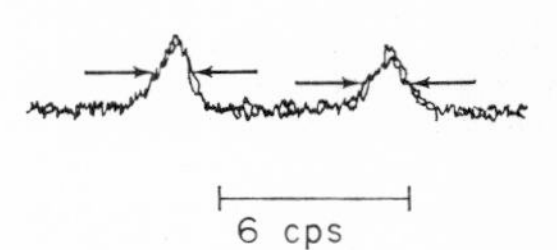

FIG. 8. The effect of liver alcohol dehydrogenase (*LADH*) and ADP-R· on the resonance of the methyl protons of isobutyramide. Conditions are as described in Fig. 2.

TABLE VI

Effect of alcohol dehydrogenase-ADP-R· on transverse relaxation rate of methyl protons of isobutyramide

Temperature, 25°, 0.03 M sodium phosphate buffer, pD = 6.45.

[Isobutyramide]	[ADP-R·]	[E]	[NADH]	$1/T_{2p}$	$1/pT_{2p}$
mM	*μM*	*μM*	*μM*	*sec⁻¹*	*sec⁻¹*
17.7	0	90	0	0	0
16.3	219	83	0	2.43	180
15.8	212	80	386	1.02	75
15.0	202	77	924	0.68	50

TABLE VII

Relative distances between unpaired electron of ADP-R· and protons of ethanol, acetaldehyde, and isobutyramide in ternary complexes with alcohol dehydrogenase

Proton	$1/T_{2M}$	r
	sec⁻¹	*A*
(CH_3)—CH_2—OH	282	3.6[a]
CH_3—(CH_2)—OH	118	4.1
CH_3—(CH)=O	770	3.1
($(CH_3)_2$)CH—C(=O)—NH_2	≥180	≤3.9

[a] Midpoint of range of distances (Table III).

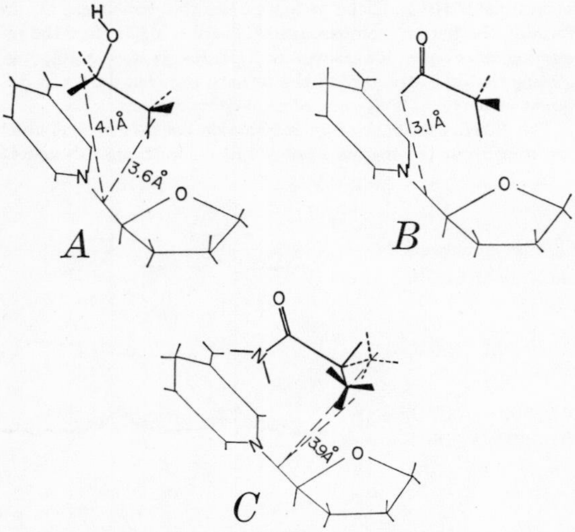

FIG. 9. Relative positions on liver alcohol dehydrogenase of the appropriate coenzymes and ethanol (*A*), acetaldehyde (*B*), and isobutyramide (*C*) as determined by the distances measured to alcohol dehydrogenase-ADP-R· (Table VII).

from tetramethyl silane with a coupling constant J = 6.9 ± 0.1 cps and a septet resulting from the β-proton 3.21 ± 0.01 ppm downfield from tetramethyl silane. The latter signal was not suitable for line broadening studies. The effects of $E(\text{ADP-R}\cdot)_2$ on the methyl resonance of isobutyramide shown in Fig. 8 reveal a broadening but no change in the chemical shift or coupling constant. The paramagnetic contribution to the relaxation rate given in Table VI indicates that *E*-ADP-R·effectively relaxes the methyl protons of isobutyramide. NADH reverses this effect with an apparent K_D of 8.1 μM, a value one-sixth as large as that observed in the absence of substrates (2). Isobutyramide is known to raise the affinity of liver alcohol dehydrogenase for NADH (21). The value of $1/pT_{2p}$ = 180 sec⁻¹ sets a lower limit on the exchange rate of isobutyramide into the environment of enzyme-bound ADP-R· and on the relaxation rate of isobutyramide in the ternary complex; *i.e.* $1/\tau_M$ (isobutyramide) ≥ 180 sec⁻¹ and $1/T_{2M}$ (isobutyramide) ≥ 180 sec⁻¹. From $1/T_{2M}$, with the use of Equation 6, the average distance from the unpaired electron of ADP-R· to the methyl protons of isobutyramide in the ternary complex is ≤3.9 A (Table VII).

In a study of the competitive inhibitor acetamide (35 mM), no change in the NMR spectrum of the methyl protons was detected in the presence of alcohol dehydrogenase (177 μM) and ADP-R· (234 μM). Hence, if a ternary complex involving acetamide exists it is either exchanging slowly on an NMR time scale ($1/\tau_M$ < 30 sec⁻¹) or the methyl protons are more distant than 5.2 A from the unpaired electron.

Table VII summarizes the average distances in the ternary complexes between the unpaired electron of ADP-R· and the protons of ethanol, acetaldehyde, and isobutyramide, based on a value of 3.6 A for the distance to the methyl protons of ethanol. Although the uncertainty in the absolute value of r is high (±28%) because of the uncertainty of τ_c, the relative distances to the methyl and methylene protons of ethanol and to the protons of other substrates are more precise (< ±6%) since they would have similar values of τ_c, and are in error by the sixth root of the error in $1/pT_{2p}$. These distances reflect the relative average distances on the enzyme from the protons of ethanol, acetaldehyde, and isobutyramide to the ribosidic bond to the pyridine nitrogen of bound NAD (Fig. 9).

DISCUSSION

Diamagnetic broadenings of the resonances of ethanol have been observed with yeast alcohol dehydrogenase (26) at much higher levels of enzyme (1.1 mM) and much lower levels of ethanol (1.0 mM) than those used here in which a significant fraction of the residual ethanol is bound to the enzyme. Such effects are two orders of magnitude smaller than the paramagnetic effects described here and are difficult to interpret in terms of structural or kinetic parameters.

The present studies of the effect of the enzyme-ADP-R· complex on the NMR spectrum of ethanol, acetaldehyde, and isobutyramide have yielded both kinetic and structural information about the respective ternary complexes. From the exchange rate of ethanol into the ternary *E*-ADP-R·-ethanol complex ($1/\tau_M$ = 610 sec⁻¹) one can calculate the rate constant for the combination of ethanol with the binary *E*-ADP-R· complex, a parameter which has thus far been unobtainable for the enzyme-coenzyme complex (27).

If $1/\tau_M$ for ethanol is limited by a first order dissociation of *E*-ADP-R·-ethanol, the rate constant for combination of ethanol with *E*-ADP-R· from the dissociation constant of ethanol (1.1 mM) (3) is 5.5 × 10⁵ M⁻¹ sec⁻¹. This scheme may be represented as follows:

$$\text{Ethanol} + E\text{—ADP—R}\cdot \underset{6.1 \times 10^2 \text{ sec}^{-1}}{\overset{5.5 \times 10^5 \text{ M}^{-1} \text{ sec}^{-1}}{\rightleftharpoons}} E\text{—ADP—R}\cdot\text{ethanol}$$

Alternatively, if $1/\tau_M$ is measuring a second order process, the rate constant for combination of ethanol with the ternary complex $(1/\tau_M)(1/\text{ethanol})$ is 1.7 × 10⁴ M⁻¹ sec⁻¹:

$$\text{Ethanol}^* + E\text{—ADP—R}\cdot\text{ethanol} \xrightarrow{1.7 \times 10^4 \text{ M}^{-1} \text{ sec}^{-1}} E\text{—ADP—R}\cdot\text{ethanol}^* + \text{ethanol}$$

Although neither mechanism can be excluded by the limited NMR data, the assumption of a first order dissociation process leads to a rate constant for dissociation of ethanol (610 sec⁻¹)

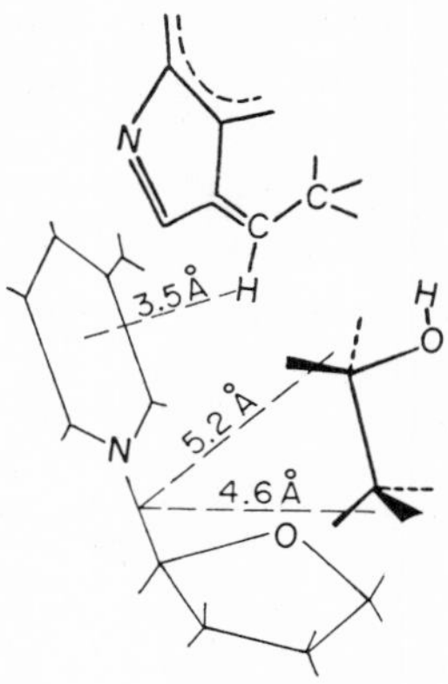

FIG. 10. Open structure of the ternary complex of liver alcohol dehydrogenase-NAD-ethanol, consistent with the upper limit of the distances measured to alcohol dehydrogenase-ADP-R· (Table III), showing the mode of interaction of tryptophan.

which is significantly slower than k_3 found by steady state kinetics at pH 7 (18) for the reaction:

$$CH_3CHO + E\text{—}NADH \underset{k'_3 = 1.22 \times 10^4 \text{ M}^{-1} \text{sec}^{-1}}{\overset{k_3 = 3.1 \times 10^5 \text{ M}^{-1} \text{sec}^{-1}}{\rightleftharpoons}} E\text{—}NAD + \text{ethanol}$$

which includes the oxidation-reduction reaction as well. Hence the first order assumption for τ_M seems unlikely. The second order assumption yields a rate constant which is in reasonable agreement with k'_3. The larger value for k_3 might be explained by assuming that aldehyde displaces water faster than ethanol displaces water.

Whichever mechanism for the combination of ethanol with the enzyme-analogue complex is operative, the rate of this process is several orders of magnitude slower than the rate of a diffusion controlled reaction. Two explanations for this may be suggested. First, the ethanol molecule may coordinate to zinc in a rate-limiting reaction. In certain sterically hindered complexes, zinc binds ligands at a rate as low as 5×10^5 M^{-1} sec^{-1} (28). Second, the rate of combination of ethanol with the enzyme may be limited by the rate of a conformational change in the protein or in the bound substrate. At present no evidence exists which can distinguish among these alternatives.

From the structural properties of the ternary complexes of the analogue one may gain information about the structure of the ternary complexes of the coenzymes. Since the unpaired electron of ADP-R· occupies a position which corresponds to the bond between the pyridine nitrogen and the C_1 ribose of NAD (3), one can map the relative positions of this ribosidic bond and the protons of ethanol, acetaldehyde, and isobutyramide in the ternary complexes (Fig. 9) from the average values of the distances (r) between the unpaired electron and the protons (Table VII).

While distances from a single reference point are insufficient to define uniquely the relative positions of the coenzymes and the substrates on liver alcohol dehydrogenase in solution, two restrictions limit the range of possible structures. First, our studies of the proton relaxation rate of water (3) (Figs. 5 and 7) suggest that ethanol, acetaldehyde, and isobutyramide bind on the "water side" of the bound coenzyme, thus halving the range of positions available to the substrate. Second, the space occupied by the pyridine and ribose rings of NAD in all conformations further delimits the range of positions available to the substrate.

In Fig. 9 it may be seen that structures which are consistent with our calculated average distances permit a single position for the oxygen atoms of ethanol, acetaldehyde, and isobutyramide. Such a position might be fixed by coordination to the zinc ion (29), although no direct evidence for such a metal bridge complex exists. It can also be seen (Fig. 9*A*) that the structure of the ternary ethanol complex permits van der Waals contact between the hydrogen which is transferred and the C-4 position of pyridine. Similarly, in the ternary aldehyde complex (Fig. 9*B*) contact between the pyridone hydrogen of NADH and the C-2 carbon of the substrate is permitted by our calculated distance. Hence, the structures of the ternary complexes in solution are consistent with the direct transfer of hydrogen between the substrates and the coenzymes (30, 31). The concept of direct transfer has recently been challenged by Schellenberg (32–34) who has detected the transfer of hydrogen to tryptophan residues of various dehydrogenases from stereospecifically labeled substrates and coenzymes. Schellenberg has therefore suggested that tryptophan might serve as a hydrogen carrier between substrates and coenzymes. As mentioned above, the structures shown in Fig. 9, while fitting our data, do not represent a unique solution of the relative positions of coenzyme, ethanol, acetaldehyde, and isobutyramide on the enzyme, since they are based on a single reference point. The most open structure for the ternary complex with ethanol which fits our values of r (Fig. 10) allows a maximum cleft of only 1.7 A between ethanol and the pyridine ring of NAD and, therefore, does not permit the insertion of the indolenine ring system of tryptophan (3.7 A thick) between them. Hence these data exclude hydrogen transfer by way of stacked pyridine and indolenine rings. However, an edge to edge structure for tryptophan and pyridine as depicted in Fig. 10 is possible and would allow hydrogen transfer by way of the amino acid residue. Hence our data cannot exclude the participation of a tryptophan residue as proposed by Schellenberg. A solution to this problem may be achieved by the use of other paramagnetic analogues of NAD with the unpaired electron in different positions. Such sets of spin-labeled coenzyme analogues will permit the determination, by triangulation, of the precise orientation of the substrates with respect to NAD on the enzyme in solution. In the crystalline state, of course, the technique of x-ray diffraction should contribute to a definitive solution of this problem. Because of the possibility of differences in structures between crystalline and dissolved proteins (35–37)[5] both techniques are necessary and complementary.

It is concluded that by studying the effects of a paramagnetic analogue-enzyme complex on the nuclear magnetic relaxation rates of the solvent, the substrates, and inhibitors, one can gain detailed information about the structural, thermodynamic, and kinetic properties of ternary complexes. Specifically, one can determine distances between individual atoms of the bound analogue and the bound substrate molecules on an enzyme in solution, as well as the binding constants and exchange rates of the substrates. The general applicability of these techniques to

[5] MILDVAN, A. S., RUMEN, N. M., AND CHANCE, B., *Abstracts of the American Chemical Society Meeting, Atlantic City, September 1968*, Biol-32.

other two-substrate enzymes is, therefore, limited only by the availability of appropriate paramagnetic substrate analogues.

Acknowledgments—We are grateful to Professors Roland G. Kallen and Mildred Cohn for making their NMR spectrometers available to us, to Professor Anders Ehrenberg and Dr. Jenny Pickworth Glusker for helpful comments, and to Miss Marie L. Smoes for capable technical assistance.

REFERENCES

1. Weiner, H., and Mildvan, A. S., *Fed. Proc.*, **28**, 533 (1969).
2. Weiner, H., *Biochemistry*, **8**, 526 (1969).
3. Mildvan, A. S., and Weiner, H., *Biochemistry*, **8**, 552 (1969).
4. Theorell, H., and Yonetani, T., *Arch. Biochem. Biophys.*, Suppl. 1, 209 (1962).
5. Taniguchi, S., Theorell, H., and Åkeson, Å., *Acta Chem. Scand.*, **21**, 1903 (1967).
6. Mildvan, A. S., and Cohn, M., *Biochemistry*, **2**, 910 (1963).
7. Mildvan, A. S., Leigh, J. S., and Cohn, M., *Biochemistry*, **6**, 1805 (1967).
8. Cohn, M., and Leigh, J. S., *Nature*, **193**, 1037 (1962).
9. Eisinger, J., Shulman, R. G., and Szymanski, B. M., *J. Chem. Phys.*, **36**, 1721 (1962).
10. Swift, T. J., and Connick, R. E., *J. Chem. Phys.*, **37**, 307 (1962).
11. Luz, Z., and Meiboom, S., *J. Chem. Phys.*, **40**, 2686 (1964).
12. Solomon, I., *Phys. Rev.*, **99**, 559 (1955).
13. Solomon, I., and Bloembergen, N., *J. Chem. Phys.*, **25**, 261 (1956).
14. Hamilton, C. L., and McConnell, H., in A. Rich and N. Davidson (Editors), *Structural chemistry and molecular biology*, W. H. Freeman and Company, San Francisco, 1968, p. 115.
15. Leffler, J. E., and Grunwald, E., *Rates and equilibria of organic reactions*, John Wiley and Sons, Inc., New York, 1963, p. 112.
16. Mildvan, A. S., and Scrutton, M. C., *Biochemistry*, **6**, 2978 (1967).
17. Lajzerowicz-Bonneteaw, J., *Acta Cryst.*, **24**, 196 (1968).
18. Theorell, H., and McKinley-McKee, J. S., *Acta Chem. Scand.* **15**, 1797 (1961).
19. Theorell, H., and McKinley-McKee, J. S., *Acta Chem. Scand.*, **15**, 1811 (1961).
20. Theorell, H., and Chance, B., *Acta Chem. Scand.*, **5**, 1127 (1951).
21. Winer, A. D., and Theorell, H., *Acta Chem. Scand.*, **14**, 1729 (1960).
22. Lombardi, E., and Sogo, P. G., *J. Chem. Phys.*, **32**, 635 (1960).
23. Evans, P. G., Miller, G. R., and Kreevoy, M. M., *J. Phys. Chem.*, **69**, 4325 (1965).
24. Bell, R. P., *Advan. Phys. Org. Chem.*, **4**, 1 (1966).
25. Dalziel, K., and Dickinson, F. M., *Nature*, **206**, 255 (1965).
26. Hollis, D. P., *Biochemistry*, **6**, 2080 (1967).
27. Sund, H., and Theorell, H., in P. Boyer, H. Lardy, and K. Myrbäck (Editors), *The enzymes, Vol. 7*, Academic Press, New York, 1963, p. 25.
28. Hertz, H. G., *Z. Electrochem.*, **65**, 36 (1961).
29. Theorell, H., and Yonetani, T., *Biochem. Z.*, **338**, 537 (1963).
30. Westheimer, F. H., Fisher, H. F., Conn, E. E., and Vennesland, B., *J. Amer. Chem. Soc.*, **73**, 2403 (1951).
31. Levy, H. R., and Vennesland, B., *J. Biol. Chem.*, **228**, 85 (1957).
32. Schellenberg, K. A., *J. Biol. Chem.*, **240**, 1165 (1965).
33. Schellenberg, K. A., *J. Biol. Chem.*, **241**, 2446 (1966).
34. Schellenberg, K. A., Chan, T., and McLean, G. W., *Fed. Proc.*, **27**, 453 (1968).
35. Chance, B., Ravilly, A., and Rumen, N. M., *J. Mol. Biol.*, **17**, 525 (1966).
36. Maricic, S., Ravilly, A., and Mildvan, A. S., in B. Chance, R. Estabrook, and T. Yonetani (Editors), *Hemes and hemoproteins*, Academic Press, New York, 1966, p. 157.
37. Theorell, H., Chance, B., and Yonetani, T., *J. Mol. Biol.*, **17**, 513 (1966).

Reprinted from THE JOURNAL OF BIOLOGICAL CHEMISTRY, Vol. 239, 1964

Direct Studies on the Electron Transfer Sequence in Xanthine Oxidase by Electron Paramagnetic Resonance Spectroscopy

I. TECHNIQUES AND DESCRIPTION OF SPECTRA*

GRAHAM PALMER, ROBERT C. BRAY,† AND HELMUT BEINERT

From the Institute for Enzyme Research, University of Wisconsin, Madison 6, Wisconsin

(Received for publication, November 8, 1963)

The unique usefulness of electron paramagnetic resonance spectroscopy in the observation of electron transfer in enzyme systems involving certain paramagnetic species has been amply demonstrated in recent years (1–4). Auxiliary instrumentation has been developed for following the kinetics, in the early stages, of changes in the EPR[1] signals, related to changes in either the substrate or in the enzyme, in enzymic reactions. Changes in substrate molecules have been followed by EPR at ambient temperatures (1, 4–6), while those in enzyme molecules have been followed both at ambient temperatures (4) and in the frozen state (7–9). Milk xanthine oxidase has been an attractive object of such investigations, because at least two components, molybdenum and flavin, and almost certainly iron as well, appeared amenable to study by EPR (2); the enzyme can be obtained in good yield and high purity; it represents a multicomponent electron transport system within 1 soluble molecule; and its catalytic constants (10) are of a convenient magnitude for experimentation with the presently available methods. Consequently, with this protein the rapid freezing method of kinetic analysis has been extended farther at this time than with any other enzyme (9). Nevertheless, it appeared promising to us to resume and extend this work for two reasons. First, EPR instrumentation is now available which is more sensitive than that used in previous work. This might allow one to identify and study in more detail signals or components of signals which were barely perceptible previously. Thus Bray, Pettersson, and Ehrenberg (2) observed a broad signal, which they tentatively attributed to iron. The kinetics of this signal could not be studied. These authors also observed changes in the structure of the molybdenum signal during reduction, but could not resolve the spectra sufficiently for a more definite identification. Second, a new type of electron carrier, presumably a nonheme iron complex, has recently been found by EPR in a number of organisms, tissues, and isolated enzymes (3, 11–15). The EPR spectrum of this type of compound appears closely related to the spectrum of that component of xanthine oxidase which is attributed to reduced iron (2). It was therefore of particular interest to study at higher resolution and in more detail the characteristics of the EPR signal of this component of xanthine oxidase, and to compare its properties with those of the signals obtained from the nonheme iron complexes present in other oxidation-reduction systems. If a close relationship between all these iron compounds could be established, the xanthine oxidase component would afford a unique opportunity to study the kinetic behavior of this type of compound in a relatively well defined system. From such a kinetic analysis, we expected the long sought answer to the question whether representatives of the family of nonheme iron complexes, characterized by an EPR signal at $g = 1.93$ to 1.95 in the reduced state, would qualify as intermediate catalysts in electron transfer.

The two main objectives of the present work were realized. The molybdenum signal of xanthine oxidase was sufficiently resolved so that four major components of the EPR signal were recognized. We attribute these to Mo(V) in the active center of the enzyme in different states of chemical binding. It is likely that only two chemical species are present, each giving rise to one set of components of the signal. We have evaluated our results on the basis of this latter assumption. The EPR characteristics of the iron component were established, and its kinetic behavior is consistent with its being a functional catalyst in the electron transfer system of xanthine oxidase. A sequence of electron transfer within the enzyme from molybdenum to flavin and finally to iron is most compatible with our results.

In this paper, details of the experimental techniques will be given, and the EPR spectra obtained from inorganic complexes of Mo(V) and from xanthine oxidase under different conditions and at different stages of the reaction will be discussed. Paper II (16) is concerned with the results of kinetic studies by the rapid mixing-freezing technique of Bray (7) and Bray and Pettersson (8).

EXPERIMENTAL PROCEDURE

Enzyme Preparation—Xanthine oxidase was prepared by a modification of a method of Gilbert and Bergel (17). Cream was separated from fresh milk, without cooling below 20°,

* This investigation was supported by the United States Public Service through Research Grants AM 02512, GM 06762, and GM 05073, by Research Career Program Award GM-K6-18,442 from the Division of General Medical Sciences to H. Beinert, and by the Atomic Energy Commission through Contract At(11-1)-909. The participation of one of us (R. C. B.) in this work was made possible through the interest of Professor F. Bergel, F.R.S., in this work and by grants from E. I. du Pont de Nemours and Company, Inc., the Research Committee of the Graduate School of the University of Wisconsin, and the National Institutes of Health, which are gratefully acknowledged.

† Permanent address, Chester Beatty Research Institute, Institute of Cancer Research: Royal Cancer Hospital, London S.W. 3, England.

[1] The abbreviation used is: EPR, electron paramagnetic resonance; often also called electron spin resonance, ESR.

and buttermilk was prepared from it by churning at 5–10°. Sodium salicylate was added to the buttermilk to a final concentration of 1 mM. It was then heated to 36° and the following were added for each liter of buttermilk: $NaHCO_3$, 17 g; pancreatin (British Drug Houses, Ltd., Poole, Dorset, England), 1.6 g; and cysteine hydrochloride, 0.3 g. The mixture was then held at 36–38°, with stirring, for 3½ hours. The digest was cooled overnight to 10°, and the following were added for each liter of digest: 1-butanol (at −11°), 143 ml; $(NH_4)_2SO_4$, 190 g; and disodium EDTA (final concentration, 1 mM). The remaining operations were carried out at approximately 5°, and from this point on all buffers and dialysis solutions were passed through Dowex A-1 chelating resin and EDTA and sodium salicylate were added to them to a final concentration of 1 mM. The mixture was centrifuged and the upper layer rejected. To each liter of the yellow lower layer, which remained somewhat cloudy, 110 g of $(NH_4)_2SO_4$ were added. The floating, bulky, brown precipitate containing the enzyme was then separated by centrifugation and dialyzed thoroughly against dilute phosphate buffer; the solution so obtained was clarified by centrifuging at 10,000 × *g*. The enzyme was then purified by chromatography twice on calcium phosphate gel. The columns (packed volume, 15 cm in height × 5 cm in diameter) consisted of a mixture of calcium phosphate gel (18) and cellulose[2] (5 g of cellulose per g of calcium phosphate) and could be used several times. Each column was suitable for chromatographing the enzyme from about 5 liters of buttermilk in a single run. The enzyme was applied in 0.1 M phosphate buffer (0.02 M Na_2HPO_4 and 0.08 M KH_2PO_4). The column was then washed with the same buffer at 0.2 M and finally eluted with 1.0 M buffer, and the brown fractions were combined; a small forerun and tail were rejected. The second chromatography was carried out in exactly the same manner, except that the final elution was done with 0.09 M pyrophosphate buffer, pH 7.0 (prepared from $Na_4P_2O_7$ and HCl), in place of the 1.0 M phosphate. The enzyme solution was finally concentrated by ammonium sulfate precipitation and dialysis against polyvinylpyrrolidone or by vacuum dialysis (2). The final yield from 17 liters of buttermilk was 5 g of purified xanthine oxidase. Xanthine oxidase concentrations were calculated from the extinction coefficient at 450 mμ (19). The product showed a high degree of purity in the ultracentrifuge, although traces of faster and slower sedimenting components could be detected (20). The proportion of the slow material was always less than 5%, but the amount of the fast sedimenting component was more variable, and in one sample as much as 10% was present. No evidence was found in the present work to suggest that this impurity had any influence on the EPR signals. Since the properties (20) of the faster sedimenting material suggested that it might be a dimer of xanthine oxidase, ultracentrifugal runs were made at different times after diluting the stock enzyme solution and at two pH values (6.0 and 8.3). However, the proportion of this faster component remained unchanged, so that no direct evidence for its being a dimer was obtained. The ratio $E_{420}:E_{450}$ for the xanthine oxidase samples used ranged from 0.89 to 1.02; most of the work was done with samples with ratios near the lower of these values. The significance of the ratio is uncertain, and it has not been commented on by previous workers. Variations did not seem to influence the EPR signals, as will be shown in Paper II (16). The enzyme was stored at 2–4° in 1 M phosphate buffer, prepared as for chromatography, which contained 30 mM salicylate and 1 mM EDTA. Under these conditions the enzyme is stable for several months (21). Shortly before use the enzyme was dialyzed thoroughly against 0.05 M pyrophosphate buffer, pH 8.2. The activity declines slowly after this dialysis, and samples were only kept under these conditions for immediate use.

Enzyme Assay—Enzyme activity was assayed with xanthine as substrate as described by Bergel and Bray (21). The preparations used had a specific activity (activity/E_{450}) of 80 to 100, corresponding, on the basis of previous assumptions (2), to 68 to 86% of xanthine oxidase *a*. For one of the samples with the highest specific activity, the extent of anaerobic bleaching by the substrate (22, 23) was measured.[3] In agreement with the high activity, most of the bleaching took place rapidly. Thus, with 12.5 moles of xanthine, the E_{450} was 48% of its initial value after 1 minute, 40% after 15 minutes, and 34% after 60 minutes at 28°. It was observed that for preparations obtained by this method, the optimal pH is not necessarily 8.3, the value indicated by earlier workers (22, 24, 25). Under the conditions of the routine assay mentioned above, we find an optimum at pH 8.9. The relative activities at different pH values were: pH 6.0, 41; pH 8.3, 90; pH 8.9, 100; and pH 9.6, 77.

Substrates, Inhibitors, and Model Compounds—In most experiments xanthine was used as substrate. Xanthine (Hoffmann-La Roche) was dissolved in NaOH to a 10 mM solution of pH 10. Other substrates used were hypoxanthine (A. D. McKay Company), salicylaldehyde (Eastman Kodak), purine (Aldrich Chemical Company), and DPNH (Pabst Laboratories); the inhibitors 8-phenylxanthine[4] and 2,6-diamino-8-phenylpurine[4] were studied in some experiments.

$K_4Mo(CN)_8 \cdot 2H_2O$ was prepared according to Furman and Miller (26), and $MoCl_5$ was a gift of Climax Molybdenum Company. Immediately before use, the cyanide complex was oxidized to the pentavalent form by addition of ceric sulfate.

Rapid Freezing Technique—The rapid freezing method of quenching the reactions (7) and the technique for obtaining the frozen samples in a form suitable for EPR measurements (8) were as described previously. Some of the following points were not made clear in the original publications, while some represent additional refinements. An improved hydraulic unit (Stephens, Smith and Company, Ltd., London E.14) and 0.5-, 1.0-, or 2.5-ml glass syringes with Teflon plungers ("gastight," Hamilton Company, Whittier, California) were employed. Polyethylene tubing with an internal diameter of 0.38 mm and a wall thickness of 0.35 mm (Intramedic PE 20, Clay Adams Company, New York) was used throughout. The tubing was connected to the syringes by cementing aluminum end pieces to the syringes with epoxy cement. Drilled out stainless steel screws fastened into the end pieces then served to press the flanged ends of the polyethylene directly onto the glass nozzles of the syringes.

The entire reaction unit, including the syringes, the mixing

[2] Cellulose powder Aku-20, from Mo och Domsjö Aktiebolag, Stockholm, was used. This powder, which gives a very good flow rate, is no longer manufactured.

[3] The experiment was carried out in EPR tubes, along with the manual EPR titration experiments (see Fig. 10 of the accompanying paper (16)).

[4] These compounds were kindly provided by Drs. F. Bergmann and G. M. Timmis, respectively.

chamber, and the connecting tubes, together with the ram of the hydraulic unit, was contained in an open Lucite box, which could be filled with water circulated from a thermostat for making measurements at other than room temperature. The jet was underneath this box and was connected through its bottom.

The conditions necessary to obtain optimum packing of the crystals in the EPR tubes were somewhat critical. The linear velocity of the stream of liquid leaving the jet was of particular importance, and in most of the present work a velocity of about 30 meters per second was used. At lower velocities (less than 20 meters per second), the crystals tended to be coarse, while at velocities above 50 meters per second they became unduly fine and packing became very difficult. However, the optimum jet velocity depended on the solutions being used and also on the contours of the individual jet. Isopentane (Matheson, Coleman, and Bell, "practical grade") at $-145 \pm 5°$ was employed as the cooling liquid, and the packing device was cooled before use. The EPR tubes were made from precision bore quartz of 3-mm internal diameter and 0.5-mm wall thickness. This facilitated both the packing and direct quantitative comparison of the EPR signal intensities.

It was found possible to reduce the volume of solutions squirted. Previously, 0.3 to 0.5 ml was used (8, 9), but in the present work this was decreased to 0.15 to 0.20 ml, plus a volume equal to the dead space of the system. Connecting tube and jet systems that provided total dead spaces of 4 to 350 μl were used. The procedure for packing the ice crystals was described previously (8). The height of the ice packed in the bottom of the EPR tubes was 2 to 3 cm; according to one set of measurements, about 42% of the packed volume in the present work was ice. Thus, usually a little less than half of the liquid leaving the jet was eventually packed. However, since the amount packed was always more than sufficient to cover the sensitive part of the EPR cavity, the *proportion* packed was irrelevant so far as signal intensities were concerned. The critical factor would be the *uniformity* of the packing, but the relative lack of scatter of the experimental points in the figures in Paper II (16) indicates that adequate uniformity of packing was, in fact, achieved.

Since, as indicated previously (7), the jet diameter might be expected to have a critical effect on the efficiency of the quenching process, parallel experiments of the single turnover type (see Paper II, Fig. 1) were carried out on xanthine oxidase.

Complete signal intensity *versus* time curves were obtained both with a 0.14-mm jet (jet velocity, approximately 40 meters per second) and with a 0.21-mm jet (jet velocity, approximately 20 meters per second). It was found that changing the jet size had a definite effect both on the intensity of the signals and on the times for their maximum development. Relative to the smaller jet, the 0.21-mm jet caused the FADH and Mo δ signal curves to be decreased in intensity by about 30%, while the Mo β signal curve was increased by about 15%. Furthermore, with the larger jet, the curves generally appeared to be displaced toward shorter times by about 10 milliseconds. This effect was apparent particularly for the Mo δ and Mo β signal curves, but it probably applied also to the radical and the iron, although here experimental errors made it difficult to be certain. This effect would be the one expected if quenching of the reaction was not instantaneous as a result of slow freezing with the larger jet. As a first approximation, the time for quenching might be expected to be related to the square of the jet diameter (7), so that in principle it might have been desirable to carry out further experiments with smaller jets and to extrapolate all results to zero jet diameter. In practice, small jets often cause difficulties owing to the high pressures which they generate in the flow system, but extrapolation, on the basis of a 10-millisecond difference between the two jets, would put the quenching times at 8 milliseconds for the smaller jet and 18 milliseconds for the larger. The fact that some of the signals increased and some decreased on changing the jet size shows that differences in the packing in the EPR tubes, resulting from the different sizes of the crystals, cannot be the whole explanation of the intensity differences caused by the different jets. Presumably the activation energies of the various reactions which are taking place (16) are not all the same, so that some of them are quenched more readily than others. One possible explanation of the above data would be that the transformation (16) of Mo δ to Mo β is a reaction which is relatively difficult to quench.

A 0.14-mm diameter jet was used for almost all of the current work, and in the light of the discussion above it seems reasonable to assume that the reactions being studied were all quenched rapidly enough for the results to be meaningful under the conditions employed. However, it should be pointed out (*cf.* Bray (7)) that the apparent time taken to quench a given chemical reaction depends on several factors and might be expected to vary widely when reactions of different types are considered. The observed apparent quenching time (in terms of milliseconds of reaction occurring at the temperature of the experiment) will be measured by the extent to which the reaction proceeds from the moment the reaction mixture leaves the jet to the time at which the EPR measurements are made. It is made up of the reaction which takes place while the solution is being cooled down from the temperature of the experiment to the freezing point of the solution, the reaction which takes place while the solution is actually freezing, the reaction (if any) in the frozen state while the sample is being cooled to liquid nitrogen temperature, and finally, the reaction (if any) during storage at this temperature. Thus the apparent quenching time is only indirectly related to the time required for the reaction mixture to freeze, under the experimental conditions, although the two are frequently likely to be of the same order of magnitude. Previous calculations (7), which predicted freezing times of the order of a few milliseconds, seem to be in fair agreement with other data in the literature on rates of cooling with isopentane (27). Furthermore, previous relatively crude tests with a model reaction indicated a reasonably short quenching time (7). However, in view of the possible variability of quenching times, it seemed desirable to try other reactions. The reaction between ferric perchlorate and thiocyanate ion appeared particularly suitable, as the kinetics has been studied in detail (28). Samples obtained by freezing the reactants at various times after mixing by the technique applied in the enzyme studies were examined by low temperature reflectance spectroscopy (3, 29). The readings so obtained were converted to concentration of product by means of a calibration curve obtained by rapid freezing of spectroscopically standardized solutions containing the red $FeSCN^{++}$ ion, which is essentially the only colored species formed under the experimental conditions of Below, Connick, and Coppel (28). The results are presented in Fig. 1. In view of the possible sources of error inherent in experiments of this nature, the agreement between the experimental points and the

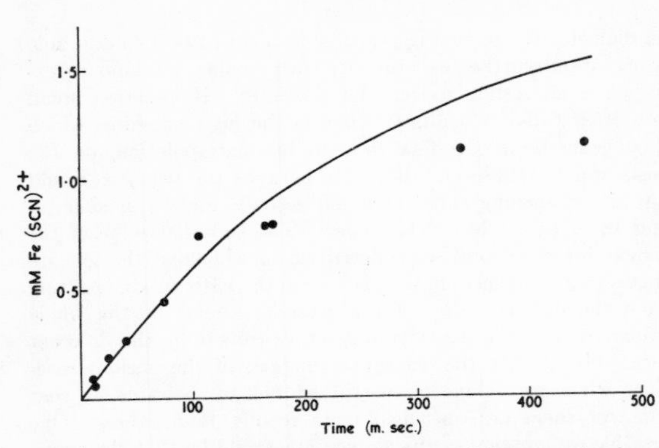

FIG. 1. Time course of reaction between 2.5 mM potassium thiocyanate and 75 mM ferric perchlorate. The solvent was an aqueous solution 0.2 M with respect to $HClO_4$ and $NaClO_4$. The temperature was 22°. The *solid circles* represent the experimental points obtained by reflectance spectroscopy of the frozen samples, converted to concentration of $Fe(SCN)^{++}$ as described in the text. The error is estimated to be ±20%. The *solid line* is the theoretical curve computed from published data (28). The jet diameter of the rapid freezing apparatus was 0.14 mm, the flow rate 0.5 ml per second, and the jet velocity 32 meters per second.

theoretical curve is suprisingly good, suggesting that the freezing time is extremely short, probably of the order of a few milliseconds.

Rapid Mixing Technique—The efficiency of the mixing chamber calls for further discussion. The simple "Y" design (7) made it possible to keep the dead space small and so obtain short reaction times without using unduly high flow rates. The efficiency of the chamber would be expected to be greater at higher flow rates; the rates used ranged from 0.15 ml per second to 1.0 ml per second, while turbulent flow in the 0.38-mm diameter polyethylene tubing would be expected at flows above about 0.6 ml per second. (Generally, the lower flow rates were only used in experiments when the shorter reaction times were not of particular interest.) Extrapolation from data quoted by Roughton and Chance (30) for simple "T" junctions suggests that mixing might take 10 milliseconds or more, even with turbulent flow. The adequacy of mixing was checked by the qualitative method described by Berger (31), which relies on visual observation of color homogeneity on mixing two liquids of markedly different color. At ram speeds considerably below those used in our experiments, color "streaking" was obvious. However, no inhomogeneity was observed during flow at our routine ram speeds, although streaking was apparent close (within 2 cm) to the mixing chamber on cessation of flow. This poorly mixed material, however, is not expelled, but remains either in the polyethylene hose or as a large drop which is invariably formed on the jet tip when flow ceases. An experiment in which the octacyanide of Mo(IV) was oxidized by ferric phenanthroline to the octacyanide of Mo(V) was carried out at the extremely low flow rate of 0.1 ml per second. A mixing time of the order of 100 milliseconds was suggested by this test. Slow mixing would affect experimental results in the direction opposite to slow quenching, and the two effects might, under certain conditions, cancel one another out. However, exact cancelation is relatively unlikely, and the agreement between experiment and theory for the data obtained on mixing ferric and thiocyanate ions (Fig. 1) is taken as evidence that the mixing chamber must, despite the above considerations, be reasonably efficient under the conditions used. Further evidence was obtained with an extremely fast reaction, the oxidation of $K_4Mo(CN)_8$, in this case by ceric ions rather than by ferric phenanthroline. Oxidations by ceric sulfate under the conditions employed are known to be very rapid. With 0.1 mM potassium cyanomolybdate (IV) and 5 mM ceric sulfate in 0.2 N HCl, the reaction appeared to be complete at the shortest reaction times which could be studied, even with a flow rate as low as 0.17 ml per second. Unfortunately, this is not conclusive proof of the efficiency of the mixing chamber, since mixing might conceivably have been taking place in the jet rather than in the chamber. However, taking all the data together and including some of the xanthine oxidase results (*cf.* Paper II (16)), there seems to be no reason for suspecting that any of the results presented are seriously distorted by either nonideal mixing or nonideal quenching.

For the anaerobic fast reaction studies, the substrate solution was freed from oxygen by gassing with purified nitrogen before it was transferred into its syringe. Similarly, the enzyme solution was rendered anaerobic by placing it in a tonometer connected directly to the enzyme syringe (33). The tonometer was partially evacuated and flushed with nitrogen repeatedly before the sample was transferred to the syringe. The jet and connecting tube of the system were flushed with argon until immediately before the shot was made. As a further precaution, the ram was advanced slightly, with the reaction tube disconnected from the mixing chamber, until sufficient solution had been expelled from the mixing chamber to ensure that the liquid was renewed in the hoses connecting the syringes. This was done just before the shot in case oxygen had diffused into the solutions through the polyethylene. For the aerobic experiments the enzyme solution was allowed to remain in equilibrium with air, while the substrate solution was equilibrated either with air or with oxygen, according to the oxygen concentration required. The solubility of pure oxygen was taken as 1.25 mM at 22° and 2.1 mM at 1°.

For studies on the reoxidation of the enzyme, after controlled periods of anaerobic reduction, the apparatus was modified so that three syringes and two mixing chambers could be employed, as shown in Fig. 2. The upper syringe contained oxygenated buffer; for control experiments this was replaced by nitrogen-saturated buffer. The other two syringes contained, respectively, anaerobic substrate and enzyme. The system was filled to the points *A* and *B*, and the remainder of the apparatus was flushed with argon. The dead spaces were so arranged that (with three equal syringes) the hold-up between *A* and *C* was just half that between *B* and *C*. The duration of the anaerobic reduction was then controlled by the *BC* (or *AC*) dead space, and that of the reoxidation, by the *CD* space.

For anaerobic experiments with manual mixing, tubes of the type described previously (32) were employed. The tubes were flushed thoroughly with specially purified argon, and the deoxygenated enzyme and substrate were added from syringes and mixed with a wire, while still flushing with argon; then the tubes were closed and the contents frozen after 1 minute.

EPR Spectroscopy—EPR spectroscopy was carried out essentially as described (34). Routinely a microwave power of 25 milliwatts was applied. With further development of instrumentation it was found that at this power saturation of the free radical signals becomes just noticeable (5 to 10%) and the molybdenum signals are approximately 20% saturated. Since the iron signals are weak at lower powers, routine work was continued at 25 milliwatts and appropriate corrections were applied for saturation of molybdenum and radical signals. Spectra recorded for integration of free radical signals were obtained at a power of 3 milliwatts. No saturation was observed under these conditions. The modulation amplitude was 3 to 5 gauss for spectra in which molybdenum and free radicals were to be determined, and 12 to 20 gauss for evaluation of the iron signal; scanning rates of 105 gauss per minute (molybdenum) and 165 to 275 gauss per minute (iron) were used. Although, at the power levels used, better resolution of the molybdenum signals was obtained at higher temperatures (optimum, −100 to −150°), samples were generally kept at −175° (±1°)[5] during routine runs to prevent difficulties from melting of the isopentane and release of dissolved gases, and in order to obtain optimal iron signals. It has been shown previously that signals of this type of compound broaden rather rapidly with rising temperature (2, 3).

For room temperature spectra, flat quartz cells (0.3 × 8 × 50 mm) were used. The magnetic field was monitored by a proton probe. Proton frequencies were counted with a Hewlett-Packard 524 D electronic counter and 525 A frequency converter, and microwave frequencies were measured with a Hewlett-Packard 532 B frequency meter. We estimate that the g values derived from these measurements are accurate to within ±0.002 for relatively narrow lines (Mo(V) and radicals) and ±0.005 for broader lines (iron). The precision is considerably better, however, since proton frequencies can be measured to within 100 cycles per second and the precision of the measurement of microwave frequency is 1 megacycle (0.01%). The separation of the various absorption lines is therefore measurable with considerable precision, and the error is estimated to be about ±0.2 gauss for the narrower lines. The increased error for the broader lines is due to the uncertainty involved in the location of the actual point of measurement.

When an estimate of the quantity of material represented by a signal was desired, a double integration of the derivative signal was carried out manually and the area under the absorption curve was compared to that of a standard. The following standards were used: a pitch sample in KCl, standardized and supplied by Varian Associates, Palo Alto, California, was used for calibration of free radical content, and copper nitrate EDTA served as standard for the calibration of molybdenum in the enzyme. The error of these estimates is unfortunately high, particularly when signals of quite different width are compared, as in the case of the molybdenum and copper signals.[6] Problems pertaining to quantitative measurements of this kind are discussed in a forthcoming paper (35).

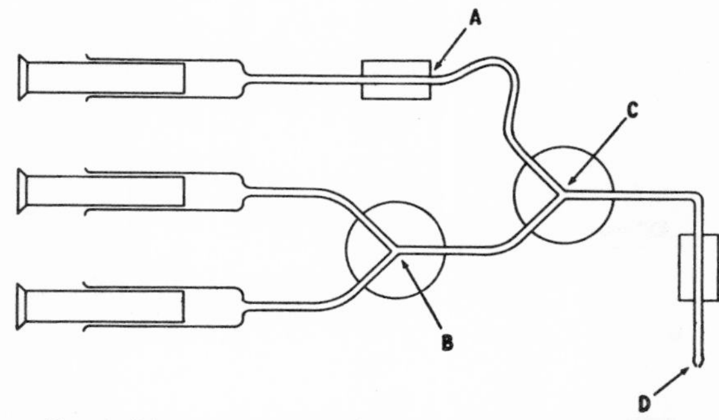

Fig. 2. Diagram illustrating the method (*cf.* Bray (7)) of bringing two solutions (from the *lower syringes*) into reaction together, then introducing a third reagent (from the *uppermost syringe*) after a predetermined reaction time. For further details, see the text.

The methods used for relative quantitative determination of the various signals are discussed below. Such a discussion appears more instructive when the features of the observed signals have been described.

RESULTS AND DISCUSSION

Description of EPR Spectra of Known Complexes of Mo(V)—Parameters of the EPR signals observed with model complexes of pentavalent molybdenum at low temperature were reported by Bray *et al.* (2). Since molybdenum is found as a constituent of several enzymes, we consider it of interest actually to present the spectra obtained at ambient and low temperature from two compounds of Mo(V). Fig. 3 shows spectra of Mo(V) complexes in solution. The spectrum of potassium octacyanomolybdate(V) as obtained at room temperature has been described previously by Weissman and Cohn (36). The intense central line is due to the molybdenum isotopes lacking a net nuclear magnetic moment. The six minor components arise through hyperfine splitting by the molybdenum isotopes with nuclear spin 5/2 (^{95}Mo and ^{97}Mo; combined abundance, 25%). Furthermore, additional structure at the main peak has been shown to be due to the naturally occurring carbon isotope, ^{13}C, of spin 1/2 (36). The spectrum of $MoCl_5$ in strong HCl shows a similar six-line spectrum with a seventh intense center line and the same intensity ratio of the central line to the hyperfine lines, but considerably larger hyperfine splitting (*cf.* Bray *et al.* (2)). The values that were measured and could be assigned to certain parameters are given in Table I. The spectra show clearly what is called a "second order effect," *i.e.* an increase in splitting with increasing field between the six less intense lines due to the isotopes of molybdenum that have a nuclear magnetic moment. This effect is expected, because the spins

[5] It was found that the actual sample temperature generally lies 1° above the temperature measured at the location of the thermocouple in the Varian Dewar system.

[6] In order to have an estimate of the error involved in comparisons of this kind, a known amount of potassium octacyanomolybdate (IV) was oxidized with ceric sulfate and its EPR spectrum was determined at −145° and integrated. Because of the great difference in signal width and g value, the potassium octacyanomolybdate (V) sample was not directly compared with copper nitrate EDTA. Nitrosyldisulfonate in 0.05 N KOH and pitch in KCl were used as standards in this case. The sample was 4.15 × 10^{-3} M with respect to molybdenum according to weight, whereas a value of 1 × 10^{-2} M was found according to the nitrosyldisulfonate standard, and 2.3 × 10^{-3} M according to the pitch standard. When 2 mM copper nitrate EDTA was compared with nitrosyldisulfonate at −70°, a 4 mM copper concentration was indicated by the latter standard. It has been our experience and that of others that nitrosyldisulfonate tends to give high values. The copper standard was therefore chosen for the estimation of molybdenum in xanthine oxidase. That under favorable circumstances the integration may be very accurate was shown in a comparison of copper nitrate EDTA and copper sulfate, in which the discrepancy amounted to less than 5%.

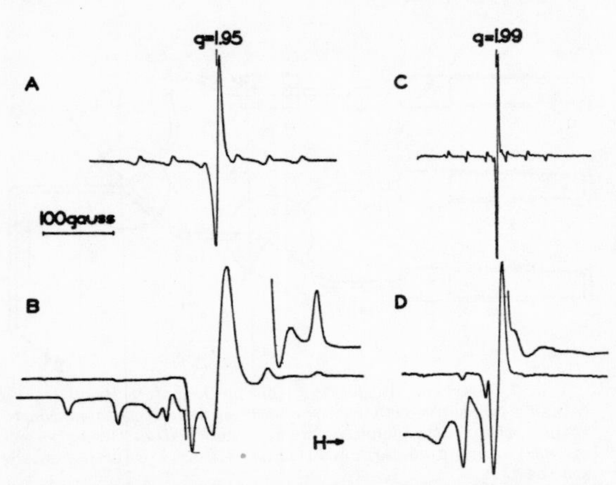

FIG. 3. EPR spectra of Mo(V) compounds at ambient and low temperature. *A*, $MoCl_5$ (2 mg per ml) in concentrated HCl at room temperature; *B*, at $-172°$. *C*, $K_3Mo(CN)_8$ (1 mg per ml) in 1 N HCl oxidized from the tetravalent form with 0.04 volume of 0.1 M ceric sulfate in 1 N H_2SO_4 at room temperature; *D*, at $-172°$. Microwave power, 25 milliwatts; modulation amplitude, 0.6 (*A*), 0.08 (*B*), and 0.02 gauss (*C* and *D*). To make visible the weaker hyperfine lines in *B* and *D*, parts of these spectra are shown in *insets* above and below the complete spectra at 12.5 times the modulation amplitude.

TABLE I

EPR parameters of molybdenum pentachloride and potassium octacyanomolybdate (V)

Compound	Solvent (HCl)	g (25°)	$g_x \approx g_y$ (−172°)	g_z (−172°)	Line width*	Hyperfine splitting*
	N				gauss	
$MoCl_5$	12	1.949	1.94_5†	1.97_1	6.6	52
$K_3Mo(CN)_8$	1	1.991	‡	‡	2.1	33

* These data apply only to the spectra recorded at 25°.
† The value of the third decimal place is uncertain.
‡ Assignment of measured values to parameters is not certain.

of these paramagnetic nuclei may be oriented so that their magnetic moment either adds (lines at low field) or subtracts (lines at high field) from the externally applied field and, as the local field increases, the splitting is likewise enhanced.

The low temperature spectra of both complexes are asymmetrical and not readily interpreted. An attempt to analyze these curves was not made in the present work. Some measured and assigned values are given in Table I. We show these spectra to demonstrate their difference from those observed at room temperature as well as the intensity ratios of the various lines, and to permit a comparison with the spectra observed with the enzyme at low temperature.

EPR Spectrum of Xanthine Oxidase—Fig. 4*A* shows part of the spectrum of the "resting" enzyme. The small signal at g = 2 is partly attributable to free radicals in the quartz container and Dewar flask used. There is no other signal in these spectra except for a rather small signal at g = 4.3 (signal *D* of Bray *et al.* (2)), which varies from preparation to preparation and which we attribute to Fe(III)-containing impurities, in agreement with Bray *et al.* (2). Small signals of this type are frequently found in biological materials and are almost as ubiquitous as small radical signals at g = 2.00. A broad signal (400 to 500 gauss) centered at g = 2.1, which was seen by Kubo *et al.* (37) in their resting xanthine oxidase preparations, could not be detected in our preparations, even under experimental conditions similar to those used by Kubo *et al., viz.* high modulation and room temperature.

The spectrum of xanthine oxidase at various levels of reduction after addition of substrate was considerably better resolved in the present than in previous work (2, 9, 19). An accurate interpretation of the many lines which were observed is not possible at this time. Not only are there several components of the enzyme that exhibit EPR signals, but in addition the environment of these paramagnetic species is unknown. However, certain regularly occurring features and relationships between components of the spectrum were recognized and some tentative interpretations were arrived at, which we hope to confirm and extend. These will be discussed below.

There is no oxidation state and hence no single spectrum which shows all the absorption lines well developed. We therefore have to explain the signals with the aid of a series of spectra (Fig. 4, *B* to *E*).

Free Radicals—The simplest component of the complex signal

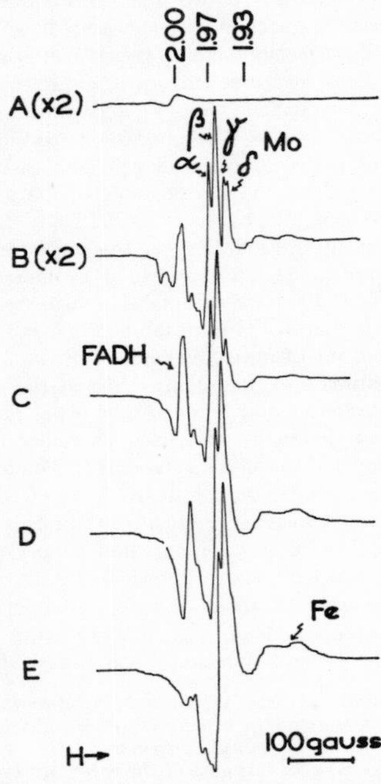

FIG. 4. EPR signals obtained from xanthine oxidase before (*A*) and during reaction with xanthine (*B* to *E*). The solutions, in 0.05 M pyrophosphate buffer, pH 8.3, were mixed in the rapid flow apparatus at 22°. The concentrations after mixing were: xanthine oxidase, 0.12 mM; xanthine, 1.25 mM; and oxygen, 0.76 mM. The mixture was frozen at the indicated times after mixing: *B*, 26 milliseconds; *C*, 77 milliseconds; *D*, 860 milliseconds; *E*, 1410 milliseconds. *Curves A* and *B* were recorded at twice the amplification of *Curves C* through *E*. The microwave power was 25 milliwatts, the modulation amplitude 5 gauss, and the temperature during recording −174°.

appearing on reduction is the free radical signal at g = 2.006, which is maximally developed at intermediate oxidation states (*cf.* Fig. 4, *B*, *C*, and *D*). It is apparent in some records (Fig. 4*B*) that this signal is made up of two components: a small, narrow radical, which probably originates from the quartz tube and Dewar flask used, and a signal of 15 gauss (peak to peak width), which we, in accordance with Bray *et al.* (2, 19), attribute to flavin semiquinones. The possibility that other free radicals, perhaps including substrate radicals, contribute to the signal under some circumstances is not excluded, although we consider it unlikely.

Molybdenum—At g = 1.95 to 1.97 we observe a complex signal made up of four peaks at early times of reaction (Fig. 4*B*) and changing in shape as the reaction proceeds (Fig. 4, *C* to *E*). In agreement with Bray *et al.* (19), this signal is attributed to Mo(V) for the following reasons: Mo(V) shows EPR absorption in this range of g values (*cf.* Fig. 3 and Table I), and the appearance of the expected hyperfine lines (*cf.* Fig. 5*A*) proves that the observed spectrum is due to a paramagnetic species of molybdenum; signals due to known Mo(V) compounds and the signals observed with the enzyme both show behavior toward variation of temperature and microwave power that is indicative of a long relaxation time; molybdenum is known to be a constituent of xanthine oxidase; and the signal appears on reduction, as would be expected if the resting enzyme contains molybdenum in the most stable, hexavalent form. Although, in principle,

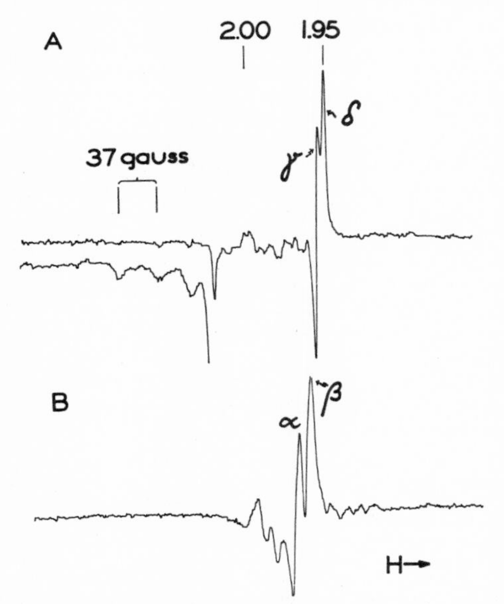

Fig. 5. EPR signals obtained from xanthine oxidase during reaction with xanthine in 0.05 M pyrophosphate, pH 6.0 and 9.6. The conditions were as follows. *A* (pH 9.6): xanthine oxidase, 0.63 mM; xanthine, 2.9 mM; oxygen, 1.0 mM; reaction time, 8 milliseconds at 1°; microwave power, 2.5 milliwatts; modulation amplitude, 2.5 gauss (5-milliwatt power and 15-gauss modulation for *inset* in *lower left corner*). *B* (pH 6.0): xanthine oxidase, 0.11 mM; xanthine, 0.10 mM; oxygen, 0.76 mM; reaction time, 71 milliseconds at 22°; microwave power, 25 milliwatts; modulation amplitude, 3 gauss. All concentrations refer to the state after mixing. The observation temperature was −173°.

EPR signals from molybdenum-containing materials could also be produced by Mo(III), this appears very unlikely in the case of xanthine oxidase from a consideration of oxidation-reduction potentials and relative stabilities (38). It is therefore assumed here that the signals centered at g = 1.95 to 1.97 are due to Mo(V).

There are at least four well separated peaks at early times of reaction. They will be called α, β, γ, and δ from left to right (increasing field). At a microwave frequency of 9115 megacycles per second, they are located at 3309, 3322, 3335.7, and 3343 gauss, respectively. It appears as if the α- and β-lines, on the one hand, and the γ- and δ-lines, on the other, are closely related. The members of each pair follow very similar if not identical kinetics (16). Because of overlap, exact measurement of their intensity is not possible in spectra (*cf.* Fig. 4, *C* to *E*) in which intensities are very different.

There are additional lines at low field (g > 2.00) associated with the Mo(V) signals at g = 1.95 to 1.97. Since we were able to obtain situations when only the α- and β- or only the γ- and δ-lines were present, we were able to determine the specific association of these lines at low field with the lines at g = 1.95 to 1.97. This could be accomplished by allowing the enzyme to react with xanthine at different pH values as shown in Fig. 5.

Fig. 5*A* shows an early reaction time at pH 9.6 when, in the g = 1.95 to 1.97 region, essentially only the γ- and δ-lines are present. It is clear from this figure that the sharp line at low field (g = 2.02_2) is associated with the γ and δ structure. Kinetic studies confirmed this (*cf.* Paper II (16)).

It is of particular interest that in the spectra of Fig. 5*A*, at alkaline pH, we were able to detect three hyperfine lines on the low field side of the sharp line at g = 2.02_2. Their number, intensity, and spacing (*A* = 37 gauss) are in perfect agreement with the interpretation given here, namely, that the observed spectra are indeed due to molybdenum. This is to our knowledge the first instance in which molybdenum hyperfine lines in EPR spectra of enzymes have been resolved. Unfortunately, the hyperfine lines at higher field are obviously overlapping with other signals stemming from Mo(V), flavin, and iron, so that a clear interpretation thus far has not been possible.

At low pH (Fig. 5*B*), practically no γ- or δ-line appears and only the α- and β-lines stand out. There is a small radical signal at g = 2.00, but the remaining structure between g = 2.00 and 1.97 and toward high field (<1.95) is associated with the α- and β-lines. The locations of all peaks in the derivative spectrum of Fig. 5*B*, observed at 9115 megacycles per second, are the following, starting at low field: 3259, 3273, 3289, 3309, 3322, 3343, 3361, 3380.7, and 3393 gauss. The peaks at 3309, 3324, and 3343 gauss are those previously designated as α, β, and δ, respectively. The γ-peak appears only as a faint shoulder. It is likely that the less intense lines at high field represent a pattern of overlapping hyperfine satellite lines of the two main lines (α and β).

At pH 8, where most experiments were done, the structure toward high field is not as pronounced although definitely present, and overlaps to some extent with the iron signal, as may be noticed on close inspection of Fig. 4. There is also some change in the low field structure associated with α and β when the pH is raised (*cf.* Fig. 4, *B* to *E*, and Fig. 5*B*). It is of interest that when hypoxanthine is used instead of xanthine, the structure associated with α and β, even at pH 8, resembles closely that seen with xanthine at pH 6.

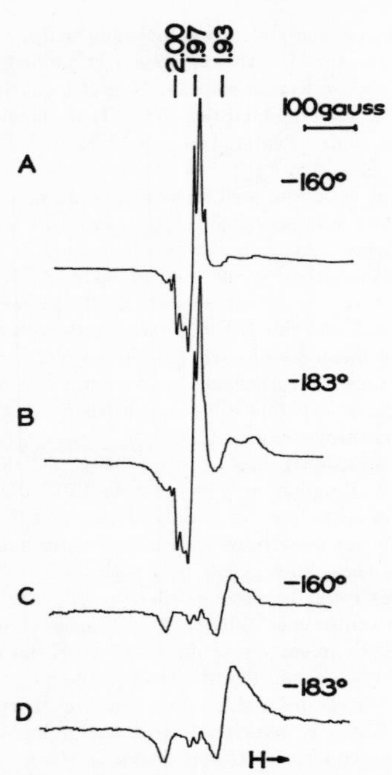

FIG. 6. EPR signals of xanthine oxidase at different temperatures after addition of xanthine (*A* and *B*) and after addition of dithionite (*C* and *D*). Xanthine oxidase in 0.05 M pyrophosphate, pH 8.3, was mixed at 22° in the rapid mixing apparatus with xanthine. The final concentrations were: xanthine oxidase, 0.12 mM; xanthine, 1.25 mM; and oxygen, 0.26 mM. The sample was frozen 1.6 seconds after mixing. The microwave power was 25 milliwatts, the modulation amplitude 3 gauss, and the observation temperature −160° for *A* and −183° for *B*. The amplification was identical for *A* and *B*. For *C* and *D*, a different preparation of xanthine oxidase (22 mg per ml, 0.04 mM, in pyrophosphate as above), which had lost most of its activity, was reduced with solid dithionite. Microwave power and modulation were as described above, and the observation temperature was −160° in *C* and −183° in *D*.

A tentative interpretation consistent with the data reported in the preceding paragraph is that the spectrum of Fig. 5*A* represents an anisotropic distributon of g values with $g_x = 1.95_3$, $g_y = 1.95_5$, and $g_z = 2.02_2$; $A_z \approx 37$ gauss; and B_x and B_y not observable. The spectrum of Fig. 5*B* can be explained in an analogous fashion, with $g_x = 1.96_6$, $g_y = 1.97_3$, and $g_z = 1.99$. These assignments are to be considered tentative, and await confirmation by means of reconstructed spectra. Computer studies to this end are in progress.

The two different spectra of Fig. 5 are tentatively attributed to two chemically different species of Mo(V) which participate in the enzymatic reaction. This will be discussed in the accompanying paper (16). The observed asymmetry of these spectra requires that, if an octahedral configuration is assumed, at least three nonequivalent pairs of ligands surround the metal.

Iron—Fig. 4, *C* to *E*, shows that, superimposed on the signals described thus far, there is a broad signal which is most conspicuous at high field (*right*). From inspection of appropriate spectra it is clear that this signal actually extends from low field (g > 2.00) to high field (g = 1.8). This is well illustrated in Fig. 6, where use is made of the high temperature dependence of the shape of this signal. The temperature dependence of this type of signal has been pointed out previously as one of its main characteristics (2, 3, 13). Fig. 6*A* shows the spectrum of xanthine oxidase after addition of xanthine, recorded at −160°. At −183° (Fig. 6*B*), the Mo(V) signal increases 12% whereas the signal at high field increases more than 200%. It is also apparent that the broad line on the low field side of the radical and Mo(V) signal (g > 2.00) shows the same change in intensity as the main iron signal. This broad line at low field is distinct from the sharper line (Fig. 5*A*) associated with the molybdenum γ and δ signal. The conclusion that this broad line is actually part of the signal at high field is supported by spectra which show essentially only the component responsible for this signal and very little Mo(V) or radical signal. Such spectra are shown in Fig. 6, *C* and *D*. These spectra were recorded on a preparation of xanthine oxidase that had lost most of its activity and showed only a poor Mo(V) signal on reduction by dithionite. The spectra were again taken at different temperatures, *C* at −160° and *D* at −183°. It is obvious that the broad line at g > 2.00 and the peak at g ≈ 1.92 respond similarly to the temperature change. The temperature-sensitive signal had been seen in xanthine oxidase by Bray *et al.* (2) (their signal *C*) and was found by Beinert *et al.* (3, 11–15) in a variety of materials. This signal is attributed to a reduced form of iron in a specific complex which does not involve porphyrin. The arguments for this assignment have been extensively discussed by Beinert *et al.* (3).[7] They are supported by magnetic susceptibility measurements on xanthine oxidase (2), which were interpreted to indicate the transition of part of the iron from a low spin ferric state to a high spin ferrous state on addition of substrate. The salient characteristics of this type of iron complex appear to be its asymmetrical EPR spectrum with a g value of 1.93 to 1.95 and a minor component at g = 2.00, and the strong temperature dependence of the signal. The iron signal of Fig. 6*B*, for instance, is no longer detectable at −120° because of broadening, indicating a relatively short relaxation time of the parent spin system.

The iron signal at high field has two distinct peaks separated by 54 gauss. We refer to the peak at low field as α and that at high field as β. When the enzyme is reduced with xanthine, both peaks appear. Their kinetics of appearance and disappearance is not completely in phase, and it was therefore concluded that two species are involved (39). Studies at various

[7] The pronounced temperature dependence and hence short relaxation time of the paramagnetic species involved, as well as the fact that maximal signal intensity is always achieved at full reduction, appear to be the strongest arguments against the hypothesis that a free radical species is involved. If it is accepted that the signal indicates a paramagnetic metal ion, there is no alternative to iron on the basis of chemical and spectroscopic analyses. It should, however, be emphasized here that there is no evidence at all, at this time, what valency states of this metal are involved. Although the presence of Fe(III) in the oxidized and Fe(II) in the reduced form of the enzyme is most likely and will be assumed here, this is purely hypothetical at the moment and other possibilities are not excluded. Higher or lower valency states of iron could be involved. It is also not rigorously excluded that the appearance and disappearance of the signal may in fact not indicate a shuttle between two valency states, but merely a change in environment of the metal at a particular fixed valency state, such as Fe(III). We consider this latter possibility unlikely.

temperatures, however, revealed that the α-peak overlaps with the high field structure of Mo(V), which was discussed above. This is seen best by comparing Fig. 4*B* with Fig. 5*B*. Since the Mo(V) signal generally has a considerably higher amplitude, the α-peak of the iron signal receives a noticeable contribution from high field structure of the Mo(V) signal. The observed differences in kinetics of the α- and β-peaks are therefore not a valid argument for two different reactive iron species. We have, therefore, considered it reasonable to assume, in the evaluation of the spectra, that both peaks, when produced by addition of substrate, are representative of a single iron species with a peak of the absorption curve located at g = 1.93_2 and closely related to previously observed iron species, characterized by a g value of 1.94 (3, 11–15). For measurements, the β-peak appeared to be the more reliable point of reference because of overlap of the α-peak with part of the Mo(V) signal.

The presence of different species of iron is, however, indicated by the difference in appearance and behavior of the signal found after addition of dithionite and that produced with substrate. This can also be seen from Fig. 6. Although the signals of Fig. 6, *A* and *B*, *C* and *D*, are very similar, the dithionite-reduced enzyme shows a predominant α-peak of iron and, more significantly, shows a less pronounced temperature dependence (Fig. 7). It is also pertinent here that on addition of dithionite to the xanthine-reduced enzyme the iron signal increases about 50% in amplitude, indicating the presence of iron which is not reduced by substrate. We conclude, therefore, that in addition to the reactive iron species, which is readily reduced by substrate, there is a related but not identical species, which is only reduced on chemical treatment. Whether this latter species might arise from the active one by damage done to the enzyme during preparation is not known.

Attempts at quantitative estimation of the iron signal are of course beset with great uncertainties, since the state of the iron in the complex is not known and simple model compounds are not available. Some assumptions have to be made, therefore,

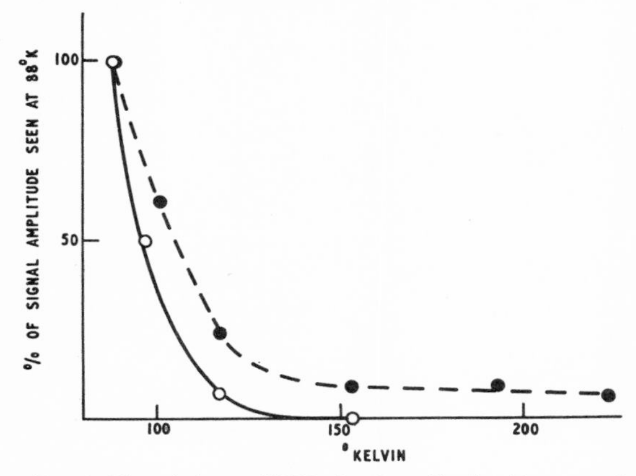

Fig. 7. Plot of observed EPR signal amplitude of iron α-peak (g = 1.912), in percentage of amplitude obtained at 88° K *versus* absolute temperature, and of xanthine oxidase reduced with xanthine (*solid line*) and with dithionite (*dashed line*). Xanthine oxidase, 0.2 mM in 0.05 M pyrophosphate buffer, pH 8.3, was reduced with xanthine (final concentration, 2.5 mM) or with solid dithionite, and frozen after 1 minute. The microwave power was 25 milliwatts, and the modulation amplitude 3 gauss.

if one wants to arrive at a tentative estimate. On the basis of several considerations (3) and the susceptibility measurements of Bray *et al.* (2), it appears reasonable to assume that the species involved could be a complex of high spin ferrous iron, in which case 4 unpaired electrons would be present per atom. When the signals were twice integrated and compared on this basis to a copper-EDTA standard, values were obtained which indicated the presence of approximately 1 equivalent of this component for every 8 equivalents of bound flavin (*cf.* Bray *et al.* (16)). Such low values have been observed for other signals of this general type.[8] One may therefore doubt the validity of the assumptions on which these estimates are based. Nevertheless the observations indicate that only a fraction of the iron of the enzyme might be represented by the signal at g = 1.94 which appears on reduction with xanthine.

Evaluation of Spectra for Relative Quantitative Determination—For comparison of relative signal size in the kinetic experiments reported in the accompanying paper (16), a base-line was drawn in the spectrum from the far left to the right end, where no absorption lines are present. The heights of the molybdenum peaks were then measured from this base-line. For evaluation generally the β- and δ-peaks were used, as they appeared least influenced by adjacent lines. The molybdenum γ-peak overlaps sufficiently with the β-peak so that it cannot be reliably measured in the presence of a strong β-peak (*cf.* Fig. 4, *D* and *E*). For evaluation of the iron, the β-peak was chosen for the reasons set forth above. Its height was measured from the right end of the base-line to avoid uncertainties from overlap with molybdenum. Radical concentration was estimated from the total height of the signal at g = 2.00, and a correction for the radical signal in the oxidized enzyme and quartz containers was applied (*cf.* Fig. 4*A*). For absolute quantitative determination (see "Experimental Procedure"), signals were chosen which showed the component under study at maximal development as compared to the other signals.

SUMMARY

Improvements and extensions in the use of the rapid mixing-freezing technique of Bray in combination with low temperature electron paramagnetic resonance (EPR) spectroscopy are described. Evidence is presented which indicates that the apparent quenching time, under the conditions used, is of the order of 1 to 8 milliseconds. This time depends on the type of reaction studied, but will frequently be of the same magnitude as the actual freezing time.

The EPR spectra of molybdenum pentachloride, octacyanomolybdate (V), and xanthine oxidase at various stages during the reaction with xanthine are shown and discussed. The different lines seen in the EPR spectra of the enzyme are attributed to distinct species of Mo(V), flavin, and reduced iron. The asymmetrical shape of the Mo(V) signals in the enzyme requires at least three nonequivalent pairs of ligands if an octahedral coordination sphere around the metal is assumed. At least two species of Mo(V) can be distinguished. These presumably differ in their ligand environment.

Acknowledgments—We are indebted to Professor H. E. Calbert of the Department of Dairy and Food Industries of the University of Wisconsin for his kind assistance in providing the buttermilk, and to Drs. F. Bergmann, G. M. Timmis, and

[8] H. Beinert, unpublished observations.

G. B. Brown for furnishing specific inhibitors. We would also like to acknowledge the kind help of Drs. R. H. Sands and T. Vänngård in the interpretation of the electron paramagnetic resonance data, the assistance and advice of Mr. R. E. Hansen with instrumentation, and the technical assistance of Mrs. R. N. Carl, I. Harris, and M. VanDeBogart.

REFERENCES

1. Yamazaki, I., Mason, H. S., and Piette, L., *J. Biol. Chem.*, **235**, 2444 (1960).
2. Bray, R. C., Pettersson, R., and Ehrenberg, A., *Biochem. J.*, **81**, 178 (1961).
3. Beinert, H., Heinen, W., and Palmer, G., in *Enzyme models and enzyme structure, Brookhaven symposia in biology, Vol. 15*, Brookhaven National Laboratory, Upton, New York, 1962, p. 229.
4. Broman, L., Malmström, B. G., Aasa, R., and Vänngård, T., *Biochim. et Biophys. Acta*, **75**, 365 (1963).
5. Chance, B., Bicking, L., and Legallais, V., in M. S. Blois, Jr., H. W. Brown, R. M. Lemmon, R. O. Lindblom, and M. Weissbluth (Editors), *Free radicals in biological systems*, Academic Press, Inc., New York, 1961, p. 101.
6. Nakamura, T., *Biochem. and Biophys. Research Communs.*, **2**, 111 (1960).
7. Bray, R. C., *Biochem. J.*, **81**, 189 (1961).
8. Bray, R. C., and Pettersson, R., *Biochem. J.*, **81**, 194 (1961).
9. Bray, R. C., *Biochem. J.*, **81**, 196 (1961).
10. Gutfreund, H., and Sturtevant, J. M., *Biochem. J.*, **73**, 1 (1959).
11. Beinert, H., and Sands, R. H., *Biochem. and Biophys. Research Communs.*, **3**, 41 (1960).
12. Sands, R. H., and Beinert, H., *Biochem. and Biophys. Research Communs.*, **3**, 47 (1960).
13. Beinert, H., and Lee, W., *Biochem. and Biophys. Research Communs.*, **5**, 40 (1961).
14. Rajagopalan, K. V., Aleman, V., Handler, P., Heinen, W., Palmer, G., and Beinert, H., *Biochem. and Biophys. Research Communs.*, **8**, 220 (1962).
15. Nicholas, D. J. D., Wilson, P. W., Heinen, W., Palmer, G., and Beinert, H., *Nature*, **196**, 433 (1962).
16. Bray, R. C., Palmer, G., and Beinert, H., *J. Biol. Chem.*, **239**, 2667 (1964).
17. Gilbert, D. A., and Bergel, F., *Biochem. J.*, **90**, 350 (1964).
18. Singer, T. P., and Kearney, E. B., *Arch. Biochem. Biophys.*, **29**, 190 (1950).
19. Bray, R. C., Malmström, B. G., and Vänngård, T., *Biochem. J.*, **73**, 193 (1959).
20. Avis, P. G., Bergel, F., Bray, R. C., James, D. W. F., and Shooter, K. V., *J. Chem. Soc.*, 1212 (1956).
21. Bergel, F., and Bray, R. C., *Biochem. J.*, **73**, 182 (1959).
22. Morrell, D. B., *Biochem. J.*, **51**, 657 (1952).
23. Gilbert, D. A., *Nature*, **198**, 1175 (1963).
24. Krebs, E. G., and Norris, E. R., *Arch. Biochem. Biophys.*, **24**, 49 (1949).
25. Bergmann, F., and Dikstein, S., *J. Biol. Chem.*, **223**, 765 (1956).
26. Furman, N. H., and Miller, C. O., in L. F. Audrieth (Editor), *Inorganic syntheses, Vol. III*, McGraw-Hill Book Company, Inc., New York, 1950, p. 160.
27. Luyet, B. J., in R. J. C. Harris (Editor), *Freezing and drying*, Institute of Biology, London, 1951, p. 77; in A. S. Parkes and A. U. Smith (Editors), *Recent research in freezing and drying*, Blackwell, Oxford, 1960, p. 3.
28. Below, J. F., Jr., Connick, R. E., and Coppel, C. P., *J. Am. Chem. Soc.*, **80**, 2961 (1958).
29. Palmer, G., and Beinert, H., *Anal. Biochem.*, **8**, 95 (1964).
30. Roughton, F. J. W., and Chance, B., in S. L. Friess and A. Weissberger (Editors), *Technique of organic chemistry, Vol. 8*, Interscience Publishers, Inc., 1953, p. 669.
31. Berger, R. L., in C. M. Herzfeld (Editor), *Temperature—its measurement and control in science and industry, Vol. 3*, Part 3, Reinhold Publishing Corporation, New York, 1963, p. 61.
32. Beinert, H., and Sands, R. H., in M. S. Blois, Jr., H. W. Brown, R. M. Lemmon, R. O. Lindblom, and M. Weissbluth (Editors), *Free radicals in biological systems*, Academic Press, Inc., New York, 1961, p. 17.
33. Beinert, H., and Palmer, G., *J. Biol. Chem.*, **239**, 1221 (1964).
34. Beinert, H., Griffiths, D. E., Wharton, D. C., and Sands, R. H., *J. Biol. Chem.*, **237**, 2337 (1962).
35. Beinert, H., and Kok, B., *Biochim. et Biophys. Acta*, in press.
36. Weissman, S. I., and Cohn, M., *J. Chem. Phys.*, **27**, 1440 (1957).
37. Kubo, H., Shiga, T., Uozumi, M., and Isomato, A., *Bull. soc. chim. biol.*, **45**, 219 (1963).
38. Williams, R. J. P., in S. Kirschner (Editor), *Advances in the chemistry of coordination compounds*, Macmillan, New York, 1961, p. 65.
39. Palmer, G., Bray, R. C., and Beinert, H., *Federation Proc.*, **22**, 466 (1963).

Reprinted from THE JOURNAL OF BIOLOGICAL CHEMISTRY, Vol. 239, 1964

Direct Studies on the Electron Transfer Sequence in Xanthine Oxidase by Electron Paramagnetic Resonance Spectroscopy

II. KINETIC STUDIES EMPLOYING RAPID FREEZING*

ROBERT C. BRAY,† GRAHAM PALMER, AND HELMUT BEINERT

From the Institute for Enzyme Research, University of Wisconsin, Madison 6, Wisconsin

(Received for publication, November 8, 1963)

This paper describes the results and interpretations of quantitative measurements of the intensities of the various electron paramagnetic resonance signals which are observed during enzymic catalysis by xanthine oxidase. The techniques used and the types of signals which were obtained are discussed in the preceding paper (1).

EXPERIMENTAL PROCEDURE

Types of Experiment

In order to study the electron transfer sequences within the enzyme, we must know the behavior of the EPR[1] signals in the approach to, during, and at the end of the steady state of the reaction. We have used four basic types of experiment. The first will be referred to as the "single turnover" technique and is the one used previously for preliminary studies (2). It consists of treating the enzyme with 1 equivalent or less (with respect to active center) of substrate in the presence of an excess of oxygen. Under these conditions no steady state is reached, since each enzyme active center reacts either once only, or else not at all. This technique has the advantage that the enzyme remains reduced for an absolute minimum of time so that the possibility of secondary changes, such as the reduction of "inactive" enzyme by "active," is effectively eliminated. This danger of secondary reduction is a very real one when the enzyme is left anaerobically in contact with the substrate for periods of minutes, as was shown by Bray, Pettersson, and Ehrenberg (3), even though their enzyme samples had high specific activities. There is also in principle a possibility of secondary changes occurring in "inactive" molecules during a prolonged steady state, although no evidence for this has in fact been obtained. From this viewpoint the single turnover technique is the preferred one. Furthermore, it has the advantage, provided that one accepts the finding of Gutfreund and Sturtevant (4), that the binding of the substrate and of oxygen is rapid and that the kinetics is simple to interpret (2).

The observation of prolonged steady states was not possible at the high enzyme concentrations used in the present work, since the solubility both of oxygen and of xanthine is not sufficient. However, useful information has been obtained from experiments in which the concentrations were such that a short "steady state" period,[2] lasting for perhaps 2 to 20 turnovers of the enzyme, was followed by its full reduction when all the oxygen was consumed.

A third technique has consisted of following the anaerobic reduction of the enzyme by the substrate, either observing the time course with a fixed amount of substrate or "titrating" the enzyme with different amounts of substrate over a fixed reaction period from 100 milliseconds up to 1 hour. Results from the anaerobic method have proved the most difficult to interpret, and reduction is undoubtedly a complex process. Hence, for the fourth type of experiment, which has been used once only, a special form of the apparatus was required. This fourth technique consisted of studying the reoxidation of enzyme that had been reduced for controlled periods which were thought to be short enough to exclude the possibility of secondary reactions. This system, which consists of three syringes and two mixing chambers, adapted for this purpose, is described in the preceding paper (1).

Methods

Details of the preparation of the enzyme, of the rapid freezing technique, and of EPR measurements are given in the preceding paper (1).

Normalized Signal Heights—The signal heights in millimeters, measured from the traces as indicated in the Paper I (1), were "normalized" to standard settings of the EPR apparatus and standard enzyme concentration, assuming, as a first approximation, direct proportionality with enzyme concentration. All values quoted in the figures, tables, and text refer to 0.10 mM xanthine oxidase with the EPR spectrometer set at 25 milliwatts

* This investigation was supported by the United States Public Health Service through Research Grants AM 02512, GM 06762, and GM 05073, by Research Career Program Award GM-K6-18,442 from the Division of General Medical Sciences to H. Beinert, and by the Atomic Energy Commission through Contract At(11-1)-909. The participation of one of us (R. C. B.) in this work was made possible through the interest of Professor F. Bergel, F.R.S., in this work and by grants from E. I. du Pont de Nemours and Company, Inc., the Research Committee of the Graduate School of the University of Wisconsin, and the National Institutes of Health, which are gratefully acknowledged.

† Permanent address, Chester Beatty Research Institute, Institute of Cancer Research: Royal Cancer Hospital, London S.W. 3, England.

[1] The abbreviation used is: EPR, electron paramagnetic resonance; often also called electron spin resonance, ESR.

[2] It is of course realized that these states of the reacting system are not true steady states, as the concentrations of the reactants undergo changes which are by no means negligible. For the purposes of this discussion, however, we shall, for the sake of simplicity, use the term steady state.

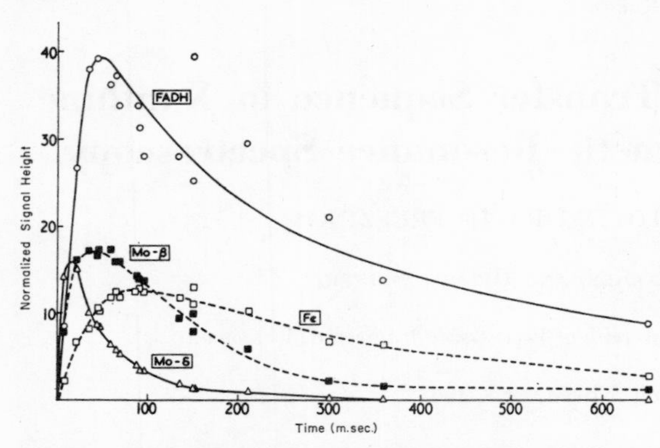

FIG. 1. Signal heights in a "single turnover" experiment. Samples were frozen at the times indicated, as described in the previous paper (1), and the signal heights were normalized as described in the text. In this experiment as in all the others described in the figures, 0.05 M pyrophosphate buffer, pH 8.2 to 8.4, was used and the temperature (except where otherwise indicated) was 22.0 ± 0.5°. The concentrations immediately after mixing were: xanthine oxidase, 0.25 mM; xanthine, 0.37 mM; and oxygen, 0.76 mM. The following symbols for the various xanthine oxidase components detectable by EPR are used in all figures of this paper: ○——○, FADH; ■---■, molybdenum β; △——△, molybdenum δ; □---□, iron.

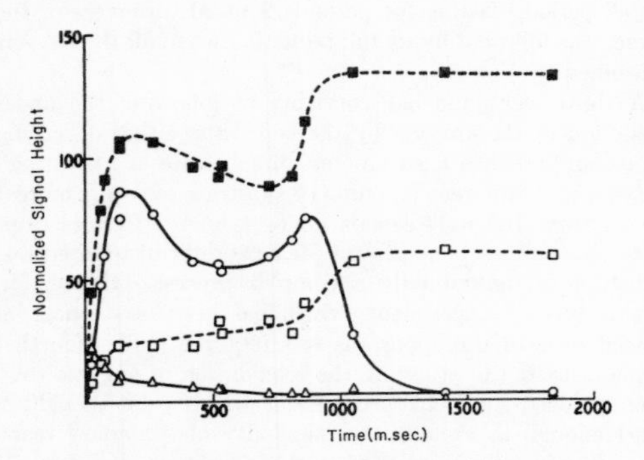

FIG. 2. Signal heights during reduction of the enzyme via a steady state. The conditions were as described in Fig. 1. The concentrations immediately after mixing were: xanthine oxidase, 0.12 mM; xanthine, 1.25 mM; and oxygen, 0.76 mM. Some of the actual traces from this experiment are reproduced in Fig. 4 of Paper I (1). For key to *symbols*, see Fig. 1.

of power incident on the sample, 5-gauss modulation amplitude, and a gain of 250 for molybdenum and radical signals and a modulation of 15 gauss and gain of 400 for the iron. In addition, small corrections for differences in diameter of the sample tubes were applied where appropriate. Except for the values referring to absolute quantitative determination, no corrections were applied for saturation (*cf.* Paper I (1)), which amounted to 20% for molybdenum and 5 to 10% for the radical at 25 milliwatts of power. For the manual mixing experiments, results were scaled down by a factor of 0.42 (see the preceding paper) to correct for the absence of isopentane. In all experiments, blank determinations were carried out on the oxidized enzyme, run under the same conditions, and the values obtained were subtracted from the experimental results. The radical and iron blanks were always reasonably small compared with the experimental values (see Fig. 4 of Paper I (1)), and the molybdenum blanks were invariably zero. At least part of the blank signal was due to the quartz tubes.

RESULTS

Single Turnover—The result of a "single turnover" experiment (Fig. 1) shows clearly that the time courses of appearance and disappearance of the four signals are quite different from one another. The molybdenum δ signal appeared[3] and disappeared very rapidly and was maximal at about 15 milliseconds. The molybdenum β, FADH, and iron signals were maximal at about 40, 45, and 100 milliseconds, respectively. The forms of the curves for these three species differed slightly. All the signals had become quite small by 650 milliseconds. As pointed out in the preceding paper (1), the kinetics of the molybdenum γ and α signals differ little from those of molybdenum δ and β, respectively. The α and γ components are therefore not specifically referred to.

Steady State—Figs. 2 and 3 show changes in signal intensities when xanthine was present in excess and the oxygen concentrations were 0.76 and 0.26 mM, respectively. In the former experiment (Fig. 2), signal changes were rapid during the first 150 milliseconds, then less rapid during a phase which lasted until about 800 milliseconds, then again more rapid in the time range from 800 to 1200 milliseconds, and after this, constant up to at least 1800 milliseconds. The calculated time for the exhaustion of the oxygen, based on the standard enzyme assay (5) and assuming a strict proportionality between enzyme concentration and activity over a 1000-fold concentration range, was 580 milliseconds. Applying a correction for temperature and for oxygen and xanthine concentrations, the calculated time becomes 950 ± 150 milliseconds.[4] The calculation thus confirms that the signal changes occurred over the time range during which the catalytic reaction must have been proceeding, and it is therefore reasonable to assume that the phases of the reaction correspond to the approach to the steady state, the steady state itself, the end of the steady state, and full reduction, respectively. The experiment shows that the iron signal was about half-maximal during the steady state, while the molybdenum β signal was about three-quarters maximal. As in the single turnover experiment, the molybdenum δ signal appeared very rapidly, but it decreased to a low level during the steady state and was absent at full reduction. Free radicals were present in the steady state but not at full reduction. In Fig. 3 the oxygen concentration was lower than in Fig. 2, so that only a very short steady state would be expected. In agreement with this, a plot of signal intensities against time shows only inflections in the region of 200 to 500 milliseconds, rather than a plateau. Fig. 4 shows the approach to steady state conditions at 1° instead of 22°, which was the temperature for all the other experiments. Since the Q_{10} is 3.0 (3), the over-all reaction would be expected to be about 10 times slower than in Figs. 2 and 3. Nevertheless, the molybdenum δ signal still appeared very rapidly, the half-time for its formation being about 10 milliseconds and the time for maximal formation being well under 100 milliseconds. The slower appearance of all the other

[3] For designation of signals, see the accompanying paper (1).

[4] The Q_{10} in this temperature range is 2.6 (3). In an assay with increased xanthine and oxygen concentrations, 70% of the activity under standard conditions was found.

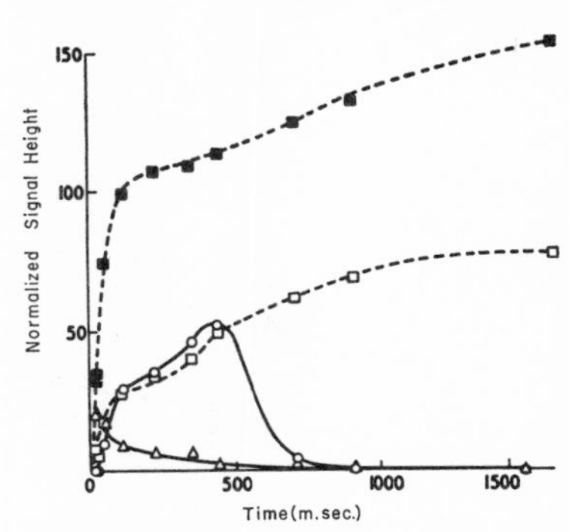

FIG. 3 (*left*). Signal heights during reduction of the enzyme via a very short steady state. The conditions were as described in Fig. 1. The concentrations immediately after mixing were: xanthine oxidase, 0.12 mM; xanthine, 1.25 mM; and oxygen, 0.26 mM. For key to *symbols*, see Fig. 1.

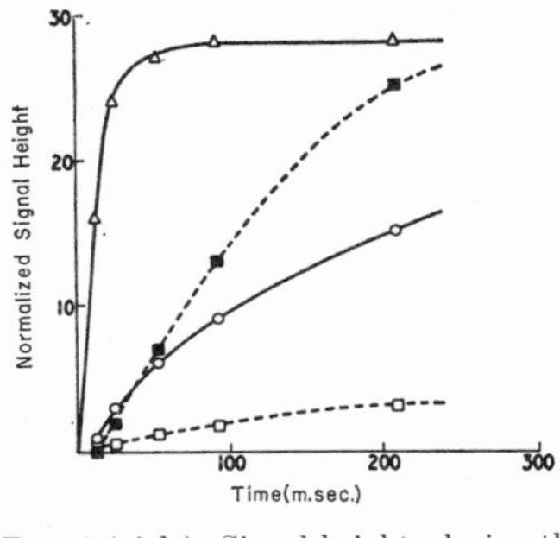

FIG. 4 (*right*). Signal heights during the early stages of reduction of the enzyme to the steady state in the cold at 0.8 ± 0.3°. The conditions were otherwise as described in Fig. 1. The concentrations immediately after mixing were: xanthine oxidase, 0.105 mM; xanthine, 1.25 mM; and oxygen, 1.27 mM. For key to *symbols*, see Fig. 1.

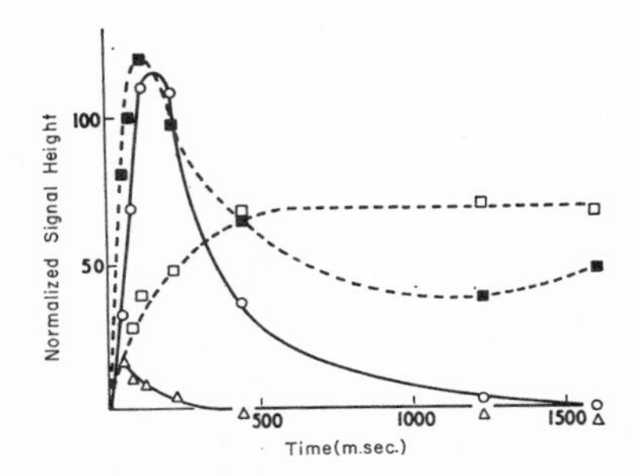

FIG. 5 (*left*). Signal heights during anaerobic reduction of xanthine oxidase (first sample, see the text) with 10.4 molecular proportions of xanthine. The conditions were as described in Fig. 1; the final xanthine oxidase concentration was 0.12 mM. For key to *symbols*, see Fig. 1.

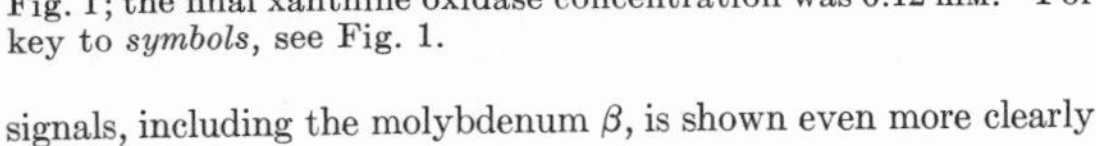

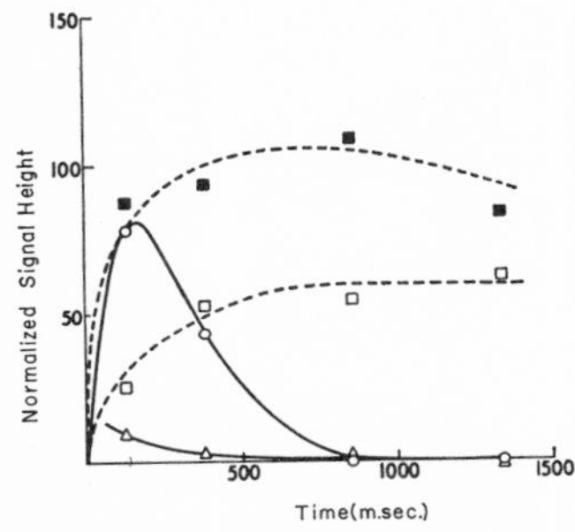

FIG. 6 (*right*). Signal heights during anaerobic reduction of xanthine oxidase (second sample, see the text) with 10.4 molecular proportions of xanthine. The conditions were as described in Fig. 1; the final xanthine oxidase concentration was 0.08 mM. For key to *symbols*, see Fig. 1.

signals, including the molybdenum β, is shown even more clearly in this experiment than in the previous ones.

Anaerobic Reduction—The time courses of the anaerobic reduction of two samples of the enzyme are given in Figs. 5 and 6. Both samples showed that the FADH free radical signal was maximal at about 150 milliseconds and disappeared by about 1000 milliseconds. Also, for both samples, the appearance of the iron signal was roughly first order, with a half-time of about 130 milliseconds, while again for both samples, and as in the other types of experiment, the molybdenum δ signal appeared rapidly, then disappeared once again. Somewhat unexpectedly, however, the behavior of the molybdenum β signal was different for the two samples. In both cases strong signals appeared within 150 milliseconds, but while in the experiment of Fig. 6 they remained roughly constant, in Fig. 5 they decreased to about one-third of their maximum height at 1000 milliseconds. This is the only point, in the work described in this paper, at which different samples of xanthine oxidase behaved differently from one another, and no explanation of the discrepancy has been obtained. The samples differed in their $E_{420}:E_{450}$ ratios, that in Fig. 5 having a ratio of 0.89, and that in Fig. 6, 1.02. A third sample, with a ratio of 0.95, also gave intermediate behavior with regard to the molybdenum β signal. However, this apparent correlation between a low $E_{420}:E_{450}$ and "fading" of the molybdenum β signal may well have been fortuitous, since there were no indications of "fading" after the end of the steady state in Figs. 2 and 3, despite the fact that samples with low absorption ratios were used.

Anaerobic experiments of the titration type were also carried out. Figs. 7 and 8 represent titrations at 100 and 550 milliseconds, respectively, on a sample of the enzyme similar to the one used for Fig. 5. As might have been expected from Figs. 5 and 6, a 100-millisecond period was not long enough for the FADH of the enzyme to be reduced to $FADH_2$, even when an excess of xanthine was present. Thus, in Fig. 7, the FADH signal increased steadily on increasing the xanthine concentration up to 1 to 2 moles of xanthine per mole of xanthine oxidase, and then remained roughly constant up to at least 13 molecular proportions. On the other hand, in the 550-millisecond titration (Fig. 8), the time was adequate for substantial reduction of the flavin to the fully reduced form provided that more than 2 moles of xanthine per mole of xanthine oxidase were present. The iron signal showed a sharp break at 2 moles, with relatively smaller increases in intensity above this xanthine concentration. The 100-millisecond iron values (not shown) were generally rather

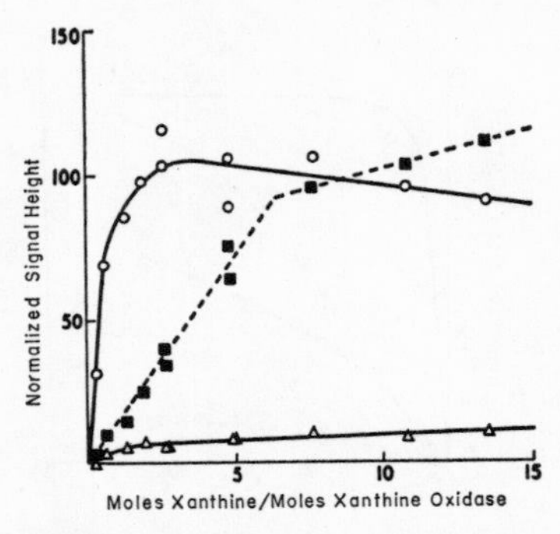

FIG. 7 (*left*). Anaerobic titration; signal heights as a function of xanthine concentration, measured after 102 ± 9 milliseconds of reaction. The conditions were as described in Fig. 1; the final xanthine oxidase concentration was 0.093 mM. For key to *symbols*, see Fig. 1.

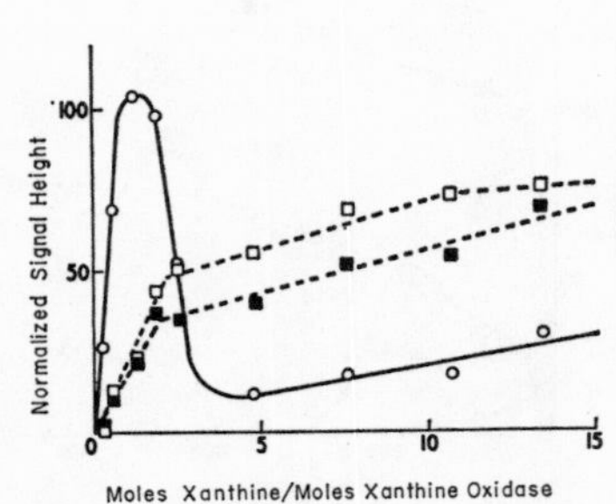

FIG. 8 (*right*). Anaerobic titration; signal heights as a function of xanthine concentration, measured after 550 ± 30 milliseconds of reaction. The conditions were as described in Fig. 1; the final xanthine oxidase concentration was 0.093 mM. For key to *symbols*, see Fig. 1.

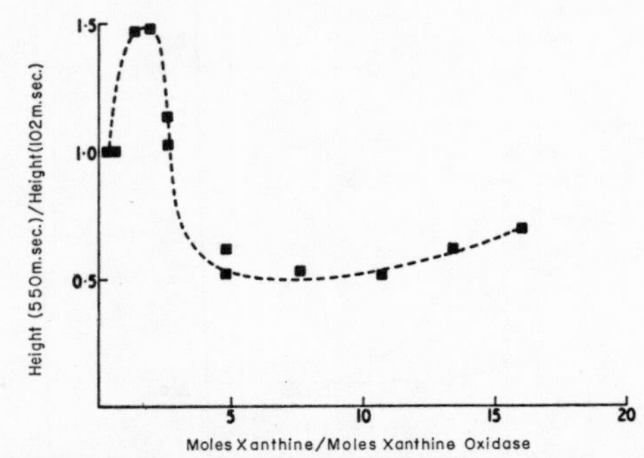

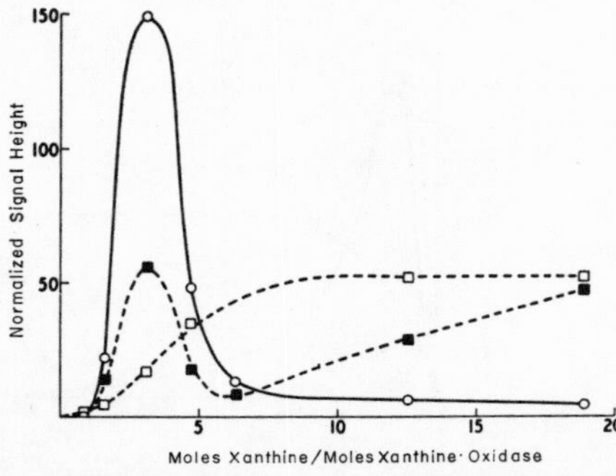

FIG. 9 (*left*). Effect of xanthine concentration on the anaerobic time course of the molybdenum β signal. The graph shows the intensity of the signal at 550 milliseconds divided by that at 102 milliseconds, at each xanthine level. The data are taken from Figs. 7 and 8.

FIG. 10 (*right*). Anaerobic titration employing manual mixing; signal heights as a function of xanthine concentration. The samples were frozen after about 1 minute of incubation at 10°; the conditions were otherwise as described in Fig. 1. The final xanthine oxidase concentration was 0.08 mM. The form of the curves suggests a slight contamination with oxygen, corresponding to about 1 molecular proportion of xanthine. For key to *symbols*, see Fig. 1.

less than half of those in Fig. 8, in agreement with the time course data given above. The molybdenum δ signal was present at 100 milliseconds but not at 550 milliseconds (not shown in Fig. 8). The behavior of the molybdenum β signal in these titration experiments was complicated by the "fading" phenomenon of Fig. 5. This can be seen most clearly in Fig. 9, in which the signal height at 550 milliseconds divided by that at 100 milliseconds is plotted as a function of the xanthine concentration, based on data from Figs. 7 and 8; it is apparent that the decrease in the molybdenum β signal was maximal with about 5 to 10 molecular proportions. In previous experiments employing manual mixing, decreased molybdenum signals in the presence of an excess of substrate were also noted and were attributed, provisionally, to reduction of the molybdenum to the tetravalent state. Fig. 10 shows a manual titration experiment in which the enzyme was frozen 1 minute after mixing with xanthine. This again shows the decrease in the molybdenum β signal, particularly in the region of 5 to 10 molecular proportions of xanthine, in agreement with the original data (compare Fig. 3*b* of Bray, Pettersson, and Ehrenberg (3)). However, since decreases in the molybdenum β signal did not occur with all batches of enzyme (Fig. 6), it must now be concluded that they are unlikely to be related to the catalytic activity. Furthermore, since the present work suggests that the decrease may be confined to intermediate concentrations of substrate and may not take place at all in the presence of a large excess (Fig. 9; see also Fig. 8), it seems that reduction to the tetravalent state is probably not the true explanation to the phenomenen. A possible explanation (3) may be some form of interaction (6, 7) between 2 molybdenum atoms within the xanthine oxidase molecule, for which there might exist an optimum xanthine concentration. The possibility that the decrease in the molybdenum β signal involved interaction between the molybdenum atoms of different xanthine oxidase molecules seems to be eliminated, since the phenomenon was not dependent on concentration. Thus, when the sample of enzyme used for the experiments in Figs. 7, 8, and 9, at twice its previous concentration, was treated with 7.7 molecular proportions of xanthine, the molybdenum β signals were as before, giving a 550 millisecond:100 millisecond ratio of 0.51, in full agreement with Fig. 9.

The failure to observe a drop in the molybdenum β signal after the end of the steady state (Fig. 2) might suggest that hydrogen peroxide or uric acid inhibited the drop. However, no evidence in favor of this could be obtained. When xanthine oxidase was treated anaerobically with 7.1 molecular proportions of xanthine plus 2.7 molecular proportions of uric acid, the 550 millisecond: 100 millisecond molybdenum β ratio was 0.62 compared with a control ratio of 0.56. Similarly, when the uric acid was replaced by 16 molecular proportions of hydrogen peroxide, a ratio of 0.60 was obtained.

Reoxidation—Table I compares the effect of reducing the enzyme anaerobically with 1 molecular proportion of xanthine for 107 milliseconds (*cf.* Fig. 7) with that of reducing it for 84 milliseconds and then treating it with an excess of oxygen for 23 milliseconds. Somewhat unexpectedly, all the signals were decreased to about the same extent by the oxygen in this experiment.

Effect of pH—All the above work was done at or near pH 8.3. The relative intensities of the various signals were markedly altered on changing the pH. Experiments described in this section were carried out both at higher and at lower pH values; but in considering the results it must be remembered that the pH optimum for the enzyme was found (1) to be 8.9 rather than 8.3, where it had been expected.[5] Table II gives the maximum intensities of the various signals obtained in single turnover experiments at three pH values. The intensities after 400 milliseconds in steady state experiments are also given; this time would presumably be somewhere toward the middle of the steady state. It is apparent from Table II that in both types of experiment, the FADH and the molybdenum δ signals were markedly increased on increasing the pH and were decreased to very low values on lowering it. At the same time, and in contrast to this, the molybdenum β signal was diminished in intensity at high pH and intensified at low pH in the single turnover experiment. In the steady state, the yield of this signal was relatively independent of pH. The iron signals were only moderately affected by change of pH, but seemed to be most intense near pH 8.3. Although the time for maximum development could not be determined precisely for the weaker signals, the time sequence of the signals did not seem to be pH-dependent in these "single turnover" experiments. The intense free radical signal obtained at pH 9.6 developed more slowly than did that at pH 8.3 (Fig. 1) and took 120 milliseconds to attain its maximum value. Similarly, the strong molybdenum β signal at pH 6.0 appeared also relatively slowly, reaching its maximum only after about 100 milliseconds. On the other hand, the time course of the molybdenum δ signal, showing its peak at an early time, was substantially the same at pH 8.2 and 9.6. When in the steady state experiment at pH 9.9 (Table II), a sample was frozen after 8 milliseconds of reaction, the molybdenum δ signal was 4 times as intense as it was at 400 milliseconds and was in fact the most intense molybdenum δ signal that we have obtained in the current work (normalized signal height of 101). Thus, the time course of the molybdenum δ signal in the steady state experiments also was probably the same at both the high and the moderate pH values (Fig. 2).

Comparison of Different Substrates—All the experiments described so far employed xanthine as substrate. Other substrates were compared with xanthine in steady state experiments at pH 8.3 (Table III). The signal intensities in the table refer to 400 milliseconds, but experiments were also carried out at 800

[5] This applies to the conditions of the routine assay. The pH optimum under the conditions of the rapid reaction experiments has not been determined, and may be slightly different.

TABLE I

EPR signals after reduction and reoxidation

The enzyme (final concentration, 0.09 mM) was reduced anaerobically with xanthine (final concentration, 0.09 mM) for 84 milliseconds and then treated with buffer saturated either with O_2 (final concentration, 0.42 mM) or with nitrogen for an additional 23 milliseconds in the three-syringe system (see Paper I (1), Fig. 2). The conditions were as described in Fig. 1.

Gas in equilibrium with buffer in third syringe	Normalized signal height of component			
	FADH	Mo β	Mo δ	Fe
Nitrogen	90	16	1	13
Oxygen	43	10	2	6

TABLE II

Effect of pH on EPR signals

The table shows normalized signal heights measured in experiments of the single turnover or steady state type. Pyrophosphate buffers were used and the concentrations of the reactants immediately after mixing were as shown; the temperature was 22°. For the single turnover experiment, maximal amplitudes are given, whereas for the steady state experiment the amplitudes refer to a reaction time of 400 milliseconds.

Type of experiment*	pH	Normalized signal height			
		FADH	Mo β	Mo δ	Fe
Single turnover (xanthine oxidase, 0.11 mM; xanthine, 0.10 mM; oxygen, 0.76 mM)	6.0	3	37	2	5
	8.2	49	10	9	6
	9.6	160	3	19	3
Steady state (xanthine oxidase, 0.05 mM; xanthine, 1.25 mM; oxygen, 0.76 mM)	6.1†	13	95	3	8
	8.4	101	111	6	38
	9.9	1030	101	25	18

* The concentrations given are final values.

† Xanthine had to be added as an unbuffered solution owing to low solubility; the pH is therefore approximate.

TABLE III

Effect of various substrates on EPR signals

The table shows normalized signal heights in steady state experiments with the following concentrations (immediately after mixing): 0.09 mM enzyme, 0.92 mM oxygen, and 1.3 mM substrates. The experiments were carried out at pH 8.3 and the signals were measured at 400 milliseconds; the temperature was 22°.

Substrate	Normalized signal height of component			
	FADH	Mo β	Mo δ	Fe
Xanthine	94	74	6	12
Hypoxanthine	129	181	4	23
Purine	69	14	0	2
Salicylaldehyde	56	7	0	3
NADH	6	0	0	0

milliseconds. The values at the two times of measurement agree quite well, so that it is reasonable to take those in the table as representative of the steady state condition. With hypoxanthine the same signals as with xanthine were obtained, but it is noteworthy that the molybdenum β signal, and to a lesser extent the FADH, were intensified, while the molybdenum δ signal was diminished (1).[6] With purine and salicylaldehyde, moderate radical signals were obtained with rather weak molybdenum β values; the molybdenum δ signal could not be detected, while iron was just visible. Addition of NADH did not produce any EPR signal, except for a very weak radical signal. These results may be compared with manual titration experiments with the various substrates reported by Bray *et al.* (3). They concluded that xanthine, purine, and salicylaldehyde under suitable anaerobic conditions all gave molybdenum and radical signals rapidly, whereas NADH only gave the signals very slowly.

The rates of oxidation of the substrates were not measured in these experiments. However, combining data in the literature from various sources (8–10), the relative rates of oxidation may be calculated as: xanthine, 100; hypoxanthine, 70; salicylaldehyde, 42; purine, 20; and NADH, 1. Since all of these rates would probably be quite sensitive to pH, as well as to substrate and oxygen concentration, they must be regarded only as a very rough indication of the actual rate of substrate oxidation in the EPR experiments. Nevertheless they do suggest some parallelism between signal intensities and rates of oxidation.

Absolute Signal Intensities—The normalized signal heights quoted throughout this paper are, as a first approximation, directly proportional in each case to the proportion of the enzyme which is present in the form which gives the signal in question. However, as explained in the preceding paper (1), it is by no means easy to obtain reliable estimates of the proportionality factors, so that any absolute values which are calculated with their aid must be interpreted wth considerable caution.

The following proportionality factors were obtained. A normalized signal height of 100 corresponded to 0.06 mole per mole of xanthine oxidase for the free radical; 0.36 mole per mole for Mo(V) represented by both the α and β signals (measured at the β-peak); 0.24 mole per mole for Mo(V) represented by the γ and δ signals (measured at the δ-peak); and 0.36 mole per mole for the iron (measured at the β-peak and on the basis of 4 unpaired electrons per iron atom; *cf.* Paper I (1)). On this basis absolute intensities may be calculated as follows. The strongest radical signal obtained (pH 9.9, steady state; Table II) corresponds to about 0.6 mole of FADH per mole of xanthine oxidase, or, allowing for 2 FAD per xanthine oxidase, to about 30% conversion to the radical. In the steady state at pH 8.3 (Fig. 2), the maximum conversion was about 3%. The highest conversion of molybdenum to the molybdenum δ form (8 milliseconds at pH 9.9) corresponded to 0.24 atom of molybdenum per mole of xanthine oxidase or, again assuming 2 atoms of the metal per enzyme molecule, a 12% conversion. In experiments at pH 8.3 (*e.g.* Fig. 2), the conversion was about 2.5%.

Since neither the FADH nor the molybdenum δ signal was present at full reduction of the enzyme, there is no reason to expect their signals to approach a 100% yield under any of the experimental conditions. In these cases, therefore, there are no grounds for questioning the above percentage conversion figures. On the other hand, ignoring those cases in which the molybdenum β signal gave the anaerobic decreases discussed above, the time courses of the iron and molybdenum β signals are such that 100% conversion might reasonably be anticipated at full reduction of the enzyme. However, the calculated conversion figures are somewhat at variance with this. Thus, the absolute intensity of the iron signal at full reduction in a number of experiments corresponded only to about 0.25 atom of iron per mole of xanthine oxidase. If one postulates—and this postulate is consistent with the magnetic susceptibility data of Bray *et al.* (3) that there are two active centers in the enzyme, each with 4 iron atoms of which only 1 is reducible—then the maximum conversion of the enzyme to the form giving the iron signal would only be 12.5%. Whether this low value is simply the result of errors arising in the integration procedure (*cf.* Paper I (1)) possibly combined with a slightly low value due to some contamination of the xanthine oxidase with the "inactive" form, or whether some other explanation, such as interactions among the iron atoms, is involved, is not certain at present. The obvious alternative explanation, *viz.* that the signals are weak because they are related to changes in impurities rather than in the enzyme itself, is rejected, first, because of the agreement between results on different batches of the enzyme (see below), and second, because of the agreement between the kinetic properties of the signals and those of the enzymic reactions, as shown above.

Similarly, in a number of experiments, the molybdenum β signal (including molybdenum α) at full reduction corresponded to about 0.5 atom of molydenum per mole of enzyme, *i.e.* (again assuming 2 atoms per mole) a 25% conversion. Again, it is not clear whether this low value is due to integration errors or to other causes; one possible interpretation, as for the iron, is interaction between molybdenum atoms. Previous work (3) indicated that on appropriate reduction "inactive" xanthine oxidase gave stronger molybdenum signals than did "active." It was claimed that the former yielded signals that approximated the theoretical maximum intensity. No attempt has been made to repeat these experiments, but if they are correct, then it would seem that the postulated interaction between molybdenum atoms must be weakened in the "inactive" enzyme.

Results with Different Xanthine Oxidase Preparations—Two samples of xanthine oxidase prepared as described in the preceding paper (1) were used in the above work. Although these samples were prepared from different milk in different countries, no differences in the signals obtained from them were noted, nor were kinetic differences observed. One experiment was also carried out with an old sample of xanthine oxidase (ratio of activity to E_{450}, 32) prepared by another method (11). Again, the same signals were obtained, although, as would be expected, they were somewhat weaker than those from the more active samples. In particular, it was noted that the ratio of the intensity of the molybdenum δ to the molybdenum β signal in the early stages of a single turnover experiment was the same with the old sample as for the others.

Effect of Inhibitors—In aerobic manual mixing experiments, xanthine oxidase was treated with 10 molecular proportions of 8-phenylxanthine or 1 proportion of 2,6-diamino-8-phenylpurine. The solutions were frozen about 1 minute after the enzyme was mixed with the inhibitors, at pH 8.3. No signals were obtained.

[6] The iron signal obtained from xanthine in this experiment was weaker than that in several other experiments under similar conditions. The reason for this is not clear. It might conceivably be due to experimental errors, although this seems unlikely.

DISCUSSION

Rapid versus Slow Reactions—The interpretation of previous EPR data employing manual mixing of the enzyme and substrate (3, 12) was complicated by the fact that changes in the signal intensities occurred anaerobically during the period from 1 minute to 1 hour after mixing. These changes were interpreted as being due to reduction of "inactive" forms of the enzyme by the reduced "active" form. However, even 1 minute at 0° is a long period compared to the turnover time of the enzyme (about 100 milliseconds), and there is, in principle, no reason why secondary reactions unrelated to the catalytic cycle might not take place within this time. It appeared necessary, therefore, to use enzyme of high activity and rapid reaction techniques in the present work. The specific activity of the xanthine oxidase preparations which we used was slightly higher than that of previous samples (3). With the possible exception of some of the slower anaerobic changes (see above), all the changes of EPR signal intensity reported are believed to be due to the "active" enzyme.

Absence of Endogenous Inhibitor—In previous rapid freezing work (2), the kinetics of the EPR signals was in agreement with the known (4) kinetics of the enzymic reaction for one sample of the enzyme, while for another sample, the signals were slow to disappear. This was presumed to be due to the presence of an inhibitor in the second preparation, perhaps identical with one discussed earlier (13), which was assumed to have a serious effect on the rate of reoxidation of the reduced enzyme, but only in the very concentrated solutions required for EPR work. In the present work no evidence suggesting complications from the presence of such an inhibitor has been obtained. As discussed above in relation to the steady state experiments, the EPR kinetics obtained with about 10^{-4} M xanthine oxidase is entirely consistent with data from the ordinary catalytic assay employing about 10^{-7} or 10^{-8} M enzyme. This indicates that the enzyme activity was directly proportional to concentration over a remarkably wide concentration range, as would, of course, be expected under ideal inhibitor-free conditions.

Internal Electron Transfer—The experiments on single turnover (*cf.* Fig. 1) and on anaerobic reduction (*cf.* Figs. 5 and 6) clearly show that the iron signal is the slowest to appear and to disappear. However, the rates of appearance and disappearance of even this component are still sufficiently rapid to fall within the limits set by the over-all reaction velocity of the enzyme (see below under "Fe(III):Fe(II)"). We assume, therefore, that the observed rapid changes of EPR signal shape and intensity of all components are directly related to the catalytic reaction of the enzyme.

In order to facilitate the consideration of various alternative interpretations of our results, we introduce at this point in Fig. 11 a tentative scheme as a guide to such a discussion.

The oxidation-reduction systems involved in the sequence from substrate to oxygen are shown in this figure in the *vertical columns* from *left to right*, and the various species within each system which participate in the reaction are placed on different horizontal levels, with the oxidized states shown on the *top line*. The species observable by EPR spectroscopy are shown in *bold print*. Electrons that become available in the turnover of a system on the *left-hand side* will then enter the neighboring system on the *right*, wherever electron uptake is indicated. Our

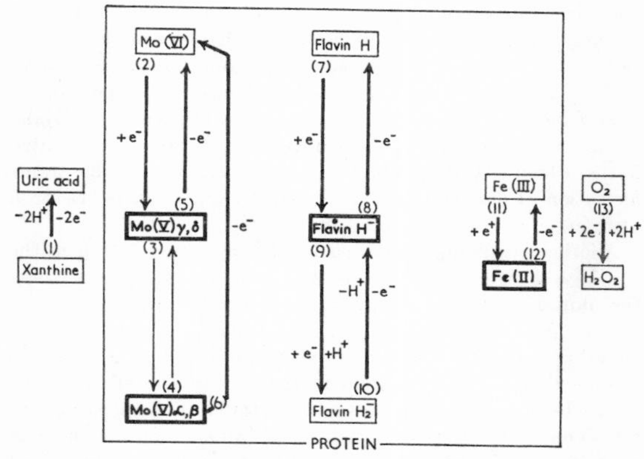

FIG. 11. Schematic representation of pathways of electron transfer in the xanthine oxidase reaction, as suggested on the basis of the present EPR studies. The various participating oxidation-reduction systems are aligned in *vertical columns* from *left* to *right*, and the various species within each system are placed on different horizontal levels within each column, with the oxidized states shown on the *top line*. Uptake or loss of electrons or protons is indicated by *heavy arrows* and corresponding *letters*. The *light arrows* (Reactions *3* and *4*) indicate no electron transfer but only a conversion without change of oxidation state. The species observable by EPR spectroscopy are indicated by *bold print* and a *heavy frame*. Concerning the Fe(III):Fe(II) oxidation-reduction system, footnote 7 of the accompanying paper (1) should be considered. The *symbols* describing the oxidation and ionization state of the flavin component are explained in footnote 9 of this paper.

work here was not concerned with the events outside the enzyme molecule. Therefore, we do not want to specify anything about the mechanism of conversion of xanthine to uric acid, or of oxygen to hydrogen peroxide; we merely indicate for these substrates the balance of protons and electrons transferred.

Mo(VI):MO(V)—The initial steps, electron transfer to and from molybdenum, are complicated by the appearance of chemically different species of the reduced metal (Mo(V)) at different reaction times and depending on pH. The Mo(V) γ,δ component is clearly, and by a considerable margin, the first reactant, except at low pH (see below). However, the events which immediately follow are less well defined. There is evidence in our data suggesting that under appropriate conditions, all of Reactions *2* to *6* can take place, but that the extent to which the two reoxidation pathways, *5* and *3* plus *6*, respectively, participate at different pH and with different substrate concentrations is not certain. It may well be that under certain conditions some of these reactions are negligibly slow. However, it seems most probable that they are all participating to some extent at pH 8.3, while at higher pH, reoxidation of the molybdenum is largely from the Mo(V) γ,δ component, and at low pH, mainly from Mo(V) α,β. Several alternative explanations for the involvement of two different species of Mo(V) may be suggested. One is that two different catalytic species of molybdenum follow each other in the electron transfer sequence, the sequence being then: molybdenum, molybdenum, FAD, iron. The stoichiometry of the catalytic components of the enzyme argues against this view. There are 2 flavins, somewhat less than 2 molybdenum, and 8 iron atoms present per molecule of enzyme (10). An even stronger argument appears to be the fact that on complete reduction of the enzyme by substrate or dithionite, only the Mo(V) α,β signal is observed and not the Mo(V) γ,δ.[7] It appears more reasonable, then, to assume that we are dealing with 1 molybdenum catalyst per sequence, which can exist in two closely related forms, which either equilibrate or undergo an irreversible transformation (from Mo(V) γ,δ to Mo(V) α,β) during electron transfer.

Another possibility suggested by the rapidity with which the Mo(V) γ,δ signal appears is that this species in fact represents the initial enzyme-substrate complex, rather than a species derived from it by electron transfer from substrate to molybdenum. However, if this were the case, it would be necessary to assume that the molybdenum in the resting as well as in the reduced enzyme is in the pentavalent state. The appearance of the signal in the presence of substrate would then not indicate reduction of Mo(VI) to Mo(V), but interference of the bound substrate with an interaction—possibly between 2 Mo(V) atoms (6, 7) in vicinal active centers—which broadens or otherwise cancels the EPR signal of Mo(V). If this were the case, one might expect that substrate analogues, which are strongly bound but cannot be oxidized, would produce the Mo(V) γ,δ signal. Two such compounds, 2,6-diamino-8-phenylpurine, a potent competitive inhibitor of xanthine oxidase (14), and 8-phenylxanthine, failed to produce any EPR signals. Although these experiments do not exclude the possibility that Mo(V) is present in the resting enzyme, they lend no support to such an assumption. A titration of the enzyme or correlation of the rate of appearance of uric acid with that of the EPR signals could lead in principle to a decision on this point. Thus far, experiments along these lines could not be brought to the degree of accuracy which would allow any conclusions. It may be added that such evidence as is available (15) for the oxidation-reduction potential of the Mo(VI):Mo(V) couple is in agreement with the concept that a reduction of Mo(VI) takes place during the oxidation of substrate.

If it is then assumed that Mo(VI) is present in the resting enzyme and that only 1 and not 2 molybdenum atoms are present per active site, we have to explain the different signals observed on the basis of a change in the environment of this molybdenum atom after reduction to Mo(V). A suggestion, similar to that discussed in the preceding paragraphs but based on the assumption that Mo(VI) is present in the resting enzyme, would be that Mo(V) γ,δ represents a complex between an intermediary or the final oxidation product of xanthine with the molybdenum site of the enzyme, after electron transfer to Mo(VI) has taken place, whereas at the Mo(V) α,β stage this complex would have dissociated. Alternatively, Mo(V) α,β and γ,δ could both represent complexes with products at different oxidation levels. An experiment bearing on this point, in which uric acid (final concentration after mixing, 0.2 and 0.3 mM) was added to the enzyme before mixing with xanthine, showed no effect on shape and kinetics of the EPR signals.

An alternative possibility is that Mo(V) α,β represents a Mo(V)-flavin complex which is formed immediately following the initial reduction of Mo(VI) to the Mo(V) γ,δ form, and preceding the reduction of flavin by molybdenum. This suggestion would make Reactions *3* and *6* (Fig. 11) obligatory steps and would exclude Reaction *5*. However, this is perhaps unlikely, since at high pH a single turnover could be accomplished with very little Mo(V) α,β appearing whereas under steady state conditions high concentrations of this species were present. This would favor the contrary supposition that Mo(V) α,β is essentially on a side path at high pH rather than an essential reactant.

Although we favor the view (*cf.* Fig. 11) that the conversion of Mo(V) γ,δ to Mo(V) α,β represents a rearrangement in the environment of the molybdenum at the pentavalent level, we have little information to offer concerning the actual nature of this process. The only clue to this interconversion we have obtained thus far is in experiments conducted at different pH values (*cf.* Fig. 5 of Paper I (1)). At low pH the conversion of Mo(V) γ,δ to Mo(V) α,β is clearly favored. One may therefore suggest that the uptake of a proton or the loss of a hydroxyl ion in the metal complex is involved. The most simple case would be that one of the ligands of the Mo(V) atom is protonated or exchanged against a more acidic ligand. It has been shown for myoglobin (16) that replacement of one ligand can lead to a significant change in the EPR spectrum, as it is observed here.

Apparently the conversion of Mo(V) γ,δ to Mo(V) α,β is relatively slow, since at all pH values the increase in the α, β signal on going from the single turnover type of experiment to the steady state ones was generally greater for this than for any other signal.[8]

The transformation of Mo(V) γ,δ to Mo(V) α,β seems to be favored when hypoxanthine rather than xanthine is used as substrate, since relative to the Mo(V) α,β the Mo(V) γ,δ signal was very weak in the steady state with this compound. Possibly this reflects some difference in the binding of the substrates.

The relatively low molybdenum signals obtained when purine and salicylaldehyde were used as substrates might be explained by higher Michaelis constants, which would tend to make the formation of the complexes slower than the subsequent reduction of molybdenum. Salicyaldehyde is known to have a relatively high Michaelis constant (9).

FAD:FADH:FADH$_2$[9]—We now consider the relevant features in the oxidation-reduction of the flavin system. A striking observation is that the concentration of semiquinone present during the steady state or single turnover is greatly enhanced by elevated pH. The pH dependence of semiquinone formation of free flavins and flavin-metal chelates has been investigated by Hemmerich (17). According to this author, the semiquinone concentration in the system flavin H + flavin $H_3 \rightleftarrows 2$ flavin$\cdot H_2$ rises steeply toward alkaline pH ("comproportionation"). The

[7] This consideration is based on the assumption that both the α,β and γ,δ signals are due to Mo(V) and that with substrate as reductant no valency state lower than Mo(V) is reached.

[8] It is suggested that even at low pH values, where the molybdenum γ,δ signal was always very weak, it was still a precursor of Mo(V) α,β. The reason for this suggestion is that no conditions have been found under which the molybdenum α,β signal has appeared before the Mo(V) γ,δ, although in the low pH, single turnover experiment the two species appeared virtually simultaneously.

[9] In the heading of this paragraph the conventional symbols for the oxidized, semiquinone, and reduced forms of flavin are used. In the succeeding discussion, however, as well as in Fig. 11, we prefer to use a designation which indicates the oxidation as well as the ionization state of the flavin. At neutral pH, the oxidized form is therefore designated as flavin H, the semiquinone form as flavin$\cdot H^-$, and the reduced form as flavin H_2^-. Below pH 6 the latter two forms would be protonated to flavin$\cdot H_2$ and flavin H_3, respectively.

pH range in which this rise is observed, however, strongly depends on the acidity of the semiquinone and reduced forms in the particular system under study and is, therefore, obviously influenced by binding to protein and neighboring metal components. In chelates of flavin semiquinone and metal, the acidity of the semiquinone form is increased and "comproportionation" (17) is found at pH values lower than in metal-free systems. The effect that we observed here with xanthine oxidase is therefore in line with Hemmerich's model studies, if we assume that with the enzyme intra- or intermolecular electron transfer between flavins is possible. More detailed comparisons and predictions are not permissible, because we know too little about the environment of the flavin in the enzyme. We consider it likely that decreased free radical signals at low pH result from increased reduction of the flavin to the fully reduced form, a situation which would, incidentally, also be favored by high substrate concentration. Steady state reduction of a significant portion of the flavin to the fully reduced form could also be responsible for the decreased over-all rate of xanthine oxidase found at low pH. It is interesting to note here that xanthine oxidase has, at low pH, a greater tendency toward inhibition by excess substrate (18).

Fe(III):Fe(II)[10]—Whatever we have learned of the Fe(III):Fe(II) oxidation-reduction system in xanthine oxidase warrants little specific comment. It can, however, safely be concluded that this system in xanthine oxidase, as is the case in other oxidizing enzymes studied more recently (aldehyde oxidase, dihydroorotic dehydrogenase, and DPNH dehydrogenase)[11] (19), has the kinetic behavior of an active catalytic component, located on the main path of electron transfer. The anaerobic reduction experiments showed that the slowest signal to appear was the iron and that the half-time for this process was about 130 milliseconds. This corresponds to a first order velocity constant of 5.3 sec^{-1}. The velocity constant for enzyme turnover under these conditions is probably about 4.3 sec^{-1}.[12] The rate of appearance of the iron signal is thus rather faster than the over-all turnover, as is, of course, necessary if it is on the main pathway. In full agreement with these figures, such data as are available on the reoxidation of the enzyme indicate that this, in contrast, is a relatively rapid process. In the case of all the enzyme systems studied, iron appears to be the electron-transferring agent closest to the external acceptor, which is oxygen in the case of xanthine oxidase. No characteristics, however, could be found from our work which would point to a unique affinity or aptness of this Fe(III):Fe(II) system for reaction with oxygen. The related systems in the dehydrogenases certainly do not do so efficiently, and it is of particular interest with regard to our xanthine oxidase work that the xanthine dehydrogenase from chicken liver (20),[13] which fails to show appreciable activity with oxygen, shows the same iron signals and similar kinetics of their appearance. There must, therefore, be some other features, possibly not amenable to study by EPR, which determine the oxygen affinity of these metallo-flavoproteins.

Substrate and Oxygen Radicals—If, as suggested earlier (3), there are two independent active centers in the xanthine oxidase molecule, each with 1 molybdenum, 1 FAD, and 4 irons, then the present scheme implies that electrons are transferred singly from the substrate to the molybdenum. Thus, the scheme in this form implies the participation of substrate and of oxygen radicals in the mechanism, though these might remain bound to the enzyme until the second electron has been transferred. Somewhat inconclusive evidence for xanthine radicals in the mechanism was presented by Ackerman and Brill (21), whereas our work has given no indication that substrate radicals are involved. We are aware, however, that failure to observe such radicals, particularly in low temperature EPR spectroscopy, is not a conclusive argument against their existence. This applies even more to the detection of oxygen radicals. They may not be detectable under our conditions, even if present, or they may be so reactive that they are not trapped by the quenching process. Fridovich and Handler (22) have concluded that oxygen radicals participate in the xanthine oxidase reaction. It is of interest here that Stauff, Schmidkunz, and Hartmann (23) observed a chemiluminescence during this reaction. These aspects of the problem were not the aim of the present investigation, however, and their clarification requires more direct experimental evidence on the formation of substrate and oxygen radicals. In addition, the meaning of the nonintegral molybdenum content of the enzyme has to be understood for the stoichiometry of its reduction to be studied properly.

Oxidation-Reduction Potentials—Finally, the anaerobic titration data will be considered. In all the experiments, the molybdenum, flavin, and iron were apparently titrated simultaneously, rather than successively, so that all the signals showed up with even the smallest addition of xanthine and their titrations all appeared to be complete at a common xanthine level. Thus the oxidation-reduction potentials of all the systems in the enzyme must be very similar to one another. A conclusion somewhat similar to this was reached earlier (3) in the case of the flavin and molybdenum of the enzyme.

Conclusions—For the electron transfer within the enzyme molecule, we consider a sequence from molybdenum to flavin and to iron most compatible with our results. This had previously been suggested by Bray (24). We see the strongest arguments in favor of this conclusion in the results of the single turnover experiments (*cf.* Fig. 1). In these experiments the sequence of components in electron transfer is indicated by the sequence in which maximal reduction is reached, *i.e.* in the presentation of Fig. 1, the sequence of peaks. It is clear from an inspection of the curves that Mo(V) γ,δ is the first component reduced and iron is obviously the last. The differences in the order of Mo(V) α,β and flavin or flavin and iron are not so pronounced, and approach the limits of accuracy of our techniques. If it is accepted, however, that Mo(V) α,β is formed from Mo(V) γ,δ merely by a rearrangement in the ligand sphere of the metal at the pentavalent oxidation level, then obviously Mo(V) α,β also precedes flavin and iron in the electron transfer sequence from substrate. Since iron was recognized as the last component in

[10] We refer here to a discussion of the state of our knowledge concerning the identification of this system in the accompanying paper, particularly in footnote 7 of that paper (1).

[11] H. Beinert and G. Palmer, unpublished work in collaboration with K. V. Rajagopalan, V. A. Aleman, and P. Handler; with R. Coleman and A. J. Merola; and with T. P. Singer and T. Cremona.

[12] For the most active samples used (ratio of activity to E_{450}, 100) the turnover, assuming two active centers, is 6.1 sec^{-1} under the standard assay conditions. Extrapolating (3) to the hypothetical ratio of activity to E_{450} of 117, the value becomes 7.1 sec^{-1}. This figure then has to be corrected from 23.5° to 22° and for the over-all slowing down of the reaction, which takes place at high xanthine concentrations. Since the latter effect depends to some extent on the oxygen concentration, accurate calculation is not easy. However, as a first approximation the correction may be made as indicated in footnote 4.

[13] We are indebted to Dr. W. J. DeAngelis and Dr. J. R. Totter for their collaboration in these experiments.

this sequence, flavin would have to be placed between molybdenum and iron.

The conclusions reached from the single turnover experiments are in agreement with those drawn from experiments in which the steady state reduction level was reached (Figs. 2 to 4). The relative position in the sequence of Mo(V) and iron is clear from the rate of approach to the steady state (Fig. 4). On the other hand again, the position of the flavin is not as clear. The relatively low absolute levels of FADH concentration and the possible complication of the presence of $FADH_2$ make it difficult to be certain that flavin participates in the electron transfer before, rather than after, iron. Attempts to obtain further information on this by studying the reoxidation kinetics did not give any helpful information.

The possibility has to be considered that flavin may act in conjunction with one of the metals, either the Mo(V) α,β species or the iron. Studies on the saturation with increasing microwave power of the substrate-induced free radical in xanthine oxidase indicate that it is interacting with a neighboring paramagnetic species.

Experiments with the related enzyme aldehyde oxidase[14] have led to conclusions similar to those presented here for xanthine oxidase. Preliminary analogue computer studies—which we hope to extend—on the single turnover type of experiments with xanthine oxidase confirmed that the proposed mechanism is compatible with the experimental data that we have obtained.

SUMMARY

The oxidation of xanthine by milk xanthine oxidase was investigated by the use of a rapid mixing-freezing technique and low temperature electron paramagnetic resonance spectroscopy, with particular emphasis on the kinetics of electron transfer between the catalytic components of the enzyme. All the electron paramagnetic resonance signals, which appear on reduction and disappear on reoxidation, do so in times less than the turnover time of the enzyme, and from detailed studies it is concluded that a sequence of elec on transfer from substrate to molybdenum, flavin (FAD, FAI ·), iron, and thence to oxygen is most compatible with the dat. It is tentatively suggested that per active site and electron transfer chain within the enzyme, 1 atom of molybdenum, 1 molecule of flavin, and most likely not all 4 iron atoms present are engaged in electron transfer, shuttling between the states Mo(VI) and Mo(V); FAD and FADH· (at pH $\geq$ 8); and Fe(III) and Fe(II). At lower pH, FADH· formation is decreased, presumably in favor of reduction to $FADH_2$. Two different chemical species of Mo(V) participate in the reaction. One appears at early times (maximal at 15 milliseconds at 22°) only and is favored by elevated pH; the other species predominates at low pH and later times. It is suggested that protonation or replacement of at least one ligand with a more acidic ligand in the coordination sphere of Mo(V) occurs after initiation of electron transfer. The implications of these findings and alternative possibilities and explanations of the data are discussed.

[14] K. V. Rajagopalan, P. Handler, G. Palmer, and H. Beinert, unpublished experiments.

Acknowledgments—We are indebted to Drs. R. H. Sands and T. Vänngård for valuable discussions on the interpretation of electron paramagnetic resonance spectra, to Mr. R. E. Hansen for his assistance and advice in instrumentation problems, and to Mrs. R. N. Carl, I. Harris, and M. Van De Bogart for technical assistance.

REFERENCES

1. Palmer, G., Bray, R. C., and Beinert, H., *J. Biol. Chem.*, **239**, 2657 (1964).
2. Bray, R. C., *Biochem. J.*, **81**, 196 (1961).
3. Bray, R. C., Pettersson, R., and Ehrenberg, A., *Biochem. J.*, **81**, 178 (1961).
4. Gutfreund, H., and Sturtevant, J. M., *Biochem. J.*, **73**, 1 (1959).
5. Avis, P. G., Bergel, F., and Bray, R. C., *J. Chem. Soc.*, 1100 (1955).
6. Sacconi, L., and Cini, R., *J. Am. Chem. Soc.*, **76**, 4239 (1954).
7. Mitchell, P. C. H., and Williams, R. J. P., *J. Chem. Soc.*, 4570 (1962).
8. Bergmann, F., and Dikstein, S., *J. Biol. Chem.*, **223**, 765 (1956).
9. Booth, V. H., *Biochem. J.*, **32**, 503 (1938).
10. Avis, P. G., Bergel, F., and Bray, R. C., *J. Chem. Soc.*, 1219 (1956).
11. Mackler, B., Mahler, H. R., and Green, D. E., *J. Biol. Chem.*, **210**, 149 (1954).
12. Bray, R. C., Malmström, B. G., and Vänngård, T., *Biochem. J.*, **73**, 193 (1959).
13. Bray, R. C., *Biochem. J.*, **73**, 690 (1959).
14. Bray, R. C., Ph.D. thesis, University of London, 1956.
15. Williams, R. J. P., in S. Kirschner (Editor), *Advances in the chemistry of coordination compounds*, Macmillan, New York, 1961, p. 65.
16. Ehrenberg, A., *Arkiv Kemi*, **19**, 119 (1962).
17. Hemmerich, P., *Helv. Chim. Acta*, **47**, 464 (1964).
18. Fridovich, I., and Handler, P., *J. Biol. Chem.*, **233**, 1581 (1958).
19. Beinert, H., Palmer, G., Cremona, T., and Singer, T. P., *Biochem. and Biophys. Research Communs.*, **12**, 432 (1963).
20. Remy, C. N., Richert, D. A., Doisy, R. J., Wells, I. C., and Westerfeld, W. W., *J. Biol. Chem.*, **217**, 293 (1955).
21. Ackerman, E., and Brill, A., *Biochim. et Biophys. Acta*, **56**, 397 (1962).
22. Fridovich, I., and Handler, P., *J. Biol. Chem.*, **236**, 1836 (1961).
23. Stauff, J., Schmidkunz, H., and Hartmann, G., *Nature*, **198**, 281 (1963).
24. Bray, R. C., in P. D. Boyer, H. Lardy, and K. Myrbäck (Editors), *The enzymes, Vol. 7*, Ed. 2, Part A, Academic Press, Inc., New York, 1963, pp. 533–556.

Reprinted from THE JOURNAL OF BIOLOGICAL CHEMISTRY, Vol. 238, 1963

α-Keto Acid Dehydrogenation Complexes

IV. RESOLUTION AND RECONSTITUTION OF THE *ESCHERICHIA COLI* PYRUVATE DEHYDROGENATION COMPLEX*

MASAHIKO KOIKE,† LESTER J. REED, AND WILLIAM R. CARROLL

From the Clayton Foundation Biochemical Institute and the Department of Chemistry, The University of Texas, Austin 12, Texas, and the Laboratory of Physical Biology, National Institute of Arthritis and Metabolic Diseases, National Institutes of Health, Bethesda 14, Maryland

(Received for publication, July 16, 1962)

Enzyme systems that catalyze the lipoic acid-mediated oxidative decarboxylation of pyruvate (Reaction 1) have been isolated from pigeon breast muscle (1, 2) and from *Escherichia coli* (3) as

$$\text{Pyruvate} + \text{Co}\bar{\text{A}}\text{-SH} + \text{DPN}^+ \rightarrow \text{acetyl-S-Co}\bar{\text{A}} + CO_2 + \text{DPNH} + \text{H}^+ \quad (1)$$

multienzyme complexes with molecular weights of approximately 4 and 4.8 million, respectively. Enzyme complexes that catalyze an analogous decomposition of α-ketoglutarate have been isolated from pig heart (4) and from *E. coli* (3). It has been presumed that as many as four separate enzymes are involved in overall Reaction 1, since the latter reaction apparently occurs in a sequence of four steps (5, 6) as shown in Equations 2 to 5, where the brackets indicate enzyme-bound compounds. However,

$$\text{Pyruvate} + [\text{thiamine-PP}] \rightarrow [\text{acetaldehyde-thiamine-PP}] + CO_2 \quad (2)$$

$$[\text{Acetaldehyde-thiamine-PP}] + [\text{lipS}_2]^1 \rightarrow [\text{acetyl-S-lipSH}] + [\text{thiamine-PP}] \quad (3)$$

$$[\text{Acetyl-S-lipSH}] + \text{Co}\bar{\text{A}}\text{-SH} \rightarrow \text{acetyl-S-Co}\bar{\text{A}} + [\text{lip(SH)}_2] \quad (4)$$

$$[\text{Lip(SH)}_2] + \text{DPN}^+ \xrightarrow{\text{(FAD)}} [\text{lipS}_2] + \text{DPNH} + \text{H}^+ \quad (5)$$

previous attempts to separate the individual enzymes and thereby confirm the proposed reaction sequence have met with limited success. No separation of the constituent enzymes of the mammalian pyruvate and α-ketoglutarate dehydrogenation complexes was obtained during purification (1, 4). Both the pyruvate and α-ketoglutarate dehydrogenation systems of *E. coli*, on the other hand, were easily separated into two fractions during purification (7, 8), Fractions A and B and A′ and B, respectively. Although complete separation was not achieved, it was apparent that Fractions A and A′ were rich in the enzymes catalyzing the decarboxylation, reductive acylation, and acyl transfer reactions (Equations 2 to 4) and that Fraction B was rich in dihydrolipoic dehydrogenase (Reaction 5) (8). Recently, Massey (9) achieved a separation of the pig heart α-ketoglutarate dehydrogenation complex into a flavoprotein (dihydrolipoic dehydrogenase) and a fraction similar in function to *E. coli* Fraction A′ by fractionation on calcium phosphate gel-cellulose in the presence of 2.5 M urea.

This paper describes a separation of the *E. coli* pyruvate dehydrogenation complex into three essential components: (*a*) pyruvic carboxylase (Reaction 2), (*b*) a component containing the bound lipoic acid and exhibiting dihydrolipoic transacetylase activity (Reactions 3 and 4), tentatively designated lipoic reductase-transacetylase, and (*c*) a flavoprotein, dihydrolipoic dehydrogenase (Reaction 5). Reassociation of the three isolated components to produce a large unit resembling the original complex in composition and enzymatic activities is demonstrated. Some of this work has been reported briefly (10).

EXPERIMENTAL PROCEDURE

Enzyme Preparations and Reagents—The pyruvate dehydrogenation complex was isolated essentially as described previously (3) from sonic extracts of *E. coli* (Crookes strain) cells that had been grown in the presence of DL-lipoic acid-S_2^{35}. Further purification of some preparations of the complex was achieved by fractionation with solid ammonium sulfate. The complex precipitated between 0.40 and 0.48 ammonium sulfate saturation. The preparations of the complex used in this investigation showed a specific activity of 800 to 900 in the pyruvate dismutation assay. Calcium phosphate gel suspended on Whatman standard grade cellulose powder was prepared as described by Price and Greenfield (11). The sources of other materials are described in previous papers (3, 12, 13). All solutions were prepared with deionized, glass-distilled water.

Assay Procedures—Pyruvate dismutation activity, based on Reaction 6, was determined as described previously (12). Specific activity is expressed as micromoles of acetyl phosphate formed per hour per mg of protein. Dihydrolipoic transacetylase and dihydrolipoic dehydrogenase activities, based on Reactions 7 and 8, respectively, were determined with substrate amounts of DL-dihydrolipoic acid essentially as described by Hager and

$$2\ \text{Pyruvate} + P_i \rightarrow \text{acetyl-P} + CO_2 + \text{lactate} \quad (6)$$

* This investigation was supported in part by a grant (RG-6590) from the National Institutes of Health, United States Public Health Service.

† Present address, Institute of Atomic Diseases, Nagasaki University School of Medicine, Nagasaki, Japan.

[1] The abbreviations used are: lipS_2, lipoic acid; acetyl-S-lipSH, *S*-acetyldihydrolipoic acid; lip(SH)_2, dihydrolipoic acid.

$$\text{Acetyl-P} + \text{lip(SH)}_2 \xrightarrow{\text{(CoA)}} \text{acetyl-S-lipSH} + \text{P}_i \quad (7)$$

$$\text{Lip(SH)}_2 + \text{pyruvate} \xrightarrow{\text{(DPN)}} \text{lipS}_2 + \text{lactate} \quad (8)$$

Gunsalus (8, 14). The latter assay was carried out at pH 7.0. Sulfhydryl groups were determined as described by Hager (14) and by Ellman (15). Specific activities are expressed, respectively, as micromoles of heat-stable thioester formed per hour per mg of protein and micromoles of dihydrolipoic acid oxidized per hour per mg of protein. The ferricyanide reduction assay for pyruvic carboxylase activity, based on Reaction 9, was carried out as described by Hager (14). Specific activity is expressed

$$CH_3COCO_2H + 2\,Fe(CN)_6^{3-} + H_2O \rightarrow CH_3CO_2H + CO_2 + 2\,Fe(CN)_6^{4-} + 2\,H^+ \quad (9)$$

as micromoles of pyruvate oxidized per hour per mg of protein. The amount of pyruvate dehydrogenation complex used in the various assays has been indicated previously (16). Protein was determined routinely by the phenol method of Lowry *et al.* (17) and, in certain cases, by the biuret method as described by Gornall, Bardawill, and David (18). The lipoic acid content of the protein fractions is based on radioactivity determinations performed as described previously (12). The lipoic acid content (calculated as (+)-lipoic acid) of several preparations also was determined, after hydrolysis, with lipoic acid-deficient *Streptococcus faecalis* cells (3, 19). Good correlation of the results was obtained (*cf.* (3)). The flavin content of the protein fractions was determined according to the method of Beinert and Page (20) by measuring the absorbancy of neutralized trichloroacetic acid extracts at 450 mμ before and after reduction with dithionite.

RESULTS

Resolution of Pyruvate Dehydrogenation Complex with Urea—In preliminary experiments, a preparation of the pyruvate dehydrogenation complex in 0.05 M potassium phosphate buffer, pH 7.5, was incubated at 0° with different concentrations of urea. Aliquots were removed at various time intervals and tested in the pyruvate dismutation assay. In the presence of 1.2 M and 2.5 M urea, the complex retained full activity for at least 4 hours. In 3.6 M urea, approximately 70% of the activity remained after 1 hour. In 5 M urea, only 10% of the activity remained after 1 hour. Under the latter conditions, approximately 90% of the pyruvic carboxylase activity (Reaction 9) and 50% of the dihydrolipoic transacetylase activity (Reaction 7) was destroyed, but the dihydrolipoic dehydrogenase activity (Reaction 8) was not affected. Examination of a solution of the complex in 5 M urea in a Spinco model E ultracentrifuge revealed the presence of several components, indicating a dissociation of the complex. However, the apparent instability of the pyruvic carboxylase component of the complex placed restrictions on the concentration of urea and period of contact that could be used in attempting to resolve the complex. After a number of trials, a satisfactory procedure was developed, involving fractionation on a calcium phosphate gel-cellulose column in the presence of 4 M urea (*cf.* (9)). A typical resolution is described below.

A slurry of calcium phosphate gel suspended on cellulose in 0.02 M phosphate buffer, pH 6.0, was poured into a column of 2.8-cm diameter. The slurry was allowed to pack by gravity to give a column approximately 3.9 cm in height, comprising about 24 ml of gel-cellulose.[2] All subsequent operations were carried out in a cold room. A solution of the pyruvate dehydrogenation complex containing 50 mg of protein in 1.1 ml of 0.05 M phosphate buffer, pH 7.0, was applied to the column. After adsorption of the protein, 3.5 ml of a solution containing 4 M urea, 2% ammonium sulfate, and 0.1 M phosphate buffer, pH 7.5, were passed into the column. This was followed by 2.5 ml of a solution containing 4 M urea,[3] 1% ammonium sulfate, and 0.1 M phosphate buffer, pH 7.5. The column was then washed with approximately 70 ml of a solution of 1% ammonium sulfate in 0.1 M phosphate buffer, pH 7.5. A colorless protein fraction was eluted, leaving a broad yellow, fluorescent band on the column. The latter band was then eluted with a solution of 4% ammonium sulfate in 0.1 M phosphate buffer, pH 7.5. The colorless fraction was dialyzed with stirring against 0.05 M phosphate buffer, pH 7.0, as soon as possible after elution. The dialyzed solution, comprising 39 mg of protein in a volume of 11 ml, was fractionated with solid ammonium sulfate. The active protein (35 mg) precipitated sharply between 0.40 and 0.44 saturation. The yellow fraction, comprising 8 mg of protein in a volume of 12 ml, was fractionated with ammonium sulfate to remove unresolved pyruvate dehydrogenation complex. The latter precipitated below 0.50 saturation. The uncomplexed flavoprotein was collected between 0.50 and 0.80 saturation. Recovery of protein in the latter fraction was 4 mg.

The composition and enzymatic activities of the colorless fraction and the flavoprotein are shown in Table I.[4] The colorless fraction contained all of the protein-bound lipoic acid of the

TABLE I

Composition and activities of pyruvate dehydrogenation complex, components from urea resolution, and reconstituted complex

Sample	Bound lipoic acid	Bound flavin	Specific activities: Dismutation	Ferricyanide reduction	Dihydrolipoic transacetylase	Dihydrolipoic dehydrogenase
	mμmoles/mg protein		*μmoles/hr/mg protein*	*μmoles/hr/mg protein*	*μmoles/hr/mg protein*	
Pyruvate dehydrogenation complex	12.2	2.8	900	11.8	123	1240
Colorless fraction	14.9	0.03	16	14.2	150	36
Flavoprotein	0	17.7	0	0	0	8400
Reconstituted complex	12.8	3.4	890	11.6	120	1480

[2] The flow rate of the column should be approximately 15 to 20 ml per hour. Use of columns with a slower flow rate resulted in loss of pyruvic carboxylase activity, apparently because of prolonged contact with urea.

[3] The total volume of 4 M urea solution used should be approximately one-fourth the volume of gel-cellulose comprising the column.

[4] No attempt was made to determine the thiamine-PP content of the pyruvate dehydrogenation complex and the colorless fraction, since data reported previously (3, 21) indicated that loss of this coenzyme occurred during isolation of the complex. The colorless fraction showed almost complete dependence on added thiamine-PP in the pyruvate dismutation and ferricyanide reduction assays. The dismutation assay was carried out in the presence of excess flavoprotein (14).

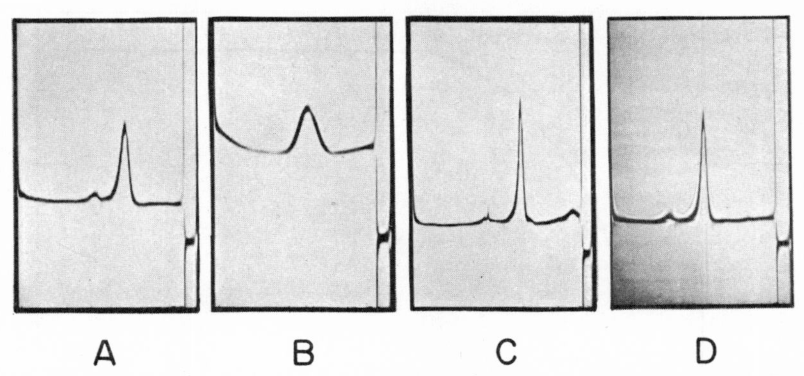

FIG. 1. Ultracentrifuge schlieren patterns obtained with the pyruvate dehydrogenation complex, the components from urea resolution of the complex, and the reconstituted complex. *Pattern A*, colorless fraction after 19 minutes at 42,040 r.p.m.; *B*, flavoprotein after 90 minutes at 59,780 r.p.m.; *C*, reconstituted complex after 16 minutes at 42,040 r.p.m.; *D*, pyruvate dehydrogenation complex after 20 minutes at 42,040 r.p.m. Sedimentation proceeds from right to left. Protein concentrations were, respectively, 5.0, 3.0, 5.0, and 4.1 mg per ml of 0.05 M potassium phosphate buffer, pH 7.0; temperature, 5°. Sedimentation coefficients given in the text correspond to the major component in each picture.

pyruvate dehydrogenation complex, catalyzed the oxidation of pyruvate with ferricyanide as electron acceptor (Reaction 9), and exhibited dihydrolipoic transacetylase activity (Reaction 7). The flavoprotein contained all of the flavin of the pyruvate dehydrogenation complex, and its dihydrolipoic dehydrogenase activity (Reaction 8) per mole of flavin was essentially the same as that found with the intact complex. These results indicate that the colorless fraction contains the enzymes that catalyze Reactions 2 through 4 of the overall sequence, and that the flavoprotein catalyzes Reaction 5.

The sedimentation pattern of the colorless fraction determined in a Spinco model E ultracentrifuge showed (Fig. 1, *Pattern A*) a major component ($s_{20,w}$ = 49.7 S) comprising approximately 90% of the total protein. The sedimentation pattern of the flavoprotein (Fig. 1, *Pattern B*) showed a single component ($s_{20,w}$ = 6.3 S). These results indicate that the colorless fraction is itself a complex of the enzymes catalyzing Reactions 2 through 4 of the overall sequence. Additional evidence supporting this conclusion was obtained by resolution of the colorless complex at pH 9.5 as described below. Although the molecular weight of the colorless complex is not yet known, its sedimentation coefficient indicates that it is a large unit. On the basis of data such as those given in Table I and the recovery of protein in the resolution experiments, it is estimated that the colorless fraction comprises approximately 84% of the protein of the pyruvate dehydrogenation complex. The remainder (16%) is the flavoprotein.

Reconstitution of Pyruvate Dehydrogenation Complex with Components from Urea Resolution—Neither the colorless fraction nor the flavoprotein alone was capable of catalyzing the DPN-linked oxidation of pyruvate (Table I).[5] When the two components were mixed, however, this activity was reconstituted. Fig. 2 shows that in the presence of a constant amount of the colorless fraction, the rate of pyruvate oxidation is directly proportional to the amount of flavoprotein added until saturation is obtained.

Direct evidence that the colorless fraction and the flavoprotein combined to reconstitute the pyruvate dehydrogenation complex was obtained in the experiments described below. Examination of a mixture of the colorless fraction and the flavoprotein (ratio of 4:1 by weight) in a Spinco model E ultracentrifuge revealed the virtual absence of peaks corresponding to the two individual components. Instead, a major, faster moving peak ($s_{20,w}$ = 56.8 S) was observed (Fig. 1, *Pattern C*), with which the boundary of the yellow color of the flavoprotein was associated. In a second experiment, a solution of 10.0 mg of the colorless fraction in 0.74 ml of 0.05 M phosphate buffer, pH 7.0, was mixed with a solution of 2.54 mg of the flavoprotein in 0.26 ml of the same buffer. The mixture was kept in an ice bath for 10 minutes, then diluted with 3 ml of 0.05 M phosphate buffer, pH 7.0, and centrifuged for $2\frac{1}{2}$ hours at 173,000 × g in the No. SW-39L rotor of a Spinco model L ultracentrifuge. A yellow pellet was obtained, which was dissolved in 1.0 ml of 0.05 M phosphate buffer, pH 7.0. The solution was clarified by centrifugation for 10 minutes at 20,000 r.p.m. in the No. SW-39L rotor. This solution contained 9 mg of protein. Its sedimentation pattern was very similar to that shown in Fig. 1, *Pattern C*, with the exception that the minor peak near the meniscus, due to uncomplexed flavoprotein, was absent. The boundary of the yellow color was associated with the main peak ($s_{20,w}$ = 57.9 S). The sedimentation pattern of the original complex is shown in Fig. 1, *Pattern D*, for comparison. The sedimentation coefficient ($s_{20,w}$) of the major component was 57.7 S. The composition and enzymatic activities of the reconstituted complex were very similar to those of the original complex (Table I).

Flavin Content and Molecular Weight of Flavoprotein Component—Preparations of the flavoprotein obtained by urea resolution of the pyruvate dehydrogenation complex showed a flavin content of 17 to 18 mμmoles per mg of protein, corresponding to a minimal molecular weight of 56,000 to 59,000. The flavin had been identified previously (22) as FAD. A diffusion study of the flavoprotein was carried out at 6.3° in a Rayleigh synthetic

[5] The colorless fraction showed approximately 2% of the pyruvate dismutation activity exhibited by the intact complex, apparently because of slight contamination with the flavoprotein. However, other preparations of the colorless fraction have been obtained that were completely inactive in the dismutation assay.

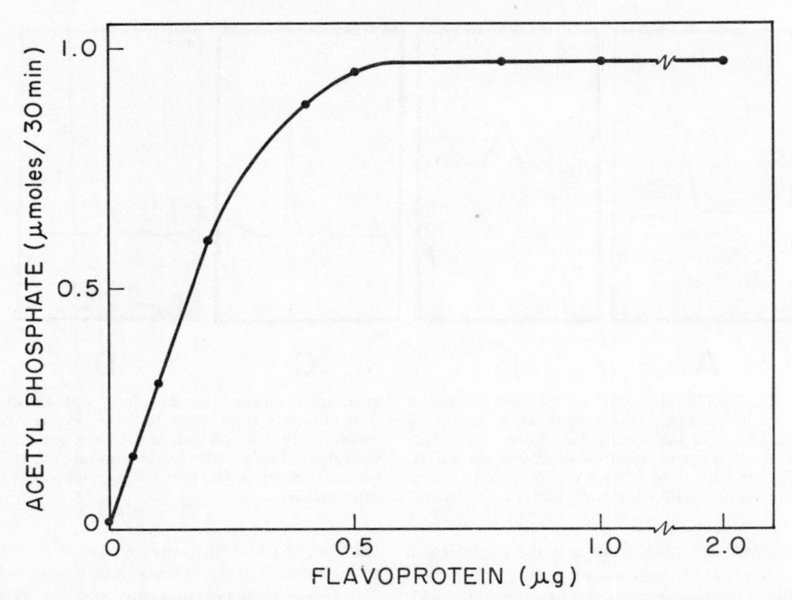

FIG. 2. Influence of flavoprotein concentration on the rate of pyruvate oxidation by the colorless fraction. Two micrograms of the colorless fraction and the indicated amounts of flavoprotein were mixed in a total volume ot 0.2 ml of 0.02 M phosphate buffer, pH 7.0, and the solutions were allowed to stand at 0° for 10 minutes. The other components required in the dismutation assay were then added to give a final volume of 1.0 ml, and the assay was carried out as described previously (12).

boundary cell in a Spinco model E ultracentrifuge. The solution used contained 8.0 mg of protein per ml of 0.05 M phosphate buffer, pH 7.0, and had been dialyzed with stirring against the same buffer for 9 hours. The rotor speed was set at 4000 r.p.m., and photographs of the fringes of the diffusing boundary were taken at intervals over a period of 5 hours. The fringe spacings were measured on a microcomparator, and the diffusion coefficient was calculated by the method of Longsworth (23). The corrected diffusion coefficient, $D_{20,w}$, of this preparation was 5.01×10^{-7} cm² sec⁻¹. The sedimentation pattern of this preparation showed a single component, $s_{20,w} = 6.24$ S. From these data and an assumed partial specific volume of 0.73 ml per g, the molecular weight was calculated to be 112,000. It is thus apparent that the flavoprotein contains 2 molecules of FAD per molecule of enzyme.

Resolution of Pyruvate Dehydrogenation Complex at pH 9.5—When the pyruvate dehydrogenation complex was allowed to stand in contact at 0° with an ethanolamine-phosphate buffer, pH 9.5, which was 0.02 M with respect to ethanolamine and approximately 0.025 M with respect to potassium phosphate,[6] only 15% of the DPN-linked pyruvate dehydrogenation activity was destroyed in 1 hour. Examination of a freshly prepared mixture in the analytical ultracentrifuge showed two components (Fig. 3) with $s_{20,w}$ values of 29.6 S and 6.6 S, respectively. The yellow color was associated with the faster moving component. A portion of the mixture was dialyzed with stirring for 5 hours against two changes of 0.05 M phosphate buffer, pH 7.0. Examination of the dialyzed preparation in the ultracentrifuge showed only one major component with a sedimentation coefficient close to that of the untreated complex. The specific activity of the dialyzed preparation in the pyruvate dismutation assay (Reaction 6) was 95% of that obtained with the untreated complex. These results indicated a reversible dissociation of the complex into two components. This dissociation did not occur below pH 9.0. At pH values above 9.5, the loss of dismutation activity increased. That pH rather than ethanolamine was responsible for the dissociation was indicated by the observation that two

[6] The ethanolamine-phosphate buffer was prepared by mixing 200 ml of 0.04 M ethanolamine and 160 ml of 0.05 M potassium phosphate buffer, pH 7.0. The pH was adjusted to 9.5 with approximately 2 ml of 1 M phosphate buffer, pH 6, and the solution was diluted to 400 ml with water.

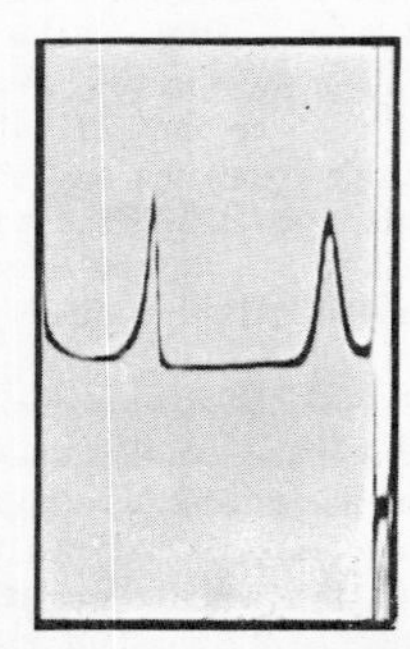

FIG. 3. Ultracentrifuge schlieren pattern obtained with the pyruvate dehydrogenation complex in ethanolamine-phosphate buffer, pH 9.5, after 27 minutes at 59,780 r.p.m.; protein concentration, 6.4 mg per ml; temperature, 6°. Sedimentation proceeds from right to left. The sedimentation pattern of the complex in 0.05 M phosphate buffer, pH 7.0, is shown in Fig. 1, *Pattern D*.

Table II

Composition and activities of components from resolution of pyruvate dehydrogenation complex at pH 9.5

Component	Bound lipoic acid	Bound flavin	Specific activities			
			Dismutation	Ferricyanide reduction	Dihydrolipoic transacetylase	Dihydrolipoic dehydrogenase
	mμmoles/mg protein		*μmoles/hr/mg protein*	*μmoles/hr/mg protein*	*μmoles/hr/mg protein*	
Pyruvic carboxylase	0	0	0	25	0	0
Yellow fraction	24.6	4.9	20	0.1	240	2200

components with $s_{20,w}$ values close to those given above were produced when a solution of the complex in 0.02 M phosphate buffer, pH 7.0, was adjusted to pH 9.5 with a dilute potassium hydroxide solution. However, the ethanolamine-phosphate buffer was retained as a convenient means of maintaining the desired pH.

Good separation of the two components was achieved by fractionation on calcium phosphate gel-cellulose in the presence of the ethanolamine-phosphate buffer. A typical resolution is described below. A column (2.8 × 4.0 cm) of calcium phosphate gel suspended on cellulose was prepared as described above. The column was washed with the ethanolamine-phosphate buffer until the pH of the effluent was approximately 9.5. All subsequent operations were carried out in a cold room. A solution of 63 mg of the pyruvate dehydrogenation complex in 5 ml of the ethanolamine-phosphate buffer was applied to the column. The column was then washed with approximately 73 ml of the ethanolamine-phosphate buffer. A colorless protein fraction was eluted, leaving a broad yellow, fluorescent band on the column. The latter band was eluted with a solution of 4% ammonium sulfate in 0.1 M phosphate buffer, pH 7.5. The two fractions were dialyzed as soon as possible after elution

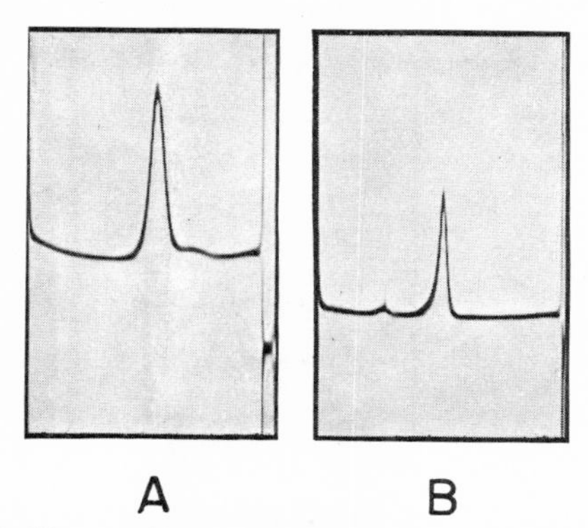

Fig. 4. Ultracentrifuge schlieren patterns obtained with the components from resolution of the pyruvate dehydrogenation complex at pH 9.5. *Pattern A*, pyruvic carboxylase after 67 minutes at 59,780 r.p.m.; *B*, yellow fraction after 52 minutes at 42,040 r.p.m. Sedimentation proceeds from right to left. Protein concentrations were, respectively, 5.2 and 4.0 mg per ml of 0.05 M phosphate buffer, pH 7.0; temperature, 6°.

against 0.05 M phosphate buffer, pH 7.0. The colorless fraction, comprising 29 mg of protein in a volume of 10 ml, was fractionated with solid ammonium sulfate. The active protein (26 mg) precipitated between 0.50 and 0.60 saturation. The yellow fraction, comprising 26 mg of protein in a volume of 6 ml, was concentrated by centrifugation for 4 hours at 173,000 × g in the No. SW-39L rotor of a Spinco model L ultracentrifuge. The yellow pellet obtained (22 mg) was dissolved in 0.05 M phosphate buffer, pH 7.0.

The composition and enzymatic activities of the two fractions are shown in Table II. The colorless fraction did not contain protein-bound lipoic acid. The only enzymatic activity exhibited by this fraction was the ferricyanide-linked oxidation of pyruvate (Reaction 9). Complete dependence on added thiamine-PP was exhibited in this reaction (*cf.* footnote 4). The specific activity of the colorless fraction was approximately twice that of the original complex. The sedimentation pattern of the colorless fraction showed (Fig. 4, *Pattern A*) a major component ($s_{20,w}$ = 9.2 S)[7] comprising over 90% of the total protein. These data indicate that the colorless fraction contains the pyruvic carboxylase component of the pyruvate dehydrogenation complex, which catalyzes Reaction 2 of the overall sequence, and that this component comprises approximately 50% of the protein of the complex.

The yellow fraction contained protein-bound lipoic acid and flavin and exhibited dihydrolipoic transacetylase and dihydrolipoic dehydrogenase activities. The sedimentation pattern of this fraction showed (Fig. 4, *Pattern B*) a major component ($s_{20,w}$ = 26.8 S),[8] comprising approximately 90% of the total protein, with which the boundary of the yellow color was associated. These results indicate that the yellow fraction is a complex of the enzymes catalyzing Reactions 3 through 5 of the overall sequence. Additional evidence supporting this conclusion was obtained by resolution of the yellow fraction with urea as described below.

Reconstitution of Pyruvate Dehydrogenation Activity with Components from Resolution at pH 9.5—Neither the colorless fraction, *i.e.* pyruvic carboxylase, nor the yellow fraction alone was capable of catalyzing the DPN-linked oxidation of pyruvate (Table II). When the two fractions were mixed, however, this activity was reconstituted. Fig. 5 shows that in the presence of a constant amount of the yellow fraction the rate of pyruvate oxidation is proportional to the amount of pyruvic carboxylase added until saturation is obtained.

Molecular Weight of Pyruvic Carboxylase—The diffusion coefficient of pyruvic carboxylase was determined at 4.9° in a Rayleigh synthetic boundary cell in the Spinco model E ultracentrifuge. The solution used contained 5.9 mg of protein per ml of 0.05 M phosphate buffer, pH 7.0, and had been dialyzed with stirring against the same buffer for 20 hours. The rotor speed was set at 5000 r.p.m. The corrected diffusion coefficient, $D_{20,w}$, of this

[7] The sedimentation pattern of the isolated pyruvic carboxylase (4.6 mg per ml) in ethanolamine-phosphate buffer, pH 9.5, showed a single component with a sedimentation coefficient ($s_{20,w}$) equal to 6.2 S. Apparently there is a significant difference in the hydrodynamic properties of the carboxylase in the two buffer systems.

[8] The sedimentation coefficients ($s_{20,w}$) of different preparations of the yellow complex varied between approximately 27 S and 32 S. This variation is due largely to differences in flavoprotein content of the preparations, which in turn are due to losses of flavoprotein incurred in purification of the pyruvate dehydrogenation complex and in the resolution on gel-cellulose at pH 9.5.

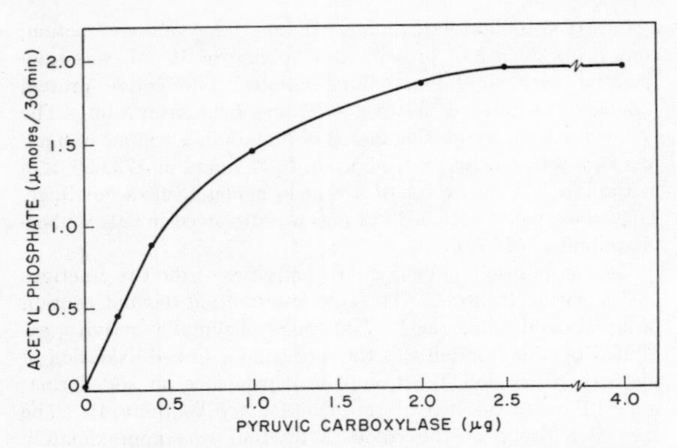

FIG. 5. Influence of pyruvic carboxylase concentration on the rate of pyruvate oxidation by the yellow fraction. Two micrograms of the yellow fraction were used. Other conditions were as in Fig. 2.

preparation was 4.54×10^{-7} cm^2 sec^{-1}. The $s_{20,w}$ value was found to be 9.24 S. A second determination at a protein concentration of 1.5 mg per ml gave a value, $s_{20,w}$ = 9.45 S, indicating little concentration dependence of the sedimentation coefficient. The molecular weight calculated from the sedimentation and diffusion coefficients, with the partial specific volume of the enzyme assumed to be 0.73 ml per g, was 183,000.

Resolution of Yellow Complex with Urea—The yellow fraction obtained by resolution of the pyruvate dehydrogenation complex at pH 9.5 (*cf.* Table II and Fig. 4, *Pattern B*) was separated into two components, a colorless fraction and the flavoprotein, by fractionation on calcium phosphate gel-cellulose in the presence of 4 M urea essentially as described above. From 25 mg of the yellow complex there were obtained 15 mg of the colorless fraction and 4 mg of the flavoprotein.

The colorless fraction contained all of the protein-bound lipoic acid and exhibited dihydrolipoic transacetylase activity (Table III). Its sedimentation pattern showed (Fig. 6) a major component with a sedimentation coefficient ($s_{20,w}$) equal to 21.2 S. This component is tentatively designated lipoic reductase-transacetylase. It comprises approximately 34% of the protein of the pyruvate dehydrogenation complex and apparently catalyzes Reactions 3 and 4 of the overall sequence. Attempts to separate this large unit into active subunits have been unsuccessful thus far.

The sedimentation pattern of the flavoprotein showed a single component ($s_{20,w}$ = 6.3 S). Its flavin content and dihydrolipoic dehydrogenase activity (Table III), together with its sedimentation characteristics, indicate that it is identical with the flavoprotein obtained by urea resolution of the pyruvate dehydrogenation complex.

Resolution of Colorless Complex at pH 9.5—The colorless fraction obtained by urea resolution of the pyruvate dehydrogenation complex (*cf.* Table I and Fig. 1, *Pattern A*) was treated at 0° with ethanolamine-phosphate buffer, pH 9.5. Examination of the solution in the analytical ultracentrifuge revealed the presence of two components (Fig. 7) with $s_{20,w}$ values of 22.6 S and 5.7 S, respectively. Separation of the two components was achieved by fractionation on a column of calcium phosphate gel-cellulose in the presence of the ethanolamine-phosphate buffer. A column of 2.8-cm diameter, comprising approximately 25 ml of gel-cellulose, was washed with ethanolamine-phosphate buffer, pH 9.5, as described above. A solution of 52 mg of the colorless fraction in 6 ml of ethanolamine-phosphate buffer was applied to the column. The column was then washed with approximately 80 ml of the same buffer to elute the slower moving component shown in Fig. 7. The faster moving component was then eluted with a solution of 4% ammonium sulfate in 0.1 M phosphate buffer, pH 7.5. The two fractions were dialyzed as soon as

TABLE III

Composition and activities of components from urea resolution of yellow complex

Component	Bound lipoic acid	Bound flavin	Specific activities: Dihydrolipoic transacetylase	Specific activities: Dihydrolipoic dehydrogenase
	mμmoles/mg protein		*μmoles/hr/mg protein*	
Lipoic reductase-transacetylase	33	0	290	0
Flavoprotein	0	18	0	8500

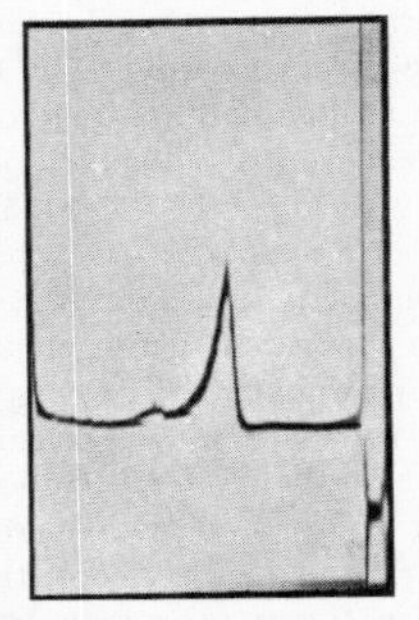

FIG. 6. Ultracentrifuge schlieren pattern obtained with a preparation of lipoic reductase-transacetylase after 52 minutes at 42,040 r.p.m. Sedimentation proceeds from right to left. The protein concentration was 3.8 mg per ml of 0.05 M phosphate buffer, pH 7.0; temperature, 5°.

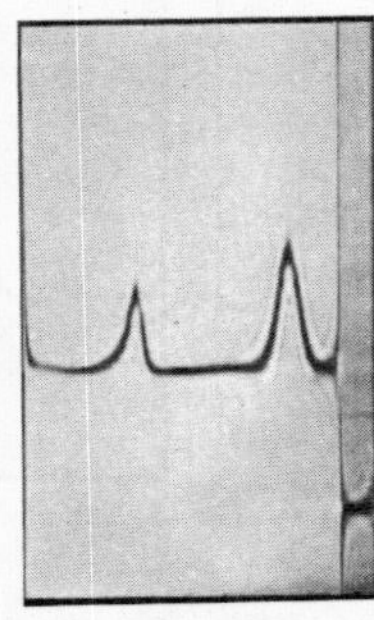

FIG. 7. Ultracentrifuge schlieren pattern obtained with the colorless complex in ethanolamine-phosphate buffer, pH 9.5, after 34 minutes at 59,780 r.p.m.; protein concentration, 5.0 mg per ml; temperature, 5°. Sedimentation proceeds from right to left. The sedimentation pattern of the colorless complex in 0.05 M phosphate buffer, pH 7.0, is shown in Fig. 1, *Pattern A*.

possible after elution against 0.05 M phosphate buffer, pH 7.0. Recovery of protein in the two fractions was 21 mg and 20 mg, respectively.

The first fraction eluted from the column did not contain bound lipoic acid, and the only enzymatic activity it exhibited was the ferricyanide-linked oxidation of pyruvate (Table IV). The active protein precipitated between 0.50 and 0.60 ammonium sulfate saturation. No significant change in specific activity was observed. Its sedimentation pattern was very similar to that shown in Fig. 4, *Pattern A*. The sedimentation coefficient ($s_{20,w}$) of the major component was 9.3 S. It is apparent from these data that this component is pyruvic carboxylase.

The second fraction eluted from the column contained all of the protein-bound lipoic acid and exhibited dihydrolipoic transacetylase activity (Table IV). It is apparent from these data that this fraction contained lipoic reductase-transacetylase.

TABLE IV

Composition and activities of components from resolution of colorless complex at pH 9.5

Component	Bound lipoic acid	Specific activities	
		Ferricyanide reduction	Dihydrolipoic transacetylase
	mμmoles/mg protein	*μmoles/hr/mg protein*	*μmoles/hr/mg protein*
Pyruvic carboxylase	0	24	0
Lipoic reductase-transacetylase	32	0	270

TABLE V

Reconstitution of pyruvate dehydrogenation activity

The individual components and combinations thereof were incubated at 0° for 10 minutes in a total volume of 0.2 ml of 0.02 M phosphate buffer, pH 7.0. Other components required for the pyruvate dismutation assay were then added in a final volume of 1.0 ml, and the assay was carried out as described previously (12). The amounts of protein used were: carboxylase, 3 μg; lipoic reductase-transacetylase, 2 μg; flavoprotein, 1 μg.

Components	Acetyl phosphate produced
	μmoles/30 min
Carboxylase	0
Lipoic reductase-transacetylase	0
Flavoprotein	0
Carboxylase + lipoic reductase-transacetylase	0.03
Carboxylase + flavoprotein	0
Lipoic reductase-transacetylase + flavoprotein	0.04
Carboxylase + lipoic reductase-transacetylase + flavoprotein	2.2

Reconstitution of Pyruvate Dehydrogenation Complex with Three Isolated Components—All three components, pyruvic carboxylase, lipoic reductase-transacetylase, and flavoprotein, were required to reconstitute the DPN-linked oxidation of pyruvate (Table V). Direct evidence that the three isolated components combined to reconstitute the pyruvate dehydrogenation complex was obtained in the experiments described below. A mixture of the three components was prepared which simulated the composition of the pyruvate dehydrogenation complex. The mixture contained 2.9 mg (45.2%) of carboxylase, 2.3 mg (36%) of lipoic reductase-transacetylase, and 1.2 mg (18.8%) of flavoprotein in a total volume of 1.4 ml of 0.05 M phosphate buffer, pH 7.0. Examination of the mixture in the analytical ultracentrifuge revealed the virtual absence of peaks corresponding to the three individual components. Instead, a major, faster moving peak ($s_{20,w}$ = 52.8 S) was observed (Fig. 8, *Pattern A*) with which the boundary of the yellow color of the flavoprotein was associated. The protein solution was recovered, combined with the remainder of the mixture, and centrifuged for 2½ hours at 173,000 × *g* in the No. SW-39L rotor of a Spinco model L ultracentrifuge. A yellow pellet was obtained and was dissolved in 0.05 M phosphate buffer, pH 7.0. Recovery of protein in the yellow pellet was approximately 84%. The composition and enzymatic activities of the yellow pellet (Pellet 1) are shown in Table VI. These values are very similar to those obtained with the original complex (*cf.* Table I). In a second experiment with different preparations of the three isolated components, 4.2 mg (50%) of carboxylase, 2.8

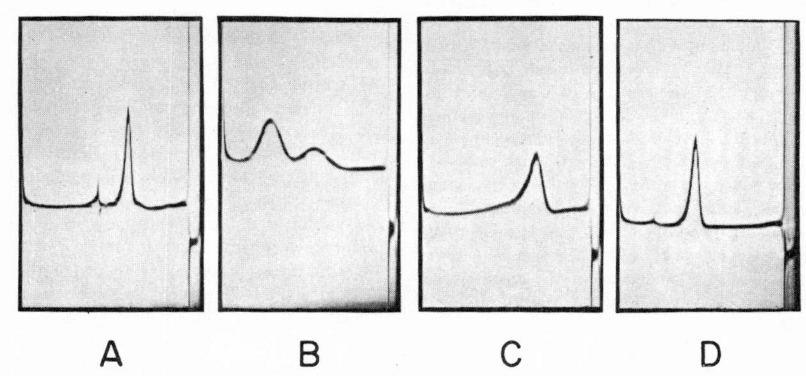

FIG. 8. Ultracentrifuge schlieren patterns obtained with mixtures of carboxylase, lipoic reductase-transacetylase, and flavoprotein. *Pattern A*, mixture of all three components after 19 minutes at 42,040 r.p.m.; *B*, mixture of carboxylase and flavoprotein after 206 minutes at 42,040 r.p.m.; *C*, mixture of lipoic reductase-transacetylase and carboxylase after 19 minutes at 42,040 r.p.m.; *D*, mixture of lipoic reductase-transacetylase and flavoprotein after 43 minutes at 42,040 r.p.m.; temperature, 6°. Protein concentrations were, respectively, 4.7, 5.0, 4.4, and 4.6 mg per ml of 0.05 M phosphate buffer, pH 7.0. The amounts of the individual components used are given in the text.

TABLE VI

Composition and activities of pyruvate dehydrogenation complex reconstituted from three isolated components

Sample	Bound lipoic acid	Bound flavin	Specific activities			
			Dismutation	Ferricyanide reduction	Dihydrolipoic transacetylase	Dihydrolipoic dehydrogenase
	mμmoles/mg protein		*μmoles/hr/mg protein*	*μmoles/hr/mg protein*	*μmoles/hr/mg protein*	
Pellet 1....	11.0	3.7	800	10.2	121	1620
Pellet 2.....	10.9	3.2	980	13.0	122	1450

mg (33.3%) of lipoic reductase-transacetylase, and 1.4 mg (16.7%) of flavoprotein were mixed in a total volume of 2.0 ml of 0.05 M phosphate buffer, pH 7.0. The sedimentation pattern of the preparation was similar to that shown in Fig. 8, *Pattern A*. The sedimentation coefficient ($s_{20,w}$) of the major component was 52.4 S. The protein solution was recovered, an excess of carboxylase and flavoprotein (0.67 mg and 0.50 mg, respectively) was added, and the mixture was centrifuged for 3 hours at 173,000 × g. The yellow pellet was dissolved in 0.05 M phosphate buffer, pH 7.0. Its sedimentation pattern was similar to that shown in Fig. 8, *Pattern A*. The sedimentation coefficient ($s_{20,w}$) of the major component was 52.0 S. The composition and enzymatic activities of the yellow pellet (Pellet 2) are shown in Table VI.

Other sedimentation studies indicated that the carboxylase and the flavoprotein did not combine with each other, but that each of these components did combine with the lipoic reductase-transacetylase fraction. Thus, the sedimentation pattern of a mixture of the carboxylase (2.9 mg) and the flavoprotein (1.2 mg) showed (Fig. 8, *Pattern B*) two components with sedimentation coefficients ($s_{20,w}$ = 9.5 S and 6.2 S, respectively) essentially the same as those found with the individual components. The boundary of the yellow color was associated with the slower moving component. The sedimentation pattern of a mixture of lipoic reductase-transacetylase (2.4 mg) and pyruvic carboxylase (3.6 mg) showed (Fig. 8, *Pattern C*) the virtual absence of peaks corresponding to the individual components and, instead, a major, faster moving component with a sedimentation coefficient ($s_{20,w}$) equal to 44.8 S. Similarly, the sedimentation pattern of a mixture of lipoic reductase-transacetylase (2.5 mg) and the flavoprotein (1.5 mg) did not show peaks corresponding to these two components. Instead, a major, faster moving component ($s_{20,w}$ = 32.7 S) was observed (Fig. 8, *Pattern D*) with which the boundary of the yellow color of the flavoprotein was associated.

That the carboxylase and the flavoprotein combined independently of each other with the lipoic reductase-transacetylase fraction is indicated by the observation that the specific activity (Reaction 6) and sedimentation coefficient of the reconstituted complex were not significantly affected by the order of mixing of the three components.

DISCUSSION

The results reported in this communication indicate that the *E. coli* pyruvate dehydrogenation complex is composed of at least three enzymatic components, pyruvic carboxylase, lipoic reductase-transacetylase, and dihydrolipoic dehydrogenase (flavoprotein). It is not yet clear whether lipoic reductase-transacetylase is a complex of two separate enzymes, a reductive acetylase (Reaction 3) and a transacetylase (Reaction 4), or whether both activities reside in the same enzyme. It is apparent from the sedimentation coefficient of lipoic reductase-transacetylase (21.2 S) that it is a large unit. It comprises approximately 34% of the protein of the pyruvate dehydrogenation complex, which indicates that its "molecular weight" is approximately 1.6 million. Its high lipoic acid content suggests that it is an aggregate of a small subunit. If it is assumed that there is 1 molecule of bound lipoic acid per subunit, the minimal molecular weight of the hypothetical subunit would be approximately 30,000. From the available data on the composition of the *E. coli* pyruvate dehydrogenation complex and the molecular weights of the complex and its constituent enzymes, the number of molecules of each constituent enzyme per molecule of complex has been calculated: for the carboxylase, approximately 12 to 14; for the lipoic reductase-transacetylase, 1; and for the flavoprotein, approximately 6 to 8.

The results demonstrate that pyruvic carboxylase and dihydrolipoic dehydrogenase can be dissociated selectively from the pyruvate dehydrogenation complex. The former enzyme is released at a pH above 9.0, leaving a complex of lipoic reductase-transacetylase and dihydrolipoic dehydrogenase. Dissociation of dihydrolipoic dehydrogenase from the pyruvate dehydrogenation complex is accomplished by fractionation on calcium phosphate gel-cellulose in the presence of 4 M urea. Release of dihydrolipoic dehydrogenase leaves a complex of lipoic reductase-transacetylase and pyruvic carboxylase. These observations are consistent with the results of the reconstitution experiments. The carboxylase and the dihydrolipoic dehydrogenase do not combine with each other, but these two components do combine independently of each other with the lipoic reductase-transacetylase component. These resolution and reconstitution experiments indicate that lipoic reductase-transacetylase possesses separate binding sites for the carboxylase and the flavoprotein. It may be inferred from this conclusion that the prosthetic groups of the latter two enzymes are oriented specifically with respect to the lipoic acid which is bound to the lipoic reductase-transacetylase component. It is visualized that the 12 to 14 molecules of carboxylase and 6 to 8 molecules of flavoprotein are embedded in a matrix of lipoic reductase-transacetylase.

The ratio of bound lipoic acid to FAD in the complex is 3:1 or 4:1. There is apparently an "excess" of bound lipoic acid. As shown in the present investigation, all of the lipoic acid in the complex is bound to the lipoic reductase-transacetylase component. That all of the lipoic acid participates in the oxidative decarboxylation of pyruvate (Reaction 1) is indicated by the observation[9] that the decrease in pyruvate dismutation activity produced by incubating the complex with lipoyl-X hydrolase (13, 16) is proportional to the amount of lipoic acid released from the complex. It is evident from the sequence of Reactions 3 through 5 that the lipoyl moiety undergoes a cyclic series of transformations, *i.e.* reductive acetylation, acetyl transfer, and reoxidation. These transformations apparently involve interaction of the lipoyl moiety, which is covalently bound to one enzyme (lipoic reductase-transacetylase), with "acetaldehyde-thiamine-PP," which is bound to a second enzyme (carboxylase),

[9] K. Suzuki and L. J. Reed, unpublished results.

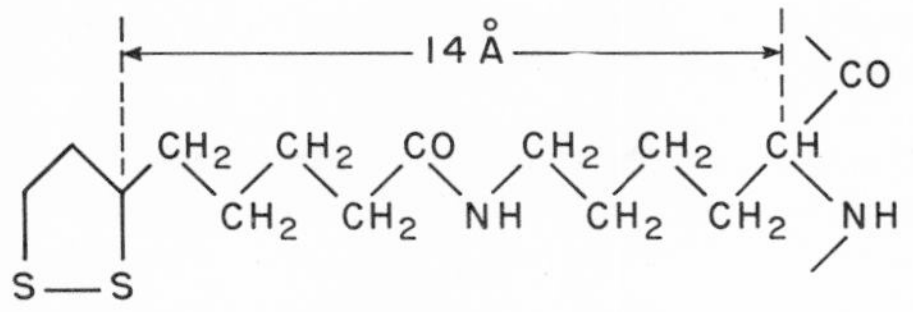

Fig. 9. N^{ϵ}-Lipoyllysine residue

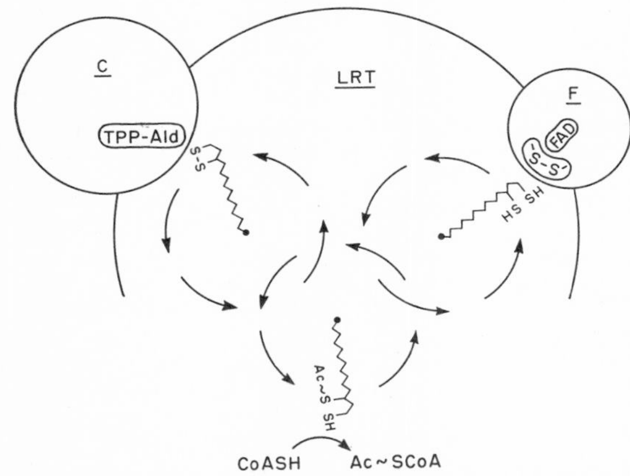

Fig. 10. A diagrammatic representation of possible interactions between several N^{ϵ}-lipoyllysyl moieties in the *E. coli* pyruvate dehydrogenation complex. The *arrows* describe the area covered by each lipoyllysyl moiety. *C*, *LRT*, and *F* represent, respectively, carboxylase, lipoic reductase-transacetylase, and flavoprotein.

and with the prosthetic group[10] of a third enzyme (dihydrolipoic dehydrogenase). The lipoyl moiety must also interact with CoA at an as yet unidentified site on the lipoic reductase-transacetylase. These interactions between prosthetic groups of separate enzymes occur within a complex in which there is apparently restricted movement of the individual enzymes and no dissociation of intermediates, with the exception of acetyl-CoA. Highly specific positioning of the three enzymatic components of the complex, and, by inference, the prosthetic groups of these components, is indicated by the reconstitution experiments discussed above. Granted this assumption, it still appears impossible for the three prosthetic groups (and the CoA) to be in close enough molecular proximity to interact. A possible solution to this enigma is provided by the discovery of Nawa *et al.* (26) that the lipoyl moieties are bound to the ε-amino groups of lysine residues. This attachment provides a flexible arm of approximately 14 A[11] for the reactive dithiolane ring (Fig. 9), conceivably permitting rotation of the latter between the bound "acetaldehyde-thiamine-PP" of the carboxylase, the site for acetyl transfer to CoA, and the prosthetic group of the flavoprotein. A similar suggestion has been made independently by Green, Bock, and Criddle (*cf.* (27)). It is also conceivable that the distances between the "acetaldehyde-thiamine-PP" of a carboxylase molecule, a site for acetyl transfer to CoA, and the prosthetic group of a flavoprotein molecule are too large to be encompassed by the rotation of a single lipoyllysyl moiety, necessitating interaction between several such moieties (Fig. 10). This possibility could account for the "excess" of bound lipoic acid in the complex. Interaction of the lipoyllysyl moieties is visualized as comprising thiol-disulfide interchange and acetyl transfer as illustrated in Equation 10 (*cf.* (13, 28)).

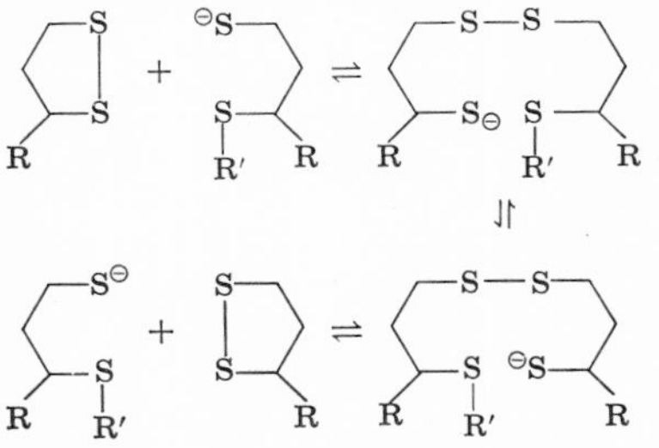

(10)

$R = (CH_2)_4CO$—enzyme; $R' = H$ or CH_3CO

These studies provide strong support for the concept of multienzyme complexes propounded by Green (*cf.* (29)). The pyruvate dehydrogenation complex appears to be "an organized mosaic of enzymes" in which each of the component enzymes is uniquely located to permit efficient implementation of a consecutive reaction sequence. Electron microscope studies of the pyruvate dehydrogenation complex and subunits thereof which are now in progress should shed further light on its structural organization.

Gounaris and Hager (30) have independently reported a requirement for three partially purified protein fractions to reconstitute the *E. coli* DPN-linked pyruvate dehydrogenation activity. Two fractions, enriched in dihydrolipoic transacetylase activity and dihydrolipoic dehydrogenase activity, respectively, were obtained from extracts of an *E. coli* acetate-requiring mutant which apparently lacks pyruvic carboxylase. A third fraction containing the latter enzyme was obtained from extracts of wild-type *E. coli*. The data of Gounaris and Hager do not permit a decision as to whether or not separate enzymes catalyze Reactions 3 and 4.

Preparations of dihydrolipoic dehydrogenase obtained from extracts of *E. coli* by modification of the procedure of Hager and Gunsalus (8, 14) showed an FAD content of 13 to 14 mμmoles per mg of protein (22), which corresponds to a minimal molecular weight of 72,000 to 77,000. Ultracentrifugation experiments indicated a homogeneously sedimenting boundary, and the molecular weight was found to be approximately 88,000, suggesting the presence of 1 molecule of FAD per molecule of enzyme. Preparations of dihydrolipoic dehydrogenase obtained in the present investigation by urea resolution of the highly purified pyruvate dehydrogenation complex showed an FAD content of 17 to 18 mμmoles per mg of protein, which corresponds to a

[10] The pig heart dihydrolipoic dehydrogenase has been shown to contain a reactive disulfide group which participates, in addition to FAD, in Reaction 5 (24, 25). *E. coli* dihydrolipoic dehydrogenase is inhibited by preincubation with DPNH and arsenite or with DPNH and cadmium ion. Dithiols are more effective than monothiols in reversing the inhibition. These observations (J. Matthews and L. J. Reed, unpublished results) indicate that the *E. coli* enzyme also contains a reactive disulfide group.

[11] The authors are indebted to Dr. Rowland Pettit, Department of Chemistry, The University of Texas, for this calculation.

minimal molecular weight of 56,000 to 59,000. This value is incompatible with the older value of 88,000. The sedimentation coefficients of the preparations of the flavoprotein obtained by the two procedures were in good agreement (approximately 6.3 S). The lower flavin content of the older preparations appeared to be due to a loss of flavin during the isolation procedure. It was suspected that the previously reported value of the diffusion coefficient ($D_{20,w}$ = 6.7 × 10^{-7} cm^2 sec^{-1}), which had been determined by using a synthetic boundary cell and schlieren optics in a Spinco model E ultracentrifuge, was too high. The diffusion coefficient of a recent preparation of the flavoprotein was then determined in a Rayleigh synthetic boundary cell and was found to be lower than the value reported previously. The value of the molecular weight obtained in the present investigation is 112,000, which correponds to 2 molecules of FAD per molecule of enzyme.

SUMMARY

1. The *Escherichia coli* pyruvate dehydrogenation complex was separated into the flavoprotein component (dihydrolipoic dehydrogenase) and a complex of lipoic reductase-transacetylase and pyruvic carboxylase by fractionation on calcium phosphate gel-cellulose in the presence of 4 M urea. These two components reassociated when mixed to produce a unit resembling the original complex in composition, enzymatic activities, and sedimentation characteristics. The complex of lipoic reductase-transacetylase and carboxylase was separated into its individual components by fractionation on gel-cellulose at pH 9.5.

2. The pyruvate dehydrogenation complex was separated into pyruvic carboxylase and a complex of lipoic reductase-transacetylase and dihydrolipoic dehydrogenase by fractionation on gel-cellulose at pH 9.5. The latter complex was separated into lipoic reductase-transacetylase and dihydrolipoic dehydrogenase by fractionation on gel-cellulose in the presence of 4 M urea.

3. The molecular weight of pyruvic carboxylase, calculated from sedimentation and diffusion measurements, is approximately 183,000.

4. The flavin (flavin adenine dinucleotide) content of dihydrolipoic dehydrogenase isolated from the complex is 17 to 18 mμmoles per mg of protein, which corresponds to a minimal molecular weight of 56,000 to 59,000. The molecular weight calculated from sedimentation and diffusion measurements is approximately 112,000, which indicates 2 molecules of flavin per molecule of enzyme.

5. The three isolated components, pyruvic carboxylase, lipoic reductase-transacetylase, and dihydrolipoic dehydrogenase, combine when mixed to produce a large unit resembling the original complex in composition and enzymatic activities. Lipoic reductase-transacetylase appears to possess separate binding sites for pyruvic carboxylase and dihydrolipoic dehydrogenase.

6. The structural organization of the pyruvate dehydrogenation complex is discussed in the light of the resolution and reconstitution experiments.

Acknowledgment—We wish to thank Mrs. Elizabeth Thompson for excellent technical assistance.

REFERENCES

1. Jagannathan, V., and Schweet, R. S., *J. Biol. Chem.*, **196**, 551 (1952).
2. Schweet, R. S., Katchman, B., Bock, R. M., and Jagannathan, V., *J. Biol. Chem.*, **196**, 563 (1952).
3. Koike, M., Reed, L. J., and Carroll, W. R., *J. Biol. Chem.*, **235**, 1924 (1960).
4. Sanadi, D. R., Littlefield, J. W., and Bock, R. M., *J. Biol. Chem.*, **197**, 851 (1952).
5. Gunsalus, I. C., in W. D. McElroy and B. Glass (Editors), *The mechanism of enzyme action*, Johns Hopkins Press, Baltimore, 1954, p. 545.
6. Reed, L. J., in P. D. Boyer, H. Lardy, and K. Myrbäck (Editors), *The enzymes, Vol. III*, Academic Press, Inc., New York, 1960, p. 195.
7. Korkes, S., del Campillo, A., Gunsalus, I. C., and Ochoa, S., *J. Biol. Chem.*, **193**, 721 (1951).
8. Hager, L. P., and Gunsalus, I. C., *J. Am. Chem. Soc.*, **75**, 5767 (1953).
9. Massey, V., *Biochim. et Biophys. Acta*, **38**, 447 (1960).
10. Koike, M., and Reed, L. J., *J. Biol. Chem.*, **236**, PC33 (1961).
11. Price, V. E., and Greenfield, R. E., *J. Biol. Chem.*, **209**, 363 (1954).
12. Reed, L. J., Leach, F. R., and Koike, M., *J. Biol. Chem.*, **232**, 123 (1958).
13. Reed, L. J., Koike, M., Levitch, M. E., and Leach, F. R., *J. Biol. Chem.*, **232**, 143 (1958).
14. Hager, L. P., Ph.D. Dissertation, University of Illinois, 1953.
15. Ellman, G. L., *Arch. Biochem. Biophys.*, **82**, 70 (1959).
16. Koike, M., and Reed, L. J., *J. Biol. Chem.*, **235**, 1931 (1960).
17. Lowry, O. H., Rosebrough, N. J., Farr, A. L., and Randall, R. J., *J. Biol. Chem.*, **193**, 265 (1951).
18. Gornall, A. G., Bardawill, C. J., and David, M. M., *J. Biol. Chem.*, **177**, 751 (1949).
19. Gunsalus, I. C., Dolin, M. I., and Struglia, L., *J. Biol. Chem.*, **194**, 849 (1952).
20. Beinert, H., and Page, E., *J. Biol. Chem.*, **225**, 479 (1957).
21. Das, M. L., Koike, M., and Reed, L. J., *Proc. Natl. Acad. Sci. U. S.*, **47**, 753 (1961).
22. Koike, M., Shah, P. C., and Reed, L. J., *J. Biol. Chem.*, **235**, 1939 (1960).
23. Longsworth, L. G., *J. Am. Chem. Soc.*, **74**, 4155 (1952).
24. Massey, V., and Veeger, C., *Biochim. et Biophys. Acta*, **48**, 33 (1961).
25. Searls, R. L., Peters, J. M., and Sanadi, D. R., *J. Biol. Chem.*, **236**, 2317 (1961).
26. Nawa, H., Brady, W. T., Koike, M., and Reed, L. J., *J. Am. Chem. Soc.*, **82**, 896 (1960).
27. Green, D. E., and Oda, T., *J. Biochem. (Tokyo)*, **49**, 742 (1961).
28. Goldman, D. S., *Biochim. et Biophys. Acta*, **45**, 279 (1960).
29. Green, D. E., *The Harvey Lectures, Series 52*, Academic Press, Inc., New York, 1958, p. 177.
30. Gounaris, A. D., and Hager, L. P., *J. Biol. Chem.*, **236**, 1013 (1961).

Reprinted from THE JOURNAL OF BIOLOGICAL CHEMISTRY, Vol. 237, 1962

Side-chain Interactions Governing the Pairing of Half-cystine Residues in Ribonuclease*

EDGAR HABER† AND CHRISTIAN B. ANFINSEN

From the Laboratory of Cellular Physiology and Metabolism, National Heart Institute, National Institutes of Health, Public Health Service, United States Department of Health, Education, and Welfare, Bethesda 14, Maryland

(Received for publication, January 29, 1962)

Fully reduced ribonuclease (RNase), devoid of demonstrable secondary or tertiary structure, may be oxidized by molecular oxygen to yield a product which is indistinguishable by physical measurements or enzymatic activity from the native enzyme (2–4). Since reduced RNase contains eight sulfhydryl groups, 105 possible arrangements with four disulfide bridges may occur (5, 6), yet the native enzyme can be recovered in nearly quantitative yields after such oxidations. It is apparent, then, that certain interactions among elements of the primary structure must serve as guides for the unique pairing of sulfhydryl groups. An attempt has been made to study these interactions by carrying out oxidations of reduced RNase in the presence of various reagents known to influence inter- or intramolecular bonding. Enzymatically inactive derivatives were produced, quite similar to native RNase in physical properties, which could, under appropriate circumstances, be rearranged to the native enzyme. These materials are believed to be a mixture representing some, if not all, of the possible isomeric configurations of disulfide bonding in the molecule. The observation that this mixture of materials can be readily converted to the native structure is taken as evidence that the unique secondary and tertiary structure of RNase is, thermodynamically, the most stable configuration.

EXPERIMENTAL PROCEDURE

Reduction—RNase (Sigma Chemical Company, chromatographic grade, Lot No. R60-B-204) was allowed to react with mercaptoethanol in the presence of 8 M urea according to the procedure described previously (7). Completeness of reduction was determined both by titration with *p*-chloromercuribenzoate (8) and by reaction with C^{14}-labeled iodoacetic acid (7). The reduced protein was separated from buffers and reagents by gel filtration on columns of Sephadex G-25 (Pharmacia, Uppsala), developed with 0.1 M acetic acid.

Oxidations—The reduced protein as obtained in the column effluent was diluted with 0.1 M acetic acid, in each instance, to a concentration of 0.1 mg per ml, determined spectrophotometrically. The pH was adjusted to the desired value by dropwise addition of a saturated solution of tris-(hydroxymethyl)aminomethane (Sigma, primary standard grade). Final pH values were determined on a Radiometer pH meter, model 22, equipped with a scale expander. In the instance of oxidations in urea, sufficient neutral methylamine acetate was added to make a 0.1 M solution. This reagent was intended to react with any cyanate ions which might be present in urea solutions (9). After addition of appropriate reagents, the solutions were permitted to stand at room temperature (approximately 23°) in a beaker exposed to air and covered with a single layer of tissue paper. The solutions were not stirred. Aliquots were taken periodically for sulfhydryl group determination by either *p*-chloromercuribenzoate titration (8) or the isotopic technique (7). The reaction was terminated when no free —SH groups could be detected, and the entire reaction mixture was concentrated by lyophilization. Gel filtration on Sephadex G-25 freed the protein from reagents. Developing solutions for this separation were either 0.1 M acetic acid or 0.1 M ammonium carbonate (Fisher, reagent grade), depending on whether the added reagents were more soluble in acidic or basic solutions. The protein peak from the column was lyophilized and thus rendered salt-free, both developing solutions being entirely volatile.

Oxidations were carried out at pH 8.0 in various concentrations of the following reagents, and the products were isolated: urea (Fisher, reagent grade, freshly recrystallized from ethanol); guanidinium chloride (Eastman, twice recrystallized from methanol); phenol (Fisher, reagent grade), purified by sublimation under reduced pressure immediately before use; and *p*-hydroxyphenylacetic acid (Aldrich), sublimed under reduced pressure. Similar oxidations were carried out in the presence of various concentrations of the following materials after purification by sublimation, but the protein derivatives were not isolated: catechol, *p*-cresol, *p*-chlorophenol, 1-naphthol, benzoic acid, *p*-hydroxybenzoic acid, phenylacetic acid, and *p*-nitrophenol (all these were Eastman products, highest purity). The purity of these substances was checked by melting point determinations in most instances. The following group of reagents was used as obtained from the manufacturer: methanol (Fisher, reagent), dioxane (Fisher, reagent), aniline (Eastman, highest purity), cyclohexanol (Fisher, reagent), bovine serum albumin (California Corporation for Biochemical Research), gelatin (commercial grade), and L-phenylalanine (Nutritional Biochemicals Corporation). A copolymer of tyrosine and glutamic acid, in the ratio of 1:1, containing approximately 30 residues of amino acids per

* A short summary (1) of this work was presented at the annual meeting of the Federated Societies for Experimental Biology and Medicine, Atlantic City, New Jersey, April 1961.

† Present address, Massachusetts General Hospital, Boston, Massachusetts.

molecule, and a polyaspartic acid sample of molecular weight of approximately 5000 were generous gifts of Dr. Michael Sela.

Assays for enzymatic activity were carried out in the presence of these reagents, which were shown, in separate experiments, not to influence the activity of native RNase at the concentrations under consideration. Because of the absorbancy of most of these reagents in the ultraviolet, sulfhydryl group determinations were, in general, followed by the isotopic method.

Assays of Enzymatic Activity—Activity against RNA, prepared according to Crestfield, Smith, and Allen (10), was estimated either by absorbancy determination in uranyl acetate filtrates (11) or by a modification of the method of Kunitz (12) in which the initial rate of change of absorbancy at 300 mμ of 0.1% solutions of RNA was compared with that produced by standard RNase solutions.

Physical Methods—Molecular weights were determined in the Spinco ultracentrifuge, model E, according to the method of Archibald (13). Optical rotation and rotatory dispersion measurements were performed in the Rudolph precision polarimeter, model 80, at 23°. Ultraviolet absorption measurements were made in a Cary recording spectrophotometer. Viscosity was determined in an Ostwald viscometer with an outflow time for water of 80 seconds, in a bath maintained at 25° ± 0.01°. Trypsin digestions were carried out for 12 hours at 37° with a Radiometer pH-Stat, model TTT1, with an Ole Dich drum recorder, and Worthington five times recrystallized trypsin which had been treated with a 10% molar equivalent of diisopropyl fluorophosphate to inactivate the bulk of the chymotrypsin-like contamination.[1]

RESULTS

Effects of Urea, Guanidinium Chloride, and Low pH—The oxidation of dilute solutions of reduced RNase at pH 8 results in the recovery of over 90% of the potential enzyme activity. However, if such oxidations are carried out in the presence of 8 M urea or 4 M guanidinium chloride at the same pH, the final products, after separation from reagents, have an enzymatic activity of only approximately 1% that of the native enzyme. This low level of activity is consistent with that predicted on the basis of a random intramolecular association of the eight —SH groups in the reduced protein to form disulfide bonds (5, 6). The experiments summarized in Table I indicate that the regain of activity is inversely related to the concentrations of urea and guanidinium chloride employed. These data also show that the effect of guanidinium chloride is not due to its strong electrolyte nature, since control oxidations in equivalent concentrations of KCl yielded essentially full reactivation. Thus, 3 M guanidinium chloride permitted only a 4% regain in activity, whereas, at this same concentration, KCl solutions yielded a 94% recovery.

Stark, Stein, and Moore (9) have demonstrated that prolonged incubations of ribonuclease in the presence of urea result in reactions between cyanate ions and amino groups. When present, sulfhydryl groups would be particularly reactive. To evaluate the significance of this reaction, oxidations were performed in the presence of C^{14}-labeled urea. After separation of the protein from reagents by Sephadex G-25 filtration, its radioactivity was estimated in the scintillation counter (7). The results indicate that approximately 1 mole of cyanate ion had reacted with each molecule of RNase. The possibility that the oxidations carried out in urea yielded inactive material because of reaction with cyanate ions becomes even less likely on the basis of experiments described later in this paper, which show that the inactive derivatives may be converted in good yield to active RNase by a procedure designed specifically to induce disulfide interchange.

[1] B. Hartley, personal communication.

TABLE I

Effect of denaturants on reduced ribonuclease oxidation

Compound	Concentration	Activity
	M	%
Urea	0.25	100
	1.0	92
	2.0	40
	4.0	25
	6.0	10
	8.0	<1
Guanidinium chloride	1.0	88
	2.0	25
	3.0	4
	4.0	<1
KCl	1.0	101
	3.0	94

Relative rates of oxidation of the —SH groups in reduced RNase were determined in aqueous and urea solutions. As shown in Fig. 1, the oxidation of reduced RNase in urea solutions at pH 8 is quite slow and, even after 100 hours of incubation, one sulfhydryl group per mole may still be detected.

To examine further the influence of hydrogen ion concentration, a series of experiments was performed in which oxidation was allowed to proceed to completion at various pH values. These experiments are summarized in Fig. 2. Above pH 6.2, the production of active material is essentially as great as at pH 8.0. Below this hydrogen ion concentration, the amount of

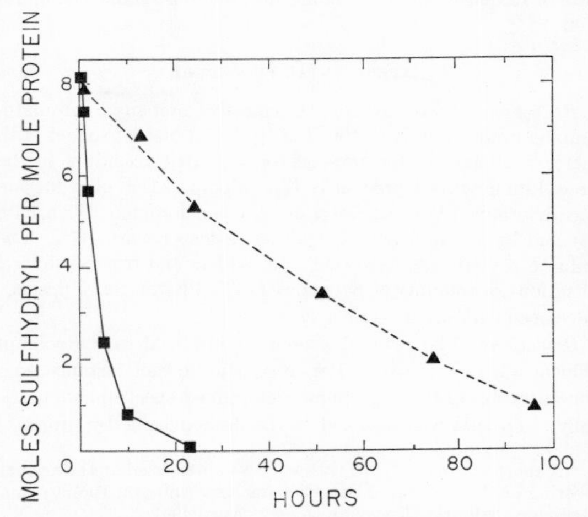

FIG. 1. Rates of oxidation of sulfhydryl groups in reduced RNase at pH 8.0 as measured by CMB titrations. ■——■, oxidation in buffer solution; ▲- - -▲, oxidation in 8 M urea.

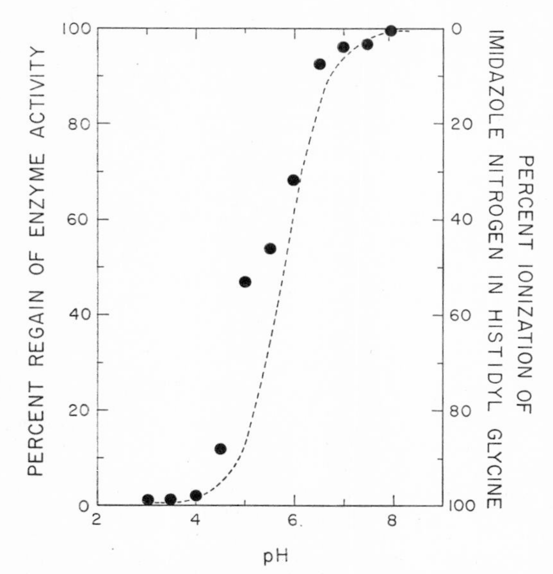

FIG. 2. Percentage of regain of enzyme activity at various pH values. ●, enzyme activity relative to native RNase; (- - -), ionization of the imidazole nitrogen in histidylglycine.

active material produced drops markedly, and, as mentioned above, only inactive products are produced below pH 4 to 4.5. The data are presented together with a curve representing the degree of ionization of the imidazole nitrogen of histidine as it exists in the peptide, histidylglycine. The close similarity between the two curves may be only coincidental, although involvement of histidine residues in the establishment of the correct three-dimensional structure of RNase would not be too surprising in view of the obvious importance of this amino acid in the active configuration of the protein (14, 15).

Tyrosine Analogues—In view of previous studies (16–19) that suggest that some of the tyrosine residues of RNase may be of critical importance in the stabilization of tertiary structure, a variety of compounds, more or less related in structure to this amino acid, were tested as possible inhibitors of the "correct" oxidation process. Tyrosine, itself, could not be tested because of its limited solubility at the pH employed. However, a number of compounds (Table II) were effective in influencing the formation of products with reduced activity (approximately 1% of native) when tested at concentrations of 0.1 M or less (mole ratios of compound to RNase of 10^4:1). A number of other substances, although in many instances structurally similar to the effective compounds, did not markedly reduce the specific activity of the products when employed at the same, or usually greater, concentration.

In the presence of equimolar, *i.e.* approximately 10^{-5} M, concentrations of a copolymer of glutamic acid and tyrosine (see "Experimental Procedure"), the products formed during oxidation had a specific activity of the order of 1%. At the pH used for oxidation, pH 8, this copolymer is not an inhibitor of RNase activity (20). Oxidations carried out in the presence of a 50-fold molar excess of polyaspartic acid yielded material with essentially native enzymatic activity.

Characterization of Oxidation Products—After separation of the products from reagents by gel filtration, followed by lyophilization, the materials formed during oxidation in urea, guanidinium chloride, at pH values below pH 4, or in the presence of the "effective" compounds listed in Table II, were very soluble in water and formed colorless solutions. Their enzymatic activities varied between 0.5 and 1.7% that of native RNase. Physical properties of many of these materials are listed in Table III. Absorption maxima were 2760 A (Fig. 3), a value which lies between that of native RNase (2775 A) and the reduced carboxymethylated protein (2750 A) and is similar to the absorption maxima of other inactive derivatives of RNase in which disulfide bonds are intact, such as pepsin-inactivated RNase (16, 21) and RNase S-protein (22).

The inactive oxidation products (Table III) have optical rotations similar to that of native RNase and, in the presence of 8 M urea, a similar increase in negative optical rotation occurs. This increase is actually not as marked as with the native protein, an observation that may indicate cross-linking that is even more rigid and interlocked than in RNase itself. The critical wavelength, λ_c, of the inactive products is, however, essentially the same as that of reduced, alkylated RNase and does not change perceptibly in 8 M urea solutions.

The inactive derivatives appear to be monomeric, since their molecular weights are essentially identical with that of native RNase. Specific viscosities lie between those of native RNase and the reduced alkylated form.

Trypsin Digestibility—Native RNase is not susceptible to trypsin digestion except above a critical temperature, in the neighborhood of which its internal structure becomes labile and partially disoriented (23, 24). The data in Table IV demonstrate

TABLE II

Effect of tyrosine analogues and other compounds on regain of activity in oxidations

Reduced ribonuclease concentration was 1.4 × 10^{-5} M in all cases. All oxidations were done at pH 8.0.

Compound	Concentration	Final activity of oxidized RNase
		%
Tyrosine glutamic copolymer (1:1)	1 × 10^{-5} M	<1
Phenol	0.1 M	<1
Catechol	0.1 M	<1
p-Cresol	0.1 M	<1
p-Chlorophenol	0.05 M	<1
1-Naphthol	0.04 M	<1
p-Hydroxyphenylacetic acid	0.1 M	<1
Polyaspartic acid (MW 5000)	5 × 10^{-4} M	67
Benzoic acid	0.2 M	89
Aniline	0.1 M	67
Phenylacetic acid	0.2 M	60
Phenylalanine	0.17 M	104
p-Hydroxybenzoic acid	0.5 M	76
p-Nitrophenol	0.1 M	45
Cyclohexanol	0.2 M	88
Benzyl alcohol	0.2 M	39
Methanol	1.0 M	88
Dioxane	1.0 M	100
Bovine serum albumin	1%	74
Gelatin	1%	67

Table III

Physical properties of native ribonuclease and some of its derivatives

Material	Absorption maxima	$[\alpha]_D$	$[\alpha]_D$ in 8 M urea	λ_c	λ_c in 8 M urea	Molecular weight	$[\eta]$
	A						*(g/100 ml)*$^{-1}$
Native RNase	2775	−74.2	−108.5	2330	2200	13,683*	0.033
Reduced carboxymethylated RNase	2750	−91.6	−92.1	2230	2050	14,155*	0.133
Urea-oxidized RNase	2760	−79.6	−94.3	2120	2120	13,800	0.060
Guanidine-oxidized RNase	2760	−79.4		2120		14,200	
RNase oxidized at pH 3.5	2760	−82.2		2120		13,700	
RNase oxidized in *p*-hydroxyphenylacetic acid	2760	−79.0				13,900	0.058

* Formula weight.

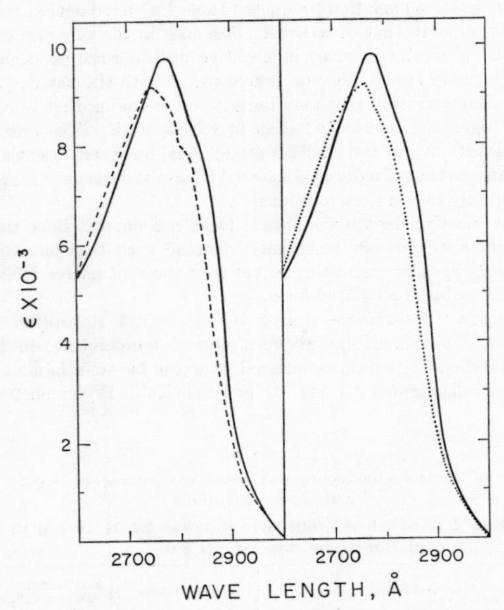

Fig. 3. Ultraviolet spectra of native RNase (——), reduced carboxymethylated RNase (- - -), and RNase oxidized in 8 M urea (. . . . .).

that, in contrast, the inactive oxidation products are as susceptible as reduced, alkylated RNase, indicating a profound change in surface topography, which now makes trypsin-susceptible peptide bonds available for cleavage. The product isolated after oxidation of reduced RNase under optimal conditions is as resistant as the native enzyme itself, an observation that adds additional evidence for the identity of native and reduced-reoxidized RNase.

Regeneration of Active Enzyme—When these relatively inactive derivatives of RNase are incubated with sulfhydryl compounds for periods of approximately 20 hours, a large fraction of the theoretical specific activity is regained (Table V). As shown in Table VI, incubation of the product obtained from oxidation in 8 M urea solutions with mercaptoethanol (100 moles per mole of RNase) yields a product with 79% the specific activity of native RNase. Regain of activity at the end of 20 hours appears to be a function of mercaptoethanol concentration up to a limit of

Table IV

Trypsin digestion of ribonuclease derivatives

Digestions were carried out at pH 9.3; the number of moles of base uptake was taken to equal the number of bonds cleaved. There are 12 peptide bonds in ribonuclease that are potentially susceptible to trypsin digestion. The maximal number of bonds cleaved, approximately 11, observed in these experiments may differ from the theoretical number for several reasons. First, the bond between the arginine residue at position 39 and the half-cystine residue at position 40 is known to be unusually resistant, and a fraction of the molecules may have still contained an intact bond at this point even after 12 hours at 37°. Furthermore, although α-amino groups would be expected to be essentially fully un-ionized at pH 9.3, there may exist some such groups with abnormally high pK values. Thus the number of bonds cleaved might be slightly greater than would be indicated by our titration.

Material	No. of bonds cleaved per mole
Native RNase	0
Reduced alkylated RNase	11
RNase oxidized in buffer	<1
RNase oxidized in guanidine	11
RNase oxidized in urea	9
RNase oxidized in *p*-hydroxyphenylacetic acid	11
RNase oxidized at pH 3.5	11

Table V

Rearrangement of ribonuclease derivatives in presence of mercaptoethanol

Material	Initial activity	Activity after 24 hours' incubation at pH 8.0	Activity after 24 hours' incubation at pH 8.0 in the presence of 100-fold molar excess of mercaptoethanol
	%	%	%
RNase oxidized at pH 5	7	7	83
RNase oxidized in urea	<1	<1	79
RNase oxidized in guanidine	<1	<1	55
RNase oxidized in phenol	<1	<1	41

100-fold molar excess. Several sulfhydryl-containing substances, including cysteine and reduced RNase itself, have been used to effect this reactivation.

Work reported in a previous paper (4) has shown that, during the early stages of reoxidation of reduced RNase, a marked lag phase occurs in which no activity appears in spite of extensive formation of disulfide bonds. As one explanation for this lag, it has been suggested that "incorrect" disulfide bonds are formed initially and that these subsequently are rearranged through a process of disulfide interchange. Although this explanation may at present be considered only as one of several alternatives, the reactivation of inactive products described in this paper are consistent with this interpretation.

DISCUSSION

The molecule produced by the reduction of the disulfide bonds of RNase does not, to the best of our present knowledge, contain *covalent* structure beyond that of the amino acid sequence itself. It has been assumed that this simple polypeptide structure exists in solution in the form of a random chain and that the entire information required for the establishment, through oxidation of —SH groups, of the correct, native pairing of —SS bonds resides in the identity and sequential order of the individual amino acid residues. It is, nevertheless, quite possible that the interactions between the functional groups of the side chains may exert, by a concerted action, a powerful set of forces that allow a significant fraction of molecules to favor a configuration resembling that of the native enzyme, even in the absence of stabilizing covalent tie-points. Under such circumstances, the population of reduced molecules would not form a completely random set of three-dimensional conformations but might exist in a broad Gaussian distribution, any single molecule being most often in a form approaching that of native RNase. The much slower disappearance of —SH groups during oxidation of reduced RNase in 8 M urea solutions (Fig. 1) supports the idea that a highly specific set of directing interactions can exist even in the absence of disulfide bonds.

Studies reported elsewhere (25, 26) have shown that approximately 8 of the 11 amino groups of RNase may be masked with polyalanine chains of considerable length without significant loss of activity or of the capacity to reform, after reduction, a set of disulfide bonds consistent with an active configuration. These experiments indicated that the spatial orientation of the amino groups themselves is not of critical importance in the reoxidation process, since, in the peptidyl derivatives, the amino groups residing on the ends of the polypeptide chains are removed from the RNase molecule by a distance of 10 A or more. The present studies enable us to extend further the evaluation of the importance of various functional groups in the determination of "correct" pairing of —SH groups. The striking effect of urea or guanidinium chloride suggests that short range, rather than electrostatic, forces must be primarily involved. The marked effect of various analogues of tyrosine on reoxidation, and the absence of significant effects with certain other compounds of related structure, leads to the conclusion that aromatic compounds possessing a phenolic hydroxyl group and bearing a charged group (when present) removed by at least one methylene group from the aromatic ring (*e.g.* *p*-hydroxyphenylacetic acid) are effective, whereas those in which the charged group resides on the ring are not (*e.g.* *p*-hydroxybenzoic acid). The results are consistent with the hypothesis that tyrosine residues in RNase are involved in intramolecular interactions that help to predetermine, or at least stabilize, the ultimate, enzymatically active configuration. These results are also in agreement with observations that establish a correlation between the active configuration of RNase and the presence of several tyrosine residues in the molecule in a form having anomalous resonance properties (17, 19) (as evidenced by a shift in the ultraviolet absorption spectrum). The observation that the inactive derivatives produced by random disulfide linking do not exhibit anomalous resonance also supports the concept that tyrosine interactions are involved in the stabilization of the native configuration.

TABLE VI

Rearrangement of ribonuclease oxidized in urea at various mercaptoethanol concentrations

All samples possessed less than 1% activity before oxidation.

Mercaptoethanol	Enzymatic activity after 24 hours
moles/mole protein	%
0	<1
1	5
5	32
10	34
100	79

The experiments on the effect of variations in hydrogen ion concentration show that, at low pH values, the requisite information for —SH pairing is no longer available. The pH dependency of "correct" oxidation appears, from the results summarized in Fig. 2, to establish a correlation with the state of ionization of histidine residues in the protein, although such a correlation is, of course, entirely hypothetical on the basis of these results alone.

SUMMARY

Reduced ribonuclease (RNase), containing eight sulfhydryl groups and devoid of covalent structure other than the primary sequence itself, was allowed to reoxidize in air in the presence of 8 M urea, 4 M guanidine, and a number of compounds bearing a structural resemblance to tyrosine. The effect of variation in pH was also studied. Concentrated urea and guanidine, as well as compounds with aromatic character and bearing a phenolic hydroxyl group but having no charged group on the aromatic ring itself, were effective inhibitors of the "correct" pairing of sulfhydryl groups, suggesting that short range forces, some possibly associated with tyrosine, are of importance in orienting the reduced molecule. At low pH values, inactive derivatives were also produced. The pH dependence of the efficiency of production of active RNase from the reduced form suggests that histidine residues may be involved as determinants in the process. Inactive derivatives were compared with native RNase by a variety of physical and chemical methods. They were found to be of similar molecular weight. They exhibited a modified, "inactive" type of ultraviolet absorption spectrum, but a relatively normal optical rotatory value. The critical wave length was quite different, however, resembling that of reduced, alkylated RNase. The inactive derivatives were quite susceptible to trypsin digestion, in contrast to the native enzyme. Their viscosities lay between those of native and reduced alkylated RNase. These findings suggest that the inactive derivatives are monomeric, symmetrical, reasonably globular molecules, with a surface topography different from that of native RNase, and that they are distinguished mainly by random disulfide pairing.

When the "incorrectly" oxidized derivatives were dissolved under conditions compatible with disulfide interchange (pH 8, in the presence of sulfhydryl compounds) they regained a large fraction of the theoretical activity. These results suggest that the native molecule is the most stable configuration, thermodynamically speaking, and that the major force in the correct pairing of sulfhydryl groups in disulfide linkage is the concerted interaction of side-chain functional groups distributed along the primary sequence.

Acknowledgments—The authors would like to thank Drs. Frank Reithel and D. Michael Young for their aid in performance of some of the physical measurements.

REFERENCES

1. Haber, E., Sela, M., and Anfinsen, C. B., *Federation Proc.*, **20**, 217 (1961).
2. Sela, M., White, F. H., Jr., and Anfinsen, C. B., *Science*, **125**, 691 (1957).
3. White, F. H., Jr., *J. Biol. Chem.*, **236**, 1353 (1961).
4. Anfinsen, C. B., Haber, E., Sela, M., and White, F. H., Jr., *Proc. Natl. Acad. Sci. U. S.*, **47**, 1309 (1961).
5. Sela, M., and Lifson, S., *Biochim. et Biophys. Acta*, **36**, 471 (1959).
6. Kauzmann, W., in R. Benesch, R. Benesch, P. D. Boyer, I. M. Klotz, W. R. Middlebrook, A. G. Szent-Gyorgyi, and D. R. Schwartz (Editors), *Symposium on sulfur in proteins*, Academic Press, Inc., New York, 1959, p. 93.
7. Anfinsen, C. B., and Haber, E., *J. Biol. Chem.*, **236**, 1361 (1961).
8. Boyer, P. D., *J. Am. Chem. Soc.*, **76**, 4331 (1954).
9. Stark, G. R., Stein, W. H., and Moore, S., *J. Biol. Chem.*, **235**, 3177 (1960).
10. Crestfield, A. M., Smith, K. C., and Allen, F. W., *J. Biol. Chem.*, **216**, 185 (1956).
11. Anfinsen, C. B., Redfield, R. R., Choate, W. L., Page, J., and Carroll, W. R., *J. Biol. Chem.*, **207**, 201 (1954).
12. Kunitz, M., *J. Biol. Chem.*, **164**, 563 (1946).
13. Archibald, W. J., *J. Phys. Chem.*, **51**, 1204 (1947).
14. Barnard, E. A., and Stein, W. D., *J. Molecular Biol.*, **1**, 339, 350 (1959).
15. Stark, G. R., Stein, W. H., and Moore, S., *J. Biol. Chem.*, **236**, 436 (1961).
16. Sela, M., and Anfinsen, C. B., *Biochim. et Biophys. Acta*, **24**, 229 (1957).
17. Sela, M., Anfinsen, C. B., and Harrington, W. F., *Biochim. et Biophys. Acta*, **26**, 502 (1957).
18. Scheraga, H. A., *Biochim. et Biophys. Acta*, **23**, 196 (1957).
19. Tanford, C., and Hauenstein, J. D., *J. Am. Chem. Soc.*, **28**, 5287 (1956).
20. Sela, M., *J. Biol. Chem.*, **237**, 418 (1962.)
21. Anfinsen, C. B., *J. Biol. Chem.*, **221**, 405 (1956).
22. Richards, F. M., *Proc. Natl. Acad. Sci. U. S.*, **44**, 162 (1958).
23. Harrington, W. F., and Schellman, J., *Compt. rend. trav. lab. Carlsberg, Sér. chim.*, **30**, 463 (1958).
24. Rupley, J. A., and Scheraga, H. A., *Biochim. et Biophys. Acta*, **44**, 191 (1960).
25. Anfinsen, C. B., Sela, M., and Cooke, J., *J. Biol. Chem.*, in press.
26. Sela, M., and Anfinsen, C. B., *Symposium on poly-L-amino acids*, Wisconsin University Press, Madison, Wisconsin, 1961, in press.

Reprinted from The Journal of Biological Chemistry, Vol. 241, 1966

Implication of an Ionizing Group in the Control of Conformation and Activity of Chymotrypsin*

(Received for publication, September 27, 1965)

Hannah L. Oppenheimer, Bernard Labouesse,‡ and George P. Hess
From the Department of Chemistry, Cornell University, Ithaca, New York

SUMMARY

1. Acetylated chymotrypsinogen was prepared by the reaction of crystallized chymotrypsinogen with acetic anhydride at pH 6.7 and 4°, and purification of the product. End group analysis and Van Slyke amino nitrogen determination indicated that all the free ϵ- and α-amino groups of the chymotrypsinogen molecule had been blocked by acetylation.

2. Activation of acetylated chymotrypsinogen by trypsin, followed by chromatography, yielded acetylated δ-chymotrypsin. Quantitative end group determination indicated that the molecule had a single amino group, the NH_2-terminal α-amino group of an isoleucine residue. In the catalytic hydrolysis of *N*-acetyl-L-tryptophan amide, the steady state kinetic parameters were found to be the same, in the pH region 6 to 10, with acetylated δ-chymotrypsin and with δ-chymotrypsin.

3. The activation of acetylated chymotrypsinogen to acetylated δ-chymotrypsin was found to be accompanied by the appearance of a single new ionizing group, titratable in the pH region 5 to 10.5, with a pK(app) of 8.3.

4. The specific rotation of acetylated δ-chymotrypsin was found to vary with pH in the pH region 6 to 10.5, the change in specific rotation following the ionization of a single group with pK(app) of 8.3. Above pH 10 the specific rotation of the zymogen and the enzyme were found to be the same at the three different wave lengths investigated.

5. Previous investigations in this and other laboratories of the pH dependence of the quotient of the steady state kinetic parameters, $k_{cat}\,[K_m(app)]^{-1}$—a function which yields the ionization constants of the catalytically important ionizing groups of the free enzyme—have indicated the importance of an ionizing group with pK(app) $\sim$ 8.5 in all α-chymotrypsin-catalyzed reactions studied, and in the δ- or acetylated δ-chymotrypsin-catalyzed hydrolysis of *N*-acetyl-L-tryptophan amide.

6. An interpretation of these results that is consistent with all available experimental information is that the conversion of catalytically inactive chymotrypsinogen to active enzyme is accompanied by the appearance of a new ionizing group (a group that has a pK(app) of 8.3 in acetylated δ-chymotrypsin) which controls the conformation and thereby the activity of the enzyme. Molecules in which this ionizing group is unprotonated are in a catalytically inactive conformation, resembling that of chymotrypsinogen, and molecules in which this ionizing group is protonated are in a catalytically active conformation. A diisopropyl fluorophosphate (DFP)-induced protein isomerization involves only the enzyme molecules that are in the catalytically active conformation. This interpretation leads to a prediction that in the presence of DFP there will be an increase in pK(app) of the ionizing group. Such an increase in pK(app) was previously observed, and accounts for the proton uptake and the change in specific rotation of the enzyme observed in its reaction with DFP.

7. All evidence obtained so far suggests that the group with pK(app) $\sim$8.5 is an NH_2-terminal isoleucyl α-amino group.

There is a considerable amount of evidence that the interaction of α-chymotrypsin with substrates or inhibitors is accompanied by a conformational change in the enzyme, and that an ionizing group in the enzyme is involved in this process (1–14). The object of this study was to investigate further the identity and function of this ionizing group.

Kinetic investigations (10, 11) of the reaction of α-chymotrypsin with diisopropyl fluorophosphate indicate that a reversible change in conformation, observed by measurements of changes in absorption spectra and in specific rotation of the enzyme, occurs prior to the formation of the covalently bonded phosphoryl enzyme and is an intermediate step in the reaction. It was found, in these studies, that an ionizing group with a pK(kin) $\sim$9 controls the rate of conformational change and thereby the rate of phosphorylation above pH 8 (11). It was also noted that as a result of the conformation change, an ionizing group with pK(app) $\sim$9 becomes perturbed and the enzyme takes up protons from solution (11, 12). Reversible potentiometric titrations (11, 12) indicated that in the pH range 7 to 10 a group with pK(app) $\sim$9.0 (13) can be titrated in α-chymotrypsin but not in diisopropyl-α-chymotrypsin, and that the difference

* We are grateful to the National Institutes of Health and the National Science Foundation for financial support. This work includes material from a thesis submitted by Hannah L. Oppenheimer to the Graduate School of Cornell University in partial fulfillment of the requirements for the degree of Doctor of Philosophy.
‡ Visiting scientist, Cornell University, 1961 to 1963. Permanent address, Faculté des Sciences, Université de Paris, Centre d'Orsay, S. et O., France.

between the titration curves accounts for the kinetically observed proton uptake in the phosphorylation reaction. Since spectrophotometric titrations (14) showed that values of pK(app) of the tyrosyl residues are the same in α-chymotrypsin and in diisopropyl-α-chymotrypsin, the suggestion arising from the titration data is that the ionizing group involved in the conformational change is an α- or ε-amino group.

In the work reported here, studies were made with a chymotrypsin derivative in which all free amino groups except one, the isoleucyl α-amino group liberated during activation, were blocked by acetylation. The preparation of this enzyme, acetylated δ-chymotrypsin, was accomplished by fully acetylating chymotrypsinogen and then activating it with trypsin (15, 16). Spectral characteristics, titration curves, and optical rotation measurements of acetylated chymotrypsinogen, acetylated δ-chymotrypsin, and acetylated diisopropyl-δ-chymotrypsin were studied in order to deduce the function of the amino-terminal isoleucyl α-amino group in the conformation and catalytic activity of chymotrypsin. A preliminary report of some of the data has appeared (17).

EXPERIMENTAL PROCEDURE

Materials

Crystallized, salt-free chymotrypsinogen, Lot CG 871, and chromatographically homogeneous chymotrypsinogen, Lot CGC 762, were obtained from Worthington. Crystallized, lyophilized trypsin and three times crystallized, salt-free α-chymotrypsin, Lot 6072, were also from Worthington. The DFP[1] and *N,O*-diacetyl-L-tyrosine were purchased from K & K Laboratories; 2,4-dinitrofluorobenzene, DNP-amino acids, *N*-acetyl-L-tyrosine ethyl ester, and *p*-tosyl-L-arginine methyl ester-HCl were from Mann. Tris was obtained from Sigma. The 2,4-dinitrophenol, *N,N*-diethylaminoethyl cellulose, and 2-octanol were purchased from Eastman Organic Chemicals. Hydrochloric acid was "Baker Analyzed" reagent. All other reagents were obtained from Mallinckrodt.

Sephadex G-50 and G-25, fine grade, and CM Sephadex C-50, medium (Pharmacia) were used for gel filtration. Visking 18/32 seamless cellulose tubing was used for dialysis. Silica Gel G was from Research Specialities.

Apparatus[2]

A Radiometer pH-stat (TTT1a) was used for pH determinations, enzyme activity assays, acetylation reactions, and protein titrations. The pH measurements were made with reference to Beckman standard buffers, and had a precision of ±0.05 pH unit. An Agla micrometer syringe, obtained from Wellcome Research Laboratories, Beckenham, England, was used for the titration experiments; the accuracy of this syringe is such that volumes as small as 1×10^{-3} ml can be measured to within $\pm 5 \times 10^{-5}$ ml. For absorbance readings, a Beckman DU spectrophotometer was used. A Cary 14 spectrophotometer was used for absorption and difference spectrum measurements, and for spectrophotometric determinations of tyrosine acetylation. A Rudolph high precision spectropolarimeter (model 200S-80Q), with a high pressure mercury light source, was used for optical rotation measurements. Centrifugations were performed with a Servall super speed RC-2 automatic refrigerated centrifuge. A Technicon automatic amino acid analyzer was used in the quantitative end group determination.

[1] The abbreviations used are: DFP, diisopropyl fluorophosphate; DNP-, dinitrophenyl-; DIP-, diisopropylphosphoryl-.

[2] We are grateful to Professor Harold A. Scheraga for use of the Rudolph high precision spectropolarimeter and to Professor Harold H. Williams for use of the Technicon automatic amino acid analyzer.

Methods

Protein Concentration Determination—Protein concentrations were determined spectrophotometrically at 280 mμ. A molar extinction coefficient of 5×10^4, reported for α-chymotrypsin (18), was used for α-chymotrypsin and chymotrypsinogen; a molar extinction coefficient of 4.7×10^4 was used for their acetylated derivatives, since the determination of protein concentration by nitrogen analysis (19) indicated that the acetylated proteins have a slightly lower molar extinction coefficient at 280 mμ than chymotrypsinogen. The nitrogen content of chymotrypsinogen was taken to be 16.5% (20), and the molecular weight as 25,000 (21). Since acetylation added 15 acetyl groups of molecular weight 43 to chymotrypsinogen, a molecular weight of 25,600 was assumed for acetylated chymotrypsinogen. Trypsin concentration in milligrams per ml was calculated by multiplying the optical density at 280 mμ by a factor of 0.72 (22).

Enzyme Assay—Enzyme activity was measured by the method of Schwert *et al.* (23). Substrates used were *N*-acetyl-L-tyrosine ethyl ester for chymotrypsin, and *p*-tosyl-L-arginine methyl ester for trypsin.

Van Slyke Amino Nitrogen Analysis—The original procedure of Van Slyke (24, 25) was used to determine amino nitrogen content of proteins. The reaction time was varied according to temperature (see Table I). The reaction vessel was covered to avoid light, for light causes extra nitrogen to be liberated from tyrosine peptides (26). A reagent blank was subtracted. Samples contained approximately 0.01 mg of amino nitrogen, and at this concentration level the Van Slyke analysis is capable of 1% accuracy (24).

Acetylation of Chymotrypsinogen—The reaction was carried out at 4° in the water-jacketed sample cup of the pH-stat. A portion, 1 g, of chymotrypsinogen was added to 100 ml of 0.01 M $CaCl_2$, and the pH was adjusted to 8.0. Approximately 1×10^{-3} M DFP was added to inhibit any chymotrypsin present. With constant stirring by means of a magnetic stirring device, acetic anhydride was added with a Gilmont micropipette-burette over a period of 40 min until the solution was 1% (v/v) in acetic anhydride. The pH was maintained at 6.7 ± 0.2 with 5 N

TABLE I

Reaction time for Van Slyke amino nitrogen analysis

Temperature	Time	
	min	*sec*
29°	10	
28	10	50
27	11	30
26	12	10
25	12	50
24	13	30
23	15	
22	15	40
21	17	10

NaOH. After the final addition of acetic anhydride, base was added by pH-stat until all acetic anhydride was hydrolyzed, as evidenced by a leveling off of the base consumption curve. The sample was dialyzed against two changes of 5 liters of 0.01 M sodium borate, pH 8.0, for 17 hours at 4°.

Fractionation of Acetylated Chymotrypsinogen with Ammonium Sulfate—Acetylated chymotrypsinogen, prepared as described above, was diluted to a concentration of approximately 1 mg per ml with a 0.02 M sodium borate buffer solution (pH 8.0, 0.01 M in $CaCl_2$). This zymogen solution was kept on ice while ice-cold saturated ammonium sulfate solution was added slowly with stirring. Two precipitable fractions were obtained. The first fraction, precipitated at 13 to 22% of ammonium sulfate saturation, was removed by centrifugation and discarded. The second fraction, precipitated at 27 to 36% of ammonium sulfate saturation, contained approximately 80% of the starting material. This second fraction, Cut II, was found to be chromatographically homogeneous.

Activation of Acetylated Chymotrypsinogen to Acetylated δ-Chymotrypsin—After fractionation with ammonium sulfate, the Cut II zymogen precipitate was dissolved in sufficient 0.02 M sodium borate buffer (pH 8.0, 0.01 M in $CaCl_2$) to make the protein concentration approximately 5 mg per ml. A trypsin solution (4 mg of trypsin per ml, 1×10^{-3} M HCl) was added so that the final trypsin concentration was 0.12 mg per ml. This activation mixture, containing approximately 200 mg of acetylated chymotrypsinogen, was dialyzed for 15 hours against two changes of 5 liters of 0.01 M sodium borate buffer solution (pH 8.0, 0.01 M in $CaCl_2$) at 4°; the dialysate was changed after 4 hours of activation. It was observed that dialysis during activation had the effect of increasing the specific activity of the product slightly. Presumably, any remaining denatured protein was either autolyzed or digested by trypsin and removed by dialysis. The presence of calcium ions in the activation mixture was found to enhance enzyme stability.

Preparation of DIP-δ-chymotrypsin and Acetylated DIP-δ-chymotrypsin—The phosphorylated enzymes were prepared according to the method of Schaffer, May, and Summerson (27). Excess DFP was removed by dialysis against distilled water at 4° or by filtration through Sephadex G-50 (10).

Removal of Trypsin from Acetylated δ-Chymotrypsin and Acetylated DIP-δ-chymotrypsin—Solutions of the proteins (3.5 mg per ml in 0.01 M sodium borate buffer, pH 8.0, 0.01 M in $CaCl_2$) were dialyzed for 6 hours against three changes of 4 liters of distilled water at 4°. Then the aqueous protein solutions were filtered through a carboxymethyl Sephadex C-50 column, 12 × 0.8 cm, at 4°. The CM-Sephadex had initially been washed with 1 M acetic acid, and the column was equilibrated with distilled water. After filtration, the protein solutions were lyophilized. All trypsin remained on the column, as indicated by enzyme assay of eluted sample with *p*-tosyl-L-arginine methyl ester-HCl as substrate.

Chromatography—Chromatographic fractionation of acetylated DIP-δ-chymotrypsin and acetylated chymotrypsinogen (both before and after ammonium sulfate fractionation) was carried out on a DEAE-cellulose column, 24 × 0.9 cm, equilibrated at pH 7.5 with 0.02 M Tris-HCl buffer. The protein solutions, prepared as described, were dialyzed against 0.02 M Tris-HCl, pH 7.5, at 4°, and samples containing about 6 mg of protein in 2 ml of solution were adsorbed onto the column. The protein was eluted at 4° with 0.02 M Tris-HCl buffer at pH 7.5, containing KCl in the concentration of 0.14 M for the phosphorylated preparation and 0.12 M for the zymogen. Protein still remaining on the column was eluted with 0.02 M Tris-HCl buffer which was 0.5 M in KCl. Fractions of 4.2 ml were collected, and protein concentration of these was determined by optical density measurements at 280 mμ.

Determination of End Groups by 2,4-Dinitrofluorobenzene Method—Samples (20 mg) of α-chymotrypsin or acetylated δ-chymotrypsin were used for these experiments. Before being allowed to react, the protein was oxidized with performic acid at 0° (28) to cleave disulfide bonds, thereby denaturing the proteins and exposing all the amino groups. The procedure of Levy and Li (29) was used for the reaction. Subsequent hydrolysis of the DNP-proteins at 105° was allowed to proceed for 40 hours; under these conditions, less than 10% of the difficultly hydrolyzable dipeptide DNP-isoleucylvaline remains (15). Thin layer chromatography was used to identify the DNP-amino acids.

Thin Layer Chromatography—The procedures used for thin layer chromatography of the DNP-amino acids are described by Randerath (30). Hydrolysate aliquots of 0.25 ml, containing approximately 0.5 μg of the ether-soluble DNP-amino acids or 1 μg of the water-soluble DNP-amino acids, were used to spot the chromatograms. In an additional test for trace amounts of ε-DNP-lysine and DNP-cysteic acid in acetylated δ-chymotrypsin hydrolysate, the amount of sample per spot was increased to about 1 mg. DNP-amino acid standards were used in each chromatogram. The presence of dinitrophenol was confirmed by testing the appropriate spots with formic acid fumes (31); bleaching occurred upon exposure to the acid, but the spots reappeared instantly when exposed to ammonia fumes.

Determination of End Groups by Cyanate Method—The procedure of Stark and Smyth (32) was used for reaction of acetylated δ-chymotrypsin with cyanate and subsequent quantitative determination of end groups with an amino acid analyzer. Synthetic carbamyl-DL-norleucine, used as the internal standard, was added to the reaction mixture just prior to the cyclization of the hydantoins (32).

Difference Spectrum Measurements—The procedure used for obtaining the difference spectrum of DIP-enzyme *versus* enzyme has been described (2).

Titration of Acetylated Proteins—Solutions of acetylated chymotrypsinogen (Cut II) and of trypsin-free acetylated δ-chymotrypsin and acetylated DIP-δ-chymotrypsin were made up to be about 4×10^{-5} M in protein and 0.15 M in KCl. A 25-ml portion of the protein solution was pipetted into the thermostated (4°) sample cup of the pH-stat. The sample was titrated first with 0.0888 N carbon dioxide-free KOH standardized with potassium acid phthalate (33), and then with standard 0.1000 N HCl. Calibrated Agla micrometer syringes were used to make a titration from pH 5.0 to pH 10.4 and back to pH 5.0. The reversibility of this titration was rechecked by rapid addition of acid and base from pH 6.0 to pH 9.0 and back again to pH 6.0. During titration, the sample was kept covered, and nitrogen was bubbled into it to avoid carbon dioxide absorption. Each protein titration curve was corrected for the solvent blank.

Spectrophotometric Determination of Tyrosine Acetylation—The extent of acetylation of tyrosyl residues in the acetylated chymotrypsinogen and acetylated δ-chymotrypsin preparations was determined from measurements of change of optical density at 244 mμ (34) upon addition of base to solutions of the proteins. The procedure was as follows. In the sample cell of the Cary spectrophotometer, 0.2 ml of protein solution was mixed with 2.0 ml of NaOH solution (see below) to make a solution of protein concentration 0.6 to 0.8 mg per ml. This was compared to a reference consisting of the same concentration of protein at pH 8.0. Difference in optical density at 244 mμ was recorded as a function of time, beginning within 10 sec after mixing. The measurements were made at 25°.

The NaOH solutions used in preparing the sample solutions were 0.001 N NaOH, 0.12 M in KCl; 0.01 N NaOH, 0.11 M in KCl; and 0.02 N NaOH, 0.1 M in KCl. The resulting pH values, after addition of the protein solution, were 11.0, 11.9, and 12.2, respectively.

Optical Rotation Measurements—Solutions of trypsin-free acetylated δ-chymotrypsin and acetylated DIP-δ-chymotrypsin (Cut II) were diluted with KCl solution so as to have a final protein concentration of approximately 1 mg per ml and a 0.15 M KCl concentration. After pH adjustments with 1 N KOH and 1 N HCl at 12°, each sample was placed in a 2-dm quartz tube with water jacket. Optical rotation was measured at 313, 334, and 366 mμ at 12°. A solvent blank was subtracted. The acetylated δ-chymotrypsin sample was brought back to pH 7 after measurements at high pH in order to check the reversibility of the changes in optical rotation.

RESULTS

Preparation and Characterization of Acetylated Derivatives of Chymotrypsinogen, δ-Chymotrypsin, and DIP-δ-chymotrypsin

Chromatography of the initial acetylated chymotrypsingen preparation (see "Experimental Procedure") on DEAE-cellulose column revealed two distinct fractions (Fig. 1). That portion of the eluate represented by the major absorbance peak at 280 mμ contained approximately 89% of the protein, and, after activation, exhibited most of the enzymic activity. As a consequence of this finding, the simple ammonium sulfate fractionation described under "Experimental Procedure" was devised. Results of this separation are shown in Table II. Upon activation, one fraction (Cut II) showed greater enzymic activity toward N-acetyl-L-tyrosine ethyl ester than did the activated starting material; the other fraction (Cut I) was relatively inactive and insoluble at pH 8.0. When chromatographed on the cellulose column, Cut II produced a single

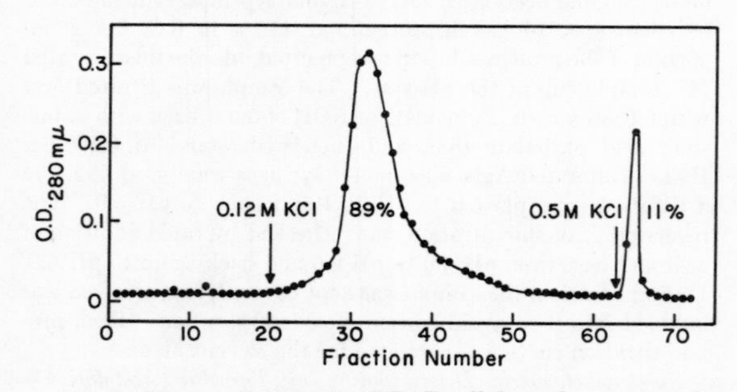

FIG. 1. Elution diagram for DEAE-cellulose chromatography of unfractionated acetylated chymotrypsinogen. Optical density at 280 mμ is plotted against fraction number. Each fraction contained 4.2 ml. A 6-mg sample of the protein was applied to the column and eluted with 0.02 M Tris-HCl buffer (pH 7.5) at 4°. The additions of KCl to the buffer are indicated.

TABLE II

Effect of ammonium sulfate fractionation of acetylated chymotrypsinogen on enzymic activity of acetylated δ-chymotrypsin

Measurements of k(obs) were made with reaction mixtures 4 to 8×10^{-7} M in enzyme, and 1.67×10^{-2} M in N-acetyl-L-tyrosine ethyl ester in buffer at pH 8 which was 0.004 M in Tris-HCl and 0.1 M in KCl. Temperature = 25°.

Enzyme preparation	Zymogen	Saturation with $(NH_4)_2SO_4$	Starting material	k(obs)
		%	%	sec^{-1}
Acetylated δ-chymotrypsin	Acetylated chymotrypsinogen			
	Unfractionated	0	100	169
	Cut I	13–22	15	97
	Cut II	27–36	80–85	200
δ-Chymotrypsin[a]	Chymotrypsinogen			246
α-Chymotrypsin	Chymotrypsinogen			174
"Reacetylated" acetylated δ-chymotrypsin				18

[a] Control. Activated from chymotrypsinogen under conditions identical with those used for the preparation of acetylated δ-chymotrypsin from acetylated chymotrypsinogen.

symmetrical absorbance peak at 280 mμ. Cut II was used for the reported experiments, and is the material referred to as acetylated chymotrypsinogen.

Acetylated chymotrypsinogen was activated with trypsin under conditions that are known to lead, in the case of chymotrypsinogen, to the δ form of the enzyme (15, 16, 25, 36). The dialyzed acetylated chymotrypsin preparation, when assayed with N-acetyl-L-tyrosine ethyl ester (1.67×10^{-2} M) as substrate, had a k(obs) value that is 80% of the k(obs) value obtained when chymotrypsinogen was activated under identical conditions (see Table II). This is the material referred to as acetylated δ-chymotrypsin.

The acetylated δ-chymotrypsin was permitted to react with DFP to form the material designated as acetylated DIP-δ-chymotrypsin. A difference spectrum with a peak at 290.5 mμ was observed when acetylated DIP-δ-chymotrypsin was compared with acetylated δ-chymotrypsin (Fig. 2). The molar extinction difference coefficient, $(\Delta\epsilon_M)_{290.5}$, was 650. This difference spectrum closely resembled that of DIP-α-chymotrypsin *versus* α-chymotrypsin, also shown in Fig. 2, for which $(\Delta\epsilon_M)_{290}$ is 590 (2). Chromatography of the DFP-treated preparation on DEAE-cellulose resulted in a chromatogram with one symmetrical peak at 280 mμ (Fig. 3).

When the acetylated δ-chymotrypsin was reacetylated, under conditions identical with those described for acetylation of chymotrypsinogen, a loss of activity occurred. The value of k(obs) decreased more than 90% (Table III) when enzymic activity was assayed with 1.67×10^{-2} M N-acetyl-L-tyrosine ethyl ester at pH 8.0. The concurrent loss of free amino groups was indicated by Van Slyke analysis (see below).

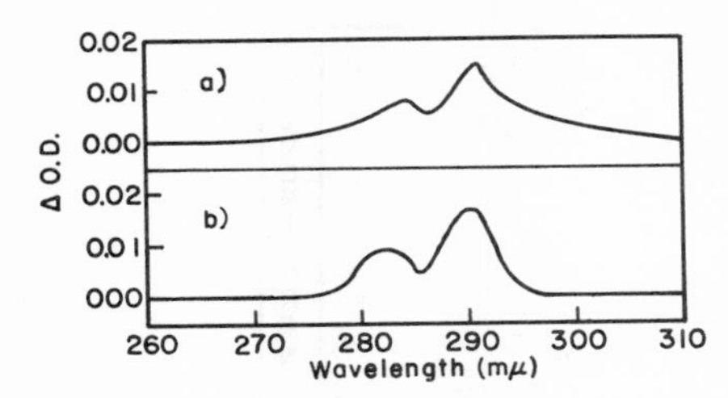

FIG. 2. Ultraviolet difference spectra of enzyme *versus* phosphorylated enzyme. The samples were made up in 0.02 M sodium borate buffer (pH 8.0), 0.01 M in $CaCl_2$. Temperature = 25°. *a*, acetylated DIP-δ-chymotrypsin *versus* acetylated δ-chymotrypsin. Protein concentration = 2.30×10^{-5} M. *b*, DIP-chymotrypsin *versus* chymotrypsin. Protein concentration = 2.85×10^{-5} M. The chymotrypsin was prepared by activation of chymotrypsinogen under conditions identical with those used for activation of acetylated chymotrypsinogen, and is presumably the δ form.

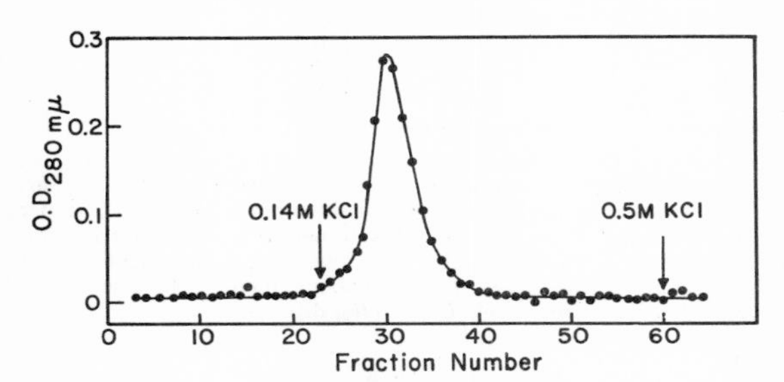

FIG. 3. Elution diagram for DEAE-cellulose chromatography of acetylated DIP-δ-chymotrypsin. Optical density at 280 mμ is plotted against fraction number. Each fraction contained 4.2 ml. A 6-mg sample of protein was applied to the column and eluted with 0.02 M Tris-HCl buffer (pH 7.5) at 4°. The additions of KCl to the buffer are indicated.

A comparison of amino nitrogen content in chymotrypsinogen and the acetylated derivatives was made by Van Slyke analysis. Experimental results are shown in Table III, along with results expected on the assumption of complete acetylation of chymotrypsinogen. It is apparent that all the amino groups in the chymotrypsinogen molecule were blocked by the acetylation process. It is also seen that the conversion of acetylated chymotrypsinogen to acetylated δ-chymotrypsin resulted in the liberation of one amino group, as expected, and that this group was lost when the acetylated δ-chymotrypsin was reacetylated.

The complete acetylation of amino groups in chymotrypsinogen was also shown by reaction of acetylated chymotrypsinogen with 2,4-dinitrofluorobenzene followed by thin layer chromatography. There was no evidence of ε-DNP-lysine or DNP-cysteic acid in the hydrolysate.

Identity of the single free amino group in acetylated δ-chymotrypsin was determined by end group analysis by two different methods. Thin layer chromatography following reaction of the enzyme with 2,4-dinitrofluorobenzene showed (Fig. 4,*a* and *b*) that the ε-DNP-lysine, DNP-alanine, and DNP-cysteic acid residues found with α-chymotrypsin were not present in the acetylated δ-chymotrypsin hydrolysate; only DNP-isoleucine was seen. In the chromatogram of water-soluble DNP-amino acids, a sufficient amount of acetylated δ-chymotrypsin hydrolysate sample was used so that even trace amounts of ε-DNP-lysine or DNP-cysteic acid would have been detected. Quantitative end group determination by the cyanate method (32) revealed 0.95 mole of isoleucine, 0.05 mole of alanine, and 0.1 mole of glycine per mole of enzyme in the acetylated chymotrypsin preparation. The small amounts of alanine and glycine can be accounted for by the fact that KNCO itself, when carried through the end group procedure, gives trace amounts of alanine and glycine in a molar ratio of 1:2 (32). A control experiment with commercial α-chymotrypsin showed not only a small amount of glycine (0.12 mole), but also alanine (0.90 mole) in greater quantity than isoleucine (0.84 mole), and traces of aspartic acid, serine, and threonine.

There is evidence that in addition to amino groups, two tyrosyl residues that are titratable in chymotrypsinogen and chymotrypsin at high pH, were acetylated in the preparations under study. It is known that there are four tyrosines in chymotrypsin (20); two titrate normally, and two abnormally (14). Spectrophotometric measurements at 244 mμ were made of alkaline solutions of acetylated chymotrypsinogen and acetylated δ-chymotrypsin in order to estimate the number of acetylated and the number of "buried" tyrosyl residues per mole of protein. Upon addition of the base to the protein solutions, a rapid increase in $O.D._{244}$, followed by a slow increase, was readily observable at pH values 11.9 and 12.2 (which yielded the same information). This rapid increase was interpreted as due to deacetylation and ionization of acetylated tyrosyl residues, for it was found that at 25° k(obs) values for the rapid increase exhibited by the experimental preparations (0.04 min^{-1} and 0.4 min^{-1} at pH values 11 and 11.9, respectively) corresponded to k(obs) values for the *O*-deacetylation of *N*,*O*-diacetyl tyrosine. The slow change was attributed to the ionization of free but "buried" tyrosine residues, which are known to become slowly unmasked at high pH (14). The total tyrosine content was calculated from the total optical density change, and extrapolation of the slow optical density change was used to calculate the

TABLE III

Extent of acetylation as determined by Van Slyke amino nitrogen analysis

Sample	No. of determinations	Amino groups per mole protein		Standard deviation
		Expected[a]	Observed[b]	
Chymotrypsinogen[c]	9	15.0	15.2	0.3
Acetylated chymotrypsinogen	11	0	0.6	0.2
Acetylated δ-chymotrypsin	11	1.0	1.6	0.3
Acetylated DIP-δ-chymotrypsin	2	1.0	1.7	0.1
"Reacetylated" acetylated δ-chymotrypsin	5	0	0.5	0.3

[a] The calculation for chymotrypsinogen is based on the data of Hartley (37). In the case of the acetylated derivatives, complete acetylation is assumed.

[b] High Van Slyke values for proteins are attributed to unspecific reaction of guanidino groups with nitrous acid (38).

[c] On the basis of two determinations, chromatographically homogeneous chymotrypsinogen (see "Materials") was found to have 15.2 amino groups per mole of protein.

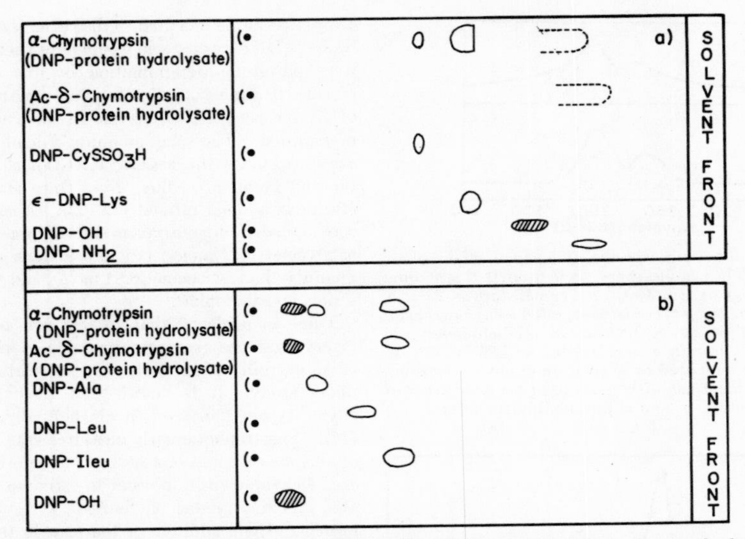

FIG. 4. Tracings of thin layer chromatograms on Silica Gel G. *a*, water-soluble DNP-amino acids. Samples were applied in 1-butanol. Concentration of DNP-amino acid was approximately 1 μg per spot; amount of hydrolysate per spot was 0.25 mg for α-chymotrypsin or 1 mg for acetylated δ-chymotrypsin. Solvent for chromatography was 1-propanol-NH_4OH (7:3), the NH_4OH containing 29.5% NH_3. Development time was about 2 hours. DNP-NH_2 was added as marker to the hydrolysate samples. *b*, ether-soluble DNP-amino acids. Samples were applied in ether. Concentration of DNP-amino acid was approximately 0.5 μg per spot; amount of protein hydrolysate per spot was 0.25 mg. Solvent for chromatography was benzene-pyridine-acetic acid (80:20:2). Development time was about 50 min. *Ac*, acetylated; *DNP-OH*, dinitrophenol; *DNP-NH₂*, dinitroaniline.

content of acetylated tyrosine in the protein. An extinction coefficient of 1.17×10^4 (34) was used in the calculation. Results are given in Table IV.

Titration of Acetylated Proteins

Titrations of solutions of acetylated chymotrypsinogen, acetylated δ-chymotrypsin, and acetylated DIP-δ-chymotrypsin, about 4×10^{-5} M in concentration, were performed at 4° between pH 5.0 and pH 10.4 (Fig. 5), and were found to be perfectly reversible. Titration below pH 5.0 was not possible because of protein insolubility.

It was found that in the acetylated δ-chymotrypsin molecule, three groups were titrated between pH 6 and pH 9. With acetylated DIP-δ-chymotrypsin, however, only two groups could be titrated. The titration curve of acetylated chymotrypsinogen, which has no free amino groups, was identical with that of the phosphoryl enzyme. Additional evidence that an ionizing group titrated in acetylated δ-chymotrypsin is not titratable in acetylated DIP-δ-chymotrypsin was obtained by titrating a sample of the phosphorylated product which had been prepared from the active enzyme immediately after this had been titrated. Table V shows that one of the three groups titrated between pH 6 and pH 9 in acetylated δ-chymotrypsin was no longer titratable after phosphorylation. It should be noticed, in Table V, that the number of amino groups per mole of protein, as determined by Van Slyke analysis, did not change during the course of the experiments. This, in addition to the reversibility of the titration curves, rules out the possibility that the "extra" group titrated in the active enzyme was a result of autolysis during titration or absorption of peptide impurities on acetylated δ-chymotrypsin but not on acetylated DIP-δ-chymotrypsin.

TABLE IV

Extent of tyrosine acetylation as determined spectrophotometrically at 244 mμ

The data were calculated from measurements at both pH 11.9 and pH 12.2. Temperature = 25°.

Sample	Storage[a]	Tyrosyl residues per mole protein		
		"Exposed"		"Buried"
		Not acetylated	Acetylated	
	days			
Acetylated chymotrypsinogen	0	0.0	2.0	2.0
	4	0.3	1.7	2.0
	8	1.1	0.9	2.0
Acetylated δ-chymotrypsin	1	0.1	1.9	2.0

[a] The zymogen stock solution was stored at pH 8 and 4°; deacetylation of tyrosyl residues occurs slowly under these conditions.

Similar results were obtained by Erlanger, Castleman, and Cooper (39), who found one less titratable group in diphenylcarbamyl-α-chymotrypsin than in α-chymotrypsin between pH 7.3 and pH 9.7.

The difference between the titration curves of acetylated δ-chymotrypsin and acetylated chymotrypsinogen or acetylated DIP-δ-chymotrypsin in the pH region 6 to 10.4 is plotted against pH in Fig. 6. From this curve, the pK(app) of the group which can be titrated in the free enzyme but not in the phosphorylated enzyme or the zymogen can be estimated (40). The midpoint

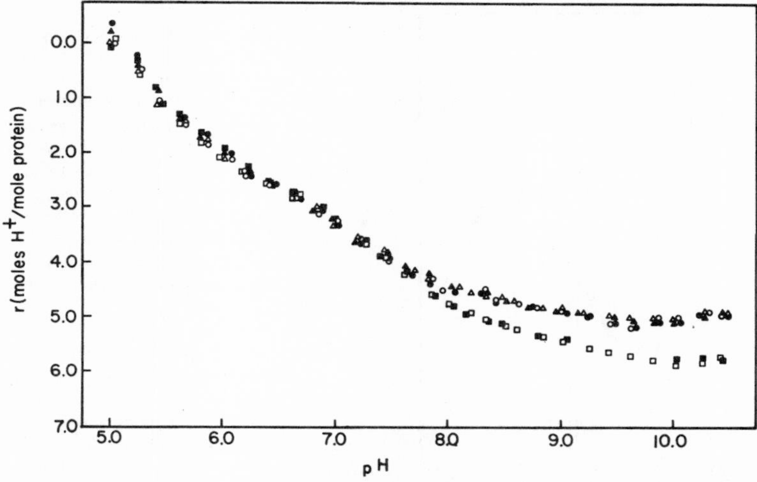

FIG. 5. Titration curves for acetylated chymotrypsinogen, acetylated δ-chymotrypsin, and acetylated DIP-δ-chymotrypsin. Samples contained 4 × 10^{-5} M protein in 0.15 M KCl. Temperature = 4°. The titration curves have been corrected for the solvent blank. ○, acetylated chymotrypsinogen, measured in the forward direction (low to high pH); ●, acetylated chymotrypsinogen, measured in the backward direction; □, acetylated δ-chymotrypsin, forward direction; ■, acetylated δ-chymotrypsin, backward direction; △, acetylated DIP-δ-chymotrypsin, forward direction; ▲, acetylated DIP-δ-chymotrypsin, backward direction.

TABLE V

Comparison of titratable groups and amino groups in acetylated δ-chymotrypsin and acetylated DIP-δ-chymotrypsin

Sample	Amino groups per mole protein (Van Slyke analysis)	Moles H+ titrated per mole protein in pH region 6 to 9
Acetylated δ-chymotrypsin		2.9
Before titration	1.4 ± 0.1	
After titration	1.4 ± 0.1	
Acetylated DIP-δ-chymotrypsin[a]	1.4 ± 0.1	1.9

[a] Prepared from the acetylated δ-chymotrypsin above, after titration.

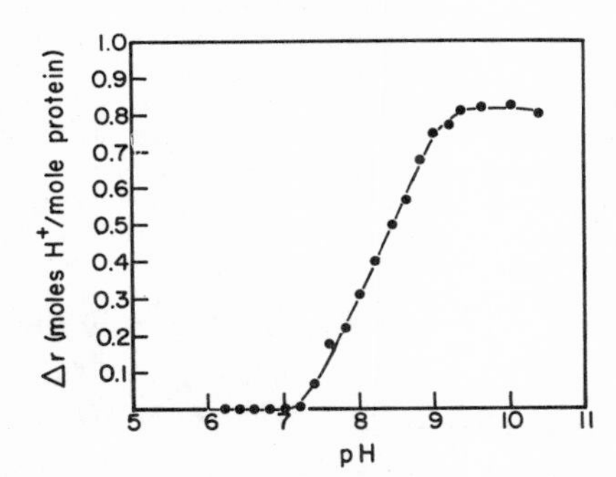

FIG. 6. Difference (Δr) between the titration curves for acetylated DIP-δ-chymotrypsin or acetylated chymotrypsinogen and acetylated δ-chymotrypsin, as a function of pH. Data for the plot were taken from Fig. 5.

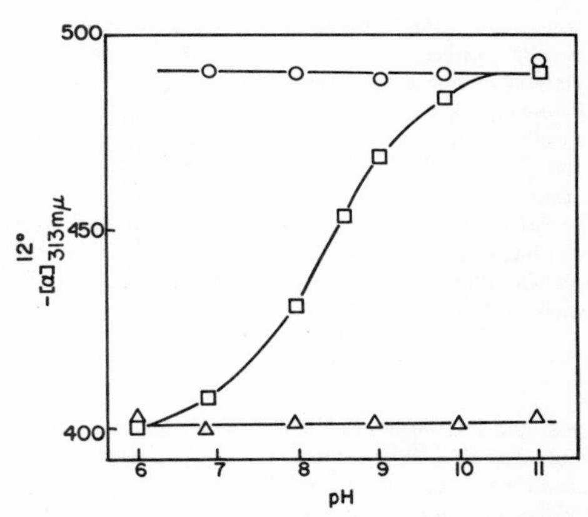

FIG. 7. Specific rotation plotted against pH at 313 mμ. Solutions contained 3.20 to 4.64 × 10^{-5} M protein in 0.15 M KCl. Temperature = 12°. The specific rotation changes of the enzyme were reversible. □, acetylated δ-chymotrypsin; △, acetylated DIP-δ-chymotrypsin; ○, acetylated chymotrypsinogen.

of the curve is at pH 8.3, indicating that the group in question has a pK(app) of 8.3.

The titration curves for all the acetylated proteins leveled off in the region between pH 9.0 and pH 10.4, suggesting that titratable tyrosyl residues had been acetylated. This possibility was confirmed spectrophotometrically, as discussed above.

pH Dependence of Optical Rotation of Acetylated Proteins

The pH dependence of optical rotation in the pH region 6 to 11 was determined for the acetylated proteins at 313, 334, and 336 mμ. The data obtained at 313 mμ are shown in Fig. 7;

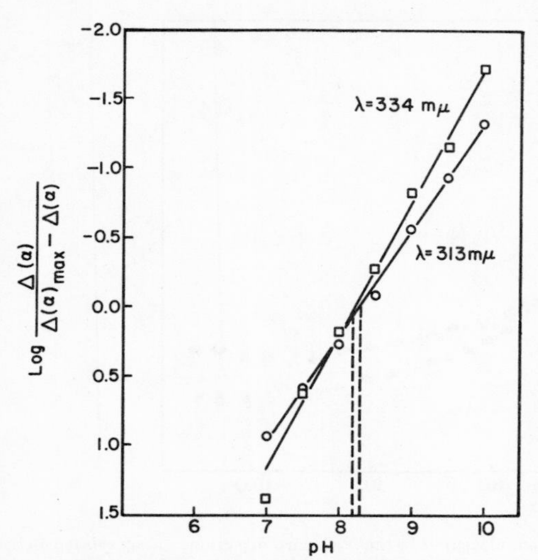

FIG. 8. Plot of log $\Delta[\alpha](\Delta[\alpha]_{max} - \Delta[\alpha])^{-1}$ against pH, for acetylated δ-chymotrypsin.

similar results were obtained at the other two wave lengths. The specific rotation values for acetylated chymotrypsinogen and acetylated DIP-δ-chymotrypsin were found to be significantly different from each other, and pH independent. With acetylated δ-chymotrypsin, reversible changes of specific rotation were observed in the pH 6 to 11 region. In the pH region 6 to 7, the specific rotation of the active enzyme was similar to that of the phosphorylated product; above pH 10, it appeared identical with that of the zymogen.

The pH-dependent change in specific rotation of acetylated δ-chymotrypsin, as related to specific rotation of its precursor and its phosphorylated derivative, is shown also in Fig. 8. The equation used as basis for this plot is

$$\text{pH} = \text{pK} + \log \frac{[E]}{[E\text{H}]} \qquad (1)$$

where E and EH represent the different ionization states of the enzyme molecule. With definition of $\Delta[\alpha]_{max}$ as the total change in specific rotation between pH 6 and the pH where values for acetylated δ-chymotrypsin and acetylated chymotrypsinogen are the same, and $\Delta[\alpha]$ defined as the change in specific rotation between pH 6 and any pH, Equation 1 becomes

$$\text{pH} = \text{pK} + \log \left(\frac{\Delta[\alpha]}{\Delta[\alpha]_{max} - \Delta[\alpha]}\right) \qquad (2)$$

Fig. 8 is a plot of pH *versus* log $\Delta[\alpha](\Delta[\alpha]_{max} - \Delta[\alpha])^{-1}$. As can be seen from the figure, Equation 2 is obeyed, indicating that the specific rotation follows the ionization of a single group. The pK(app) of the group controlling the change in specific rotation may be obtained from the abscissa of the Fig. 8 plot at the point where log $\Delta[\alpha](\Delta[\alpha]_{max} - \Delta[\alpha])^{-1} = 0$. This value is pK (app) = 8.3.

DISCUSSION

There are a number of indications that the observations made with acetylated proteins apply also to the corresponding nonacetylated proteins. The pH dependence of specific rotation of acetylated chymotrypsinogen, acetylated δ-chymotrypsin, and acetylated DIP-δ-chymotrypsin appears to be the same as that of the corresponding nonacetylated proteins (4, 41). The steady state kinetic parameters for the catalytic hydrolysis of *N*-acetyl-L-tryptophan amide in the pH region 8 to 10.5 appear to be the same for δ-chymotrypsin and acetylated δ-chymotrypsin (42). The difference between the titration curves of acetylated δ-chymotrypsin and acetylated DIP-δ-chymotrypsin is similar to the difference in titration curves between α-chymotrypsin and DIP-α-chymotrypsin (11, 13). The difference spectra between free enzyme and phosphorylated enzyme (2) appear indistinguishable for the acetylated and nonacetylated forms of the enzyme.

An interpretation of experiments reported here, together with some previously published findings with nonacetylated enzymes, is represented in Fig. 9. The interpretation is that the conversion of catalytically inactive chymotrypsinogen to active enzyme is accompanied by the appearance of a new ionizing group which controls the conformation and thereby the activity of the enzyme. Molecules in which this ionizing group is unprotonated are in a catalytically inactive conformation (designated *I* in Fig. 9), resembling that of the zymogen; molecules in which this ionizing group is protonated are in a catalytically active conformation (designated *A* in Fig. 9). Evidence for this interpretation is summarized below.

Evidence that a conformational change accompanies the conversion of inactive zymogen to enzyme has been previously published. Neurath, Rupley, and Dreyer (41) suggested that the change in specific rotation observed during the conversion of chymotrypsinogen to chymotrypsin at neutral pH is due to a conformational change that converts the catalytically inactive zymogen to catalytically active enzyme. The suggestion of conformational change was confirmed by optical rotatory dispersion measurements of Imahori, Yoshida, and Hashizume (43), and Fasman and Foster (44).

A decrease in the catalytic properties of chymotrypsin above pH 8.0 was originally reported by Northrop, Kunitz, and Herriot (45). More detailed information has recently been obtained from determinations of the pH dependence of the steady state kinetic parameters for the chymotrypsin-catalyzed hydrolysis of specific substrate amides and esters. Investigations of the pH dependence of k_{cat} $[K_m(\text{app})]^{-1}$, which gives information about catalytically important ionizing groups of the free enzyme (46) have revealed the importance of an ionizing group with pK(app) ~8.5 in all α-chymotrypsin-catalyzed reactions investigated (47) and in the δ- or acetylated δ-chymotrypsin-catalyzed hydrolysis of *N*-acetyl-L-tryptophan amide (42).

In the experiments reported here, it was observed that the activation of acetylated chymotrypsinogen to acetylated δ-chymotrypsin is accompanied by the appearance of a single new

C=O, NH —Trypsin→ (I) + H^+ $\underset{}{\overset{k_H}{\rightleftharpoons}}$ (A) + H^+

K_I ⇅ ⇅ K_A

(I)H $\underset{}{\overset{k_L}{\rightleftharpoons}}$ (A)H

chymotrypsinogen

FIG. 9. Scheme for the activation of chymotrypsinogen (or acetylated chymotrypsinogen), showing equilibria between active (*A*) and inactive (*I*) conformational forms of the enzyme molecule, and ionization equilibria. The constants k_H and k_L apply to high (10.5) and low (6 to 7) pH conditions.

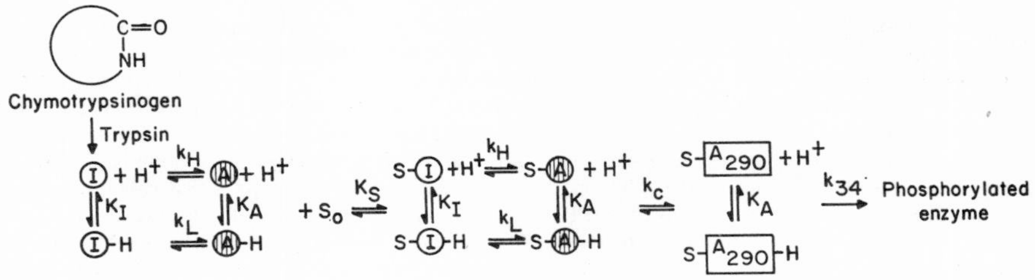

Chymotrypsinogen

Trypsin

Phosphorylated enzyme

FIG. 10. Scheme for the activation of chymotrypsinogen and the reaction of chymotrypsin with DFP, showing ionization equilibria and equilibria between active (A) and inactive (I) forms of the enzyme in free state and in complexes with DFP. Enzyme conformations include a DFP-induced form (SA_{290}) characterized by spectral changes at 290 mμ; k_c is the equilibrium constant pertaining to the DFP-induced protein isomerization. The equilibrium constants k_H and k_L apply to high (10.5) and low (6 to 7) pH conditions.

ionizing group with pK(app) of 8.3 (Figs. 5 and 6). It was also observed that the specific rotation of acetylated δ-chymotrypsin varies with pH and that the change in specific rotation follows the ionization of a single group with pK(app) of 8.3 (Figs. 7 and 8). It can be seen in Fig. 7 that the specific rotation of acetylated δ-chymotrypsin becomes the same as that of acetylated chymotrypsinogen above pH 10.

Results of the experiments with acetylated δ-chymotrypsin suggest a correlation of (*a*) the ionization of the group of the enzyme with pK(app) of 8.3, (*b*) the pH-dependent changes in specific rotation of the enzyme, and (*c*) enzymic activity. These relationships are incorporated in the interpretation shown graphically in Fig. 9, which is taken to be applicable also to nonacetylated chymotrypsinogen and chymotrypsin. The specific rotation of the enzyme observed at high pH (pH 10.5) is identical with that of the zymogen and is considered to be characteristic of the catalytically inactive conformation, designated I in the figure; and the specific rotation of the enzyme observed in the low pH region (pH 6 to 7) is considered to be characteristic of the catalytically active conformation, designated A in the figure.

The general case of conformational changes associated with titratable groups has been treated by Tanford (48). Equations pertaining to the cyclic equilibria of Fig. 9 are given below.

$$E_0 = E + EH = I + A + IH + AH \quad (3)$$

$$k_L = \frac{IH}{AH} = <1 \quad (4)$$

$$k_H = \frac{I}{A} = >1 \quad (5)$$

$$K_I = \frac{(I)(H)}{IH} \quad (6)$$

$$K_A = \frac{(A)(H)}{AH} \quad (7)$$

$$K_I k_L = K_A k_H \quad (8)$$

$$K(\text{app})_E = \frac{(E)(H)}{EH} = \frac{(I + A)H}{IH + AH} = K_A \frac{(1 + k_H)}{(1 + k_L)} = K_I \frac{k_L}{k_H} \frac{(1 + k_H)}{(1 + k_L)} \quad (9)$$

where $K(\text{app})_E$ represents the apparent acid dissociation constant of the free enzyme. Designation of the specific rotation observed at high pH as characteristic of I requires $k_H > 1$, and designation of the specific rotation observed in the pH region 6 to 7 as characteristic of A requires $k_L < 1$.

As can be seen from Equations 1, 2, and 9, $pK(\text{app})_E$ obtained from titration data and from the analysis of the pH dependence of the specific rotation must be the same, provided that the conformational changes follow the ionization of the same ionizing group. Therefore, the experiments shown in Figs. 7 and 8 provide an excellent correlation between the ionization state of the group with $pK(\text{app})_E = 8.3$ and the change in specific rotation, and are consistent with the cyclic equilibria of Fig. 9.

The pH-dependent equilibria between active and inactive conformations of acetylated δ-chymotrypsin (Fig. 9) are also consistent with results of a kinetic investigation of the reaction of chymotrypsin with DFP (10, 11). These studies indicated that a reversible DFP-induced protein isomerization, characterized by spectral changes of the enzyme at 290 mμ, precedes the phosphorylation of the enzyme, and that the observed decrease in phosphorylation rate above pH 8.0 is controlled by an ionizing group with pK(kinetic) of about 9.0 which exerts its effect on the DFP-induced protein isomerization. It was also observed that phosphorylation of the enzyme stabilizes the DFP-induced protein isomerization, and that the bond-breaking step itself causes no further changes in the spectral or optical rotational characteristics of the enzyme. We assume that DFP binds equally well to active and inactive conformations, and that the pH dependence of the DFP-induced protein isomerization arises because only the active conformations of the enzyme participate in this process. The same conclusions are reached, however, if the assumption is that only the forms of the enzyme in which the group with pK(app) = 8.3 is protonated (IH and AH) participate in the DFP-induced protein isomerization, and that DFP binds better to the catalytically active conformations. The particular assumptions we have made make Equations 3 to 7 applicable not only to the cyclic equilibria of free enzyme but also to those of DFP-bound enzyme, shown in Fig. 10. Additional equations for the cyclic equilibria involving DFP bound to enzyme are

$$k_c = \frac{SA}{SA_{290}} = \frac{SAH}{SAH_{290}} = <1 \quad (10)$$

$$K(\text{app})_{ES} = \frac{(SI + SA + SA_{290})(H)}{(SIH + SAH + SAH_{290})} = K_A(1 + k_H k_c) \quad (11)$$

where S is DFP, A and I are the active and inactive conformational forms of the enzyme, A_{290} is the enzyme conformational form induced by substrate and characterized by optical changes at 290 mμ, and $K(app)_{ES}$ is the apparent acid dissociation constant of substrate-bound enzyme.

Provided that k_c is small compared to 1, the DFP-induced conformational change has the effect of increasing pK(app): while in the free enzyme $K(app)_E \cong K_A (1 + k_H)$, in the DFP-bound enzyme $K(app)_{ES} \cong K_A (1 + k_H k_c)$. From this increase in pK(app) as a result of the DFP-induced conformational change, it is predicted that the reaction of chymotrypsin with DFP should be accompanied by hydrogen ion uptake by the enzyme, and that the onset of the pH-dependent change in specific rotation due to the conversion of conformation A to conformation I should occur at a higher pH in the phosphorylated enzyme. Experimentally it was found that the reaction of α-chymotrypsin with DFP is accompanied by a proton uptake by the enzyme which is accounted for by the difference between titration curves for the enzyme and for its phosphorylated derivative (11, 13). A similar difference in titration curves is observed between the analogous acetylated enzymes (Figs. 5 and 6). Furthermore, the kinetic investigation of the reaction of α-chymotrypsin with DFP indicated that the proton uptake occurs as a result of the DFP-induced conformational change. The prediction about pH-dependent change in specific rotation is also borne out by experiment. It can be seen in Fig. 7 that the specific rotation of acetylated DIP-δ-chymotrypsin is pH-independent throughout the pH region 6 to 10, and is similar to the specific rotation of the free enzyme at pH 6 where its group with pK(app) = 8.3 is protonated. Thus, the previously reported observations of the DFP-chymotrypsin reaction are consistent with the present experimental results which indicate pH-dependent equilibria between active and inactive conformations of the enzyme in the pH region 6 to 10.5.

There are several indications that the group with pK(app) = 8.3 is the NH_2-terminal isoleucyl α-amino group that is liberated in the conversion of acetylated chymotrypsinogen to acetylated δ-chymotrypsin. Van Slyke analysis (Tables III and V) and qualitative (Fig. 4) and quantitative end group determinations (see "Results") indicate that a single amino group is liberated during activation of the enzyme, and that this group is the NH_2-terminal α-amino group of an isoleucine residue. Titration experiments (Table III, Figs. 5 and 6) show that the conversion of the zymogen to the enzyme is accompanied by the appearance of a single new titratable group in the pH region 5 to 10.5. Reacetylation of acetylated δ-chymotrypsin decreases the Van Slyke amino nitrogen by one amino group, and leads to a corresponding decrease in the catalytic activity (Tables II and III). On the supposition that the group with pK(app) = 8.3 and this α-amino group are the same, the reacetylation experiments are consistent with (*a*) the experimental results reported in this paper, which indicate that the catalytically active conformation of the enzyme has a protonated group with pK(app) = 8.3; and (*b*) the well known fact that chymotrypsinogen, in which the isoleucyl α-amino group is in peptide linkage and cannot be protonated, is catalytically inactive and has a specific rotation that is pH independent.

The interpretation (Fig. 9) based on the premise that the amino group liberated during the activation of acetylated chymotrypsinogen to acetylated δ-chymotrypsin is also the single new titratable group that appears during the activation process, is consistent with all experimental data available so far. This scheme is not inconsistent with the supposition that the group with pK(app) = 8.3 is the NH_2-terminal isoleucyl α-amino group. This scheme merely requires that $pK_I < pK(app) \leq pK_A$, and that $k_L < 1$ and $k_H > 1$. These conditions would be satisfied, for instance, by a pK_I value of 7.7 and a k_L value of 0.1. Assigning the pK(app) = 8.3 value to a different ionizing group would make it difficult, at present, to account for the observations that acetylation of the NH_2-terminal isoleucyl α-amino group leads to a loss in catalytic activity and that only a single new ionizing group appears in the pH region 6 to 10.5 when acetylated chymotrypsinogen is converted to acetylated δ-chymotrypsin.

Results obtained in these studies, and their interpretations, compare well with previous reports (41). Desnuelle (49) reported that neochymotrypsin, which is devoid of catalytic properties, has the same specific rotation as chymotrypsinogen. Neochymotrypsin has the NH_2-terminal alanyl α-amino group that is present in α-chymotrypsin, but the isoleucyl α-amino group is in amide linkage. In some chymotrypsin-catalyzed reactions it has been observed (47) that above pH 8.0 the rate of formation of chymotrypsin-substrate intermediates is pH-dependent, but their decomposition is pH-independent. This observation is consistent with the supposition that there is an increase in pK of the ionizing group responsible for the pH dependence at alkaline pH during formation of the chymotrypsin-substrate intermediate. The data are also consistent with observations of Chervenka and Wilcox (50) which indicate that neither the α-amino group of the NH_2-terminal half-cysteine residue nor any of the lysyl ϵ-amino groups are involved in chymotrypsin activity.

REFERENCES

1. Havsteen, B. H., and Hess, G. P., *J. Am. Chem. Soc.*, **84**, 491 (1962).
2. Wootton, J. F., and Hess, G. P., *J. Am. Chem. Soc.*, **84**, 440 (1962).
3. Labouesse, B., Havsteen, B. H., and Hess, G. P., *Proc. Natl. Acad. Sci. U. S.*, **48**, 2137 (1962).
4. Havsteen, B. H., and Hess, G. P., *J. Am. Chem. Soc.*, **85**, 791 (1963).
5. Havsteen, B. H., Labouesse, B., and Hess, G. P., *J. Am. Chem. Soc.*, **85**, 796 (1963).
6. Oppenheimer, H. L., Mercouroff, J., and Hess, G. P., *Biochim. Biophys. Acta*, **71**, 78 (1963).
7. Mercouroff, J., and Hess, G. P., *Biochem. Biophys. Res. Commun.*, **11**, 283 (1963).
8. Parker, H., and Lumry, R., *J. Am. Chem. Soc.*, **85**, 483 (1963).
9. Sigler, P. B., Skinner, R. C. W., Coulter, C. D., Kalos, J., Braxton, H., and Davies, D. R., *Proc. Natl. Acad. Sci. U. S.*, **51**, 1146 (1964).
10. Moon, A. Y., Mercouroff, J., and Hess, G. P., *J. Biol. Chem.*, **240**, 717 (1965).
11. Moon, A. Y., Sturtevant, J. M., and Hess, G. P., *J. Biol. Chem.*, **240**, 4204 (1965).
12. Moon, A. Y., Sturtevant, J. M., and Hess, G. P., *Federation Proc.*, **21**, 229 (1962).
13. Havsteen, B. H., and Hess, G. P., *Biochem. Biophys. Res. Commun.*, **14**, 313 (1964).
14. Havsteen, B. H., and Hess, G. P., *J. Am. Chem. Soc.*, **84**, 448 (1962).
15. Bettelheim, F. R., and Neurath, H., *J. Biol. Chem.*, **212**, 241 (1955).
16. Rovery, M., Poilroux, M., Curnier, A., and Desnuelle, P., *Biochim. Biophys. Acta*, **16**, 590 (1955).
17. Labouesse, B., Oppenheimer, H. L., and Hess, G. P., *Biochem. Biophys. Res. Commun.*, **14**, 318 (1964).

18. Dixon, G. H., and Neurath, H., *J. Biol. Chem.*, **225**, 1049 (1957).
19. Ferrari, A., *Ann. N. Y. Acad. Sci.*, **87**, 792 (1960).
20. Wilcox, P. E., Cohen, E., and Tan, W., *J. Biol. Chem.*, **228**, 999 (1957).
21. Dixon, G. H., Neurath, H., and Pechere, J.-F., *Ann. Rev. Biochem.*, **27**, 489 (1958).
22. Liener, I. E., *Arch. Biochem. Biophys.*, **88**, 217 (1960).
23. Schwert, G. W., Neurath, H., Kaufman, S., and Snoke, J. W., *J. Biol. Chem.*, **172**, 221 (1948).
24. Van Slyke, D. D., *J. Biol. Chem.*, **83**, 425 (1929).
25. Fraenkel-Conrat, H., in D. M. Greenberg (Editor), *Amino acids and proteins*, Charles C Thomas, Springfield, Illinois, 1951, p. 532.
26. Fraenkel-Conrat, H., *J. Biol. Chem.*, **148**, 453 (1943).
27. Schaffer, N. K., May, S. C., and Summerson, W. H., *J. Biol. Chem.*, **202**, 67 (1953).
28. Hirs, C. H. W., *J. Biol. Chem.*, **219**, 611 (1956).
29. Levy, A. L., and Li, C. H., *J. Biol. Chem.*, **213**, 487 (1955).
30. Randerath, J., *Thin-layer chromatography*, Academic Press, Inc., New York, 1963.
31. Biserte, F., Holleman, J. W., Holleman-Dehove, J., and Sautière, P., *Chromatog. Rev.*, **2**, 59 (1960).
32. Stark, G. R., and Smyth, D. G., *J. Biol. Chem.*, **238**, 214 (1963).
33. Kolthoff, J. M., *Z. Anal. Chem.*, **61**, 48 (1922).
34. Wetlaufer, D. B., *Advan. Protein Chem.*, **17**, 303 (1962).
35. Jacobsen, C. F., *Compt. Rend. Trav. Lab. Carlsberg, Sér. Chim.*, **25**, 325 (1947).
36. Chervenka, C. H., *J. Biol. Chem.*, **237**, 2105 (1962).
37. Hartley, B. S., *Nature*, **201**, 1284 (1964).
38. Greenstein, J. P., and Winitz, M., *Chemistry of the amino acids, Vol. 2*, John Wiley and Sons, Inc., New York, 1961, p. 1332.
39. Erlanger, B. F., Castleman, H., and Copper, A. G., *J. Am. Chem. Soc.*, **85**, 1872 (1963).
40. Scheraga, H. A., *Protein structure*, Academic Press, Inc., New York, 1961, p. 265.
41. Neurath, H., Rupley, J. A., and Dreyer, W. J., *Arch. Biochem. Biophys.*, **65**, 243 (1956).
42. Himoe, A., Parks, P. C., and Hess, G. P., *Federation Proc.*, **24**, 473 (1965).
43. Imahori, K., Yoshida, A., and Hashizume, H., *Biochim. Biophys. Acta*, **45**, 380 (1960).
44. Fasman, G. D., and Foster, R. J., *Federation Proc.*, **24**, 472 (1965).
45. Northrop, J. H., Kunitz, M., and Herriot, R. M., *Crystalline enzymes*, Ed. 2, Columbia University Press, New York, 1948.
46. Peller, L., and Alberty, R. A., *J. Am. Chem. Soc.*, **81**, 5907 (1959).
47. Bender, M. L., Clement, G. E., Kézdy, F. J., and Heck, H. d'A., *J. Am. Chem. Soc.*, **86**, 3680 (1964).
48. Tanford, C., *J. Am. Chem. Soc.*, **83**, 1628 (1961).
49. Desnuelle, P., *Enzymes*, **4**, 93 (1960).
50. Chervenka, C. H., and Wilcox, P. E., *J. Biol. Chem.*, **222**, 635 (1956); **222**, 621 (1956).

Reprinted from NATURE, Vol. 206, 1965

STRUCTURE OF HEN EGG-WHITE LYSOZYME

A Three-dimensional Fourier Synthesis at 2 Å Resolution

By DR. C. C. F. BLAKE, DR. D. F. KOENIG*, DR. G. A. MAIR, DR. A. C. T. NORTH†, DR. D. C. PHILLIPS†, and DR. V. R. SARMA

Davy Faraday Research Laboratory, Royal Institution, 21 Albemarle Street, London, W.1

THE X-ray analysis of the structure of hen egg-white lysozyme[1] (*N*-acetylmuramide glycanohydrolase, *EC* 3.2.1.17), which was initiated in this Laboratory in 1960 by Dr. R. J. Poljak, has now produced a Fourier map of the electron density distribution at 2 Å resolution in which the structure of the molecule can be clearly seen.

Comparison of this map with the primary structure, determined by Jollès *et al.*[2,3], by Canfield[4,5] and in part by Brown[6], has made possible the location of each of the 129 amino-acid residues of which the molecule is composed, many of which, including the four disulphide bridges, can be identified unambiguously in the X-ray image. The arrangement of groups in the molecule is very complicated, and many interactions between them will be revealed only by detailed analysis of the map. This preliminary account of the work is intended, therefore, to indicate the quality of the results, and to show the general structure of the molecule. Further work which is described in the following article[7] has also revealed the location of a site at which competitive inhibitor molecules are bound in the crystal structure so that it is possible, even at this stage, to identify amino-acid residues which may be involved in the binding of substrates.

X-ray Methods

Crystals of hen egg-white lysozyme chloride grown at *p*H 4·7 (ref. 8) are tetragonal with unit cell dimensions, $a = b = 79{\cdot}1$, $c = 37{\cdot}9$ Å, and space group $P4_12_12$ or $P4_32_12$ (ref. 9). Each unit cell contains eight lysozyme molecules (one per asymmetric unit), molecular weight about 14,600, together with 1 M sodium chloride solution which constitutes about 33·5 per cent of the weight of the crystal[10].

A preliminary investigation[11], in which the method of isomorphous replacement was used, showed that the space group is in fact $P4_32_12$ and made possible a calculation of the electron density distribution at 6 Å resolution.

In the extension of this analysis to higher resolution it has been necessary to find new and better isomorphous heavy-atom derivatives. Those originally used contained: (1) mercuri–iodide ions; (2) chloropalladite ions; and (3) *ortho*-mercuri hydroxytoluene *para*-sulphonic acid (MHTS) and of these only the MHTS derivative was found to be suitable for work at higher resolution. An extensive search, using methods already described[11], revealed, however, that derivatives containing (4) $UO_2F_5^{\equiv}$ (ref. 12); and (5) an ion derived from $UO_2(NO_3)_2$, probably $UO_2(OH)_n^{(n-2)-}$, were very suitable for this purpose. In addition derivatives including (6) *para*-chloromercuri benzene sulphonic acid (PCMBS) and (7) $PtCl_6^{=}$ (ref. 13) were found to be satisfactory for use at low resolution. All these derivatives except (1) and (2) were used in a re-examination of the structure at 6 Å. The necessary diffraction measurements were made with the Hilger and Watts, Ltd., linear diffractometer[14] adapted to measure three reflexions at a time[15,16] and the data were processed and the other calculations made on the Elliott 803 *B* computer at the Royal Institution. The phases were calculated by the phase probability method[17], adapted to make proper allowance for anomalous scattering of the copper $K\alpha$ X-rays by the various heavy atoms[18]. The availability of uranium derivatives was particularly fortunate in that the anomalous scattering effect is large for uranium and indeed a particular search was made for them with that in mind. The mean figure of merit obtained in these calculations was 0·97 as compared with the 0·86 obtained before, and the greater part of this improvement has been shown to arise from the improved treatment of anomalous scattering rather than from the inclusion of additional heavy-atom derivatives.

Fig. 1 shows a model of the electron-density distribution at 6 Å resolution obtained in this new work. The outline of the molecule is certainly clearer in the new map than in the old, and within the molecule there is improved continuity suggesting the course of a folded polypeptide chain; but the two maps are, in fact, very similar (root mean square difference in electron density 0·12 eÅ^{-3}) and it remains impossible to determine the structure from them in any detail. Nevertheless a tentative general interpretation was possible. The model can be divided roughly into two parts; on the left-hand side of Fig. 1 the density has the appearance of a single chain without cross-connexions while to the right the arrangement appears to be much more complex. In general agreement with the picture, the primary structure, shown in Fig. 2, includes some relatively long lengths of polypeptide chain free from cross-connexions together with a region in which two disulphide bridges and two proline residues are close together.

2 Å Resolution

In preparation for the high-resolution investigation, the various heavy-atom derivatives were investigated in the centro-symmetric $hk0$ and $h0l$ projections at 3 and 2 Å resolution. Comparison of the various sign predic-

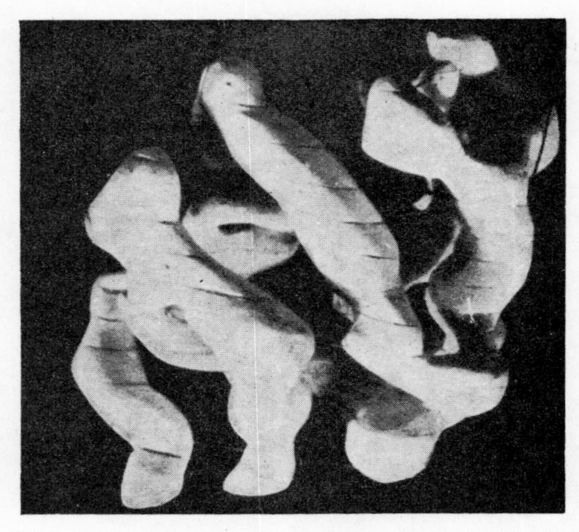

Fig. 1. Solid model of the lysozyme electron-density greater than about 0·5 electrons/Å³ at 6 Å resolution

* Present address: Brookhaven National Laboratory, Upton, Long Island, New York.

† Medical Research Council External Staff.

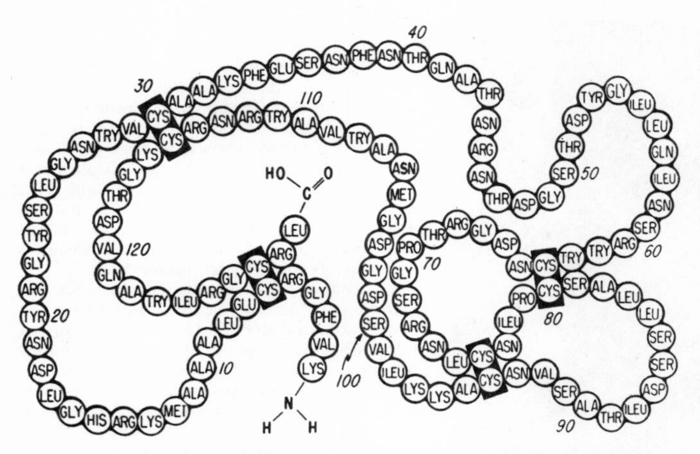

Fig. 2. Sequence of amino-acid residues in hen egg-white lysozyme reproduced from Canfield and Liu (ref. 5)

tions and examination of the radial distributions of the intensity differences[19] showed that only derivatives (3), (4) and (5) were suitable for high-resolution work. Accordingly 2 Å data were collected from the native crystals and for crystals containing these heavy atoms. The same experimental methods were used; but in order to minimize the effects of irradiation damage each crystal was exposed for only 20 h or less to the β-filtered X-ray beam from a copper tube operated at 40 kV, 20 m.amp. About 16 crystals were used to obtain each set of measurements which included generous overlaps for scaling[20] and separate measurement of the reflexions in the Bijvoet pairs *hkl* and *khl*. Semi-empirical corrections were made for absorption[21]. The data-processing system was designed at every stage to detect errors and to avoid their introduction and to eliminate any need for comprehensive manual checking.

The complete set of 2 Å data comprises some 9,040 reflexions of which 1,640 are centric, a property of this space group which greatly facilitates the determination and refinement of the heavy-atom parameters. The main heavy-atom positions were all found from $(\Delta F)^2$ maps in projection, and subsidiary sites were found from ΔF maps in two and three dimensions when preliminary signs and phases had been found. The best parameters were determined by the use of the quasi-three-dimensional set of centric reflexions, *hk*0, *k*0*l*, *hhl*, 0*kl*, in an adaptation of Hart's method[22]. The final parameters are shown in Table 1. The large proportion of centric reflexions also made possible an examination of the variation in apparent occupancy with scattering angle, thus revealing the shape of the appropriate scattering factor curve for each heavy-atom. These curves agree well with those expected for a heavy central atom surrounded by a shell of light atoms replacing a sphere of uniform electron density.

The temperature factors shown in Table 1 most deserve comment. Those obtained for the most important sites (of which there are no more than two for any derivative)

Table 1. Heavy-atom Parameters

	Site	X	Y	Z	O	B	E	N	R
MHTS	I	0·2068	0·6138	0·0507	39·2	17·8	58	1247	0·60
	II	0·2415	0·6393	0·9326	8·8	14·9			
UO_2F_5	I	0·1783	0·5849	0·7204	55·5	21·0	74	1277	0·52
	II	0·0974	0·8976	0·4650	29·3	24·2			
$UO_2(OH)_n$	I	0·0961	0·8938	0·4664	47·1	19·2	80	1140	0·57
	II	0·1898	0·5901	0·7168	42·1	124·8			
	III	0·0446	0·7266	0·5515	9·0	190·2			
	IV	0·0869	0·8976	0·4866	11·4	68·4			
	V	0·2024	0·6388	0·6781	28·6	42·8			

O, Occupancy of heavy-atom site—electrons; B, isotropic temperature factor constant—$Å^2$; E, root mean square difference between observed and calculated heavy-atom changes for centric reflexions—electrons; N, number of centric reflexions in the range $0·01 < \sin^2\theta < 0·15$ used in refinement; R, reliability index for observed and calculated heavy-atom changes of centric reflexions.

are all comparable with the overall value for the protein crystals themselves. The very large values obtained for other sites of (5) show that these sites are of little importance at high angles and may not represent true sites of heavy-atom attachment. The minor site of (3) MHTS is clearly the SO_3^- group of this molecule: it occupies the same position as the equivalent group in the not strictly isomorphous PCMBS derivative[23].

The refinement programme gave the root mean square 'lack of closure' errors required in the isomorphous phase determination[17] that are shown in Table 1. Analysis of the agreement obtained between measurements of symmetry equivalent reflexions gave the corresponding values required for the anomalous-scattering phase determination[18]. These were found to vary as a function of sin θ having the values $19 + 455 \sin^2\theta/\lambda^2$ for derivatives (3) and (4) and $26·5 + 485 \sin^2\theta/\lambda^2$ for derivative (5).

The phases of the 9,040 reflexions were calculated on the Elliott 803 *B* computer and had a mean figure of merit of 0·60. The variation with angle was very similar to that obtained in the comparable analysis of sperm-whale myoglobin[24]; the mean value fell to 0·38 at the 2 Å limit. The 'best'[17] electron-density distribution, unsharpened, was calculated from these phases and 'figures-of-merit' on the University of London *Atlas* computer. The density was calculated at 1/120ths of the cell edge along *a* and *b* and 1/60ths along *c*, and was tabulated in a form suitable for immediate contouring to a scale of 0·75 in. ≡ 1 Å. The Elliott 803 *B* computer was used to generate that part of the map which was expected, from the 6 Å resolution work, to include one molecule, from the arbitrary asymmetric unit calculated by *Atlas*. The absolute scale of the measurements is known only roughly (from statistical arguments and reasonable occupancies for the heavy-atoms) and the map was contoured at estimated 0·25 $eÅ^{-3}$ intervals.

The Fourier Map

The electron-density corresponding to a single molecule extends over the full length of the *c*-axis of the unit cell and in the other axial directions lies within the ranges $-\frac{1}{4} < x < +\frac{1}{4}$, $0 < Y < \frac{1}{2}$. The contour map was plotted on 60 sections of constant Z so as to extend safely beyond these ranges of X and Y. Two separate blocks of ten sections each, which are representative of the whole map, are shown in Figs. 3 and 4. No $F(000)$ term was included in the Fourier synthesis, so that density values are all relative to the average crystal density. Only contours of higher than average density are shown in the maps, but there are no patches of high negative density even at the heavy atom positions, and it is immediately apparent that the general background is satisfactorily uniform. In regions which must be occupied by water and salt[9], such as those near the two-fold rotation axes, there are many small peaks, which are consistent with some degree of order in the liquid of crystallization. Within the molecular boundary the density indicates clearly the course of the polypeptide chain and the nature of many side-chains.

At this resolution atoms are not expected to appear in the image as separate peaks of electron density, but groups of atoms connected only by ionic or van der Waals' interactions or by hydrogen bonds are expected to be resolved. It is satisfactory therefore to find a continuous ribbon of high density with characteristic features at regular intervals to represent the main polypeptide chain with its carbonyl groups and with side-chains protruding

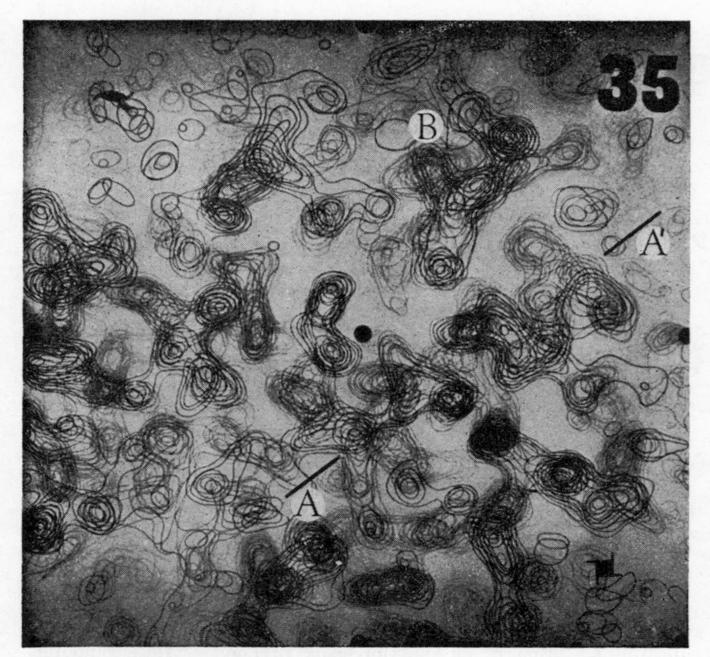

Fig. 3. Sections $Z = 35$—44/60 of the 2 Å resolution electron density map. Contours at intervals of 0·25 eÅ^{-3}. *AA'* shows the course of a length of α-helix lying in the plane of the sections. *B* indicates a length of helix more nearly normal to the sections

from it. In some regions the conformation of the main chain is clearly α-helical. For example, Fig. 3 includes the density corresponding to two lengths of α-helix, one of which runs approximately parallel to the sections (*AA'*) while the other is more nearly perpendicular to the sections (*B*). Inspection of the density in comparison with models shows immediately that the helices are right-handed running towards the terminal amino end of the polypeptide chain in the direction *AA'* and at *B* down into the map.

It is in such helical regions that the side-chains are most easily identified. Thus the peak of high density a little below the helix *AA'* and connected back to it clearly represents a disulphide bridge between the helix and more extended chain near the bottom of the picture. To the left of the disulphide bridge, density characteristic of another side-chain, a phenylalanine, can be seen emerging from the helix at an α-carbon position four residues removed from the cystine towards the terminal carboxyl end of the chain. This observation, when combined with the sequence of Fig. 2, immediately identifies the disulphide bridge as that between residues 30 and 115 and was, in fact, the starting-point of a detailed correlation between map and sequence. The helix at *B* corresponds to a part of the chain nearer the terminal amino end and has features clearly consistent with the sequence Arg-Lys-Met-Ala-Ala-Ala of residues 14–9.

In the non-helical regions, the courses of the main chain and the side-chains are rather less easily recognized by mere inspection of the map. Nevertheless, no real difficulty was encountered in following the main chain and nowhere did the density corresponding to it fall to the general background-level. As in the corresponding study of myoglobin[25], however, it was often simpler to recognize the density corresponding to particular well-marked side-chains than to follow the main-chain density directly. Some of these clearly identifiable side-chains in a non-helical region are shown in Fig. 4. They include most prominently the indole rings of tryptophan residues 28, 108 and 111, and the sulphur atom of methionine 105 which come together with tyrosine 23 in this part of the structure to form a curiously regular hydrophobic box.

Of course, not all the residues are as clearly recognizable as this; but using the criteria formulated in the examination of myoglobin[25] it is possible to identify many of them with considerable confidence. Nevertheless, the existence of almost complete knowledge of the primary structure has been of enormous value, and no attempt was made to analyse the X-ray image without recourse to it. No disagreements with the chemical information were obvious at this stage of analysis except in those few small regions in which the chemical studies[2,4] do not agree. These are discussed here.

General Description of the Molecule

The molecule is roughly ellipsoidal with dimensions about 45 × 30 × 30 Å. The general arrangement of the polypeptide chain is indicated in the schematic diagram of Fig. 5. It is clearly even more complicated than that of myoglobin, and a comprehensive description cannot yet be given. Some important features, however, are immediately apparent. First, the α-helical content is relatively low. Detailed in-

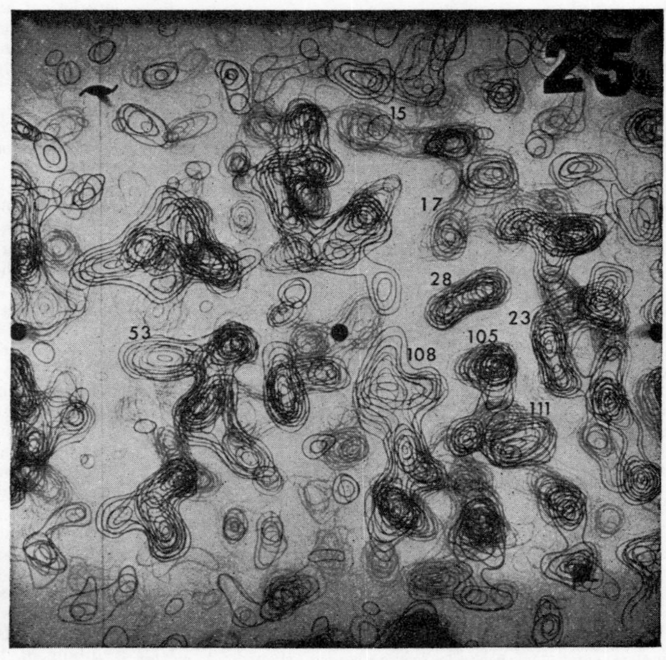

Fig. 4. Sections $Z = 25$—34/60 of the 2 Å resolution electron density map. The densities corresponding to a number of amino-acid side-chains are numbered in accordance with Fig. 2

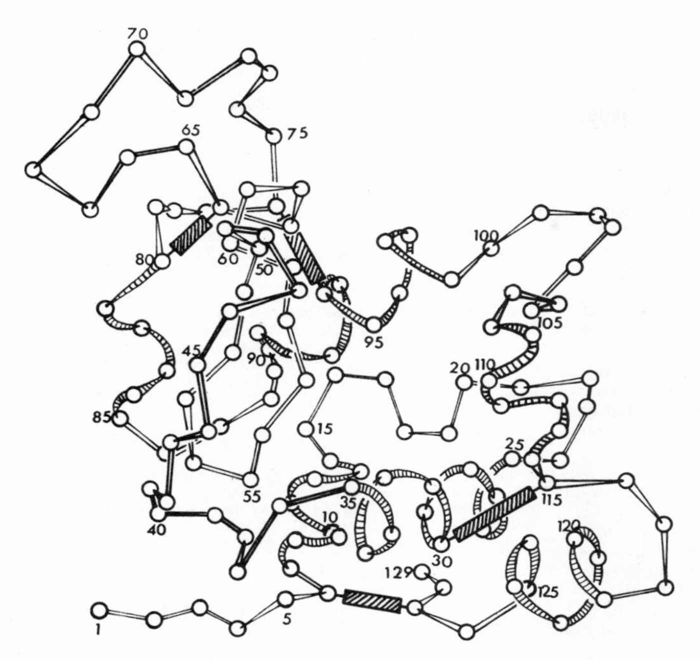

Fig. 5. Schematic drawing of the main chain conformation (by W. L. Bragg)

vestigations of the helical parameters have not yet been made, but about six lengths of helix have been recognized, some of them very short, as indicated in Figs. 5 and 6. If a residue is regarded as being in a helix if one or other of its possible hydrogen bonds is properly made, then about 55 of the 129 residues appear at this stage to be included in helices. This suggests a maximum helix content of about 42 per cent, in fair agreement with the predictions of optical rotatory dispersion[26,27].

The remainder of the molecule is less easily described, the folding of residues 35–80 being particularly complex. In this region three lengths of chain are roughly anti-parallel, two of them, residues 41–54, forming a nearly closed loop. Nearby are the two closely related disulphide bridges connected by short lengths of chain which include the proline residues (70 and 79).

The disulphide bridges each have a helix on at least one side, although the half-cystine is generally the last helical residue or at least in the last turn of its helix. Only one bridge, 30–115, has a helix on both sides and 115 is a terminal residue.

The inside and outside of the molecule are less easily defined than they are for myoglobin, in which the interior of the molecule consists almost exclusively of hydrophobic side-chains[28]. The lysozyme molecule appears to have a hydrophobic spine consisting mainly of the six tryptophan side-chains and including the hydrophobic box already described (Fig. 4). Three of these tryptophan side-chains, 62, 63 and 123, protrude, however, beyond the molecular boundary and there are in addition a number of strongly hydrophobic side-chains clearly on the molecular surface (for example, Val 2, Phe 3, Phe 34, Leu 17).

On the other hand, the parts of the molecule most shielded from contact with surrounding liquid appear to include Ser(91) and Gln(57). In addition, residue 58, which may be asparagine[2], can be regarded as internal. All the lysine and arginine side-chains are external.

Comparison with Low Resolution Results

It is interesting to compare this molecule with the model obtained at 6 Å resolution. The boundary of the molecule was correctly chosen in the revised 6 Å work (Fig. 1) and was incorrect only in minor details in the earlier investigation[11]. Furthermore, the very general interpretation was correct in that the complex folding indicated by the right-hand part of Fig. 1 does indeed include the two disulphide bridges 64–80 and 76–94. On the other hand, it is somewhat surprising to discover that the 6 Å maps contain no significant indications of the positions of three of the disulphide bridges, possibly because they lie in diffraction minima associated with neighbouring helices. Clearly it is very unlikely indeed that the 6 Å model could have been interpreted correctly in detail, even in the light of the primary structure, and this is true also of a model at 5 Å resolution which is not described here.

Comparison with Chemical Data

The independent investigations of the primary structure[2,4,5] now agree in all but a few details on the sequence of amino-acid residues in the lysozyme molecule. Two of the remaining doubts[5] are concerned with the distinction between aspartic acid and asparagine, which are not readily distinguishable by the X-ray method. The other three difficulties, which are more serious, can be resolved to some extent by examination of the present map. Thus, residues 92 and 93 are included in a length of helix and their side-chains have strongly contrasted appearances. That of 92 is clearly forked close to the helix and is surrounded by a clear region of low electron density, while the side-chain of residue 93 is longer and approaches much more nearly to neighbouring density. It is probable, therefore, that the sequence 92–93 is valine–asparagine[4] rather than asparagine–valine[2]. In the second case, the side-chains associated with residues 40–42, which are in a non-helical region, appear to be much more consistent

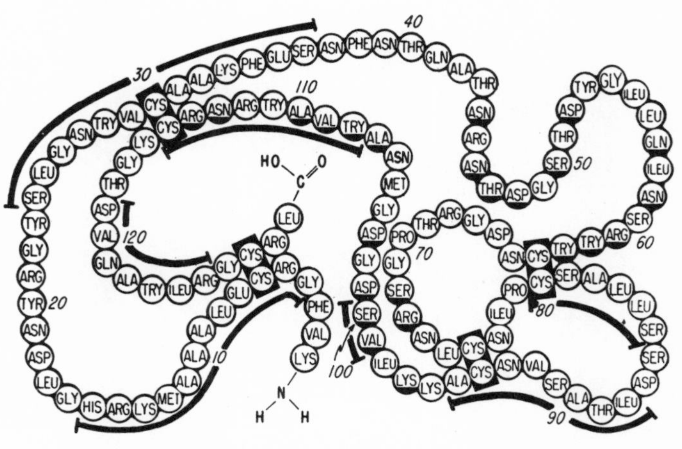

Fig. 6. Primary structure (ref. 5) showing (continuous line) the residues found in α-helical conformation and those in the apparent binding site (underlined)

(by similar arguments) with the sequence Thr-Gln-Ala[4] than with Gln-Ala-Thr[2]. The third ambiguity, between isoleucine and asparagine in positions 58 and 59, has already been referred to. In this case the differences between the two densities are less clear, but on balance that associated with residue 58 seems more consistent with asparagine[2] than with isoleucine[4] even though this residue can be regarded as being inside the molecule. Its environs include the disulphide bridge 76–94 and Leu 83, but also serine 91, and there is an unexplained neighbouring patch of density which could represent an included water molecule. It is hoped that the detailed analysis of side-chain interactions which is now in progress will resolve this uncertainty.

In addition to the primary structure there are, of course, very many chemical data on the physical and chemical properties of lysozyme which cannot be discussed here, but which are expected to be interpretable eventually in terms of the detailed structure. Something must now be said, however, about amino-acid residues probably concerned in the enzymatic activity, a subject on which there is so far rather little definite chemical evidence.

Inhibitor Binding Site

The following article describes the location at low resolution of the site of attachment of *N*-acetyl-glucosamine and its dimer, chitobiose, both of which molecules act as competitive inhibitors of lysozyme. They attach themselves, at least in this crystal structure, in a cleft in the surface of the molecule which runs nearly vertically down the middle of Fig. 5 and which can be seen clearly in the low-resolution model (Fig. 1, ref. 7). The amino-acid residues bounding this portion of the molecular surface come from various parts of the polypeptide chain and are indicated in Fig. 6. Clearly, further examination is required to establish whether this site represents the binding site of substrate molecules, but for the present it seems a reasonable hypothesis that it does so and that the properties of these residues may well repay detailed investigation. In accordance with chemical evidence[29–32], they include three tryptophans, now shown to be 62, 63 and 108, but not the single histidine which is located on the opposite side of the molecule.

We thank Sir Lawrence Bragg, who drew the diagram of Fig. 5, and Prof. R. King for their advice, and the Managers of the Royal Institution for their support; the Director of the London University Computer Unit for facilities on *Atlas*; and Mrs. R. Arthanari, Mrs. W. Browne, Mrs. S. J. Cole, Mrs. J. A. Conisbee, Miss M. Hibbs, Mrs. K. Sarma and Messrs. A. R. Knott, S. B. Morris and J. E. T. Thirkell for their assistance. We also thank the Medical Research Council and the U.S. National Institutes of Health for financial support.

1 Fleming, A., *Proc. Roy. Soc.*, B, **93**, 306 (1922).
2 Jollès, J., Jauregui-Adell, J., and Jollès, P., *Biochim. Biophys. Acta*, **78**, 68 (1963).
3 Jollès, P., Jauregui-Adell, J., and Jollès, J., *C.R. Acad. Sci., Paris*, **258**, 3926 (1964).
4 Canfield, R. E., *J. Biol. Chem.*, **238**, 2698 (1963).
5 Canfield, R. E., and Liu, A. K., *J. Biol. Chem.* (in the press).
6 Brown, J. R., *Biochem. J.*, **92**, 13P (1964).
7 Johnson, L. N., and Phillips, D. C. (following article).
8 Alderton, G., and Fevold, J., *J. Biol. Chem.*, **164**, 1 (1946).
9 Palmer, K. J., Ballantyre, M., and Galvin, J. A., *J. Amer. Chem. Soc.*, **70**, 906 (1948).
10 Steinrauf, L. K., *Acta Cryst.*, **12**, 77 (1959).
11 Blake, C. C. F., Fenn, R. H., North, A. C. T., Phillips, D. C., and Poljak, R. J., *Nature*, **196**, 1173 (1962).
12 Holmes, K. C., and Leberman, R., *J. Mol. Biol.*, **6**, 439 (1963).
13 Poljak, R. J., *J. Mol. Biol.*, **6**, 244 (1963).
14 Arndt, U. W., and Phillips, D. C., *Acta Cryst.*, **14**, 807 (1961).
15 Phillips, D. C., *J. Sci. Instrum.*, **41**, 123 (1964).
16 Arndt, U. W., North, A. C. T., and Phillips, D. C., *J. Sci. Instrum.*, **41**, 421 (1964).
17 Blow, D. M., and Crick, F. H. C., *Acta Cryst.*, **12**, 794 (1959).
18 North, A. C. T., *Acta Cryst.*, **18**, 212 (1965).
19 Crick, F. H. C., and Magdoff, B. S., *Acta Cryst.*, **9**, 901 (1956).
20 Hamilton, W. C., Rollett, J. S., and Sparks, R. A., *Acta Cryst.*, **18**, 129 (1965).
21 North, A. C. T., Phillips, D. C., and Mathews, F. S. (in preparation).
22 Hart, R. G., *Acta Cryst.*, **14**, 1188 (1961).
23 Blake, C. C. F., Fenn, R. H., and Phillips, D. C. (in preparation).
24 Kendrew, J. C., Dickerson, R. E., Strandberg, B. E., Hart, R. G., Davies, D. R., Phillips, D. C., and Shore, V. C., *Nature*, **185**, 422 (1960).
25 Kendrew, J. C., Watson, H. C., Strandberg, B. E., Dickerson, R. E., Phillips, D. C., and Shore, V. C., *Nature*, **190**, 663 (1961).
26 Urnes, P., and Doty, P., *Adv. Prot. Chem.*, **16**, 401 (1961).
27 Hamaguchi, K., and Imahori, K., *J. Biochem.* (Japan), **55**, 388 (1964).
28 Kendrew, J. C., *Brookhaven Symp. Quant. Biol.*, **15**, 216 (1962).
29 Kravchenko, N. A., Kleopina, G. V., and Kaverzneva, E. D., *Biochim. Biophys. Acta*, **92**, 412 (1964).
30 Bernier, I., and Jollès, P., *C.R. Acad. Sci., Paris*, **253**, 745 (1961).
31 Ramachandran, L. K., and Rao, G. J. S., *Biochim. Biophys. Acta*, **59**, 507 (1962).
32 Hartdegen, F. J., and Rupley, J. A., *Biochim. Biophys. Acta*, **92**, 625 (1964).

Reprinted from NATURE, Vol. 206, 1965

STRUCTURE OF SOME CRYSTALLINE LYSOZYME-INHIBITOR COMPLEXES DETERMINED BY X-RAY ANALYSIS AT 6 Å RESOLUTION

By MISS LOUISE N. JOHNSON and DR. D. C. PHILLIPS*

Davy Faraday Research Laboratory, Royal Institution, 21 Albemarle Street, London, W.1

A RECENT investigation of the azide derivative of sperm-whale myoglobin[1] has shown that the interactions of proteins with small molecules can be examined in crystals when the phases of reflexions from isomorphous unsubstituted-protein crystals are known. The difference Fourier method is used in which the changes in structure amplitude caused by the introduction of the small molecule are combined with the known phases to give a map showing the change in electron density. The work on myoglobin[1] was done at 2 Å resolution; but we have now shown that interesting results can be obtained by this method at 6 Å resolution even when the substituted molecules comprise only light atoms. The method depends on the existence of precise phase information for the protein crystals, information which is now available for lysozyme[2], and by its means we have been able to investigate the structural relationship between lysozyme and a number of compounds which are related to the substrate of the enzyme and which act as competitive inhibitors of its action. In this way we have been able to locate a part of the molecule which may be responsible for its enzymatic activity.

* Medical Research Council External Staff.

Lysozyme has a β(1–4) glucosaminidase activity with the ability to hydrolyse a mucopolysaccharide component of some bacterial cell walls releasing *N*-acetyl amino sugars derived from glucosamine and muramic acid[3]. The proposed structure of a tetrasaccharide[4] isolated from the cell wall of *Micrococcus lysodeikticus*, which is made up of alternate *N*-acetylglycosamine and *N*-acetyl muramic acid units, is shown in Fig. 1 in which the bond probably hydrolysed by lysozyme is indicated by an arrow. Lysozyme will also hydrolyse chitin[5], the (1–4) linked linear chain polymer of *N*-acetylglucosamine, and Wenzel *et al.*[6] have shown that lysozyme will promote the cleavage of the trimer, tri-*N*-acetyl chitotriose,

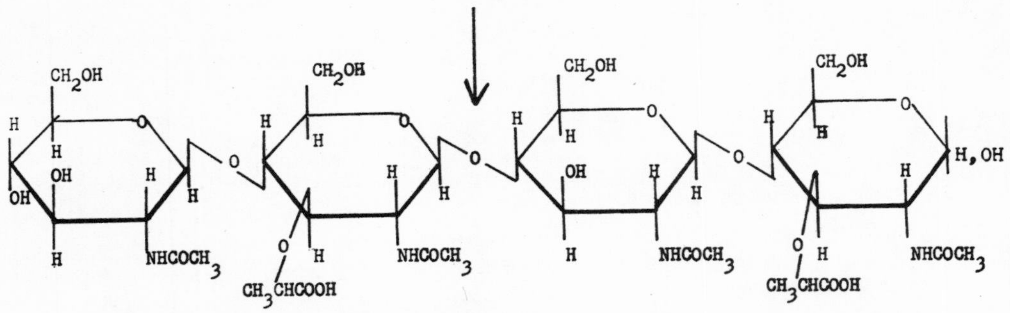

Fig. 1. Tetrasaccharide from *Micrococcus lysodeikticus* with the bond hydrolysed by lysozyme ref. 4) shown by an arrow

releasing dimer and monomer molecules. They reported that *N*-acetylglucosamine inhibits the activity of lysozyme but that glucosamine does not, and recently Rupley[7] has reported that the dimer, di-*N*-acetylchitobiose, is also an inhibitor and that it is not hydrolysed.

With this background we have examined the binding to lysozyme in the crystal structure[2] of five compounds: *N*-acetylglucosamine, di-*N*-acetylchitobiose, 6-iodo α-methyl *N*-acetylglucosaminide, glucosamine (hydrochloride) and muramic acid. The first three of these, which are competitive inhibitors of lysozyme, bind specifically to one and the same site on the enzyme whereas the last two which do not inhibit do not bind specifically to this site either.

Crystallographic Methods

Crystals of lysozyme chloride were grown in the usual way at *p*H 4·7 (ref. 2) and then soaked for 24–48 h in a solution of the inhibitor with concentration about 0·6 M. A precession photograph was then taken of the *hk*0 reflexions to 3 Å resolution and the intensities were compared with those of the native protein crystals. If there were changes complete three-dimensional data were collected to 6 Å resolution, processed in the usual way and scaled to the native data. (An outline of the experimental methods is included in the previous article[2].) Typically the average F value was about 255 (arbitrary units) while the mean absolute difference in amplitude between the derivative and the native protein was about 30 with some differences as large as 86. About 400 ΔF values were used with associated phases to calculate each three-dimensional difference map at 6 Å resolution. The calculations were done on the Elliott 803 *B* computer using a programme written by Mr. R. M. Simmons.

Results

(*a*) *N-acetylglucosamine*. The three-dimensional difference synthesis gave a remarkably clear map with only one peak which could be interpreted as representing the *N*-acetylglucosamine (NAG) molecule. The peak height was 0·16 e/Å^3 and the peak contained 48 electrons distributed throughout a volume of 460 Å^3. Luzzatti[8] has shown that in an electron-density synthesis the height of peaks representing atoms not taken into account in the phase determination approaches 50 per cent of their true height when the ratio of known to unknown atoms is large. Furthermore, in the present situation a reduction in peak heights is expected due to the replacement of water molecules. The peak height calculated for a molecule of NAG placed in the position of the peak was found to be 0·5 e/Å^3 so that the observed value is in satisfactory agreement with the value expected for a molecule of NAG in this position displacing some water.

(*b*) 6-*Iodo α-methyl N-acetylglucosaminide*. This derivative was synthesized for us in this laboratory by Dr. J. W. H. Oldham. It gave rise to a single peak in the same position as that occupied by NAG itself and with a peak height of 0·22 e/Å^3. The peak contained 64 electrons distributed throughout a volume of 530 Å^3. The peak was significantly higher than that obtained for the uniodinated compound, but the iodine atom was not resolved.

(*c*) *Di-N-acetylchitobiose*. This dimer, in which two NAG molecules are joined by a β(1–4) linkage, was kindly provided by Dr. J. A. Rupley. The difference synthesis showed that the dimer binds to the same place as the monomer, giving rise to a peak of height 0·2 e/Å^3 with 100 electrons distributed throughout 960 Å^3.

The position occupied by chitobiose in relation to the lysozyme molecule is shown in Fig. 2. The model of the lysozyme molecule is based on the structure analysis at 6 Å resolution[2], while the shaded peak representing chitobiose was constructed from the difference synthesis. The peak due to the single NAG molecule is shown as the dark region in the peak due to the dimer.

(*d*) *Glucosamine hydrochloride*. The addition of glucosamine gave rise to significant changes in the reflexion intensities, but difference maps showed no single peak where the NAG had bound but rather a large number of smaller peaks scattered over the unit cell. From this result it appears that glucosamine will bind to lysozyme, perhaps through its amino-group, but that it does not bind specifically in one place.

Fig. 2. Photograph of the model of a lysozyme molecule obtained by X-ray analysis at 6 Å resolution together with the increase in electron density observed in the presence of di-*N*-acetylchitobiose (hatched). The increase in electron density due to *N*-acetylglucosamine is shown as the darker part of the chitobiose

(*e*) *Muramic acid.* Muramic acid, kindly supplied to us by Dr. R. H. Gigg, likewise showed no signs of specific binding to lysozyme.

Inhibition experiments. The effects of NAG, 6-iodo α-methyl NAG and glucosamine on the activity of lysozyme were investigated by means of the usual method of assay[9]. The results confirmed the earlier conclusion[6] that NAG inhibits competitively while glucosamine does not, and showed further that 6-iodo α-methyl NAG has a very similar effect to that of NAG.

Discussion

Although it is only when this work is repeated at higher resolution that detailed information concerning the interactions between the protein and inhibitor can be obtained, these results at 6 Å resolution are highly encouraging and the following conclusions can already be drawn.

There is good evidence that the site indicated by the inhibitors is indeed the region of the enzyme responsible for its activity. The inhibitors are structurally analogous to the substrate and inhibit competitively. The inhibitors bind specifically to this one site whereas closely similar molecules which are not inhibitors do not. The site is different from those to which the heavy atoms bind.

It is possible to make preliminary statements concerning the nature of the groups on the substrate which are involved in the binding. It appears the *N*-acetyl group is essential. Molecules such as glucosamine and muramic acid which do not contain this group do not bind and do not inhibit. The fact that 6-iodo α-methyl *N*-acetylglucosaminide binds and inhibits suggests that neither the reducing group nor the oxygen in the six position is essential for binding. Studies of other derivatives of NAG, in which different groups have been blocked[10], are now in progress and it is hoped that they will show, even in advance of high-resolution analysis, which groups are involved in the binding.

The results described so far do not reveal the site at which the *N*-acetylmuramic acid (NAM) moiety of the cell-wall substrate is bound. Unfortunately, this compound, though synthesized many times, is not readily available, and we have not yet obtained any for examination. However, a preliminary investigation of the interaction of lysozyme with penicillin V (ref. 11), probably a structural analogue of *N*-acetylmuramic acid[12], suggests that it is bound at a site adjacent to that occupied by NAG and above it in Fig. 2. Experiments with *N*-acetylmuramic acid itself and the NAG–NAM dimer[13] are now being planned.

From Fig. 2 it can be seen that the inhibitor molecules lie well embedded in the enzyme molecule in a crevice in its surface. This is interesting in view of the many theories of enzyme action which envisage just such a situation. At no point does the density due to the inhibitor molecules penetrate the density representing the enzyme, but in two regions they are very close. The amino-acid residues in these regions can now be identified in the image of the enzyme at 2 Å resolution and they are shown in the previous article[2]. It is encouraging that they do not include the single histidine, which is not now believed to play a direct part in the enzyme-substrate interaction[14], and that they do include three of the tryptophans, since it has been reported[15,16] that the maintenance, intact, of four of the six indole residues is essential for enzymatic activity. Clearly the high-resolution investigations of enzyme-inhibitor complexes which are now being started are needed to show in detail which amino-acid residues are involved in the binding. Similar experiments with substrate analogues resistant to hydrolysis and with true substrates such as tri-*N*-acetylchitotriose[7], using methods pioneered by Doscher and Richards[17], may well reveal any structural changes in the enzyme that take place during the reaction that it controls.

We thank our colleagues in this laboratory for their help in these experiments; Sir Lawrence Bragg and Prof. R. King for advice; the Managers of the Royal Institution for the provision of facilities; and the Medical Research Council and the U.S. National Institutes of Health for financial support. One of us (L. N. J.) thanks the Department of Scientific and Industrial Research for a research studentship.

[1] Stryer, L., Kendrew, J. C., and Watson, H. C., *J. Mol. Biol.*, **8**, 96 (1964).
[2] Blake, C. C. F., Koenig, D. F., Mair, G. A., North, A. C. T., Phillips, D. C., and Sarma, V. R., *Nature* (preceding article).
[3] Salton, M. R. J., *Biochim. Biophys. Acta*, **22**, 495 (1956).
[4] Jeanloz, R. W., Sharon, N., and Flowers, H. M., *Biochim. Biophys. Res. Comm.*, **13**, 20 (1963).
[5] Berger, L. R., and Weiser, R. S., *Biochim. Biophys. Acta*, **26**, 517 (1957).
[6] Wenzel, M., Lenk, H. P., and Schutte, E., *Z. Physiol. Chem.*, **327**, 13 (1962)
[7] Rupley, J. A., *Biochim. Biophys. Acta*, **83**, 245 (1964).
[8] Luzzatti, V., *Acta Cryst.*, **6**, 142 (1953).
[9] Shugar, D., *Biochim. Biophys. Acta*, **8**, 302 (1952).
[10] Oldham, J. W. H. (to be published).
[11] Johnson, L. N. (to be published).
[12] Collins, J. F., and Richmond, M., *Nature*, **195**, 145 (1962).
[13] Sharon, N., paper presented at the Third Intern. Symp. on Fleming's Lysozyme, Milan, April 3–5 (1964).
[14] Kravchenko, N. A., Kleopina, G. V., and Kaverzneva, E. D., *Biochim. Biophys. Acta*, **92**, 412 (1964).
[15] Bernier, I., and Jollès, P., *C.R. Acad. Sci., Paris*, **253**, 745 (1961).
[16] Rao, G. J. S., and Ramachandran, L. K., *Biochim. Biophys. Acta*, **59**, 507 (1962).
[17] Doscher, M. S., and Richards, F. M., *J. Biol. Chem.*, **238**, 2399 (1963).

Reprinted from PROCEEDINGS OF THE NATIONAL ACADEMY OF SCIENCES, Vol. 57, 1967

THE STRUCTURE OF CARBOXYPEPTIDASE A, IV. PRELIMINARY RESULTS AT 2.8 Å RESOLUTION, AND A SUBSTRATE COMPLEX AT 6 Å RESOLUTION

BY M. L. LUDWIG, J. A. HARTSUCK, T. A. STEITZ, H. MUIRHEAD, J. C. COPPOLA, G. N. REEKE, AND W. N. LIPSCOMB

DEPARTMENT OF CHEMISTRY, HARVARD UNIVERSITY

On this occasion of a symposium on three-dimensional structure of macromolecules of biological origin we would like to present two new findings in a continuing study of the molecular structure and function of carboxypeptidase A (CPA). In an electron density map at 2.8 Å resolution we can distinguish clearly resolved regions of right-handed helix. In addition, difference electron density maps at 6 Å resolution, based on data from isomorphous crystals of apoenzyme-substrate complexes, show peaks which locate a substrate binding site.

Helical Regions.—X-ray diffraction intensities have been measured to a resolution of 2.8 Å for the following: (*a*) the native enzyme CPA_{α}[1, 3] (7208 reflections), (*b*) a lead citrate derivative with Pb bound at two sites per molecule of CPA, (*c*) a $PtCl_4^{-2}$ derivative with platinum at one major and three minor binding sites, and (*d*) a mercury derivative in which mercury has replaced Zn at the active site.[2]

Scale, occupancy, positional and thermal parameters were refined by methods described previously.[1] Criteria which were used to follow the refinement are summarized as a function of interplanar spacing in Table 1. The lead atoms still contribute significantly to the scattering from the lead derivative in the 2.8–3.3 Å region. On the other hand, the average heavy atom contribution in this region is less than the mean triangle closure error, *E*, for the platinum and mercury derivatives. The rapid decrease on the root mean square structure factor for platinum with $\sin^2\theta/\lambda^2$ (Table 1) suggests that this derivative may not be satisfactory for high resolution work. Moreover, we have recently found that heavy atom occupancies of the Pt derivative vary from crystal to crystal. In the mercury derivative the heavy atom scattering is small because the effective occupancy is equal to Hg minus Zn. In order to improve the phasing we are therefore taking X-ray data to 2.8 Å resolution on a mercury derivative containing several mercury atoms per molecule.

Both regions (*A-A′* and *B-B′* in Fig. 1) of the 2.8 Å resolution electron density map which have been examined contain right-handed helix whose backbone is clearly resolved. This detailed result is in accord with the low resolution interpretation that the *A-A′* and *B-B′* regions of Figure 1 are helical. The number of residues per turn and the pitch of the helical section *A-A′* agree well with α-helical parameters. An analysis of the *A-A′* portion of the map shows that the amino to carboxyl direction is from bottom to top in Figure 2, as was suggested by our interpretation of the 6 Å map. The direction at which the side chains attach to the helical backbone and the elongations along the backbone, most probably corresponding to carbonyl oxygens, indicate that the C-terminal end of this segment is at the top of Figure 2. A serious effort to fit the few published portions of the sequence will be made upon completion of data collection on the multiple-site mercury derivative and subsequent calculation of a new map.

An Apoenzyme Substrate Complex.—Intensity changes were observed in X-ray

TABLE 1

SUMMARY OF REFINEMENT OF 2.8 Å DATA FOR CARBOXYPEPTIDASE A

Spacing (Å)		10	10–5	5–3.3	3.3–2.8
E_j [a]	Pb	7.6	5.8	6.7	6.6
	Pt	8.9	6.4	8.0	8.1
	Hg	6.4	4.6	6.5	8.3
RMS f [b]	Pb	21.5	16.2	14.9	14.4
	Pt	20.1	15.7	9.8	7.0
	Hg	10.4	9.5	7.8	6.5
R_H [c]	Pb	0.08	0.09	0.09	0.11
	Pt	0.09	0.08	0.09	0.11
	Hg	0.07	0.07	0.09	0.13
$<m>$ [d]		0.76	0.73	0.59	0.52

Values of R [e] are 0.55 for the Pb, 0.70 for the Pt, and 0.68 for the Hg derivatives.

[a] $E_j = [\sum_{hkl}(|F_H|_{obs} - |F_H|_{calc})^2/n_j]^{1/2}$ is the closure error for phase triangles where n_j is the number of observed amplitudes of scattering $|F_H|_{obs}$ for derivative j, and the sum is taken over all reflections hkl.

[b] The RMS f is the root mean square contribution of the heavy atoms for each derivative.

[c] In the calculation of $R_H = \sum_{hkl}||F_H|_{obs} - |F_H|_{calc}|/\sum_{hkl}|F_H|_{obs}$, the $|F_H|_{calc} = ||F_P|e^{i\alpha} + f_c|$ where $|F_P|e^{i\alpha}$ is the vector scattering from the protein and f_c is the heavy atom vector.

[d] The figure of merit $<m>$ is the mean value of $\cos(\alpha - \alpha_0)$ where $\alpha - \alpha_0$ is the error in the phase angle α. The average value of $<m>$ is 0.586.

[e] $R = \sum|F_H - F_P|_{calc}/\sum||F_H| - |F_P||_{obs}$.

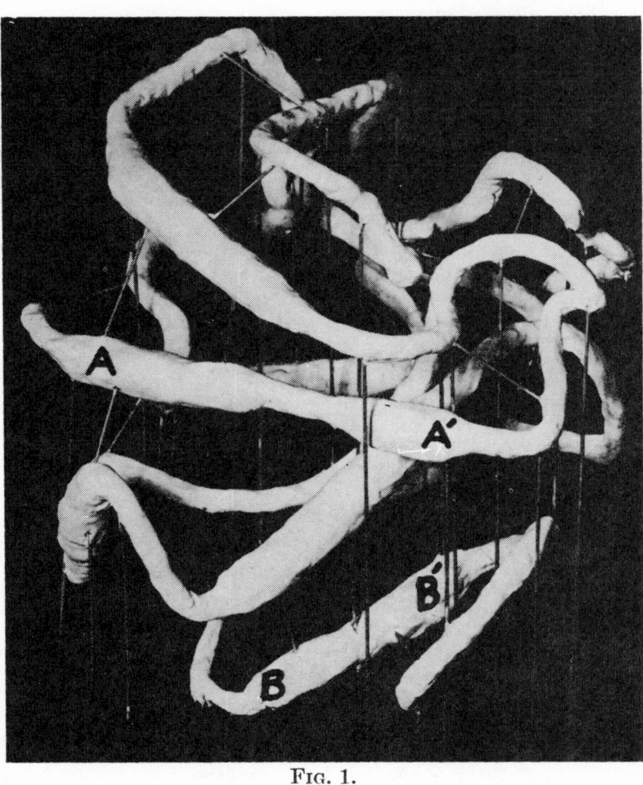

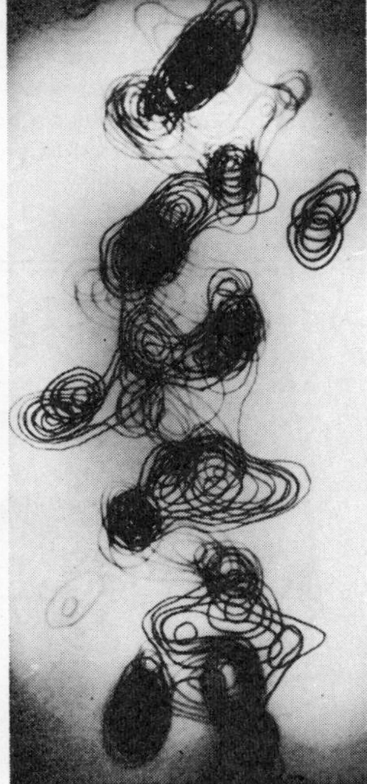

FIG. 1. FIG. 2.

FIG. 1.—View of the major helical regions (thickened) in the interpretation of the low resolution map of carboxypeptidase A. The b axis is in the vertical direction. Regions A-A' (Fig. 2) and B-B' have been examined at a resolution of 2.8 Å.

FIG. 2.—Part of the A-A' helix of carboxypeptidase A. The view is down the b axis and the helix end at the top corresponds to A of Fig. 1.

photographs of apoenzyme crystals after immersion in solutions of glycyl-L-tyrosine or chloroacetyl-L-phenylalanine. A three-dimensional difference electron density map at 6 Å resolution was calculated using data from apoenzyme crystals and from apoenzyme crystals soaked for 12 hours in 0.01 *M* glycyl-L-tyrosine, 0.1 *M* LiCl, 0.02 *M* Tris at pH 7.5. The difference peak, which is about twice the largest background variation, has its maximum 4–5 Å from the zinc position and extends into the adjacent pocket in the molecule. In Figures 3 and 4 we show the relation of the glycyl-L-tyrosine difference peak (the black area in Fig. 4) to the enzyme molecule. Present results are not inconsistent with an orientation of the substrate in which the aromatic side chain lies in the pocket, as we had surmised earlier.[1] Structural studies on the glycyl-L-tyrosine and related complexes at higher resolution should provide definitive information on the mode of substrate binding and hopefully on the mechanism of catalysis.

Note added in proof, February 10, 1967: Following the presentation of the above material, we have made substantial further progress, which is briefly summarized here.

Our final three-dimensional electron density map at 2.8 Å has now been computed, following completion of data collection on a triple site mercury derivative. Present refinement criteria (Table 2) are considerably improved over those of Table 1. The over-all figure of merit is now 0.70, substantially higher than the value of 0.59 in Table 1. Although the $PtCl_4^{-2}$ derivative contributes significantly only to phasing of reflections with resolution less than approximately 3.5 Å, the other derivatives are excellent to 2.8 Å. Detailed interpretation of this new map has just begun. It is already apparent that both the continuity of electron density, particularly in nonhelical regions, and the connection of side chains are improved over these same features in the earlier map at 2.8 Å resolution. Data from the triple site mercury derivative and the double site lead derivative are currently being collected between 2.0 and 3.0 Å on reflections found to be large in our 2.0 Å native data set.

The reduced enzymatic activity[4] of CPA in the solid state, as compared with the activity in solution, has permitted three-dimensional X-ray diffraction studies at 6 Å resolution of an enzyme-substrate complex in crystals which are isomorphous with those of the native enzyme. The

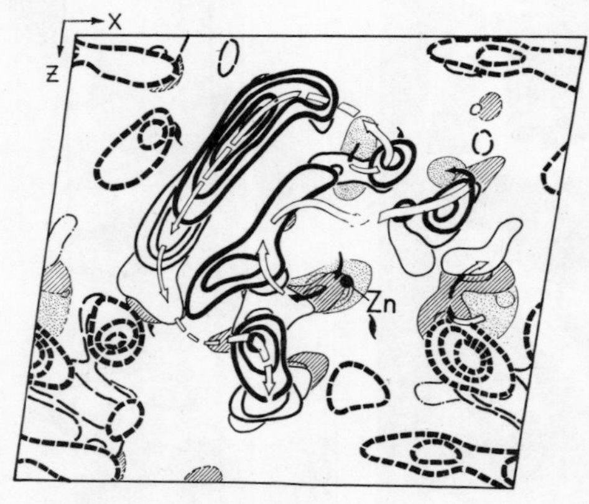

FIG. 3.

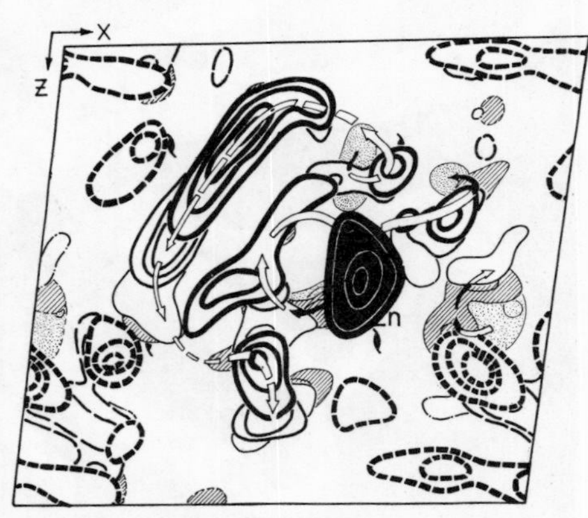

FIG. 4.

FIG. 3.—Sections 13/30, 14/30, ..., 17/30 in increasing levels perpendicular to the *b* axis, taken from the map at 6 Å resolution. The upper left corner is at $x = 1/4$ and $z = 1/4$; the lower right corner is at $x = 3/2$ and $z = 3/2$. Dotted contours represent parts of neighboring molecules. The interpretation of the smoothed trace of the polypeptide chain is suggested by arrows. The Zn atom is at 13/30 in *y*, near a pocket in the structure at the center of the figure.

FIG. 4.—The glycyl-L-tyrosine difference peak (*black area*) from section $y = 14/30$, oriented with respect to the sections through the active site of CPA.

TABLE 2

SUMMARY OF REFINEMENT OF 2.8 Å DATA FOR CARBOXYPEPTIDASE A (JANUARY 1967)

Spacing, Å		>10	10–5	5–3.3	3.3–2.8
E_j*	Pb(2)	7.1	5.0	5.9	5.7
	Hg(1)	4.7	4.5	6.0	6.7
	Hg(3)	7.4	6.0	6.5	6.7
	Pt(4)	10.3	7.8	8.8	8.8
RMS, f	Pb(2)	20.8	14.8	12.7	11.6
	Hg(1)	10.6	9.7	8.4	7.2
	Hg(3)	16.0	14.7	12.0	9.9
	Pt(4)	20.1	15.6	8.6	5.1
R_H	Pb(2)	0.08	0.08	0.07	0.10
	Hg(1)	0.05	0.07	0.08	0.12
	Hg(3)	0.08	0.09	0.09	0.12
	Pt(4)	0.10	0.09	0.10	0.12
$\langle m \rangle$		0.84	0.78	0.70	0.66

Values of R are 0.47 for the Pb(2), 0.64 for the Hg(1), 0.56 for the Hg(3), and 0.76 for the Pt(4) derivative, where the number of heavy atom sites is indicated in parentheses.
Figure of merit = 0.70.
* For definitions of symbols see footnotes to Table 1.

molecules of enzyme in these crystals were cross-linked with glutaraldehyde, and glycyl-L-tyrosine was then added as the substrate. The position of this substrate is only a few angstroms from that of this same substrate when it is bound to the apoenzyme. A detailed study at high resolution of the interaction between this enzyme and its substrate is expected to be feasible, and is now under way.

We wish to acknowledge financial support by the National Institutes of Health and the Advanced Research Projects Agency. We also thank Heather Halsey for valuable assistance throughout this study. Support of GNR by a National Science Foundation predoctoral fellowship is also acknowledged.

[1] Lipscomb, W. N., J. C. Coppola, J. A. Hartsuck, M. L. Ludwig, H. Muirhead, J. Searl, and T. A. Steitz, *J. Mol. Biol.*, **19**, 423 (1966).

[2] Hartsuck, J. A., M. L. Ludwig, H. Muirhead, T. A. Steitz, and W. N. Lipscomb, these PROCEEDINGS, **53**, 396 (1965).

[3] Sampath Kumar, K. S. V., K. A. Walsh, J.-P. Bargetzi, and H. Neurath, *Biochemistry*, **2**, 1475 (1963).

[4] Quiocho, F. A., and F. M. Richards, *Biochemistry*, **5**, 4062 (1966).